CW01111259

Persistent organic pollutants and toxic metals in foods

Related titles:

Chemical contaminants and residues in food
(ISBN 978-0-85709-058-4)

Animal feed contamination: Effects on livestock and food safety
(ISBN 978-1-84569-725-9)

Endocrine-disrupting chemicals in food
(ISBN 978-1-84569-218-6)

Details of these books and a complete list of titles from Woodhead Publishing can be obtained by:

- visiting our web site at www.woodheadpublishing.com
- contacting Customer Services (e-mail: sales@woodheadpublishing.com; fax: +44 (0) 1223 832819; tel.: +44 (0) 1223 499140 ext. 130; address: Woodhead Publishing Limited, 80 High Street, Sawston, Cambridge CB22 3HJ, UK)
- in North America, contacting our US office (e-mail: usmarketing@woodhead-publishing.com; tel.: (215) 928 9112; address: Woodhead Publishing, 1518 Walnut Street, Suite 1100, Philadelphia, PA 19102–3406, USA)

If you would like e-versions of our content, please visit our online platform: www.woodheadpublishingonline.com. Please recommend it to your librarian so that everyone in your institution can benefit from the wealth of content on the site.

We are always happy to receive suggestions for new books from potential editors. To enquire about contributing to our food series, please send your name, contact address and details of the topic/s you are interested in to nell.holden@woodheadpublishing.com. We look forward to hearing from you.

The team responsible for publishing this book:

Commissioning Editor: Sarah Hughes
Publications Coordinator: Anneka Hess
Project Editor: Rachel Cox
Editorial and Production Manager: Mary Campbell
Production Editor: Mandy Kingsmill
Project Manager: Newgen Knowledge Works Pvt Ltd
Copyeditor: Newgen Knowledge Works Pvt Ltd
Proofreader: Newgen Knowledge Works Pvt Ltd
Cover Designer: Terry Callanan

© Woodhead Publishing Limited, 2013

Woodhead Publishing Series in Food Science, Technology and Nutrition: Number 247

Persistent organic pollutants and toxic metals in foods

Edited by
Martin Rose and
Alwyn Fernandes

WP
WOODHEAD
PUBLISHING

Oxford   Cambridge   Philadelphia   New Delhi

© Woodhead Publishing Limited, 2013

Published by Woodhead Publishing Limited,
80 High Street, Sawston, Cambridge CB22 3HJ, UK
www.woodheadpublishing.com
www.woodheadpublishingonline.com

Woodhead Publishing, 1518 Walnut Street, Suite 1100, Philadelphia, PA 19102-3406, USA

Woodhead Publishing India Private Limited, G-2, Vardaan House, 7/28 Ansari Road,
Daryaganj, New Delhi – 110002, India
www.woodheadpublishingindia.com

First published 2013, Woodhead Publishing Limited
© Woodhead Publishing Limited, 2013; except Chapters 1, 2, 7, 8, 11 and 15 which are
© Crown Copyright, 2013. Note: the publisher has made every effort to ensure that permission for copyright material has been obtained by authors wishing to use such material. The authors and the publisher will be glad to hear from any copyright holder it has not been possible to contact. The authors have asserted their moral rights.

This book contains information obtained from authentic and highly regarded sources. Reprinted material is quoted with permission, and sources are indicated. Reasonable efforts have been made to publish reliable data and information, but the authors and the publisher cannot assume responsibility for the validity of all materials. Neither the authors nor the publisher, nor anyone else associated with this publication, shall be liable for any loss, damage or liability directly or indirectly caused or alleged to be caused by this book.

Neither this book nor any part may be reproduced or transmitted in any form or by any means, electronic or mechanical, including photocopying, microfilming and recording, or by any information storage or retrieval system, without permission in writing from Woodhead Publishing Limited.

The consent of Woodhead Publishing Limited does not extend to copying for general distribution, for promotion, for creating new works, or for resale. Specific permission must be obtained in writing from Woodhead Publishing Limited for such copying.

Trademark notice: Product or corporate names may be trademarks or registered trademarks, and are used only for identification and explanation, without intent to infringe.

British Library Cataloguing in Publication Data
A catalogue record for this book is available from the British Library.

Library of Congress Control Number: 2013934963

ISBN 978-0-85709-245-8 (print)
ISBN 978-0-85709-891-7 (online)
ISSN 2042-8049 Woodhead Publishing Series in Food Science, Technology and Nutrition (print)
ISSN 2042-8057 Woodhead Publishing Series in Food Science, Technology and Nutrition (online)

The publisher's policy is to use permanent paper from mills that operate a sustainable forestry policy, and which has been manufactured from pulp which is processed using acid-free and elemental chlorine-free practices. Furthermore, the publisher ensures that the text paper and cover board used have met acceptable environmental accreditation standards.

Typeset by Newgen Knowledge Works Pvt Ltd, Chennai, India
Printed by MPG Printgroup, UK

Cover image © Perica Dzeko, Shutterstock

Contents

Contributor contact details xi
Woodhead Publishing Series in Food Science, Technology
and Nutrition xv
Foreword xxiii
Preface xxv

Part I Regulatory control and environmental pathways 1

**1 Persistent organic pollutants in foods: science, policy
and regulation** 3
D. N. Mortimer, Food Standards Agency, UK
1.1 Introduction 3
1.2 Dietary exposure and total diet studies (TDSs) 4
1.3 Risk assessment, policy making and regulatory limits 6
1.4 Enforcement and implications for food businesses 12
1.5 Analytical methods and their influence on policy 14
1.6 Future trends and conclusions 15
1.7 References 17

**2 Regulatory control and monitoring of heavy metals and trace
elements in foods** 20
K. D. Hargin and G. J. Shears, Food Standards Agency, UK
2.1 Introduction 20
2.2 Risk assessment and policy making 21
2.3 Monitoring of foodstuffs 34
2.4 Impact of legislation on industry and enforcement 37
2.5 Suitability of analytical methods 40
2.6 Future trends 41
2.7 Sources of further information 43
2.8 References 44

vi Contents

3 Screening and confirmatory methods for the detection of dioxins and polychlorinated biphenyls (PCBs) in foods 47
J.-F. Focant and G. Eppe, University of Liège, Belgium
- 3.1 Introduction 47
- 3.2 Biological versus physico-chemical screening for dioxins and PCBs in food and feed 51
- 3.3 Specific analytical requirements for biological and physico-chemical tools 63
- 3.4 Quantitative versus semi-quantitative approach 66
- 3.5 Validation QA/QC 67
- 3.6 Confirmatory methods for dioxins and PCBs in food and feed 67
- 3.7 Future trends 75
- 3.8 Sources of further information and advice 76
- 3.9 References 76

4 Screening and confirmatory methods for the detection of heavy metals in foods 81
F. Arduini and G. Palleschi, Università di Roma Tor Vergata, Italy
- 4.1 Introduction 81
- 4.2 Screening methods for heavy metal detection in foods 84
- 4.3 Confirmatory methods for heavy metal detection in foods 95
- 4.4 Quality assurance and method validation 99
- 4.5 Future trends 102
- 4.6 References 102

5 Responding to food contamination incidents: principles and examples from cases involving dioxins 110
C. Tlustos, W. Anderson and R. Evans, Food Safety Authority of Ireland, Ireland
- 5.1 Introduction 110
- 5.2 The risk analysis paradigm 112
- 5.3 Food traceability 115
- 5.4 Food recall and withdrawal 117
- 5.5 Risk communication strategies 120
- 5.6 Future trends 125
- 5.7 Sources of further information 126
- 5.8 References 127

6 Uptake of organic pollutants and potentially toxic elements (PTEs) by crops 129
C. Collins, University of Reading, UK
- 6.1 Introduction 129
- 6.2 Uptake of organic pollutants by plants 131

6.3	Uptake of PTEs by plants	135
6.4	*In situ* monitoring of plant available pollutants	138
6.5	Conclusions	139
6.6	References	139

7 Transfer and uptake of dioxins and polychlorinated biphenyls (PCBs) into sheep: a case study — 145
S. W. Panton, F. Smith and A. Fernandes, *Food and Environment Research Agency (FERA), UK* and C. Foxall, *University of East Anglia, UK*

7.1	Introduction	145
7.2	Uptake pathways and sources	147
7.3	Transfer of PCBs and polychlorinated dibenzo-*p*-dioxins and dibenzofurans (PCDD/Fs) into animal tissues	151
7.4	Experimental rearing, sampling and analysis	152
7.5	Results and discussion for PCDD/Fs, dioxin-like PCBs (DL-PCBs) and ICES6 PCBs	157
7.6	Conclusions and future trends	169
7.7	Acknowledgements	169
7.8	References	170

8 Risk assessment of chemical contaminants and residues in foods — 173
D. J. Benford, *Food Standards Agency, UK*

8.1	Introduction	173
8.2	Risk assessment	176
8.3	Role of risk assessment in risk management	184
8.4	Sources of further information	185
8.5	References	185

Part II Particular persistent organic pollutants, toxic metals and metalloids — 189

9 Dioxins and polychlorinated biphenyls (PCBs) in foods — 191
D. Schrenk and M. Chopra, *University of Kaiserslautern, Germany*

9.1	Introduction	191
9.2	Properties and occurrence of polychlorinated dibenzo-*p*-dioxins and dibenzofurans (PCDD/Fs)	192
9.3	Toxicity of PCDD/Fs	195
9.4	Toxic effects of PCDD/Fs in humans and experimental animals	198
9.5	Properties and occurrence of PCBs	205
9.6	Toxicity of PCBs	209
9.7	References	213

viii Contents

10 Non dioxin-like polychlorinated biphenyls (NDL-PCBs) in foods: exposure and health hazards 215
L. E. Elabbas, E. Westerholm, R. Roos and K. Halldin, Karolinska Institutet, Sweden, M. Korkalainen, National Institute for Health and Welfare, Finland, M. Viluksela, National Institute for Health and Welfare, Finland and University of Eastern Finland, Finland and H. Håkansson, Karolinska Institutet, Sweden

10.1	Introduction	215
10.2	Sources, occurrence in foods, limit values and monitoring methods	217
10.3	Human exposure and tissue levels	220
10.4	Toxicokinetics and metabolism	223
10.5	Classification of PCB congeners	226
10.6	NDL-PCB regulatory status	228
10.7	ATHON R&D project dedicated to generating NDL-PCB toxicity data for regulatory use	229
10.8	Cell regulation and metabolism	240
10.9	Classification of NDL-PCB congeners	247
10.10	Conclusions and future trends	247
10.11	Acknowledgements	249
10.12	References	250

11 Brominated flame retardants in foods 261
R. J. Law, The Centre for Environment, Fisheries and Aquaculture Science, UK

11.1	Introduction	261
11.2	Sources, occurrence in foods and human exposure	262
11.3	Methods of analysis and monitoring of brominated flame retardants in foods	267
11.4	Toxicity of brominated flame retardants	268
11.5	Major incidences of brominated flame retardant contamination of foods	269
11.6	Implications for the food industry and policy makers for prevention and control of contamination	270
11.7	Future trends	270
11.8	Sources of further information and advice	271
11.9	References	272

12 Human dietary exposure to per- and poly-fluoroalkyl substances (PFASs) 279
R. Vestergren and I. T. Cousins, Stockholm University, Sweden

12.1	Introduction	280
12.2	Analytical methods for PFASs in foods	281
12.3	Levels of PFASs in foods	284

12.4	Pathways of food contamination	290
12.5	Estimated exposure from food and other exposure media	293
12.6	Conclusions and future trends	298
12.7	Acknowledgements	299
12.8	References	299

13 Polycyclic aromatic hydrocarbons (PAHs) in foods 308
L. Duedahl-Olesen, National Food Institute – Technical University of Denmark, Denmark

13.1	Introduction	308
13.2	Sources and formation of PAHs in foods	309
13.3	Methods of analysis of PAHs in foods	314
13.4	Human dietary exposure to PAHs from foods	319
13.5	Risk assessment of PAHs	322
13.6	Food scandals	324
13.7	Legislation of PAHs in foods within the EU	324
13.8	References	326

14 Phthalates in foods 334
T. Cirillo and R. Amodio Cocchieri, University of Naples Federico II, Italy

14.1	Introduction	334
14.2	Human exposure to phthalates	335
14.3	Sources and occurrence in foods	339
14.4	Studies of the effects of phthalates on humans	344
14.5	Methods of phthalate analysis and monitoring in foods	349
14.6	Implications for the food industry and policy making for prevention and control of contamination	352
14.7	Future trends	353
14.8	Sources of further information and advice	355
14.9	References	356

15 Polychlorinated naphthalenes (PCNs) in foods: sources, analytical methodology, occurrence and human exposure 367
A. Fernandes, The Food and Environment Research Agency (FERA), UK

15.1	Introduction	367
15.2	Sources of PCNs	368
15.3	Toxicology	370
15.4	Methods of analysis of PCNs in foods	372
15.5	Occurrence in foods	375
15.6	PCN occurrence in humans	383
15.7	Conclusions and future trends	386
15.8	References	387

16 Mercury in foods 392
E. M. Sunderland and M. Tumpney, Harvard University School of Public Health, USA

16.1 Introduction 392
16.2 Concentrations of mercury in foods 396
16.3 Mercury exposures and risks from major food categories 403
16.4 References 407

17 Arsenic in foods: current issues related to analysis, toxicity and metabolism 414
K. A. Francesconi and G. Raber, University of Graz, Austria

17.1 Introduction 414
17.2 Sources and occurrence in foods 415
17.3 Methods for determining arsenic in foods 418
17.4 Toxicity of arsenic 421
17.5 Implications for the food industry and policy makers ... 425
17.6 References 426

18 Organotin compounds in foods 430
E. Rosenberg, Vienna University of Technology, Austria

18.1 Introduction 431
18.2 Technical, agricultural and industrial uses of organotin compounds 432
18.3 Physical and chemical properties of organotin compounds 444
18.4 Analysis of organotin compounds in foods 445
18.5 Human dietary exposure to organotin compounds from foods 451
18.6 Human exposure to organotin compounds from food packaging material 457
18.7 Health risks and toxicity of organotin compounds 461
18.8 Conclusions and future trends 466
18.9 References 468
18.10 Appendix: abbreviations 474

Index 476

Contributor contact details

(* = main contact)

Editors

Martin Rose* and Alwyn Fernandes
The Food and Environment
   Research Agency
Sand Hutton
York YO41 1LZ
UK

Email: Martin.Rose@fera.gsi.gov.
   uk; alwyn.fernandes@fera.gsi.
   gov.uk

Chapter 1

Dr D. N. Mortimer
Agricultural, Process and
   Environmental Contaminants
Chemical Safety Division
Food Standards Agency
125 Kingsway
London WC2B 6NH
UK

Email: david.mortimer@
   foodstandards.gsi.gov.uk

Chapter 2

Dr K. D. Hargin* and G. J. Shears
Agricultural, Process and
   Environmental Contaminants
Chemical Safety Division
Food Standards Agency
125 Kingsway
London WC2B 6NH
UK

Email: kevin.hargin@foodstandards.
   gsi.gov.uk

Chapter 3

Jean-François Focant* and Gauthier
   Eppe
CART, Organic and Biological
   Analytical Chemistry
University of Liège
Allée de la Chimie 3
B-6c Sart-Tilman
B-4000 Liège
Belgium

Email: JF.Focant@ulg.ac.be

Chapter 4

Fabiana Arduini* and Giuseppe Palleschi
Department of Science and Chemical Technology
Tor Vergata University
Via della Ricerca Scientifica
00133 Rome
Italy

Email: fabiana.arduini@uniroma2.it

Chapter 5

Christina Tlustos
Food Safety Authority of Ireland
Abbey Court/Lower Abbey Street
Dublin 1
Ireland

Email: ctlustos@fsai.ie

Chapter 6

Chris Collins
Soil Research Centre
University of Reading
Reading RG6 6DW
UK

Email: c.d.collins@reading.ac.uk

Chapter 7

Sean H. W. Panton* and Frankie D. Smith
Food Contaminants and Integrity Group
Food and Environmental Research Agency (FERA)
Sand Hutton
York YO41 1LZ
UK

Email: sean.panton@fera.gsi.gov.uk

C. Foxall
School of Environmental Sciences
University of East Anglia
Norwich NR4 7TJ
UK

Chapter 8

Dr Diane J. Benford
Chemical Safety Division
Food Standards Agency
Aviation House
125 Kingsway
London WC2B 6NH
UK

Email: Diane.Benford@foodstandards.gsi.gov.uk

Chapter 9

D. Schrenk
Food Chemistry and Toxicology
University of Kaiserslautern
Erwin-Schroedinger Strasse 56
D-67663 Kaiserslautern
Germany

Email: schrenk@rhrk.uni-kl.de

Chapter 10

H. Håkansson
Institute of Environmental Medicine (IMM)
Karolinska Institutet
Nobels Väg 13
171 77 Stockholm
Sweden

Email: helen.hakansson@ki.se

Chapter 11

Robin J. Law
The Centre for Environment,
 Fisheries and Aquaculture Science
Cefas Lowestoft Laboratory
Pakefield Road
Lowestoft
Suffolk NR33 0HT
UK

Email: robin.law@cefas.co.uk

Chapter 12

I. T. Cousins
Department of Applied
 Environmental Science (ITM)
Stockholm University
SE-106 91 Stockholm
Sweden

Email: ian.cousins@itm.su.se

Chapter 13

Lene Duedahl-Olesen
National Food Institute – Technical
 University of Denmark
Mørkhøj Bygade 19
DK-2860 Søborg
Denmark

Email: lduo@food.dtu.dk

Chapter 14

Teresa Cirillo* and Renata Amodio
 Cocchieri
Department of Agraria
University of Naples Federico II
Via Università 100
80055 Portici
Naples
Italy

Email: tcirillo@unina.it

Chapter 15

Alwyn Fernandes
The Food and Environment
 Research Agency
Sand Hutton
York YO41 1LZ
UK

Email: alwyn.fernandes@fera.gsi.
 gov.uk

Chapter 16

Elsie M. Sunderland
Harvard University School of Public
 Health
Department of Environmental
 Health
Landmark Center West, 4th Floor
PO Box 15677
Boston
MA 02215
USA

Email: elsie_sunderland@harvard.
 edu

Chapter 17

Kevin A. Francesconi and Georg Raber*
Institute of Chemistry–Analytical Chemistry
University of Graz
Universitaetsplatz 1
8010 Graz
Austria

Email: kevin.francesconi@uni-graz.at; georg.raber@uni-graz.at

Chapter 18

Erwin Rosenberg
Vienna University of Technology
Institute of Chemical Technologies and Analytics
Getreidemarkt 9/164 AC
A-1060 Vienna
Austria

Email: E.Rosenberg@tuwien.ac.at

Woodhead Publishing Series in Food Science, Technology and Nutrition

1 Chilled foods: A comprehensive guide *Edited by C. Dennis and M. Stringer*
2 Yoghurt: Science and technology *A. Y. Tamime and R. K. Robinson*
3 Food processing technology: Principles and practice *P. J. Fellows*
4 Bender's dictionary of nutrition and food technology Sixth edition *D. A. Bender*
5 Determination of veterinary residues in food *Edited by N. T. Crosby*
6 Food contaminants: Sources and surveillance *Edited by C. Creaser and R. Purchase*
7 Nitrates and nitrites in food and water *Edited by M. J. Hill*
8 Pesticide chemistry and bioscience: The food-environment challenge *Edited by G. T. Brooks and T. Roberts*
9 Pesticides: Developments, impacts and controls *Edited by G. A. Best and A. D. Ruthven*
10 Dietary fibre: Chemical and biological aspects *Edited by D. A. T. Southgate, K. W. Waldron, I. T. Johnson and G. R. Fenwick*
11 Vitamins and minerals in health and nutrition *M. Tolonen*
12 Technology of biscuits, crackers and cookies Second edition *D. Manley*
13 Instrumentation and sensors for the food industry *Edited by E. Kress-Rogers*
14 Food and cancer prevention: Chemical and biological aspects *Edited by K. W. Waldron, I. T. Johnson and G. R. Fenwick*
15 Food colloids: Proteins, lipids and polysaccharides *Edited by E. Dickinson and B. Bergenstahl*
16 Food emulsions and foams *Edited by E. Dickinson*
17 Maillard reactions in chemistry, food and health *Edited by T. P. Labuza, V. Monnier, J. Baynes and J. O'Brien*
18 The Maillard reaction in foods and medicine *Edited by J. O'Brien, H. E. Nursten, M. J. Crabbe and J. M. Ames*
19 Encapsulation and controlled release *Edited by D. R. Karsa and R. A. Stephenson*
20 Flavours and fragrances *Edited by A. D. Swift*
21 Feta and related cheeses *Edited by A. Y. Tamime and R. K. Robinson*
22 Biochemistry of milk products *Edited by A. T. Andrews and J. R. Varley*
23 Physical properties of foods and food processing systems *M. J. Lewis*
24 Food irradiation: A reference guide *V. M. Wilkinson and G. Gould*

25 Kent's technology of cereals: An introduction for students of food science and agriculture Fourth edition *N. L. Kent and A. D. Evers*
26 Biosensors for food analysis *Edited by A. O. Scott*
27 Separation processes in the food and biotechnology industries: Principles and applications *Edited by A. S. Grandison and M. J. Lewis*
28 Handbook of indices of food quality and authenticity *R. S. Singhal, P. K. Kulkarni and D. V. Rege*
29 Principles and practices for the safe processing of foods *D. A. Shapton and N. F. Shapton*
30 Biscuit, cookie and cracker manufacturing manuals Volume 1: Ingredients *D. Manley*
31 Biscuit, cookie and cracker manufacturing manuals Volume 2: Biscuit doughs *D. Manley*
32 Biscuit, cookie and cracker manufacturing manuals Volume 3: Biscuit dough piece forming *D. Manley*
33 Biscuit, cookie and cracker manufacturing manuals Volume 4: Baking and cooling of biscuits *D. Manley*
34 Biscuit, cookie and cracker manufacturing manuals Volume 5: Secondary processing in biscuit manufacturing *D. Manley*
35 Biscuit, cookie and cracker manufacturing manuals Volume 6: Biscuit packaging and storage *D. Manley*
36 Practical dehydration Second edition *M. Greensmith*
37 Lawrie's meat science Sixth edition *R. A. Lawrie*
38 Yoghurt: Science and technology Second edition *A. Y. Tamime and R. K. Robinson*
39 New ingredients in food processing: Biochemistry and agriculture *G. Linden and D. Lorient*
40 Benders' dictionary of nutrition and food technology Seventh edition *D. A. Bender and A. E. Bender*
41 Technology of biscuits, crackers and cookies Third edition *D. Manley*
42 Food processing technology: Principles and practice Second edition *P. J. Fellows*
43 Managing frozen foods *Edited by C. J. Kennedy*
44 Handbook of hydrocolloids *Edited by G. O. Phillips and P. A. Williams*
45 Food labelling *Edited by J. R. Blanchfield*
46 Cereal biotechnology *Edited by P. C. Morris and J. H. Bryce*
47 Food intolerance and the food industry *Edited by T. Dean*
48 The stability and shelf-life of food *Edited by D. Kilcast and P. Subramaniam*
49 Functional foods: Concept to product *Edited by G. R. Gibson and C. M. Williams*
50 Chilled foods: A comprehensive guide Second edition *Edited by M. Stringer and C. Dennis*
51 HACCP in the meat industry *Edited by M. Brown*
52 Biscuit, cracker and cookie recipes for the food industry *D. Manley*
53 Cereals processing technology *Edited by G. Owens*
54 Baking problems solved *S. P. Cauvain and L. S. Young*
55 Thermal technologies in food processing *Edited by P. Richardson*
56 Frying: Improving quality *Edited by J. B. Rossell*
57 Food chemical safety Volume 1: Contaminants *Edited by D. Watson*
58 Making the most of HACCP: Learning from others' experience *Edited by T. Mayes and S. Mortimore*
59 Food process modelling *Edited by L. M. M. Tijskens, M. L. A. T. M. Hertog and B. M. Nicolaï*

Woodhead Publishing Series in Food Science, Technology and Nutrition   xvii

60 EU food law: A practical guide *Edited by K. Goodburn*
61 Extrusion cooking: Technologies and applications *Edited by R. Guy*
62 Auditing in the food industry: From safety and quality to environmental and other audits *Edited by M. Dillon and C. Griffith*
63 Handbook of herbs and spices Volume 1 *Edited by K. V. Peter*
64 Food product development: Maximising success *M. Earle, R. Earle and A. Anderson*
65 Instrumentation and sensors for the food industry Second edition *Edited by E. Kress-Rogers and C. J. B. Brimelow*
66 Food chemical safety Volume 2: Additives *Edited by D. Watson*
67 Fruit and vegetable biotechnology *Edited by V. Valpuesta*
68 Foodborne pathogens: Hazards, risk analysis and control *Edited by C. de W. Blackburn and P. J. McClure*
69 Meat refrigeration *S. J. James and C. James*
70 Lockhart and Wiseman's crop husbandry Eighth edition *H. J. S. Finch, A. M. Samuel and G. P. F. Lane*
71 Safety and quality issues in fish processing *Edited by H. A. Bremner*
72 Minimal processing technologies in the food industries *Edited by T. Ohlsson and N. Bengtsson*
73 Fruit and vegetable processing: Improving quality *Edited by W. Jongen*
74 The nutrition handbook for food processors *Edited by C. J. K. Henry and C. Chapman*
75 Colour in food: Improving quality *Edited by D. MacDougall*
76 Meat processing: Improving quality *Edited by J. P. Kerry, J. F. Kerry and D. A. Ledward*
77 Microbiological risk assessment in food processing *Edited by M. Brown and M. Stringer*
78 Performance functional foods *Edited by D. Watson*
79 Functional dairy products Volume 1 *Edited by T. Mattila-Sandholm and M. Saarela*
80 Taints and off-flavours in foods *Edited by B. Baigrie*
81 Yeasts in food *Edited by T. Boekhout and V. Robert*
82 Phytochemical functional foods *Edited by I. T. Johnson and G. Williamson*
83 Novel food packaging techniques *Edited by R. Ahvenainen*
84 Detecting pathogens in food *Edited by T. A. McMeekin*
85 Natural antimicrobials for the minimal processing of foods *Edited by S. Roller*
86 Texture in food Volume 1: Semi-solid foods *Edited by B. M. McKenna*
87 Dairy processing: Improving quality *Edited by G. Smit*
88 Hygiene in food processing: Principles and practice *Edited by H. L. M. Lelieveld, M. A. Mostert, B. White and J. Holah*
89 Rapid and on-line instrumentation for food quality assurance *Edited by I. Tothill*
90 Sausage manufacture: Principles and practice *E. Essien*
91 Environmentally-friendly food processing *Edited by B. Mattsson and U. Sonesson*
92 Bread making: Improving quality *Edited by S. P. Cauvain*
93 Food preservation techniques *Edited by P. Zeuthen and L. Bøgh-Sørensen*
94 Food authenticity and traceability *Edited by M. Lees*
95 Analytical methods for food additives *R. Wood, L. Foster, A. Damant and P. Key*
96 Handbook of herbs and spices Volume 2 *Edited by K. V. Peter*
97 Texture in food Volume 2: Solid foods *Edited by D. Kilcast*
98 Proteins in food processing *Edited by R. Yada*
99 Detecting foreign bodies in food *Edited by M. Edwards*

100 Understanding and measuring the shelf-life of food *Edited by R. Steele*
101 Poultry meat processing and quality *Edited by G. Mead*
102 Functional foods, ageing and degenerative disease *Edited by C. Remacle and B. Reusens*
103 Mycotoxins in food: Detection and control *Edited by N. Magan and M. Olsen*
104 Improving the thermal processing of foods *Edited by P. Richardson*
105 Pesticide, veterinary and other residues in food *Edited by D. Watson*
106 Starch in food: Structure, functions and applications *Edited by A.-C. Eliasson*
107 Functional foods, cardiovascular disease and diabetes *Edited by A. Arnoldi*
108 Brewing: Science and practice *D. E. Briggs, P. A. Brookes, R. Stevens and C. A. Boulton*
109 Using cereal science and technology for the benefit of consumers: Proceedings of the 12th International ICC Cereal and Bread Congress, 24th–26th May, 2004, Harrogate, UK *Edited by S. P. Cauvain, L. S. Young and S. Salmon*
110 Improving the safety of fresh meat *Edited by J. Sofos*
111 Understanding pathogen behaviour: Virulence, stress response and resistance *Edited by M. Griffiths*
112 The microwave processing of foods *Edited by H. Schubert and M. Regier*
113 Food safety control in the poultry industry *Edited by G. Mead*
114 Improving the safety of fresh fruit and vegetables *Edited by W. Jongen*
115 Food, diet and obesity *Edited by D. Mela*
116 Handbook of hygiene control in the food industry *Edited by H. L. M. Lelieveld, M. A. Mostert and J. Holah*
117 Detecting allergens in food *Edited by S. Koppelman and S. Hefle*
118 Improving the fat content of foods *Edited by C. Williams and J. Buttriss*
119 Improving traceability in food processing and distribution *Edited by I. Smith and A. Furness*
120 Flavour in food *Edited by A. Voilley and P. Etievant*
121 The Chorleywood bread process *S. P. Cauvain and L. S. Young*
122 Food spoilage microorganisms *Edited by C. de W. Blackburn*
123 Emerging foodborne pathogens *Edited by Y. Motarjemi and M. Adams*
124 Benders' dictionary of nutrition and food technology Eighth edition *D. A. Bender*
125 Optimising sweet taste in foods *Edited by W. J. Spillane*
126 Brewing: New technologies *Edited by C. Bamforth*
127 Handbook of herbs and spices Volume 3 *Edited by K. V. Peter*
128 Lawrie's meat science Seventh edition *R. A. Lawrie in collaboration with D. A. Ledward*
129 Modifying lipids for use in food *Edited by F. Gunstone*
130 Meat products handbook: Practical science and technology *G. Feiner*
131 Food consumption and disease risk: Consumer-pathogen interactions *Edited by M. Potter*
132 Acrylamide and other hazardous compounds in heat-treated foods *Edited by K. Skog and J. Alexander*
133 Managing allergens in food *Edited by C. Mills, H. Wichers and K. Hoffman-Sommergruber*
134 Microbiological analysis of red meat, poultry and eggs *Edited by G. Mead*
135 Maximising the value of marine by-products *Edited by F. Shahidi*
136 Chemical migration and food contact materials *Edited by K. Barnes, R. Sinclair and D. Watson*

Woodhead Publishing Series in Food Science, Technology and Nutrition   xix

137 **Understanding consumers of food products** *Edited by L. Frewer and H. van Trijp*
138 **Reducing salt in foods: Practical strategies** *Edited by D. Kilcast and F. Angus*
139 **Modelling microorganisms in food** *Edited by S. Brul, S. Van Gerwen and M. Zwietering*
140 **Tamime and Robinson's Yoghurt: Science and technology Third edition** *A. Y. Tamime and R. K. Robinson*
141 **Handbook of waste management and co-product recovery in food processing Volume 1** *Edited by K. W. Waldron*
142 **Improving the flavour of cheese** *Edited by B. Weimer*
143 **Novel food ingredients for weight control** *Edited by C. J. K. Henry*
144 **Consumer-led food product development** *Edited by H. MacFie*
145 **Functional dairy products Volume 2** *Edited by M. Saarela*
146 **Modifying flavour in food** *Edited by A. J. Taylor and J. Hort*
147 **Cheese problems solved** *Edited by P. L. H. McSweeney*
148 **Handbook of organic food safety and quality** *Edited by J. Cooper, C. Leifert and U. Niggli*
149 **Understanding and controlling the microstructure of complex foods** *Edited by D. J. McClements*
150 **Novel enzyme technology for food applications** *Edited by R. Rastall*
151 **Food preservation by pulsed electric fields: From research to application** *Edited by H. L. M. Lelieveld and S. W. H. de Haan*
152 **Technology of functional cereal products** *Edited by B. R. Hamaker*
153 **Case studies in food product development** *Edited by M. Earle and R. Earle*
154 **Delivery and controlled release of bioactives in foods and nutraceuticals** *Edited by N. Garti*
155 **Fruit and vegetable flavour: Recent advances and future prospects** *Edited by B. Brückner and S. G. Wyllie*
156 **Food fortification and supplementation: Technological, safety and regulatory aspects** *Edited by P. Berry Ottaway*
157 **Improving the health-promoting properties of fruit and vegetable products** *Edited by F. A. Tomás-Barberán and M. I. Gil*
158 **Improving seafood products for the consumer** *Edited by T. Børresen*
159 **In-pack processed foods: Improving quality** *Edited by P. Richardson*
160 **Handbook of water and energy management in food processing** *Edited by J. Klemeš, R.. Smith and J.-K. Kim*
161 **Environmentally compatible food packaging** *Edited by E. Chiellini*
162 **Improving farmed fish quality and safety** *Edited by Ø. Lie*
163 **Carbohydrate-active enzymes** *Edited by K.-H. Park*
164 **Chilled foods: A comprehensive guide Third edition** *Edited by M. Brown*
165 **Food for the ageing population** *Edited by M. M. Raats, C. P. G. M. de Groot and W. A. Van Staveren*
166 **Improving the sensory and nutritional quality of fresh meat** *Edited by J. P. Kerry and D. A. Ledward*
167 **Shellfish safety and quality** *Edited by S. E. Shumway and G. E. Rodrick*
168 **Functional and speciality beverage technology** *Edited by P. Paquin*
169 **Functional foods: Principles and technology** *M. Guo*
170 **Endocrine-disrupting chemicals in food** *Edited by I. Shaw*
171 **Meals in science and practice: Interdisciplinary research and business applications** *Edited by H. L. Meiselman*

172 Food constituents and oral health: Current status and future prospects *Edited by M. Wilson*
173 Handbook of hydrocolloids Second edition *Edited by G. O. Phillips and P. A. Williams*
174 Food processing technology: Principles and practice Third edition *P. J. Fellows*
175 Science and technology of enrobed and filled chocolate, confectionery and bakery products *Edited by G. Talbot*
176 Foodborne pathogens: Hazards, risk analysis and control Second edition *Edited by C. de W. Blackburn and P. J. McClure*
177 Designing functional foods: Measuring and controlling food structure breakdown and absorption *Edited by D. J. McClements and E. A. Decker*
178 New technologies in aquaculture: Improving production efficiency, quality and environmental management *Edited by G. Burnell and G. Allan*
179 More baking problems solved *S. P. Cauvain and L. S. Young*
180 Soft drink and fruit juice problems solved *P. Ashurst and R. Hargitt*
181 Biofilms in the food and beverage industries *Edited by P. M. Fratamico, B. A. Annous and N. W. Gunther*
182 Dairy-derived ingredients: Food and neutraceutical uses *Edited by M. Corredig*
183 Handbook of waste management and co-product recovery in food processing Volume 2 *Edited by K. W. Waldron*
184 Innovations in food labelling *Edited by J. Albert*
185 Delivering performance in food supply chains *Edited by C. Mena and G. Stevens*
186 Chemical deterioration and physical instability of food and beverages *Edited by L. H. Skibsted, J. Risbo and M. L. Andersen*
187 **Managing wine quality Volume 1: Viticulture and wine quality** *Edited by A. G. Reynolds*
188 Improving the safety and quality of milk Volume 1: Milk production and processing *Edited by M. Griffiths*
189 Improving the safety and quality of milk Volume 2: Improving quality in milk products *Edited by M. Griffiths*
190 Cereal grains: Assessing and managing quality *Edited by C. Wrigley and I. Batey*
191 Sensory analysis for food and beverage quality control: A practical guide *Edited by D. Kilcast*
192 **Managing wine quality Volume 2: Oenology and wine quality** *Edited by A. G. Reynolds*
193 **Winemaking problems solved** *Edited by C. E. Butzke*
194 Environmental assessment and management in the food industry *Edited by U. Sonesson, J. Berlin and F. Ziegler*
195 Consumer-driven innovation in food and personal care products *Edited by S. R. Jaeger and H. MacFie*
196 Tracing pathogens in the food chain *Edited by S. Brul, P. M. Fratamico and T. A. McMeekin*
197 Case studies in novel food processing technologies: Innovations in processing, packaging, and predictive modelling *Edited by C. J. Doona, K. Kustin and F. E. Feeherry*
198 Freeze-drying of pharmaceutical and food products *T.-C. Hua, B.-L. Liu and H. Zhang*
199 Oxidation in foods and beverages and antioxidant applications Volume 1: Understanding mechanisms of oxidation and antioxidant activity *Edited by E. A. Decker, R. J. Elias and D. J. McClements*

200 Oxidation in foods and beverages and antioxidant applications Volume 2: Management in different industry sectors *Edited by E. A. Decker, R. J. Elias and D. J. McClements*
201 Protective cultures, antimicrobial metabolites and bacteriophages for food and beverage biopreservation *Edited by C. Lacroix*
202 Separation, extraction and concentration processes in the food, beverage and nutraceutical industries *Edited by S. S. H. Rizvi*
203 Determining mycotoxins and mycotoxigenic fungi in food and feed *Edited by S. De Saeger*
204 Developing children's food products *Edited by D. Kilcast and F. Angus*
205 Functional foods: Concept to product Second edition *Edited by M. Saarela*
206 Postharvest biology and technology of tropical and subtropical fruits Volume 1: Fundamental issues *Edited by E. M. Yahia*
207 Postharvest biology and technology of tropical and subtropical fruits Volume 2: Açai to citrus *Edited by E. M. Yahia*
208 Postharvest biology and technology of tropical and subtropical fruits Volume 3: Cocona to mango *Edited by E. M. Yahia*
209 Postharvest biology and technology of tropical and subtropical fruits Volume 4: Mangosteen to white sapote *Edited by E. M. Yahia*
210 Food and beverage stability and shelf life *Edited by D. Kilcast and P. Subramaniam*
211 Processed Meats: Improving safety, nutrition and quality *Edited by J. P. Kerry and J. F. Kerry*
212 Food chain integrity: A holistic approach to food traceability, safety, quality and authenticity *Edited by J. Hoorfar, K. Jordan, F. Butler and R. Prugger*
213 Improving the safety and quality of eggs and egg products Volume 1 *Edited by Y. Nys, M. Bain and F. Van Immerseel*
214 Improving the safety and quality of eggs and egg products Volume 2 *Edited by F. Van Immerseel, Y. Nys and M. Bain*
215 Animal feed contamination: Effects on livestock and food safety *Edited by J. Fink-Gremmels*
216 Hygienic design of food factories *Edited by J. Holah and H. L. M. Lelieveld*
217 Manley's technology of biscuits, crackers and cookies Fourth edition *Edited by D. Manley*
218 Nanotechnology in the food, beverage and nutraceutical industries *Edited by Q. Huang*
219 Rice quality: A guide to rice properties and analysis *K. R. Bhattacharya*
220 Advances in meat, poultry and seafood packaging *Edited by J. P. Kerry*
221 Reducing saturated fats in foods *Edited by G. Talbot*
222 Handbook of food proteins *Edited by G. O. Phillips and P. A. Williams*
223 Lifetime nutritional influences on cognition, behaviour and psychiatric illness *Edited by D. Benton*
224 Food machinery for the production of cereal foods, snack foods and confectionery *L.-M. Cheng*
225 Alcoholic beverages: Sensory evaluation and consumer research *Edited by J. Piggott*
226 Extrusion problems solved: Food, pet food and feed *M. N. Riaz and G. J. Rokey*
227 Handbook of herbs and spices Second edition Volume 1 *Edited by K. V. Peter*
228 Handbook of herbs and spices Second edition Volume 2 *Edited by K. V. Peter*
229 Breadmaking: Improving quality Second edition *Edited by S. P. Cauvain*

230 **Emerging food packaging technologies: Principles and practice** *Edited by K. L. Yam and D. S. Lee*
231 **Infectious disease in aquaculture: Prevention and control** *Edited by B. Austin*
232 **Diet, immunity and inflammation** *Edited by P. C. Calder and P. Yaqoob*
233 **Natural food additives, ingredients and flavourings** *Edited by D. Baines and R. Seal*
234 **Microbial decontamination in the food industry: Novel methods and applications** *Edited by A. Demirci and M.O. Ngadi*
235 **Chemical contaminants and residues in foods** *Edited by D. Schrenk*
236 **Robotics and automation in the food industry: Current and future technologies** *Edited by D. G. Caldwell*
237 **Fibre-rich and wholegrain foods: Improving quality** *Edited by J. A. Delcour and K. Poutanen*
238 **Computer vision technology in the food and beverage industries** *Edited by D.-W. Sun*
239 **Encapsulation technologies and delivery systems for food ingredients and nutraceuticals** *Edited by N. Garti and D. J. McClements*
240 **Case studies in food safety and authenticity** *Edited by J. Hoorfar*
241 **Heat treatment for insect control: Developments and applications** *D. Hammond*
242 **Advances in aquaculture hatchery technology** *Edited by G. Allan and G. Burnell*
243 **Open innovation in the food and beverage industry** *Edited by M. Garcia Martinez*
244 **Trends in packaging of food, beverages and other fast-moving consumer goods (FMCG)** *Edited by N. Farmer*
245 **New analytical approaches for verifying the origin of food** *Edited by P. Brereton*
246 **Microbial production of food ingredients, enzymes and nutraceuticals** *Edited by B. McNeil, D. Archer, I. Giavasis and L. Harvey*
247 **Persistent organic pollutants and toxic metals in foods** *Edited by M. Rose and A. Fernandes*
248 **Cereal grains for the food and beverage industries** *E. Arendt and E. Zannini*
249 **Viruses in food and water: Risks, surveillance and control** *Edited by N. Cook*
250 **Improving the safety and quality of nuts** *Edited by L. J. Harris*
251 **Metabolomics in food and nutrition** *Edited by B. C. Weimer and C. Slupsky*
252 **Food enrichment with omega-3 fatty acids** *Edited by C. Jacobsen, N. Skall Nielsen, A. Frisenfeldt Horn and A.-D. Moltke Sørensen*
253 **Instrumental assessment of food sensory quality: A practical guide** *Edited by D. Kilcast*
254 **Food microstructures: Microscopy, measurement and modelling** *Edited by V. J. Morris and K. Groves*
255 **Handbook of food powders: Processes and properties** *Edited by B. R. Bhandari, N. Bansal, M. Zhang and P. Schuck*
256 **Functional ingredients from algae for foods and nutraceuticals** *Edited by H. Domínguez*
257 **Satiation, satiety and the control of food intake: Theory and practice** *Edited by J. E. Blundell and F. Bellisle*
258 **Hygiene in food processing: Principles and practice Second edition** *Edited by H. L. M. Lelieveld, J. Holah and D. Napper*

Foreword*

The safety of our food is a primary concern. However, with increased industrialization and globalization of world economies and food supplies, ensuring the safety of our food presents huge regulatory challenges. Chemical contamination of the food supply is perhaps one of the more difficult challenges. Persistent organic pollutants and toxic metals are ubiquitous environmental pollutants. Following their release into the air, water or soil, these chemicals slowly degrade and bioaccumulate in the food chain. The bioaccumulation results in low-level contamination of our food supply. Because of the long biological half-life of these chemicals in humans, even contamination at parts per trillion in the food can result in human body burdens approaching those in which adverse effects are observed in experimental studies in animals, and in observational studies in people. At present, detecting most chemical contaminants in food at parts per trillion levels can only be done with the most sophisticated, and costly, analytical techniques. In addition, because of the perishable nature of food, most analytical techniques are too time consuming to allow data generation in a real-time manner. The development of cost effective screening and intervention approaches for these chemicals are subjects of intense scientific and regulatory debates.

Low-level contamination of the food supply with persistent organic pollutants and toxic metals is an excellent example of the interaction of science and public policy. The ubiquitous microcontamination of our environment with these chemicals presents complex scientific and regulatory problems. In contrast to microbiological contaminates in food, typical hygienic practices

---

* This article may be the work product of an employee of the National Institute of Environmental Health Sciences (NIEHS) or the National Cancer Institute, National Institutes of Health (NIH), however, the statements, opinions or conclusions contained therein do not necessarily represent the statements, opinions or conclusions of NIEHS, NIH or the United States Government.

and thermal processes are often ineffective in preventing or removing chemical contaminants. Because of their ubiquitous presence in the environment, these chemicals get into the food on our tables through complex and often unknown pathways. Thus, in order to understand the level of contamination of our food, our only alternative is to measure these chemicals in the food. However, methods of detecting chemical contaminates are chemical-class specific and due to the large number of potential environmental chemicals that may get into our food, we must prioritize which chemicals we evaluate. In addition, as interest in a specific chemical class increases, analytical chemists develop better and more sensitive methods of detection, resulting in the discovery that these chemicals are present in a greater percentage of our food than previously understood. Analytical methods have increased sensitivity by orders of magnitude over the past several decades. Chemicals that were once 'not present' in our foods are now commonly found. To complement these advances in analytical chemistry, we must refine our risk assessment methods to better interpret the chemical contaminants occurrence and concentration data. Finally, given the wide range of foods, food sources and chemical contaminants, we would need much larger screening programs to ensure the safety of our food supply, with respect to chemical contaminants. However, the sizes of these screening programs are limited by the costs associated with them. Thus, developing cost effective screening programs that combine statistical approaches to sample selection, and analytical sensitivity and specificity are necessary.

In their book, '*Persistent organic pollutants and toxic metals in food*', editors, Dr Martin Rose and Dr Alwyn Fernandes, present both the regulatory and scientific challenges in ensuring our food supply is safe from these chemicals. This book is one of the few that presents both of these important issues. The first section of the book covers regulatory efforts to screen and control for persistent organic pollutants and toxic metals. It also includes case studies on regulatory responses to accidental contamination incidences for dioxins. These examples nicely set the stage for the second section of the book that describes the occurrence, exposure and toxicity of individual chemicals and chemical classes. This book provides the scientific and policy foundations for those interested in chemical contaminants and food safety.

Linda S. Birnbaum and Michael J. DeVito
National Institute of Environmental Health Sciences, USA

Preface

It has been our great pleasure to work with so many distinguished experts to put together a book that deals with the immensely important subject of chemical contamination of food. The fundamental concept behind this work was to bring together a range of perspectives on this subject – to include analytical, scientific, risk assessment and regulatory issues – and collate a useful resource for those with interests crossing these perspectives. We wanted to demonstrate how the best scientific evidence gets used for risk assessment and in turn translated into regulation. The wealth of experience and different backgrounds of our contributors has enabled us to collect together topics that cover not only a range of inorganic and organic contaminants, but also a range of viewpoints on their consequences in terms of exposure and human health.

The topics covered in this book also demonstrate two further aspects of the subject: the diversity of disciplines that are fundamental to our understanding of food safety, and the diversity of contaminants – some arising inadvertently, some through our purposeful anthropogenic activity – that analytical chemists, toxicologists, risk assessors and regulators are increasingly having to deal with. And although new and emerging issues surface regularly, new findings, improvements in measurement techniques and toxicological updates on older contaminants continue to provide insights into the interaction and influence of these chemicals on human health. Thus for example, decades after originally being alerted to the immediate toxicity of PCBs experienced by victims of acute exposure, we learn latterly about more subtle and long-term effects such as endocrine-disrupting activity and the dioxin-like toxicity of these food contaminants. Similarly, evolving knowledge of the toxicity of the different species of arsenic and mercury, combined with the advances in analytical methodology that allow speciation of these elements, has allowed refinement of the risk assessment and regulation.

We hope that the breadth of topics will also provide a useful introduction to students, a helpful resource for regulators and a general text for all those seeking further knowledge on the complex subject of food safety and human health. All that remains is for us to thank both the contributors and the staff at Woodhead for making the project so interesting and rewarding for us. We hope that you enjoy reading the book as much as we enjoyed the editing process.

Martin Rose and Alwyn Fernandes

Part I

Regulatory control and environmental pathways

1
Persistent organic pollutants in foods: science, policy and regulation

D. N. Mortimer, Food Standards Agency, UK

DOI: 10.1533/9780857098917.1.3

Abstract: This chapter provides a broad overview of the significance of persistent organic pollutants (POPs) in foods and illustrates how scientific evidence is used to formulate policy and implement regulations. It explains how analytical methodology and toxicological studies might influence decisions about further investigation. It also includes brief examples of how new scientific information might be used to revise a regulation and how the potential risks from new and emerging POPs might be addressed.

Key words: evidence, policy, regulation, emerging, risk.

1.1 Introduction

Persistent organic pollutants (POPs) comprise a wide range of environmental chemicals that can occur in food. The most widely characterised and studied are undoubtedly the polychlorinated dibenzo-p-dioxins (PCDD) and dibenzofurans (PCDDF) and polychlorinated biphenyls (PCB), both of which were included in the first listing under the Stockholm Convention. Also of concern and the subject of significant research, particularly in the last decade, are the brominated flame retardants (BFR), such as polybrominated diphenyl ethers (PBDE), listed under the Stockholm Convention since 2009 (UNEP, 2009), and hexabromocyclododecanes (HBCDD), currently under consideration for listing. A further group of recently-recognised halogenated POPs are the perfluorinated alkyl substances, of which perfluorooctane sulphonate (PFOS), its salts and perfluorooctane sulphonyl fluoride were also added to the Stockholm list in 2009. Other categories of POPs are emerging, many with dioxin-like properties, including brominated and mixed halogenated dioxins, furans and biphenyls, polychlorinated naphthalenes and chlorinated and brominated polycyclic aromatic hydrocarbons (PAH).

Most POPs are anthropogenic in origin, although non-anthropogenic dioxins have been found in clay deposits from diverse geographical regions (Ferrario *et al.*, 2000). Due to their persistent nature and ability to undergo long-range transport, POPs are ubiquitous in the environment. They are found in soil, lake and river sediment, benthos, water columns and the atmosphere. They enter and accumulate in the food chain through various pathways, such as deposition onto crops, soil ingestion by grazing animals, and bioaccumulation up through trophic levels. At concentrations typically found in food, the adverse health effects caused by POPs are almost entirely chronic and include cancer, disruption of the endocrine system, neurotoxicity, and damage to the developing foetus. Recent reports also suggest that some POPs may be obesogens (La Merrill and Birnbaum, 2011).

1.2 Dietary exposure and total diet studies (TDSs)

Where a POP is known or suspected to be present in food, there are two general areas in which information is required in order to assess the risk of adverse health impacts through consumption: toxic effects (acute and chronic) and levels of dietary exposure (of the general population, high consumers and sensitive groups). Detailed toxicological characterisation of POPs is rarely available, with most information coming from animal testing, quantitative structure–activity relationship (QSAR) models and epidemiological studies. The development and application of analytical methods to measure the typically low concentrations of POPs in foods is challenging and can be costly, although the cost is likely to be around an order of magnitude lower than that of large toxicological studies. Consequently, to provide evidence of the need for the latter it is generally preferable to begin with an assessment of levels in the diet. In simple terms, this can be achieved in one of two ways – either by measuring levels in individual food samples, targeting those in which the compounds of interest are most likely to be found at significant levels and then estimating exposure through consumption of typical portions, or by carrying out a total diet study (TDS), which attempts to estimate the exposure to the compound(s) of interest through the whole diet. For a TDS, large numbers of samples are required to account as far as possible for the full range of variables that might affect the level of the compound of interest in an individual sample (Peattie *et al.*, 1983). To give an example, variables for a sample of beef may include the age, breed and gender of the source animal, its geographical origin, and the time of year it was slaughtered (due to seasonal differences in feeding). The latter is important as localised variations in levels of contamination are possible, for instance where cattle are grazed on flood plains (Lake *et al.*, 2006), feeding regimen and general health, as well as the possibility of further contamination between slaughter and reaching the eventual retail outlet. The contaminant levels to which the consumer would be exposed would then be further influenced by the methods of preparation, e.g. trimming fat from a portion of meat and cooking (grilling or frying might remove further fat, and therefore lipophilic compounds). Similar variables would exist for other food types. Testing large numbers of samples of every possible food type individually would be highly resource-intensive, especially if different methods of preparation and cooking also had to be taken into account. Generally, therefore, testing is carried out on composites of large numbers of individual samples, representing consumption over a significant period of time.

In the UK, the preferred approach has been to purchase samples from a random selection of different locations (cities, large and small towns, rural centres) in England, Scotland, Wales and Northern Ireland, weighted according to population distribution, over a 12-month period. This is intended to take account of geographical differences and differences in production scale as well as possible climatic and seasonal variation. Contaminants susceptible to such variation include agricultural contaminants such as nitrates, for which levels in leafy greens tend to be higher when there is less sunlight, and mycotoxins, which are likely to be higher when climate conditions favour growth of field and storage fungi. Seasonal variation may also impact sources of imported food, notably between the Northern and Southern hemispheres. Levels of environmental contaminants in food would not normally be affected by climate, although seasonal differences could be associated with changes to feeding and husbandry practices, for example in winter when the main source of exposure changes from pasture to feed. Even then, short term changes would only be expected to be seen in milk and possibly free range eggs. Geochemical sources may account for geographical differences in metal concentrations in food where, for example, lead ores may lie just below or even extrude through topsoil. In the case of POPs, however, whilst they are ubiquitous at low levels, localised hotspots are also possible and these would be difficult to take account of in a randomised, widespread sampling programme. The long half-life of most POPs in animals also means that seasonal fluctuations in exposure are unlikely to be associated with significant variations in levels in the animal. Consequently, a TDS for POPs may not require full 12-month sampling.

Once individual samples have been collected they are prepared and cooked using a typical range of methods, then they are homogenised and grouped into composites. The choice of groups will be influenced by the nature and detail of the consumption information available. For instance, the UK uses 20 food groups (Table 1.1).

The individual food groups can be tested for the POPs of interest and population exposure may be calculated based on information about the quantities of each food group in the typical diet, gathered as part of a rolling programme of National Diet and Nutrition Surveys (NDNS, Department of Health, 2011). Adjustments can be made for high consumers of specific food

6   Persistent organic pollutants and toxic metals in foods

Table 1.1   Food groups used for TDSs in the United Kingdom

Bread	Oils and fats	Fresh fruit
Miscellaneous cereals	Eggs	Fruit products
Carcass meats	Sugars and preserves	Beverages
Offal	Green vegetables	Milk
Meat products	Potatoes	Dairy products
Poultry	Other vegetables	Nuts
Fish	Canned vegetables	

groups (for example, those who consume a lot of fish) as well as different eating habits across age and special interest groups (vegetarians, vegans etc.).

The information generated by a TDS has a number of uses. For contaminants newly identified as a risk to the food chain, TDS provides a means of estimating the scale of any problem and assessing the need for further investigations, whether toxicological or analytical. Alternatively, once a need for reduction or control measures has been recognised, TDS provides a baseline against which the effectiveness of any future measures can be assessed. Subsequent TDSs then allow downward trends in levels in food groups to be verified and reductions in dietary exposure to be quantified.

TDS provide other valuable information as well, such as identifying the food groups that constitute the main dietary sources for the contaminants of interest or those with the highest levels of contamination (which need not be the same). For the former, particularly where the scope for reducing the levels of contamination present may be limited, control measures may include the provision of precautionary dietary advice. The latter, on the other hand, may be the subject of more targeted investigations to understand the reasons for the higher contamination. Foods in these groups may require regulatory limits for contaminant levels as a risk management tool, providing a simple basis for removing the most contaminated products from the food chain or placing the onus on producers to seek ways of reducing contamination levels.

1.3   Risk assessment, policy making and regulatory limits

This section briefly examines the means by which any risk to health may be assessed and, where a potential risk is identified, how it might be mitigated. It also considers the possible impact on mitigation measures when new scientific information becomes available.

1.3.1   Risk to health

TDSs and the measurement of levels of contaminants in individual food samples provide one part of the information needed for risk assessment.

The other essential part is toxicological information i.e. the potential adverse health effects of exposure to a particular contaminant or group of contaminants. When there is a known and quantified acute effect, it is relatively straightforward to estimate the risk, which can be associated with a single dose. However, for the levels at which most POPs are found in food, the effects of concern are generally long-term and chronic (cancer, disruption of the endocrine system), and are associated with continuous exposure and/or gradual accumulation in the body. For some POPs, it has been possible to identify effect thresholds i.e. levels of exposure below which, when averaged over a long period of time, there should be no adverse health effects. This is the case for dioxins, for which a provisional tolerable monthly intake (PTMI) of 70 pg WHO-TEQ/kg bodyweight (i.e. toxic equivalence based on toxic equivalency factors established by the World Health Organization) was established by the Joint FAO/WHO Expert Committee on Food Additives (JECFA) in 2001 (JECFA, 2001). This value was considered to be protective for the most sensitive effect, that of developmental defects in the male foetus. Based on much the same toxicological evidence, the UK Committee on Toxicity of Chemicals in Food, Consumer Products and Environment (COT) recommended a tolerable daily intake (TDI) for dioxins of 2 pg WHO-TEQ/kg bodyweight (Food Standards Agency, 2001). Again, in 2008 the Panel on Contaminants in the Food Chain (CONTAM) panel of the European Food Safety Authority (EFSA) assigned a TDI of 150 ng/kg bodyweight to PFOS, based on effects on the thyroid hormone system and a TDI of 1.5 µg/kg bodyweight to perfluorooctanoic acid (PFOA) based on adverse liver effects (EFSA, 2008a). On a cautionary note, however, it is important to bear in mind that threshold effect levels may change in the light of new toxicological information.

In February 2012, the US Environmental Protection Agency (USEPA) published a reassessment of dioxin toxicity, establishing a reference dose of 0.7 pg WHO-TEQ/kg bodyweight (USEPA, 2012). This was reviewed by the UK COT, which concluded that there was no significant new evidence and that the derived value was broadly in agreement with the TDI of 2 pg WHO-TEQ/kg bodyweight applied in the UK and Europe (Food Standards Agency, 2012). On the other hand, recent reassessments of evidence for cadmium and lead by EFSA led to a roughly threefold reduction of the tolerable weekly intake (TWI) for cadmium, from 7.0 to 2.5 µg/kg bodyweight, and the removal of the TWI for lead because it was not possible to identify a lower threshold for neurodevelopmental effects in children (EFSA, 2009; EFSA, 2010a). Such changes can have a major impact on the strategies adopted to protect the consumer from the adverse health effects of contaminants in food, making it essential that policy makers are aware of and understand the changes.

In contrast to contaminants for which thresholds can be identified, genotoxic carcinogens i.e. chemicals that cause cancer by damaging DNA, for example certain PAHs, can have no threshold since it is not possible to define a safe level of exposure. Risk assessment may then be based on a Margin of Exposure (MoE) approach. This involves calculating the difference between the benchmark dose level (BMDL), which is the lower confidence limit on the level of exposure leading to a specific adverse outcome in laboratory studies, and the estimated dietary exposure. The magnitude of the difference is used to determine the level of risk to health. In the case of PAHs, MoEs in the order of 7–10 000 were considered of low concern for health (EFSA, 2008b). This approach may also be used for chemicals that have been shown in laboratory studies to exert toxic effects other than carcinogenicity but for which there is insufficient evidence for a 'no observable adverse effect level' (NOAEL). This was the approach taken by EFSA, for example, in their opinion on PBDEs in food, in which they estimated the MoE for neurodevelopmental effects. For most congeners there was no indication of a risk to health but, in the case of BDE-99, the low MoEs for normal and high-consuming children in the 1–3 year age group did indicate a concern for health through dietary exposure (EFSA, 2011a).

The toxicology and risk to health of POPs is discussed in greater detail in later chapters of this book.

1.3.2 Controlling the risk

Where health concerns through dietary exposure to POPs are identified, or there is insufficient information to be confident that no health concern exists, precautionary advice may be given for consumers to limit their intake of certain types of food, for example oily fish contains higher levels of POPs, or advice may be directed at particular subgroups to avoid certain foods e.g. pregnant women not to eat large predatory fish due to mercury content. Such advice may provide adequate protection when set against the risks to health but, in some cases, it may be necessary to go further and introduce regulatory limits for the purposes of risk management. European food law prohibits the placing of unsafe food on the market and provides scope for enforcement action, for instance where a batch of food is found to be contaminated, but, in the absence of defined limits, this action must be taken on the basis of a risk assessment. Risk assessments are subjective and different interpretations by various European Union (EU) Member States can lead to disagreement over the appropriate measures to take in the event of a contamination incident. Regulatory limits provide a much more straightforward tool for managing risk. However, because adverse effects due to POPs arise from low level exposure over a long period, it is not possible to associate a limit directly with a safe level of exposure, as might be the case when there is an acute risk.

Where there is not a direct link to safety, limits for contaminants such as POPs effectively have a two-fold function. Firstly, they provide a simple basis on which to prevent entry into the food chain of any food containing contamination above the relevant limit and to withdraw from sale or, if necessary, recall (i.e. advise consumers to return products already purchased) contaminated food that has already entered the food chain. Secondly, they allow pressure to be exerted on food businesses to take measures to monitor and, if necessary, reduce the level of contaminants in their products. Nevertheless, limits must not be disproportionate, nor must they place an unnecessary burden on food businesses, notably primary producers who have little control over the adventitious contamination of their produce from the environment, and they must therefore still be justifiable on public health grounds. In principle, this is done by attempting to take account of the full range of background levels of contamination found throughout the EU (and in third countries, if possible) and establishing the limit at a level that would exclude only the most contaminated foods. This is a sound approach in general, but it is reliant on having an adequate and representative set of data and this may not always be the case. It is therefore also important to be able to reassess and modify regulatory limits as fresh evidence becomes available. Within Europe, this is embraced in Article 7 of Regulation 178/2002 on the General Principles of Food Law, which stipulates that risk management measures must be proportionate and should be reviewed when new scientific information, which could include further data on levels in food as well as fresh information for risk assessment, becomes available (Regulation 178/2002).

Within Europe, limits for dioxins and PCBs are currently set out in Commission Regulation (EC) No 1881/2006 as amended by Commission Regulation (EC) No 1259/2011. These are shown in Table 1.2 (Commission Regulation 1259/2011).

1.3.3 Reviewing a limit

A good example of the process by which new scientific evidence can lead to a reassessment of a limit is provided by the case of dioxins in sheep liver. When the limit for dioxin in liver was originally set in 2002 at a level of 6.0 pg WHO-TEQ/g fat, there were only very limited data available covering all relevant species. A survey for dioxins and PCBs in offal, reported by the UK in 2006, highlighted a non-compliance rate of almost 50% in sheep liver (Fernandes et al., 2010). There was less concern about liver from other species, with the exception of venison liver, which was much more contaminated (but not covered by the regulatory limit). There was no evidence of a specific reason for the apparently high contamination. All of the samples had been purchased at random from different locations around the UK and there were differences in congener profiles. It was also noted that the dioxin/PCB ratios were much higher than were normally seen in ovine meat. It was postulated that, as well as being dissolved in fat, dioxins were binding to liver protein and that the original limit had, in fact, not only been set at an incorrect level but also wrongly on a fat basis. The Commission called for more data and these were supplied by a number of Member States over several years. Germany in particular reported a high rate of non-compliance with much average higher levels even than those reported by the UK. The UK itself produced further data, this time for paired samples of meat and liver taken at slaughter. It was possible to demonstrate from the levels of dioxins and PCBs in the shoulder meat that the animals had been exposed only to normal background levels.

Table 1.2 European limits for dioxins, furans and PCBs, effective from 1 January 2012

Foodstuffs		Maximum levels		
		Dioxins and furans (WHO-TEQ)	Sum of dioxins, furans and dioxin-like PCBs (WHO-TEQ)	Sum of CBs 28, 52, 101, 138, 153 and 180
5.1	Meat and meat products (excluding edible offal) of the following animals:			
	- bovine animals and sheep	2.5 pg/g fat	4.0 pg/g fat	40 ng/g fat
	- poultry	1.75 pg/g fat	3.0 pg/g fat	40 ng/g fat
	- pigs	1.0 pg/g fat	1.25 pg/g fat	40 ng/g fat
5.2	Liver of terrestrial animals referred to in 5.1 and derived products thereof.	4.5 pg/g fat	10.0 pg/g fat	40 ng/g fat
5.3	Muscle meat of fish and fishery products and products thereof, with the exemption of: - wild caught eel - wild caught fresh water fish, with the exception of diadromous fish species caught in fresh water - wild caught char originating in the Baltic region - wild caught river lamprey originating in the Baltic region - wild caught trout originating in the Baltic region - fish liver and derived products - marine oils The maximum level for crustaceans applies to muscle meat from appendages and abdomen. In case of crabs and crab-like crustaceans (*Brachyura* and *Anomura*) it applies to muscle meat from appendages.	3.5 pg/g wet weight	6.5 pg/g wet weight	75 ng/g wet weight

5.4	Muscle meat of wild caught fresh water fish, with the exception of diadromous fish species caught in fresh water and products thereof.	3.5 pg/g wet weight	6.5 pg/g wet weight	125 ng/g wet weight
5.5	Muscle meat of wild caught eel (*Anguilla anguilla*) and products thereof.	3.5 pg/g wet weight	10.0 pg/g wet weight	300 ng/g wet weight
5.6	Fish liver and derived products thereof with the exception of marine oils referred to in point 5.7.	–	20.0 pg/g wet weight	200 ng/g wet weight
5.7	Marine oils (fish body oil, fish liver oil and oils of other marine organisms intended for human consumption).	1.75 pg/g fat	6.0 pg/g fat	200 ng/g fat
5.8	Raw milk and dairy products, including butter fat.	2.5 pg/g fat	5.5 pg/g fat	40 ng/g fat
5.9	Hen eggs and egg products.	2.5 pg/g fat	5.0 pg/g fat	40 ng/g fat
5.10	Fat of the following animals:			
	- bovine animals and sheep	2.5 pg/g fat	4.0 pg/g fat	40 ng/g fat
	- poultry	1.75 pg/g fat	3.0 pg/g fat	40 ng/g fat
	- pigs	1.0 pg/g fat	1.25 pg/g fat	40 ng/g fat
5.11	Mixed animal fats.	1.5 pg/g fat	2.50 pg/g fat	40 ng/g fat
5.12	Vegetable oils and fats.	0.75 pg/g fat	1.25 pg/g fat	40 ng/g fat
5.13	Foods for infants and young children.	0.1 pg/g wet weight	0.2 pg/g wet weight	1.0 ng/g wet weight

12   Persistent organic pollutants and toxic metals in foods

Due to the sensitivity associated with potentially relaxing contaminant limits, the Commission sought an opinion from EFSA as to the risk to health from the consumption of sheep liver. This opinion was published in July 2011 (EFSA, 2011b). Although identifying a possible risk in the event of high consumption by sensitive consumer subgroups, EFSA concluded that the risk to health was generally low and therefore there was no need for a ban on sheep liver consumption. To further reinforce the scientific understanding, the Commission also asked the European Union Reference Laboratory (EURL) to investigate different extraction methods. A surprising outcome of this was that, for the same sheep liver sample, dioxin concentrations varied significantly when expressed on a fat basis but much less when expressed on a whole weight basis (EURL, 2012). Following the EFSA opinion and EURL report, there was progress towards a consensus with the Commission and Member States that the limit for dioxin in sheep liver should, indeed, be amended.

At the time of writing, the limit for sheep liver is still under review.

Regulations establishing limits for contaminants in food are accompanied by regulations laying down the criteria for sampling and analysis for Official Controls. In the case of dioxins and PCBs, these are set out in Commission Regulation (EC) No 252/2012. Although legally required only for Official Controls, it is always recommended that they be followed for informal and commercial samples in order to maintain a good and reliable quality of results.

1.4   Enforcement and implications for food businesses

The authority to set limits for contaminants in food is conferred on the Commission through Article 2.3 of Council Regulation (EEC) No 315/93 of 8 February 1993 laying down Community procedures for contaminants in food (Council Regulation 315/93). Article 5.1 of the same regulation states that *'Member States may not prohibit, restrict, or impede the placing on the market of foods which comply with this Regulation or specific provisions adopted pursuant to this Regulation for reasons relating to their contaminant levels'*. This is important as it illustrates how the establishment of regulatory limits for contaminants in food can ensure smooth trade and provides food businesses with a level playing field on which to operate. Regulatory limits can even facilitate production and trade. For example, fish liver often contains a very high level of dioxins and PCBs, particularly in low-oil species such as cod, which have very little body fat. Canned fish liver is an important fisheries product for a number of EU Member States, especially in Eastern Europe. Before 2008, high dioxin levels in fish liver was one of the most frequent contamination incidents reported on the European Commission's Rapid Alert System. In the absence of a regulatory limit, all of the enforcement action carried out was done on the basis of a risk assessment. The fisheries industry itself asked the Commission if it would be possible to set a limit, even if this would lead to a significant rate of exclusion from the market, simply so that the industry was able to continue to produce and sell fish liver. With the limit in place, the industry would have confidence that no enforcement action would be taken against product that complied with the established limit. Accordingly, in June 2008 the Commission published Commission Regulation (EC) No 565/2008 as an amendment to Regulation 1881/2006, setting a maximum level for dioxins and PCBs in fish liver (Commission Regulation 565/2008). Rapid Alerts for non-compliance of canned fish liver are now much less frequent.

General food law puts the onus on food business operators not to place unsafe food on the market, and food not complying with regulatory limits is deemed unsafe. This means that, for the purposes of due diligence, food businesses must be able to demonstrate that they have taken appropriate measures to ensure that they do not produce or supply non-compliant food. This may include in-house or contract testing of higher-risk raw materials, the frequency of which would also need to be justified, as well as placing strict product specifications on suppliers.

Although the burden is placed on food businesses to ensure the compliance of their products with food safety regulations, the EU provides for the carrying out of Official Controls through Regulation No 882/2004 of the European Parliament and Council (Regulation 882/2004). This regulation covers food and feed as well as animal health and animal welfare. It provides for random testing to be carried out by the competent authority to ensure that food products are, indeed, compliant with regulations. Article 3.1 of Regulation 882/2004 stipulates that Official Controls should be carried out on a risk basis. It allows for the past performance of food business operators and the results of their own testing to be taken into account, but also indicates that account should be taken of information about suspected non-compliance. In the case of dioxins and PCBs, the view taken in the UK, where food law enforcement is delegated to several hundred local authorities, is that random Official Controls are of limited benefit. Because the limits are based at the higher end of the range of normal data, the likelihood of detecting a non-compliance through random testing is very low. Combined with the slow sample turnaround time (typically 3–4 weeks), this is not regarded as an effective use of limited resource to detect problems, particularly since the risk to health in the event of a single non-compliance is likely to be low. Alternative approaches include targeting higher-risk food categories with intensive surveys, directing resource to better controls on animal feed, which has been the main source of contamination in some of the largest previous dioxin incidents, or studying environmental contamination pathways to gain a better understanding of uptake into food. An example of the latter is the investigation of contaminant levels in food produced from animals raised on flood-prone land, which has been researched in a number of countries, including the UK (Lake *et al.*, 2006; 2013).

This view of Official Controls is not shared by all EU Member States. In the Netherlands, for instance, large-scale testing of milk is carried out at State level using bioassay screening techniques. Individual German landers also operate monitoring programmes at the State level. At the time of writing, the regulations on Official Controls are no more prescriptive. However, they are under review.

1.5 Analytical methods and their influence on policy

It is not the intention to discuss analytical methods for POPs in detail in this section. Nevertheless, when Official Controls are carried out it is essential to have effective and robust analytical methods, since results may be open to challenge in the event of enforcement action. Good and reliable methods are similarly important when investigating emerging risks, since the results will be used for making important policy decisions about the direction of further work. It is therefore useful to have a brief overview of analytical methodology for POPs.

Measurement of the concentration of POPs in food by chemical means generally requires multi-step methods comprising extraction, clean-up, separation using gas or liquid chromatography and detection by low or high-resolution mass spectrometry. Food is a complex matrix and, as individual compounds or congeners are measured at part per billion or even sub-part per billion levels, extracts must be very clean in order to avoid chromatographic interference and contamination of equipment. A typical method will involve fortification with internal standards, extraction into non-polar organic solvent, purification using activated silica and activated carbon, possibly followed by alumina and, finally, concentration. Purified extracts will be further separated using high performance gas or liquid chromatography and quantification carried out using low- or high-resolution mass spectrometry. For such complex analyses, reliable results can generally be achieved only by using internal standards labelled with isotopes such as ^{13}C, the synthesis of which is also technically very challenging and consequently adds significantly to the cost of the analysis. The availability of suitable standards is also a limiting factor on the scope of an analysis and this is of particular concern in the case of POPs identified as emerging risks, since there may not be sufficient demand for commercially-available standards. These therefore have to be custom-synthesised. The difficulties this presents are most notably highlighted in the case of mixed halogenated dioxins, furans and biphenyls, for which there are many more possible congeners than is the case for chlorinated dioxins and PCBs. Even so, analytical standards for mixed halogenated compounds are now becoming commercially-available (Fernandes *et al.*, 2011)

Bioassay techniques are available for some POPs. An example is the DR CALUX® (Chemical-Activated LUciferase gene eXpression) method for the measurement of dioxins and dioxin-like PCBs (Bovee *et al.*, 1998). This relies on binding to the aryl hydrocarbon receptor of a transporter protein, which subsequently triggers a luminescent response, enabling quantification.

Such techniques offer certain advantages over conventional analysis, including simpler sample preparation, greater automation, faster turnaround and lower cost for a large sample throughput. However, they do not provide full congener profiles and they lack the precision of the conventional chemical techniques. Consequently, confirmation by GC-HRMS is needed for all suspect results, together with a proportion of normal results. This is reflected, for example, in Commission Regulation 252/2012, which permits the use of bioassay screening techniques for Official Control of dioxins and dioxin-like PCBs (Regulation 252/2012). Immunoassay techniques based on polyclonal antibodies have also been marketed for dioxins in food, although to date sufficiently adequate and consistent performance has not been demonstrated for use in Official Controls (Harrison and Carlson, 2000).

1.6 Future trends and conclusions

Improvements in analytical capability have undoubtedly been a factor in the increased interest and concern about POPs in food. Over the course of the past three decades in particular, it has become increasingly common for analytical laboratories to achieve detections in the sub-nanogram per kilogram (part per trillion) range. Contrast this with the action level initially established by the US Food and Drug Administration (FDA) during the serious polybrominated biphenyl (PBB) contamination incident in Michigan in 1974 (Reich and Amer, 1983).This was set at 1.0 ppm on a fat basis in milk and meat from affected animals, which was, at the time, at the limit of analytical capability. It was only when it became apparent that adverse health effects were occurring in animals at well below this level that the regulators and analysts were driven to reduce the action level and therefore improve analytical capability. However, the use of much more sensitive analytical methods has drawbacks of its own. It has become significantly more common to find stable, lipophilic and bioaccumulative compounds at very low levels in food samples but there is often a shortage of toxicological information on which to base any assessment of the associated risk to health. This leaves policy makers and regulators with further knowledge gaps that need to be filled.

Brominated flame retardants provide a good example of the dilemma that policy makers and regulators are faced with and the process under which work on emerging risks might progress. After a ban on the use of PBBs in the 1970s, their main replacement compounds were the polybrominated diphenyl ethers (PBDE). Concerns about the presence of PBDEs in the environment were first raised in the early 1980s, and were investigated on a significant scale in the USA and in Europe from the mid-1990s. They were also reported to be present in human body fluids such as breast milk (Meironyte *et al.*, 1999). Measurement of PBDE levels in food began in the USA and Europe in about 2000. Toxicological studies also began in earnest around 2000. Food was recognised as likely to be a significant source of human exposure and investigations into PBDE levels have been ongoing in Europe, the United States and elsewhere since 1999. The first UK study involved levels of PBDEs and hexabromocyclododecanes (HBCDD) in trout and eels downstream from a BFR production facility in the north-east of England, which was followed by a first attempt to carry out a TDS (Food Standards Agency, 2004).

The manufacture and use of penta- and octa-PBDE formulations were banned in Europe from 2004, and use has been phased out on a voluntary basis in the USA. As noted previously, certain PBDE congeners were listed under the Stockholm Convention in 2009 (UNEP, 2009). Despite bans and controls, PBDEs are now widespread in the environment and remain present in many consumer products such as textiles and furniture from which they can escape during use and at the time of disposal. They therefore continue to be of concern, not only as a contaminant of the human diet and a potential contaminant of the animal feed chain, but also through exposure from other sources such as household dust (Harrad et al., 2008). In order to assess the risk to health from PBDEs and other categories of BFRs and the need for any additional regulatory controls, in 2005 the European Commission approached the EFSA Scientific Panel on Contaminants in the Food Chain (CONTAM) for a preliminary view on a suitable approach. CONTAM published an advisory note in February 2006 recommending the collection of data for the occurrence in food of PBDEs, together with PBBs, HBCDDs, tetrabromobisphenol A (TBBPA) and three emerging BFRs: decabromodiphenyl ethane, hexabromobenzene and bis (2,4,6-tribromophenoxy)ethane (EFSA, 2006). Over the course of the following three years, Member States provided existing data and collected fresh data, which was submitted to EFSA and the Commission. EFSA received a further formal request for a scientific opinion on these compounds, splitting the work into six sections: PBBs, PBDEs, HBCDDs, TBBPA, polybrominated phenols and novel and emerging BFRs. The first five of these opinions were published between October 2010 and April 2012, with the sixth being prepared for publication at the time of writing (EFSA 2010b; 2011a, 2011c, 2011d; 2012). With the exception of PBBs, which require no further consideration, EFSA identified significant gaps in data and knowledge, as well as a potential health concern for a specific PBDE congener, BDE-99. The final opinion is likely to identify further knowledge gaps as little is known of newer BFRs.

The picture is further complicated because global trade and long-distance environmental transport mechanisms mean that the European environment and food chain are not necessarily protected by regulatory controls on the use of chemicals, such as the Registration, Evaluation, Authorisation and Restriction of Chemicals (REACH) regulations (Regulation 1907/2006). A novel BFR, details of which are withheld on the grounds of commercial confidentiality, may be developed in one country, for instance the USA, manufactured in bulk in a second country, e.g. India or Japan, transported to a third country, such as China, for use in the manufacture of consumer goods which are then exported to further countries, possibly in Europe. Controls at this stage would be focussed on meeting fire safety regulations rather than the chemicals they contain. Finally, at the end of their useful life, the same consumer goods may find their way to Africa for disposal.

Despite difficulties in discovering the identity of individual new-generation BFRs, the great power and sensitivity of current analytical methods means that more and more new and emerging POPs are being detected in environmental, food and tissue samples without a full understanding of their significance. In the face of this, policy makers and regulators must decide where to direct increasingly limited resources, after which the science and policy cycle begins again: measure dietary exposure, assess the risk to health, investigate levels in food, identify the necessary health protection measures, ensure if robust methodology is in place, implement controls and periodically review their effectiveness.

1.7 References

BOVEE TF, HOOGENBOOM LA, HAMERS AR, TRAAG WA, ZUIDEMA T, AARTS JM, BROUWER A and KUIPER HA. (1998). Validation and use of the CALUX-bioassay for the determination of dioxins and PCBs in bovine milk. *Food Additives and Contaminants*, **15**(8), 863–875.

COMMISSION REGULATION 1259/2011. Commission Regulation (EU) No 1259/2011 of 2 December 2011 amending Regulation (EC) No 1881/2006 as regards maximum levels for dioxins, dioxin-like PCBs and non dioxin-like PCBs in foodstuffs. OJ L320, 3 December 2011, p.18.

COMMISSION REGULATION 252/2012. Commission Regulation (EU) No 252/2012 of 21 March 2012 laying down methods of sampling and analysis for the official control of levels of dioxins, dioxin-like PCBs and non-dioxin-like PCBs in certain foodstuffs and repealing Regulation (EC) No 1883/2006. OJ L84, 23 March 2012, p.1.

COMMISSION REGULATION 565/2008. Commission Regulation (EC) No 565/2008 of 18 June 2008 amending Regulation (EC) No 1881/2006 setting maximum levels for certain contaminants in foodstuffs as regards the establishment of a maximum level for dioxins and PCBs in fish liver. OJ L160, 9 June 2008, p.20.

COUNCIL REGULATION 315/93. Council Regulation (EEC) No 315/93 of 8 February 1993 laying down Community procedures for contaminants in food. OJ L37, 13 February 1993, p.1.

DEPARTMENT OF HEALTH (2011). National Diet and Nutrition Survey, available at http://www.dh.gov.uk/en/Publicationsandstatistics/PublishedSurvey/ListOfSurvey Since1990/Surveylistlifestyle/DH_128165

EFSA (2006). Advice of the Scientific Panel on Contaminants in the Food Chain (CONTAM) on a Request from the Commission Related to Relevant Chemical Compounds in the Group of Brominated Flame Retardants for Monitoring. *Feed and Food EFSA Journal*, **328**, 1–4.

EFSA (2008a). Opinion of the Scientific Panel on Contaminants in the Food Chain (CONTAM) on perfluorooctane sulfonate (PFOS), perfluorooctanoic acid (PFOA) and their salts. *EFSA Journal*, **653**, 1–131.

EFSA (2008b). Scientific Opinion of the Panel on Contaminants in the Food Chain (CONTAM) on a request from the European Commission on Polycyclic Aromatic Hydrocarbons (PAHs). *Food EFSA Journal*, **724**, 1–114.

EFSA (2009). Scientific Opinion of the Panel on Contaminants in the Food Chain (CONTAM) on a request from the European Commission on cadmium in food. *EFSA Journal*, **980**, 1–139.

EFSA (2010a). Panel on Contaminants in the Food Chain (CONTAM); Scientific Opinion on Lead. *Food EFSA Journal*, **8**(4), 1570.

EFSA (2010b). Panel on Contaminants in the Food Chain (CONTAM); Scientific Opinion on Polybrominated Biphenyls (PBBs). *Food EFSA Journal*, **8**(10), 1789.

EFSA (2011a). Panel on Contaminants in the Food Chain (CONTAM); Scientific Opinion on Polybrominated Diphenyl Ethers (PBDEs). *Food EFSA Journal*, **9**(5), 2156.

EFSA (2011b). Panel on Contaminants in the Food Chain (CONTAM); Scientific Opinion on the risk to public health related to the presence of high levels of dioxins and dioxin-like PCBs in liver from sheep and deer. *EFSA Journal*, **9**(7), 2297.

EFSA (2011c). Panel on Contaminants in the Food Chain (CONTAM); Scientific Opinion on Hexabromocyclododecanes (HBCDDs). *Food EFSA Journal*, **9**(7), 2296.

EFSA (2011d). Panel on Contaminants in the Food Chain (CONTAM); Scientific Opinion on Tetrabromobisphenol A (TBBPA) and its derivatives in food. *EFSA Journal 2011*, **9**(12), 2477.

EFSA (2012). Panel on Contaminants in the Food Chain (CONTAM); Scientific Opinion on Brominated Flame Retardants (BFRs) in Food: Brominated Phenols and their Derivatives. *EFSA Journal 2012*, **10**(4), 2634.

EURL (2012). Application of different extraction methods for determination of PCDD/Fs and PCBs in sheep liver. Ref.: 5477.10–15 EURL-Dioxin.

FERNANDES A, MORTIMER D, ROSE M and GEM M. (2010). Dioxins (PCDD/Fs) and PCBs in offal: Occurrence and dietary exposure. *Chemosphere*, **81**(4), 536–540.

FERNANDES A, ROSE MD, MORTIMER DN, CARR M, PANTON S and SMITH F. (2011). Mixed brominated/chlorinated dibenzo-p-dioxins, dibenzofurans and biphenyls: Simultaneous congener-selective determination in food. *Journal of Chromatography A*, **1218**, 9279–9287.

FERRARIO JB, BYRNE CJ and CLEVERLY DH. (2000). 2,3,7,8-Dibenzo-p-dioxins in mined clay products from the United States: evidence for possible natural origin. *Environmental Science and Technology*, **34**, 4524–4532.

FOOD STANDARDS AGENCY (2001). COT statement on the tolerable daily intake for dioxins and dioxin-like polychlorinated biphenyls. Available at http://cot.food.gov.uk/cotstatements/cotstatementsyrs/cotstatements2001/dioxinsstate

FOOD STANDARDS AGENCY (2004). Brominated flame retardants in trout and eels from the Skerne-Tees river system and total diet study samples. Food Survey Information Sheet 52/04, available at http://www.food.gov.uk/multimedia/pdfs/fsis5204pdf.pdf

FOOD STANDARDS AGENCY (2012). Minutes of the meeting of the Committee on Toxicity of Chemicals in Food, Consumer Products and the Environment held on Tuesday, 20 March 2012. Available at: http://cot.food.gov.uk/cotmtgs/cotmeets/cotmeet2012/cotmeet20march2012/cotmin20mar2012

HARRAD S, IBARRA C, ABDALLAH MA, BOON R, NEELS H and COVACI A. (2008). Concentrations of brominated flame retardants in dust from United Kingdom cars, homes, and offices: Causes of variability and implications for human exposure. *Environment International*, **34**, 1170–1175.

HARRISON, RO and CARLSON, RE. (2000). Simplified sample preparation methods for rapid immunoassay analysis of PCDD/Fs in foods. *Organohalogen Compounds*, **45**, 192–195.

JECFA (2001). Joint FAO/WHO Expert Committee on Food Additives, Fifty-seventh meeting, Rome, 5–14 June 2001. Summary and Conclusions. Available at http://www.who.int/pcs/jecfa/jecfa.htm

LA MERRILL M and BIRNBAUM LS. (2011). Childhood obesity and environmental chemicals. *Mount Sinai Journal of Medicine*, **78**(1), 22–48.

LAKE IR, FOXALL CD, LOVETT AA, FERNANDES A, DOWDING A, WHITE S and ROSE MD. (2006). Effects of river flooding on PCDD/F and PCB levels in cows' milk, soil, and grass. *Environmental Science and Technology*, **39**(23), 9033–9038.

LAKE IR, FOXALL CD, FERNANDES A, LEWIS M, ROSE M, WHITE O, and DOWDING A (2013). Seasonal variations in the levels of PCDD/Fs, PCBs and PBDEs in cow's milk. *Chemosphere*, **90**, 72–79.

MEIRONYTE D, NOREN K and BERGMAN AJ. (1999). Analysis of polybrominated diphenyl ethers in Swedish human milk, a time-related trend study, 1972–1997. *Journal of Toxicology and Environmental Health Part A*, **58**(6), 329–341.

PEATTIE ME, BUSS DH, LINDSAY DG and SMART GQ. (1983). Reorganisation of the British Total Diet Study for monitoring food constituents from 1981. *Food and Chemical Toxicology*, **21**, 503–507.

REGULATION 178/2002. Regulation (EC) No 178/2002 of the European Parliament and of the Council of 28 January 2002 laying down the general principles and requirements of food law, establishing the European Food Safety Authority and laying down procedures in matters of food safety. OJ L31, 1 February 2002, p.1.

REGULATION 882/2004. Regulation (EC) No 882/2004 of the European Parliament and of the Council of 29 April 2004 on official controls performed to ensure the verification of compliance with feed and food law, animal health and animal welfare rules. OJ L165, 30 April 2004, p.1.

REGULATION 1907/2006. Regulation (EC) No 1907/2006 of the European Parliament and of the Council of 18 December 2006 concerning the Registration, Evaluation, Authorisation and Restriction of Chemicals (REACH), establishing a European Chemicals Agency, etc.

REICH MR and AMER (1983). Environmental Politics and Science: The Case of PBB Contamination in Michigan. *Journal of Public Health*, **73**(3), 302–313.

UNEP (2009). Stockholm Convention on Persistent Organic Pollutants (POPs) as amended in 2009, available at: http://chm.pops.int/Convention/ConventionText/tabid/2232/Default.aspx

USEPA (2012). EPA's Reanalysis of Key Issues Related to Dioxin Toxicity and Response to NAS Comments, Volume 1 (CAS No. 1746–01–6). Environmental Protection Agency, February 2012.

2
Regulatory control and monitoring of heavy metals and trace elements in foods

K. D. Hargin and G. J. Shears, Food Standards Agency, UK

DOI: 10.1533/9780857098917.1.20

Abstract: The difficulties of setting maximum legal limits for heavy metals and trace elements in foods are highlighted. This chapter discusses ways of determining and ensuring consumer safety and the significant roles played by risk assessment and risk management in the development of appropriate and proportionate food legislation. The importance of, and continued improvements in, analytical techniques are also discussed.

Key words: legislation, Total Diet Studies, risk assessment, heavy metals, policy.

2.1 Introduction

The monitoring and control of heavy metals and trace elements in food is important in helping to ensure that consumer health is not unduly compromised. While some trace elements, e.g. zinc, iron, etc., also have nutritional benefits many more are harmful and this may vary according to their oxidation state, for instance chromium. It is the responsibility of governments to ensure that a population's diet is not exposed unnecessarily to harmful substances, while producers and manufacturers are responsible for ensuring foods are safe to eat. The processes and procedures undertaken by regulators to ensure the safety of the nation's diet are necessarily complex and risk assessments and risk management options have to be determined through appropriate sampling and analyses of foodstuffs.

The large number of potentially toxic chemicals in the diet seems to be ever increasing and challenges constantly face regulators in determining human exposure. In addition to routine and targeted surveillance programmes, an important tool for governments is what is known as Total Diet Studies (TDS).

TDS are a cost-effective means for assuring that populations are not exposed to unsafe levels of toxic chemicals in food (see Section 2.3.2). They also help to indicate environmental contaminants, e.g. persistent organic pollutants (POP), and help develop and assess the effectiveness of risk management programmes.

Surveys of biomarkers, for example in blood or urine, provide another tool to assess the impact on populations of trace elements or other toxins in the diet and to assess the actual body burden that these contaminants inflict. This along with TDS and consumption databases provide a mechanism to both establish whether current levels of trace elements in food are of concern (taking account of acceptable levels of exposure) and to measure policy impact in reducing consumer exposure at the population level.

The setting of legal maximum limits for contaminants, where this is possible, is a complicated task and involves many apparently conflicting interests. For example, it is of little value setting a limit for, say, a heavy metal in a staple foodstuff which would effectively ban the food from sale if it meant that the population would suffer in some other way by its exclusion from their diet. In such instances other risk management options would have to be considered. These might include providing consumer advice, e.g. about washing and/or peeling of produce prior to consumption, or advising consumers on the maximum number of portions of a particular foodstuff that should be eaten per week. The regulators' role is a fine balancing act in an increasingly scientifically and economically diverse world.

2.2 Risk assessment and policy making

Good policy making requires that many different aspects, for example financial, social, scientific, etc., are taken into account to ensure proportionality and practicality. When considering heavy metals and trace elements, an important consideration is the need to determine the risk to public health, which is normally established through a risk assessment. This provides for sound, evidence-based legislation and will form an important component of any policy in this area.

2.2.1 Legislative background

In the European Union (EU), where possible, controls on trace elements as chemical contaminants in food are harmonised under EU food law, although national standards can apply where EU-wide limits have not been agreed. In England, for instance, the Arsenic in Food (England) Regulations 1959 are still in force at the time of writing, although they are under review.

European Commission Regulation (EC) No 178/2002, laying down the general principles and requirements of food law, and establishing the European Food Safety Authority (EFSA) and procedures in matters of food safety, sets out the basis for this harmonisation. The Regulation defines food safety requirements and balances a precautionary approach with the facilitation of free and unhindered trade, one of the founding principles of the Treaty establishing the European Economic Community. This is made clear early in the recitals of the Regulation, 'A high level of protection of human life and health should be assured in the pursuit of Community policies' and 'The free movement of food and feed within the Community can be achieved only if food and feed safety requirements do not differ significantly from Member State to Member State'.

The Regulation goes on to define risk, hazard, risk analysis, risk assessment and risk management in Article 3.

Food safety requirements are stipulated in Article 14, thus: 'food shall not be placed on the market if it is unsafe' and 'food shall be deemed to be unsafe if it is considered to be: (a) injurious to health; (b) unfit for human consumption'. This provides for the competent authorities of individual Member States to take action to protect consumers where maximum limits do not exist for a particular contaminant in a foodstuff but where, following a risk assessment, it is considered necessary to act to safeguard public health.

European Council Regulation (EEC) No 315/93 laying down Community procedures for contaminants in food defines a 'contaminant' as 'any substance not intentionally added to food which is present in such food as a result of the production (including operations carried out in crop husbandry, animal husbandry and veterinary medicine), manufacture, processing, preparation, treatment, packing, packaging, transport or holding of such food, or as a result of environmental contamination'. But it excludes 'Extraneous matter, such as, for example, insect fragments, animal hair, etc'. This Regulation also sets out the basis under which maximum limits and variants thereof may be established.

Commission Regulation (EC) No 1881/2006 (as amended) sets out in its annex maximum limits for certain contaminants in foodstuffs, and among other contaminants it lists limits for the metals lead, cadmium, mercury/methylmercury and inorganic tin in a range of foods (Table 2.1). The limits in this Regulation are directly applicable in all Member States (except where temporary derogations are agreed) and to all imports and are regularly reviewed and added to by the European Commission through their normal structure of Expert Working Groups and Committees. It is worth noting that there are also regulatory residue limits for some trace element contaminants in gelatine and collagen set out in Regulation (EC) No 853/2004 of the European Parliament and of the Council (as amended), laying down specific hygiene rules for the hygiene of foodstuffs.

The Directorate General for Health and Consumers (DG SANCO) of the EU Commission holds regular 'expert meetings', chaired by a Commission representative as part of the Commission's comitology procedures and is attended by national representatives. Final discussion and voting by Member States on legislation to introduce implementing measures that fall within the

Table 2.1 Summary of heavy metals and trace elements

Element	Symbol	Occurrence	Toxicity	EU regulatory limit* (mg/kg wet weight)	References
Aluminium	Al	One of the most abundant elements on earth. Widely found in soils and may accumulate in plants and in animals feeding on such plants. Found in many foods, with the highest concentrations in tea leaves, herbs, cocoa and cocoa products and spices.	The water-soluble forms of aluminium, e.g. aluminium chloride, can cause harmful effects. Uptake can be via oral, nasal or direct skin contact and significant levels can lead to damage to the central nervous system and kidneys, among other symptoms. Due to its cumulative effect, EFSA has recommended that a tolerable weekly (as opposed to daily) intake should be established.	n/a	Ščančar and Milačič, 2006; EFSA, 2008
Arsenic	As	Widely distributed in the environment, mostly from the burning of fossil fuels and mining activities but also from volcanoes and microbial activity. Arsenic is readily absorbed by most plants.	Can be highly toxic, particularly in the inorganic form, and is known to cause cancer of the bladder and lung. High level consumers in Europe are within the range of the $BMDL_{01}$ values identified by EFSA, and therefore there is little or no MOE. The possibility of a risk to some consumers cannot be excluded thus levels in food should be as low as possible.	n/a	Mandal and Suzuki, 2002; EFSA, 2009a
Barium	Ba	Occurs generally as barium sulphate or barium carbonate. Barium sulphate is present in soils but only a limited amount accumulates in plants. Main exposure routes are through drinking water and food, principally nuts.	Can cause gastroenteritis, hypokalemia and hypertension.	n/a	WHO, 2001

(Continued)

Table 2.1 Continued

Element	Symbol	Occurrence	Toxicity	EU regulatory limit* (mg/kg wet weight)	References
Cadmium	Cd	Cadmium can be found in the environment from the weathering of rocks, volcanoes and forest fires, though much is of anthropomorphic origin.	Cadmium collects in and damages the kidneys, and has a long biological half life (10–30 years). It may also cause cancer of the lungs, endometrium, bladder and breast.	From 0.05 for meat, vegetables and most fish to 1.0 for bivalve molluscs, cephalopods and muscle meat from certain fish.	Reeves and Chaney, 2008; EFSA 2009b
Chromium	Cr	Chromium exists in several different forms. It is mined as chromite ore and is used in alloys such as stainless steel, chrome plating and in metal ceramics. The levels of chromium in the air and water are generally low. Eating food with chromium III is the main source for most people, as it occurs naturally in many vegetables, fruits, meat, yeasts and grains.	Health effects are dependent on its oxidation state. Chromium III is an essential element, with shortages leading to heart conditions. However, excess will cause skin rashes. Chromium VI is a carcinogen and may also cause genetic damage.	n/a	Pechova and Pavlata, 2007
Copper	Cu	Widely distributed within the environment, released by human activity such as mining, metal production, etc., and by natural decay of vegetation and forest fires. Most copper compounds will bind to water sediment or soil particles. Water-soluble copper compounds are generally released into the environment as a result of agricultural uses.	As a trace element, copper is essential for health. However, chronic copper poisoning results in Wilson's Disease (hepatic cirrhosis, brain damage, renal disease and copper deposition in the cornea). Soluble copper compounds pose the greatest risk to human health.	n/a	Flemming and Trevors, 1989

Lead	Pb	Lead is found naturally in the environment, though mostly through human activities and occurs primarily in the inorganic form. Foods such as fruit, vegetables, meats, grains, seafood, soft drinks and wine may contain lead. Lead from pipes can contaminate drinking water.	Lead is highly toxic and poisoning can result in disruption to the nervous system, brain/kidney damage, and diminished learning abilities in children. It can be particularly harmful to the foetus, causing serious damage to the brain and nervous system. There is no evidence for a threshold of critical lead-induced effects.	From 0.020 for milk and milk products and infant formulae to 1.5 for bivalve molluscs.	Papanikolaou et al., 2005; EFSA, 2010
Manganese	Mn	Manganese is widely distributed in the environment and is one out of the three toxic essential elements. Main sources include grains, soya beans, eggs, nuts, olive oil, green beans and oysters, with spinach, tea and herbs being particularly important.	Manganese poisoning affects mainly the respiratory tract and the brain, leading to hallucinations, memory loss and nerve damage. Being an essential element, deficiency can also cause problems.	n/a	Aschner and Aschner, 2005; WHO, 2006b
Mercury	Hg	Mercury can be found in the environment as mercury salts or organic mercury compounds. Not naturally found in foodstuffs but can be found in the food chain as smaller organisms are consumed by larger ones. Certain large predatory fish can be significant sources of mercury in the human food chain. Not commonly found in plants (though may accumulate in mushrooms), vegetables and other crops can have traces due to agricultural applications.	Mercury causes disruption to the nervous system, damage to brain functions, DNA and chromosomal damage and may have adverse effects on the foetus and reproductive functions. Methylmercury toxicity has been demonstrated at low exposure levels and therefore should be minimised.	From 0.50 for most fishery products to 1.0 for certain species of fish.	Risher et al., 2002; EFSA, 2004

(*Continued*)

Table 2.1 Continued

Element	Symbol	Occurrence	Toxicity	EU regulatory limit* (mg/kg wet weight)	References
Selenium	Se	Selenium occurs naturally in the environment, though not abundantly. Selenium is often added to phosphate fertilisers and therefore levels in well-fertilised agricultural soil can be higher than normal. Main food sources are cereal grains and meat.	Accumulation of high levels of selenium can lead to reproductive failure and birth defects.	n/a	EVM, 2003; Navarro-Alarcon and Cabrera-Vique, 2008
Tin	Sn	There are few tin-containing ores but organic tin compounds are widespread and very persistent in the environment. They originate mainly from applications in the paint, plastics and agricultural industries.	The most toxic form of tin is the organic form, of which many different compounds exist. Their effects vary widely, depending on the actual substance, but generally result in eye irritations, sickness and dizziness. Long-term effects can include liver damage, impairment of the immune system and chromosomal damage.	Inorganic from 50 for canned infant foods to 200 for canned foods other than beverages.	Blunden and Wallace, 2003; EVM, 2003; WHO, 2006b
Zinc	Zn	Zinc is common and naturally occurring in soil and water. Soils heavily contaminated with zinc can be found close to zinc mining or refining areas, or where sewage sludge from industrial areas has been used as a fertiliser.	Zinc is an essential element, necessary for human health. However, too much zinc in the diet can cause stomach cramps, vomiting, skin irritations and anaemia. At very high levels protein metabolism and pancreatic function can be adversely affected. Zinc can also be harmful to unborn and newly born babies.	n/a	EVM, 2003; Frassinetti *et al.*, 2006; WHO, 2006b

* Commission Regulation (EC) No 1881/2006 of December 2006 setting maximum levels for certain contaminants in foodstuffs. OJ L 364, 20.12.2006, p5. MOE, Margin of Exposure. $BMDL_{01}$, Benchmark Dose – Lower Confidence Limit.

scope of Article 291 of the Treaty on the Functioning of the European Union (TFEU – 'Lisbon' Treaty) takes place at meetings of the Standing Committee on the Food Chain and Animal Health (SCFCAH), an Examination Committee (previously a 'Regulatory Committee', pre Lisbon). Delegated acts (under which regulations amending or introducing maximum limits are expected to fall) are dealt with under Article 290 of the TFEU and these measures include a period of scrutiny by the European Parliament before publication in the Official Journal of the EU and notification to Member States. The previous system of regulatory procedure with scrutiny (with voting at Regulatory Committees and transmission to the Council and Parliament) remains ongoing, while alignment to the TFEU takes place.

2.2.2 Setting maximum limits

When considering appropriate standards for food safety maximum limits, the Commission and Member States are required to take account of international standards, such as those produced by Codex Alimentarius Commission (CAC), where they exist (as directed in Article 4 of Commission Regulation (EC) No 178/2002).

Maximum limits in food are established by taking account of expert risk assessment. In the EU, the EFSA fulfils this function; internationally for Codex, the Joint FAO/WHO Expert Committee on Food Additives (JECFA) of the World Health Organization (WHO) and Food and Agriculture Organization (FAO) of the United Nations fulfils a similar function. Within the EU, when a risk is identified as a result of new data or information becoming available, it will be raised at the relevant EU expert group and preliminary discussions will take place. If it is considered to be worthwhile, a request will be made to EFSA by the EU Commission to undertake an assessment and provide an opinion on which risk managers can formulate policy. Individual Member States and the European Parliament are permitted as well to make direct requests to EFSA, and EFSA may also be self-tasking.

Internationally, many countries will observe the CAC maximum *levels* where they exist, particularly in regard to international trade. However, these do not carry legislative weight in the same way that EU maximum *limits* do in regard to food produced in or imported into the EU, the CAC has no power of enforcement. Of course, many national standards can exist also, either that are more stringent than CAC standards or where there are no CAC standards. This may be to reflect a local need or to address local consumption patterns for particular foods. Of course, where international trade is concerned local limits that are more stringent than existing CAC standards can lead to trade disputes and accusations of economic protectionism.

Of note, the United States have not set specific national legislative limits for trace elements such as cadmium, lead (except in certain ingredients such as sugar) or arsenic. The Food and Drug Administration (FDA) approach is to issue guideline levels in some cases (lead in juice, 50 ppb), or action levels (such as methylmercury in fish 1 ppm) or to act on the basis of risk assessments from the Center for Food Safety and Applied Nutrition (CFSAN) where high levels are found.

In Japan the administration of food safety within the Ministry of Health, Labour and Welfare is under the jurisdiction of the Department of Food Safety under the Pharmaceutical and Food Safety Bureau and is based on the Food Safety Basic Law, enacted in May 2003 (see Japanese Ministry of Health, Labour and Welfare).

Food safety in Australia and New Zealand is administered by a bi-national organisation 'Food Standards Australia New Zealand (FSANZ)', but is governed at local level. Maximum permitted levels for certain trace elements (arsenic, cadmium, lead, mercury and tin) are set out in the 'Food Standards Code' (FSANZ, 2008).

The State Food and Drug Administration (SFDA) is a good place to start for any enquiry into the Chinese regulation of trace elements (see Section 2.7), but there are multiple governmental departments involved in food safety in China. It may also be prudent to consider contacting the General Administration of Quality Supervision, Inspection and Quarantine of the People's Republic of China (AQSIQ) which is responsible for all exported commodities, and the Ministry of Agriculture (MoA), which is principally responsible for the control and supervision of the domestic market.

As is clear, international trace element food safety regulation is a complicated issue. We have attempted to give a brief selection of information to illustrate. It is beyond the scope of this chapter to highlight all the major international legislation and organisations governing trace elements in foods, and these are likely to change over time. If you require specific national advice we recommend contacting the relevant competent authority(ies) for up-to-date information on permitted levels and other controls that may be in place.

At the time of writing, the European Commission is undertaking a comprehensive review of the maximum limits for cadmium in food, which provides a current example of the processes and rationale behind the risk assessment and policy making involved in the control of trace element contaminants in foods.

At the request of the Commission, EFSA reviewed the latest data on cadmium and reported an updated opinion in 2009 (EFSA, 2009b), which is driving the current review. This report indicates that the provisional tolerable weekly intake (PTWI) of 7 μg/kg body weight (bw) per week previously established by JECFA and endorsed by the Scientific Committee for Food (which provided this function in the EU before EFSA) is no longer considered protective of public health. As a result, EFSA's Panel on Contamination in the Food Chain (CONTAM Panel) has put forward a revised figure of 2.5 μg/kg body weight (bw) per week. EFSA concluded that:

> The mean exposure for adults across Europe is close to, or slightly exceeding, the TWI of 2.5 μg/kg bw. Subgroups such as vegetarians, children, smokers and people living in highly contaminated areas may exceed the TWI by about 2-fold.

Although the risks for adverse effects on kidney function at an individual level at dietary exposures across Europe is very low, the CONTAM Panel concluded that the current exposure to Cd at the population level should be reduced.

This is in contrast to JECFA's revised opinion, based on the same data set, of 25 µg/kg bw *per month*. JECFA concluded that 'The estimates of exposure to cadmium through the diet for all age groups, including consumers with high exposure and subgroups with special dietary habits (e.g. vegetarians), examined by the Committee at this meeting are below the provisional tolerable monthly intake (PTMI)'. JECFA chose to use a tolerable monthly intake, as the effects of cadmium on human health are due to long-term exposure.

This is a good illustration of how two distinct expert risk assessment bodies can come to significantly different opinions due to subtle differences in approach. As a result of the discrepancy, the Commission and the Member States requested a statement from EFSA clarifying the differences and justifying their more conservative assessment (EFSA, 2011). Essentially, EFSA concluded, the difference was that they had based their TWI on the measurement of a biomarker in urine rather than on an observable effect level.

2.2.3 Considering risk

Cadmium, like all trace elements, is an environmental contaminant (although in some cases there may also be other sources of contamination such as food contact materials). Trace elements may be present in the environment through natural geochemical occurrence or in some cases to a great extent through industrial or agricultural deposition, emission and disturbance. Plants may bioaccumulate the trace elements from the soil or contaminated water, or they may be deposited on the plants through climatic processes. Livestock may consume contaminated plants or feed, including the ingestion of quantities of soil. Food may also become contaminated with certain trace elements through other forms of human activity – for instance, point source contamination of poultry/eggs from consumption of lead shot from clay pigeon shooting, or the strange predilection that cows seem to have for lead paint and lead acid batteries (these are often used in fields by farmers to charge electric fences) leading to localised contamination of offal, meat and dairy produce. The UK Food Standards Agency has produced advice to reduce on-farm lead food safety incidents (FSA, 2008).

Once a risk assessment and a tolerable intake have been adopted, it is a relatively straightforward matter to determine risk-based maximum limits that ought to sufficiently protect the majority of consumers, including those in particularly vulnerable groups such as those with restricted diets, or infants and children, due to their relative mass to consumption ratios. Population food consumption data is utilised, though the contribution of other sources of exposure, such as smoking, drinking water and air pollution, may need to be considered. Biomarker studies may also be useful in determining the actual body burden of trace elements for consumers. For cadmium, EFSA advises a critical level of 1.0 µg Cd/g creatinine for urine. This can be compared to the actual levels in consumers, and it is possible that this may be a more realistic indicator of exposure than can be achieved through TDS alone. A pilot study including UK urinary cadmium levels was published in 2007 (Levy *et al.*, 2007), with the full study also now in press (Bevan *et al.*, 2013); similar work has been undertaken internationally. Going forward, it is likely that such studies may increasingly be used as a part of risk assessment and to inform proportionate risk management.

There will always be individuals of course who, due to extreme consumption of particular foods, may put themselves at extra risk, which is why it is appropriate to consider the 97.5th percentile consumer (or a similar approach) for most risk assessments. Someone who eats brown crabmeat every day might well be considered an extreme consumer and may be exposing themselves to undesirable levels of cadmium, but this would be unusual. It is perfectly normal, however, for consumers to eat wheat, potatoes, rice or other dietary staples multiple times per day and in large quantities. Certain demographics or populations worldwide may eat rice as the main part of every meal of the day, which is a particular issue for arsenic exposure (Heikens *et al.*, 2007; Meharg *et al.*, 2009), and such customs must be taken into consideration.

2.2.4 Risk management considerations

Practicality is another crucial aspect that must be taken into account when determining maximum limits for environmental contaminants in food. As risk managers it may be desirable to set very challenging maximum limits, but if this seriously impacts food production and has a considerable economic impact then it may not be sustainable, particularly in the case of dietary staples. However, this is of course exactly where, on a population basis, it may be possible to have a worthwhile impact on exposures. It is therefore important to have a good understanding of the occurrence of the contaminant in different foods so that the impact of setting a maximum limit can be assessed. It may be that the risk-based limit(s) would have a devastating effect on the availability and costs of key foods or it may just preclude the outliers, above the 90th percentile or higher. In the latter case, risk managers should ask themselves what is to be gained from setting the limit. In other words, will it actually have a significant benefit for consumers, will it shift the mean exposure level from that source downwards significantly, will it affect overall exposure, and does the current level of risk demand action? If not, then it has to be considered whether a limit is necessary at all.

Risk managers also need to ask themselves whether the increased due diligence costs for producers and enforcement costs for authorities are justifiable. These costs are all in the end passed on to the consumer one way or another through the higher cost of food or through taxation. Even in such circumstances there can, nevertheless, still be very good economic reasons to have a harmonised limit in place, such as the trade issues mentioned earlier.

Furthermore, it is not reasonable to ignore appropriate risk-based proposals and instead set limits solely on the basis of occurrence data, for instance at the 90th percentile, or even higher, purely to ensure that producers and trade are not impacted adversely. Instead it is beholden on risk managers to liaise and consult carefully with all relevant stakeholders in light of the risk assessment advice and the occurrence data and to put forward measured, practicable and proportionate proposals, whether they be maximum limits or other measures such as consumer advice. In the case of genotoxic carcinogens, such as arsenic, limits should be set as low as reasonably achievable (ALARA), as stated in Recital 4 of Commission Regulation (EC) No 1881/2006.

To further illustrate this point, the UK independent Committee of the Toxicity of Chemicals in Food, Consumer Products and the Environment (COT), endorsed EFSA's 2009 cadmium TWI, but went on to state, 'Given the conservative manner in which the TWI was derived, and that exceedances from dietary exposure are modest (generally less than 2-fold) and only for a limited part of the lifespan, they do not indicate a major concern', but added 'Nevertheless, in view of the uncertainties, it would be prudent to reduce dietary exposures to cadmium at the population level where this is *reasonably practical* [italics added by authors] (COT, 2009).

For example, cadmium levels in the environment can be affected by fertiliser use (Finnish Environment Institute, 2000), and also from industrial sources from their use in modern electronics. Within the EU, the Commission (DG for Enterprise and Industry, 2011) is considering limits for cadmium in fertilisers to control this, and Directive 2006/66/EC places controls on batteries. Lead pollution often results from human activities such as mining and smelting, and arsenic, though ubiquitous in many forms in the environment, can increase due to mining or through boreholes being driven into geochemical sources. Identifying these sources and working to support producers and producing countries to grow crops in appropriate areas, providing technological solutions to remove these contaminants from water supplies and developing strains of crops that do not bioaccumulate the trace elements of concern must ultimately be part of the long lasting solution that will reduce consumer exposures worldwide.

Maximum limits set for contaminants in food in the EU are chiefly for the benefit of EU citizens. International Codex standards may help protect consumers worldwide if the balance between risk and occurrence is reasonable and proportionate. These could also mean that two-tier supply systems develop, resulting in, for example, a concentration factor of higher arsenic-containing rice for the world's poorest and subsistence producers in producing countries; meanwhile, the richer, more developed economies benefit from importing less contaminated product.

Looking forward, it is expected that lead limits will be reviewed by the EU Commission and the expert committee in 2012 or shortly thereafter. Both EFSA and JECFA have recently withdrawn their previously advised tolerable intakes as it is not possible to determine a safe level of exposure. EFSA's CONTAM Panel concluded that 'the current PTWI of 25 µg/kg bw is no longer appropriate as there is no evidence for a threshold for critical lead-induced effects. In adults, children and infants the margins of exposures were such that the possibility of an effect from lead in some consumers, particularly in children from 1–7 years of age, cannot be excluded. Protection of children against the potential risk of neurodevelopmental effects would be protective for all other adverse effects of lead, in all populations'.

Of note, there are, at the time of writing, no EU maximum limits for arsenic, neither total nor inorganic, and indeed for many other trace elements (see Table 2.1 for a summary of key trace elements). These will all be assessed on a case by case basis in due course. EU limits for inorganic arsenic in rice and rice products, and perhaps in certain algae/seaweeds, are expected to be discussed and agreed in the near future. Inorganic arsenic has been detected in quite high levels in hijiki seaweed, but not in others (FSA, 2004; Rose et al., 2007) and as such the UK Food Standards Agency has issued consumer advice against its consumption (FSA, 2010). Furthermore, CAC standards for arsenic in rice are under discussion at the Codex Committee on Contaminants in Food (CCCF). The WHO has produced a provisional guideline value for inorganic arsenic in drinking water, in the WHO Guidelines for Drinking-water Quality (GDWQ), and currently this is 0.01 mg/L (WHO, 2011). A WHO guideline value normally represents the concentration of a constituent that does not result in any significant risk to health over a lifetime of consumption. However, that does not apply to the guideline level for inorganic arsenic in water, as it could still contribute to cancer at the population level given long-term exposure. It is listed in the WHO guidelines for drinking water quality as provisional, and is constrained by water treatment techniques and analytical method performance. Therefore, in areas of high arsenic in drinking water it may not be enough to bring down the arsenic content to just under 0.01 mg/L. Furthermore, much of the world's population is exposed to much higher levels in water.

The EFSA CONTAM Panel has recommended that 'dietary exposure to inorganic arsenic should be reduced'. The CONTAM Panel noted that 'since the PTWI of 15 µg/kg bw was established by JECFA, new data had established that inorganic arsenic causes cancer of the lung and urinary tract in addition to skin, and that a range of adverse effects had been reported at exposures lower than those reviewed by the JECFA'.

They were concerned that 'dietary exposures to inorganic arsenic for average and high-level consumers in Europe are within the range of the $BMDL_{01}$ (Benchmark Dose – *Lower Confidence Limit*) values identified by the CONTAM Panel, and therefore there is little or no Margin of Exposure (MOE) and the possibility of a risk to some consumers cannot be excluded'. Therefore the CONTAM Panel concluded that 'the JECFA PTWI of 15 µg/kg bw is no longer appropriate' (EFSA, 2009a).

Clearly, for lead there is urgent need to reduce young people's total exposure and also for inorganic arsenic it would be appropriate to have harmonised EU limits in key foods (where suitable methodology exists for its analysis in different food matrices). However, it is vital that the effects of setting maximum limits on both producers and consumers are well appreciated so that proportionate measures are taken. A review of CAC lead standards is also being considered to assess appropriateness and discussions are underway at the CCCF.

Furthermore, setting regulatory limits for environmental contaminants in foods is a small part of a larger effort and, to effectively benefit consumers and reduce exposures meaningfully, a holistic approach is required. It is equally important to begin to reduce the amount of these contaminants that are in the environment in the first place, working with industry to produce and develop good agricultural practice and good manufacturing practice and to regulate how these contaminants are used or controlled in industrial and agricultural processes.

Another tool open to policy makers, particularly for more niche products, where consumption will vary greatly, is consumer advice. In the EU, for example, Member States are producing bespoke advice appropriate to their consumers (European Commission, 2011) for brown crabmeat (the internal organs and associated types of flesh as opposed to the white muscle tissue). Due to the huge variability of cadmium levels in the brown crabmeat, the EC had decided that it was not possible to set a practicable maximum limit for brown crab meat. However, there is a limit set for the white muscle meat of crabs in Commission Regulation (EC) No 1881/2006 (as amended).

Similarly, a policy need has been identified in the UK to provide regular consumers of lead-shot game with advice. EFSA has identified high consumers of lead-shot game meat as a group with higher exposure to lead (EFSA, 2010), and this prompted the FSA in Scotland to undertake a soon to be published study titled '*Habits and behaviours of high-level consumers of lead-shot wild-game meat in Scotland*', to investigate levels of consumption. As such, the Food Standards Agency is developing targeted advice for all UK consumers who regularly eat wild-game meat that has been shot using lead ammunition, and is likely to include risk management suggestions for hunters and cooks as well. This approach is preferred, as this is not an issue for the population in general or for farmed game produce.

Even where maximum limits exist, consumer advice may still be needed to augment them, for example, methylmercury in some fish. The additional advice is necessary and is targeted at those at risk, such as young children, pregnant women and those who may become pregnant (NHS Choices, 2011). The problem, of course, with consumer advice is keeping it current and ensuring the message gets to those who it will benefit. When the UK Food Standards Agency produced advice on rice milk drink consumption by infants this was also disseminated to all healthcare advisors, so that the message would be delivered appropriately (FSA, 2009).

In the UK, it has also been necessary to produce advice to consumers who may be indulging in geophagy for perceived health benefits (FSA, 2003, 2011; National Archives, 2009). Anthropologically, it is no doubt an ancient practice and part of traditional practice in some cultures, particularly in some countries in Africa and in Bangladesh, but many of the dried clays (sikor/shikor mati), clay drinks (sold to the health conscious as 'detox' products) and products such as calabash chalk can have elevated levels of lead or arsenic. Many of these products are favoured by pregnant women from particular communities, which is particularly problematic due to the effects on the developing foetus.

In summary, limit setting for foods can play an important role for consumer protection, as can consumer advice. However, particularly for environmental contaminants and trace elements of concern, a joined-up approach to minimisation in the environment and a reduction of exposure from all sources is needed. This is not straight forward as it requires coordinated action between different government departments within countries, in the EU between different Directorate Generals, and also internationally between states and the competing internal interests, not to mention developments in technology and plant breeding, analytical methods and data collection.

2.3 Monitoring of foodstuffs

Governments are accountable to the consumer for ensuring a nation's health and the monitoring of the national food supply is a vital component of this responsibility. Monitoring, as distinct from food control, is generally carried out with a view to developing policies and not for enforcement purposes. Broadly, this monitoring is split into two categories (i) surveillance, and (ii) TDS. It is important to recognise the differences between surveillance and TDS and to understand their respective strengths and weaknesses. It is also crucial to appreciate that in Europe it is the producer or processor who has the ultimate responsibility for placing safe food on the market (Regulation EC/852/2004).

2.3.1 Surveillance

Most national governments engage in food surveillance programmes designed, essentially, for food safety assurance and to control and prevent food hazards. They will be used to check compliance with current legislative requirements. Generally, food will be sampled on import, at wholesale and retail levels for chemical contaminants, such as natural toxins, food additives and contaminants.

Surveillance data are used to promote public awareness, as well as providing information on risk-based sampling and emerging or potential problems. Additionally, surveillance exercises will normally include seasonal monitoring, e.g. some produce will only be available at certain times of the year, as well as, within the EU, responding to requests from the EFSA if they have been asked to provide a risk assessment on a particular contaminant or type of food.

An important part of surveillance in the EU is the Rapid Alert System for Food and Feed (RASFF). This was put in place in Europe to provide the competent authorities within each Member State (and a few other countries, e.g. Norway) with an effective tool to exchange information about the results emanating from their surveillance programmes (Regulation EC/178/2002). This allows Member States to act more rapidly and in a coordinated manner in response to any health threat. Analyses of the outputs of the RASFF can help in identifying potential problems before they become widespread, or in aiding competent authorities on the targeting of products or sectors in the face of ever-decreasing public resources.

The Commission is responsible for managing the system, through designated contact points in each country, by providing a knowledge and technological platform to facilitate the transmission and handling of the RASFF notifications. The Commission has a responsibility to inform a non-member of RASFF if a product subject to a notification has been exported to that country, or when a product originating from that country has been the subject of a notification. The effectiveness of the RASFF relies on its simplicity of operation and the legal obligation of a member to notify the Commission immediately when it has information about, or taken action in response to, a serious health risk related to a food or feed product.

Thus, the purpose of food surveillance exercises is to develop ways of targeting specific contamination problems and investigating ways of monitoring and improving the food chain. What surveillance cannot do is provide an indication of the nation's exposure to a particular contaminant. That is the role of the total diet study.

2.3.2 Total diet studies

There are many human health risks associated with chemical contaminants when exposure exceeds safety thresholds (Table 2.1). Such conditions include cancer, kidney or liver dysfunction, birth defects, reproductive disorders, learning disabilities and impeded development (WHO, 2006a). However, how does a responsible government go about determining what the nation is being exposed to through their diet? This, of course, is an extremely difficult and complex matter, since a population generally consists of numerous subcultures or ethnic groups which may vary widely in their dietary habits and consumption patterns.

In practice, many countries rely on programmes known as 'Total Diet Studies (TDS)', sometimes referred to as a 'market basket study'. These involve the analysis of groups of foods that reflect the average food consumption patterns of a given population (Peattie *et al.*, 1983). Results of the analysis can then be used in conjunction with national food consumption data to estimate the average exposure of the general population to chemicals in foods. The data can also be used to identify changes or trends in exposure and to make assessments on the quality and safety of the food supply. The WHO supports TDS as one of the most cost-effective means for assuring that people are not exposed to unsafe levels of toxic chemicals through food, while also recognising the importance of TDS to the development of Codex standards and international trade (WHO, 2003).

The main purpose of a TDS is to protect consumers from chemical contaminants by monitoring exposure levels of the general population over time (WHO, 2006a). TDS programmes can be set with varying levels of complexity and sophistication but usually have some degree of sampling to take account of geographic region and seasonality. The extent to which ethnic groups are taken into account may depend on the resources available and the diversity of the population. It is also important to include in a TDS provision to differentiate age and sex groups, to be able to give a full picture of the exposure of different categories of consumers to the substances of concern.

TDS data differ from other chemical surveillance programmes because they focus on chemicals in the diet and not on individual foods. Additionally, the foods are processed for consumption in the home, thus they take into account the impact of cooking on the decomposition of less stable chemicals and the formation of new ones, e.g. acrylamide. As such, it is background levels of chemicals that are sought, not regulatory compliance.

TDS should not be viewed in isolation but as an additional tool for the policy makers, to be used in conjunction with other surveillance and monitoring programmes. For example, TDS used in conjunction with national food consumption data can lead to estimates of dietary exposure for the general population, highlighting trends or changes in consumption patterns. Additionally, TDS can indicate where there are needs for more targeted surveillance, e.g. metals in infant foods.

Limitations of TDS
It is very difficult to define the eating habits of a nation, particularly in a multi-cultural population; how close are the data to actual consumption habits of specific niche foods? For example, resources generally will not permit sampling all types of cheese and perhaps different risks are associated with cheeses from cow's milk compared to those made from buffalo's milk. This is where targeted surveillance may be more appropriate than TDS. Also, because of the large sample size due to the compositing of foods in TDS, there may be sensitivity issues for some contaminants present in very small quantities, or where contaminants, e.g. mycotoxins, are distributed unevenly throughout a foodstuff.

Selection of foods
A nation's diet and food habits are ever changing, not least the sources of different foodstuffs, and it can be very difficult to ensure that these evolutions are properly accounted for in government policy and legislation. It is important when embarking on a TDS that the foods chosen for analysis are as representative as possible of the typical diet, with respect to age, sex, ethnicity, etc. The samples then need to be prepared as they would for consumption

(e.g. washed, peeled, cooked, etc.) and analysed, usually as composite samples within certain food groups, for specific contaminants. This should provide accurate data about levels of contaminants as well as the relative contribution of each composite group.

It should be noted that the distribution of contaminants in foods is typically non-Gaussian, which impacts on the sampling plan and the number of samples required for statistical robustness. Thus, depending on the objectives of the study and the contaminants to be examined, the range of foods selected and the level of sampling can vary enormously. Tsukakoshi (2011) showed that by grouping food samples from different outlets in the same city the representativeness of results can be improved.

2.4 Impact of legislation on industry and enforcement

Good legislation must be practical, proportionate and enforceable. Therefore regulators, when developing legislation for metals and trace elements, have to consider not only what is regarded as the limit which is protective of public health, but also how that limit can be enforced (e.g. is there suitable methodology to detect the limit?) and whether the limit places unnecessary burdens on industry, either in having to conduct many tests or by having to reject too much product. This 'balancing act' is the very core of the regulators' work.

2.4.1 Implications of setting limits for consumers, food business operators and enforcement

As previously discussed, maximum limits for environmental contaminants in food have an important role to play, not only for consumer protection but also to ensure a harmonised approach to controls within trading groups. However, they can also have unintended consequences. If set too stringently or brought in too rapidly, they may cause severe upset to global food supplies and economies, which is clearly undesirable. However, if not stringent enough, they might only have the effect of legitimising contaminated and potentially unsafe food. Or, indeed, they may just place an extra burden on producers and manufacturers through additional testing, without having any real effect on population exposure levels.

When a limit is adopted, to some extent it removes the pressure on producers and suppliers to continue to reduce levels of contamination, which could potentially lead to consumers being subject to higher mean levels in some foods, increasing overall exposure. Where it is not possible for legislators to determine a safe level of exposure for a contaminant, this is of even more concern, as it is important that levels in food are ALARA. This is why limits must be regularly reviewed in light of the latest data, to assess their appropriateness and seek to lower them where there will be a demonstrable and proportionate consumer benefit.

A phased approach to implementing increasingly challenging limits over time can be a way to allow producers to implement strategies to ensure that they can produce compliant product while also protecting trade, but importantly progressively improving food safety. This is particularly appropriate for contaminants where the concern is chronic exposure. Thus, gradually reducing a limit over, say 15–20 years, to a stringent end-point may allow for the development of improved farming practices or the development of alternative varieties of crops that are resistant to the accumulation of the contaminant of concern.

Taking a zero tolerance policy for environmental contaminants is not really an option. Even where this may be possible (with illegal veterinary medicines, for instance), it can lead to inconsistencies in risk assessment and enforcement action by differing competent authorities that do not necessarily help protect consumers. Importers could possibly then take advantage of the most lenient ports of entry, which could result in trade distortions, thus creating a very difficult situation for producers, importers and distributors, as well as enforcement authorities.

Maximum limits, of course, are not required, as previously discussed, where the risk assessment does not indicate a need, and this may vary according to local conditions. For instance, in an area of high selenium contamination, this may be an issue of public health concern, but in other areas selenium deficiency may be a concern.

Other implications of setting limits should also be considered; for instance, should they be set for the bulk raw commodity, the products made from this, retail products or a combination? The requirement to do this will be driven by the risk assessment, but the effects of multi-layered limits throughout the food chain can pose particular problems for enforcement and due diligence by food business operators when they are not rational to each other. For instance, cadmium is more concentrated in the bran, and even more so in the germ (around three times) when compared to the raw wheat, so you may have compliant grain but non-compliant bran or germ. Furthermore, if limits are set separately for both retail products and raw materials (as is done for the fusarium mycotoxins), it could be possible to have a non-compliant product, due to a high inclusion rate, but made from a compliant ingredient.

2.4.2 Sampling

When conducting monitoring for trace elements in foods sampling protocols are a very important consideration. Unless the food has undergone an extreme level of processing as bulk, or perhaps is liquid, contaminants in food tend to be heterogeneously distributed, even on a local level where you may take the production from one small field; uptake or surface contamination across the field may vary and this can be represented in the bulk commodity by areas of greater and lesser levels of trace elements. This effect can be very pronounced for natural contaminants, such as mycotoxins where quite extreme variation through a consignment can often be seen. Also, the way trace elements are distributed within an individual raw food item such as a cereal grain will vary, for instance arsenic distribution and speciation among husk, bran and endosperm of rice (Lombi *et al.*, 2009). To make matters more complex, this can vary according to the variety of grain.

In the EU, the methods for sampling consignments or 'lots' of food for trace element contaminants for official control purposes are set out in Commission Regulation (EC) No 333/2007 (as amended). The intention is to ensure a good indication of the true level of contamination within a given consignment and to minimise the effect on the analytical result of error introduced by sampling. It is also intended to minimise damage to the lot being sampled, be practical for enforcement officers to perform, and ensure a decent level of consistency of approach. The Regulation also sets out methods of analysis that should be used for the various different contaminants.

The primary approach to sampling is to take a number of incremental samples from throughout the batch, which can then be homogenised to gain a representative sample. When sampling for official control purposes, this homogenised aggregate sample is usually split into three portions: one for enforcement, one for defence and the third for referee purposes, in case of dispute of the official analytical result. The number of incremental samples required depends on the size and nature (bulk or in subdivided packets, etc.) of the lot (or sub-lot in case of very large lots), as set out in the Regulation.

Where maximum limits are set, it is also important to only include in the sample that part of the food to which the limit applies. For instance, the EU cadmium limit for potatoes applies only to peeled potatoes and that for crab applies only to the white meat (i.e. the meat from appendages, not the cephalothorax).

Good risk assessment and risk management require accurate and reliable data to enable up-to-date and comparable information on hazards found in the food chain and on food consumption. Within the EU, EFSA is responsible for the collation of data collected by the Member States. This information, combined with reliable information on food consumption within each of the Member States, makes it possible for risk assessors to assess consumer exposure to a certain hazard both across the EU and within a country, allowing assessments to be made and recommendations for the prevention, reduction and monitoring of these hazards in the food chain.

EFSA is legally required to collect and analyse data in its fields of expertise (Regulation 178/2002), which includes, among others, on the occurrence of contaminants and chemical residues. It does this by collection of EU-wide data required by EU regulations on an ongoing basis and through calls for specific data. The calls for data are published on the EFSA website, together with details of their specific reporting requirements for data submission.

2.5 Suitability of analytical methods

A crucial consideration when setting legal limits is that appropriate methods are available to determine the trace element in the food matrix in question. Also, increasingly it is important to have methods that can give speciation data for the trace element, as although most maximum limits have been set on the basis of the total amount of the trace element present so far (Berg, 2006), not all these species necessarily have the same toxicity or bioavailability.

A good example is arsenic where, in many foods such as seafood, it can be present in quite high amounts, but predominantly present in organic species such as arsenobetaine (Berg, 2006). It is, however, the inorganic species of arsenic that are of most toxicological concern, and therefore it is appropriate to consider maximum limits for key foods on this basis, as is anticipated in the EU in the near future and is also under discussion by the Codex Committee for Contaminants in Food. Nevertheless, methods must be available and reasonably achievable for control laboratories to implement, and have the required level of sensitivity (appropriate limit of detection and/or limit of quantification) to allow enforcement authorities to take proper action, otherwise the legislation is inappropriate. It may be desirable to control hexavalent chromium, Cr (VI), compounds in foods, but as yet no reliable methods for determination in food matrices have been accepted.

An attempt to introduce international maximum levels for inorganic arsenic in food in the Codex Alimentarius General Standard for Contaminants and Toxins in Food was shelved in 1998, as no cheap and robust method of analysis was available (Berg, 2006). However, now a validated intralaboratory method is available; an interlaboratory report commissioned by the EU shows that inorganic arsenic measurements are robust, even for different methodologies, to an extent where they conclude that methodology should not preclude standard setting. The report concluded:

> In answering a request from the Directorate General for Health and Consumers of the European Commission, the European Union Reference Laboratory for Heavy Metals in Feed and Food, with the support of the International Measurement Evaluation Program, organised a proficiency test (PT), IMEP-107, on the determination of total and inorganic arsenic (As) in rice. The main aim of this PT was to judge the state of the art of analytical capability for the determination of total and inorganic As in rice. For this reason, participation in this exercise was open to laboratories from all over the world. Some 98 laboratories reported results for total As and 32 for inorganic As.
>
> The main conclusions of IMEP-107 were that the concentration of inorganic As determined in rice does not depend on the analytical method applied and that the introduction of a maximum level for inorganic As in rice should not be postponed because of analytical concerns (de la Calle *et al.*, 2011).

Also crucial is that the expanded measurement uncertainty (MU) is known and reported, and the recovery for any extraction step is calculated. This is a requirement under the EU legislation. When reporting the result of an official control sample, this is taken into account and the benefit of the doubt is given generally to the food business operator, with respect to the MU.

2.5.1 Due diligence implications

It is the responsibility of the food business operator (FBO) to ensure that food is safe and compliant with the legislation. To do this, they must have suitable processes and hazard analysis and critical control points (HACCP) in place, according to risk and means. What a small or medium enterprise (SME) can achieve compared, with a large international corporation, will be different. There are supply chain practicalities that must be considered for due diligence in regard to trace elements.

For trace elements you may require certificates of analysis from your suppliers, or you may undertake sample analysis of materials yourself (or both). But a lack of suitable rapid tests and the difficulties in undertaking representative sampling can be an issue. The bulk wheat at a grain merchant may be compliant but a single 30 tonne truck of it arriving at a mill may not be. It is all very well to send a sample off for analysis at a laboratory but chances are that by the time you receive the result the food will have long gone. This creates difficulties for FBOs and also for enforcement, particularly where you are considering a rate of < 5% marginal non-compliance across an entire food sector for a trace element with chronic but not acute effects.

2.6 Future trends

Undoubtedly, speciation of contaminants and trace elements will play a greater part in legislation. However, this will be dependent on the development of robust methodology and improved toxicological understanding. Lack of reliable toxicological data has slowed progress in this area but this should improve in coming years.

Additionally, the interaction between trace elements will become more important as the mechanisms of biological activities are better understood. This, of course, could lead to more complex legislation and risk management options.

Laser ablation technology has the potential to provide a wealth of new information regarding the location of heavy metals and trace elements within whole foods, such as cereal grains, fruit and vegetables. This could enable risk management options for the minimisation of certain undesirable elements entering the food chain, e.g. by peeling fruit and vegetables, or by selective milling processes. Additionally, the technique may be used as an aid in the development of grains, or fruit and vegetable varieties, more resistant to the uptake of certain contaminants, e.g. low arsenic rice varieties.

2.6.1 Speciation

Certainly within Europe there will be legislative amendments to lower legal maximum limits where these are already set, and to introduce new limits for additional foodstuffs and for other elements which currently are not included in regulatory requirements. Further, there is little doubt that future legislative limits will be based more on speciation levels in foodstuffs as the toxicological knowledge increases (Proust *et al.*, 2005). Thus there needs to be a concerted effort by researchers to better refine and develop the methodologies capable of delivering this change. This will require not only increased sensitivity but also improved selectivity. Not least among these issues is the lack of certified reference materials (Lagarde *et al.*, 1999, EFSA, 2009a), which is currently hindering progress in method development. The complexity of food matrices also offers a challenge to researchers, as it is far more difficult to quantify and speciate elements from many food matrices than to detect the total level present. Better methods of sample preparation, which allow for full extraction of elements without loss or degradation of the target element, are required if legislators are to take the speciation route. It is pointless legislators setting limits that cannot be reasonably and practically detected.

2.6.2 Methodologies

The development of analytical techniques is ever evolving and space does not permit a comprehensive review. Since speciation will become increasingly more important, techniques such as high performance liquid chromatography inductively coupled plasma mass spectrometry (HPLC-ICP-MS) will play a more crucial role in the analysts' armoury. However, the application of one technique that has the potential to aid legislators enormously with respect to perhaps providing more risk management options is Synchrotron-based X-ray fluorescence (S-XRF). This has been used successfully for the speciation and localisation of arsenic in rice (Meharg *et al.*, 2008). For certain foodstuffs, e.g. grains, vegetables and perhaps fruits, S-XRF could impart a level of new knowledge about the distribution of elements within a food. This could help with consumer advice, for instance on the necessity to peel fruit and vegetables, or for processors about utilisable parts of cereal grains. This more precise information could lead to better focussed and more pragmatic legislation. However, as with laser ablation technology, one of the challenges facing researchers is the lack of calibration standards.

2.6.3 Toxicology

As has already been mentioned, legislators face a very difficult task in setting legal limits when not all branches of science can progress at the same pace. So, while a particular element may be shown to be toxic at a certain level by one set of workers, the methodology to detect it at an adequately low level in all appropriate food matrices may not be sufficiently well developed to enable the setting of legal limits. This is no more evident than with the issue of speciation, creating a vicious circle between toxicologists and analysts as each tries to match the progress of the other.

Recent years have seen an increase in our knowledge and understanding of toxicological issues and bioavailability. This has been helped to a large extent by work on chemical speciation, which should continue to aid our ability to better focus the toxicological impact of different elements. For example, many consumer groups in Europe exceed the exposure levels set for cadmium, the adverse health effects of which occur at lower levels than had previously been thought. The extent to which legislators will be able to adapt to this way of thinking will largely depend on the rate of progress of the toxicologists and the analysts in developing new and practical methodologies. However, each part of the cycle, namely the legislators, the toxicologists and the analysts, is responsible for providing impetus to the others and forcing the developments that will ultimately lead to safer foods and a lowering of consumers' exposure to heavy metals and undesirable trace elements.

2.6.4 Total diet studies

As global trade increases, and countries need to be able to comply with international standards and ensure product safety, TDS can play an important role in helping to achieve these goals. Thus, there is a need for TDS to become more widespread internationally and for a greater consistency in approach in order that appropriate comparisons can be made, though developing countries, particularly, may need help in designing and conducting TDS. Nevertheless, it is important also to recognise that each study should reflect the health concerns and resources of the country in which it is conducted and that those differences need to be taken into account when comparing data.

2.7 Sources of further information

In a world where technology and legislative requirements are forever changing, the following websites may help provide up-to-date sources of information:

- Codex Alimentarius – http://www.codexalimentarius.net/web/index_en.jsp
- EFSA – http://www.efsa.europa.eu/en/aboutefsa/efsawhat.htm
- European legislation – http://eur-lex.europa.eu/RECH_legislation.do?ihmlang=en
- EU – http://europa.eu/index_en.htm
- Food Standards Australia New Zealand – http://www.foodstandards.gov.au/
- Health Canada – http://www.hc-sc.gc.ca/index-eng.php
- SFDA, China – http://eng.sfda.gov.cn/WS03/CL0755/
- The Joint FAO/WHO Committee on Food Additives – http://www.codexalimentarius.net/web/jecfa.jsp

44 Persistent organic pollutants and toxic metals in foods

- UK Food Standards Agency – http://www.food.gov.uk/
- US FDA – http://www.fda.gov/

2.8 References

ARSENIC IN FOOD (ENGLAND) REGULATIONS (1959), Food and Drugs. Composition and Labelling. SI No. 831.
ASCHNER, J. L. and ASCHNER, M. (2005), Nutritional aspects of manganese homeostasis, *Molecular Aspects of Medicine*, **26**(4–5): 353–362.
BERG, TORSTEN (2006), 2nd IUPAC Symposium on Trace Elements in Food: An introduction. *Pure and Applied Chemistry*, **78**(1): 65–68.
BEVAN, R., JONES, K., COCKER, J., ASSEM, F. L. and LEVY, L. S. (2013), Reference ranges for key biomarkers of chemical exposure within the UK population. *International Journal of Hygiene and Environmental Health*, **216**(2): 170–174.
BLUNDEN, S. and WALLACE, T. (2003), Tin in canned food: a review and understanding of occurrence and effect. *Food and Chemical Toxicology*, **41**(12): 1651–1662.
COT (2009), Committee on Toxicity of Chemicals in Food, Consumer Products and the Environment. Annual Report. http://cot.food.gov.uk/pdfs/cotsection2009.pdf
DE LA CALLE, M. B., EMTEBORG, H., LINSINGER, T. P. J., MONTORO, R., SLOTH, J. J., RUBIO, R., BAXTER, M. J., FELDMANN, J., VERMAERCKE, P. and RABER, G. (2011), Does the determination of inorganic arsenic in rice depend on the method? *Trends in Analytical Chemistry*, **30**(4): 641–651.
DG FOR ENTERPRISE AND INDUSTRY (2011), Cadmium in fertilisers. http://ec.europa.eu/enterprise/sectors/chemicals/documents/specific-chemicals/fertilisers/cadmium/index_en.htm [Accessed 12 October 2011]
DIRECTIVE 2006/66 (2006), EC Directive 2006/66 of the European Parliament and of the Council of 6 September 2006 on batteries and accumulators and waste batteries and accumulators and repealing Directive 91/157/EEC. *Official Journal of the European Union L* **266**/1.
EFSA (2004), Opinion of the Scientific Panel on Contaminants in the Food Chain on a request from the Commission related to mercury and methylmercury in food. *EFSA Journal*, **34**: 1–14
EFSA (2008), Scientific Opinion of the Panel on Food Additives, Flavourings, Processing Aids and Food Contact Materials on a request from European Commission on Safety of aluminium from dietary intake. *EFSA Journal*, **754**: 1–34.
EFSA (2009a), Panel on Contaminants in the Food Chain (CONTAM); Scientific Opinion on Arsenic in Food. *EFSA Journal*, **7**(10): 1351.
EFSA (2009b), Scientific Opinion of the Panel on Contaminants in the Food Chain on a request from the European Commission on cadmium in food. *EFSA Journal*, **980**: 1–139.
EFSA (2010), Panel on Contaminants in the Food Chain (CONTAM); Scientific Opinion on Lead in Food. *EFSA Journal*, **8**(4): 1570.
EFSA (2011), Scientific Opinion on tolerable weekly intake for cadmium. *EFSA Journal*, **9**(2): 1975.
EUROPEAN COMMISSION (2011), Consumption of brown crab meat. Available from: http://ec.europa.eu/food/food/chemicalsafety/contaminants/information_note_cons_brown_crab_en.pdf
EVM (2003), Safe upper levels for vitamins and minerals. Report of the Expert Group on Vitamins and Minerals. Food Standards Agency, May 2003. ISBN 1-904026-11-7.
FINNISH ENVIRONMENT INSTITUTE (2000), Cadmium in Fertilizers – Risks to Human Health and the Environment. Study report for the Finnish Ministry of Agriculture

and Forestry October 2000. http://ec.europa.eu/enterprise/sectors/chemicals/files/reports/finland_en.pdf
FLEMMING, C. A. and TREVORS, J. T. (1989), Copper toxicity and chemistry in the environment: a review. *Water, Air and Soil Pollution*, **44**(1–2): 143–158.
FRASSINETTI S, BRONZETTI G, CALTAVUTURO L, CINI M and CROCE C. D. (2006), The role of zinc in life: a review. *Journal of Environmental Pathology, Toxicology and Oncology*, **25**(3): 597–610.
FSA (2003), Calabash chalk warning. Available from: http://www.food.gov.uk/foodindustry/imports/banned_restricted/calabashchalk
FSA (2004), Arsenic in Seaweed. Food Survey Information Sheet 61/04. http://www.food.gov.uk/multimedia/pdfs/arsenicseaweed.pdf
FSA (2008), Help stop on-farm lead poisoning. Available from: http://www.food.gov.uk/multimedia/pdfs/publication/leadpoison0209.pdf
FSA (2009), Arsenic in rice research published. http://www.food.gov.uk/news/newsarchive/2009/may/arsenicinriceresearch [Accessed 12 October 2011]
FSA (2010), Consumers advised not to eat hijiki seaweed. http://www.food.gov.uk/news/newsarchive/2010/aug/hijikiseaweed
FSA (2011), FSA issues warning about eating clay. Available from: http://www.food.gov.uk/news/newsarchive/2011/june/clay
FSANZ (2008), Australia New Zealand Food Standards Code – Standard 1.4.1 – Contaminants and Natural Toxicants. Available from: http://www.comlaw.gov.au/Series/F2008B00618
HEIKENS, A., PANAULLAH, G. M. and MEHARG, A. A. (2007), Arsenic behaviour from groundwater and soil to crops: impacts on agriculture and food safety. *Reviews of Environmental Contamination and Toxicology*, **189**: 43–87.
JAPANESE MINISTRY OF HEALTH, Labour and Welfare. Administration of food safety. Available from: http://www.mhlw.go.jp/english/policy/health-medical/food/index.html
LAGARDE, F., AMRAN, M. B., LEROY, M. J. F., DEMESMAY, C., OLLE, M., LAMOTTE, A., MUNTAU, H., THOMAS, P., CAROLI, S., LARSEN, E., BONNER, P., RAURET, G., FOULKES, M., HOWARD, A., GRIEPINK, B. and MAIER, A. (1999), Certification of total arsenic, dimethylarsinic acid and arsenobetaine contents in tuna fish powder (BCR-CRM 627). *Fresenius Journal of Analytical Chemistry*, **363**: 18–22.
LEVY, L. S., JONES, K., COCKER, J., ASSEM, F. L. and CAPLETON, A. C. (2007), Background levels of key biomarkers of chemical exposure within the UK general population – pilot study. *International Journal of Hygiene and Environmental Health*, **210**(3–4): 387–91.
LOMBI, E., SCHECKEL, K. G., PALLON, J., CAREY, A. M., ZHU, Y. G. and MEHARG, A. A. (2009), Speciation and distribution of arsenic and localisation of nutrients in rice grains. *New Phytologist*, **184**: 193–201.
MANDAL, B. K. and SUZUKI, K. T. (2002), Arsenic round the world: a review. *Talanta*, **58**: 201–235.
MEHARG, A. A., LOMBI, E., WILLIAMS, P. N., SCHECKEL, K. G., FELDMANN, J., RAAB, A., ZHU, Y. and ISLAM, R. (2008), Speciation and localisation of arsenic in white and brown rice grains. *Environmental Science and Technology*, **42**: 1051–1057.
MEHARG, A. A., WILLIAMS, P. N., ADOMAKO, E., LAWGALI, Y. Y., DEACON, C., VILLADA, A., CAMBELL, R. C. J., SUN, G., ZHU, Y-G., FELDMANN, J., RAAB, A., ZHAO, F-J., ISLAM, R., HOSSAIN, S. and YANAI, J. (2009), Geographical variation in total and inorganic arsenic content of polished (white) rice. *Environmental Science and Technology*, **43**(5): 1612–1617.
NATIONAL ARCHIVES (2009). Agency warns consumers to avoid certain clay-based drinks. Available from: http://webarchive.nationalarchives.gov.uk/20120206100416 and http://food.gov.uk/news/newsarchive/2009/aug/clay?view=printerfriendly
NAVARRO-ALARCON, M. and CABRERA-VIQUE, C. (2008), Selenium in food and the human body: A review. *Science of the Total Environment*, **400**(1–3): 115–141.

NHS CHOICES (2011), http://www.nhs.uk/Planners/pregnancycareplanner/pages/Carewithfood.aspx. [Accessed 12 Oct 2011].

PAPANIKOLAOU, N. C., HATZIDAKI, E. G., BELIVANIS, S., TZANAKAKIS, G. N. and TSATSAKIS, A. M. (2005), Lead toxicity update. A brief review. *Medical Science Monitor*, **11**(10): RA329–336.

PEATTIE, M. E., BUSS, D. H., LINDSAY, D. G. and SMART, G. A. (1983), Reorganization of the British total diet study for monitoring food constituents from 1981. *Food and Chemical Toxicology*, **21**(4): 503–507.

PECHOVA, A. and PAVLATA, L. (2007), Chromium as an essential nutrient: a review. *Veterinarni Medicina*, **52**(1): 1–18.

PROUST, N., BUSCHER, W. and SPERLING, M. (2005), *Speciation and the emerging legislation*, in: RITA CORNELIS, JOE CARUSO, HELEN CREWS, KLAUS HEUMANN, eds., Handbook of Elemental Speciation II, John Wiley & Sons, Chichester, 737–744.

REEVES, P. G. and CHANEY, R. L. (2008), Bioavailability as an issue in risk assessment and management of food cadmium: A review. *Science of the Total Environment*, **398**(1–3): 13–19.

Regulation EC/178/2002 laying down the general principles and requirements of food law, establishing the European Food Safety Authority and laying down procedures in matters of food safety. *Official Journal of the European Union*, No L 31 of 1 February 2002, Articles 50, 51 and 52.

Regulation (EEC) No 315/93 laying down Community procedures for contaminants in food. *Official Journal of the European Union*, L37 of 8 February 1993, p. 1.

Regulation (EC) No 333/2007 laying down the methods of sampling and analysis for the official control of the levels of lead, cadmium, mercury, inorganic tin, 3-MCPD and benzo(a)pyrene in foodstuffs. *Official Journal of the European Union* L 88/29 of 28 March 2007.

Regulation EC/852/2004 on the hygiene of foodstuffs. *Official Journal of the European Union*, N° L **139**/1 of 29 April 2004.

RISHER, J. F., MURRAY, H. E. and PRINCE, G. R. (2002), Organic mercury compounds: human exposure and its relevance to public health. *Toxicology and Industrial Health*, **18**: 109–160.

ROSE, M., LEWIS, J., LANGFORD, N., BAXTER, M., ORIGGI, S., BARBER, M., MACBAIN, H. and THOMAS, K. (2007), Arsenic in Seaweed — forms, concentration and dietary exposure. *Food and Chemical Toxicology*, **45**: 1263–1267.

ŠČANČAR, J. and MILAČIČ, R. (2006), Aluminium speciation in environmental samples: a review. *Analytical and Bioanalytical Chemistry*, **386**(4): 999–1012.

Treaty of Lisbon amending the Treaty on European Union and the Treaty establishing the European Community, signed at Lisbon, 13 December 2007. *Official Journal of the European Union*, C306 of 17 December 2007.

TSUKAKOSHI, Y. (2011), Sampling variability and uncertainty in total diet studies. *Analyst*, **136**: 533–539.

WHO (2001), *Concise International Chemical Assessment Document 33: Barium and barium compounds*. World Health Organization, Geneva.

WHO (2003), *GEMS/Food regional diets: regional per capita consumption of raw and semi-processed agricultural commodities. Global Environment Monitoring System/Food Contamination Monitoring and Assessment Programme (GEMS/Food)*. Rev. ed. World Health Organization, Geneva. ISBN 92 4 159108 0.

WHO (2006a), *Total Diet Studies: A Recipe for Safer Food*. (INFOSAN Information Note No. 06/2006 – Total Diet Studies). World Health Organization, Geneva.

WHO (2006b), *Elemental speciation in human health risk assessment. Environmental Health Criteria 234, International Programme on Chemical Safety*. World Health Organization, Geneva.

WHO (2011), *Guidelines for drinking-water quality, fourth edition*. World Health Organization, ISBN: 978 92 4 154815 1.

3
Screening and confirmatory methods for the detection of dioxins and polychlorinated biphenyls (PCBs) in foods

J.-F. Focant and G. Eppe, University of Liège, Belgium

DOI: 10.1533/9780857098917.1.47

Abstract: This chapter reports on the use of the European screening-confirmatory strategy for detection of dioxins and PCBs in food and feed. The introduction explains the need for using such an approach for large monitoring programs and highlights the origin of the European Union (EU) pillars strategy and its basis of action. The various biological and physico-chemical screening approaches are described and further compared against each other and versus the confirmatory method. Quantitative and semi-quantitative aspects are discussed from the perspective of specific analytical requirements. The state-of-the-art GC-HRMS (gas chromatography in combination with high resolution mass spectrometry) confirmatory method is also presented and specific quality assurance/quality control (QA/QC) aspects are discussed. A discussion of future trends completes the chapter.

Key words: dioxins, furans, polychlorinated biphenyls (PCBs), foods, mass spectrometry, isotopic dilution (ID) technique, bioassays, screening and confirmatory methods, quality assurance (QA), quality control (QC).

3.1 Introduction

In order to properly understand the screening-confirmatory strategy for the detection of dioxins and PCBs in foods, it is important to consider the general context in which the strategy has been developed.

3.1.1 Context

Risks to human health from polychlorinated biphenyls (PCBs) and dioxins namely polychlorinated dibenzo-p-dioxins and polychlorinated dibenzofurans – PCDD/Fs) are mainly related to consumption of food of animal origin. During the last decade, repeated cases of contamination of feedingstuffs have highlighted the importance of feed as a potential contamination medium. Reducing human uptake of dioxins is thus highly dependent on actions taken to minimize the contamination of all feed materials, including raw materials, recycled products and ingredients (e.g. citrus pulp pellets, recycled fats, mineral clays, choline chloride component, hydrochloric acid related to gelatin production, guar gum thickener, biodiesel-related fatty acids, etc.).

Despite such action, isolated cases of contamination might still arise. The implementation of continuous monitoring strategies, the enforcement of the maximum-action-target level strategy (now simplified in maximum-action level only), as well as the availability of a Rapid Alert System for Food and Feed (RASFF) (Commission Regulation (EC) No 1881/2006, 2006), nowadays allows actions to be taken more rapidly and in a coordinated manner in order to reduce potential human exposure to a minimum if a contamination event is reported. This normally translates into so-called 'food crises', which now often have as much impact on our economies as on our health. Such a continuous monitoring strategy is part of a proactive approach pursued by the EU in order to obtain comprehensive reliable data on the presence of PCBs and dioxins in food and feed (Commission Regulation (EC) No 1881/2006; Commission Regulation (EC) No 1883/2006, 2006; Commission Directive 2002/70/EC, 2002). The idea is to be able to select samples with significant levels of dioxins using a screening method of analysis. The levels of dioxin in these samples would then have to be determined by a confirmatory method of analysis (Commission Regulation (EC) No 152/2009, 2009).

In order to enable a timely response to health threats caused by contaminated food or feed, laboratories have to be endowed with large and efficient analytical capacity. The entire screening-confirmatory approach of the EU relies on the responsiveness of such expert laboratories. They have to be able to handle continuous flows of samples for screening (and potential confirmation) as part of regular monitoring programs, but also to quickly go on the alert and handle large numbers of suspected samples, in the event of a contamination episode being reported. Such expert laboratories have to use state-of-the-art technology, including both sample preparation and measurement aspects. One of the major challenges is to combine a high level of QA/QC with fast turnover and large sample throughput. Depending on whether a method is being used for screening or for confirmation, analytical instructions and guidelines are somewhat different but, in each case, the requirements are very stringent and specific to the highest level.

3.1.2 Screening strategy

The concept of screening systems in analytical chemistry has been around for several years. It is based on the use of rapid analytical tools providing a binary yes/no response that indicates whether the target analytes are above or below a pre-set concentration threshold (Valcarcel *et al.*, 1999). The main objectives of screening systems are to be fast, inexpensive and reliable, while minimizing the need for the continuous use of costly instruments. Sometimes, one can also reduce the sample pretreatment steps to a minimum when using screening methods. This is, however, not the case with dioxin analyses because of the complexity of food and feed matrices, as well as the need for fractionation of dioxins from PCBs. Although the binary response approach might seem very simple, it still requires proper analytical treatment of important parameters such as specificity, limits of detection (LOD) and quantification (LOQ), threshold values, uncertainties, etc., in order to reduce the risk of false positive and false negative responses.* Reducing false negative rates close to the residue limits is actually one of the major issues for the proper setting of performance criteria for screening tools. By definition, such screening tools should always be complemented by confirmatory tools to both confirm and investigate positive samples and also verify and ensure that the false negative rate remains under acceptable levels.

3.1.3 The need for efficient monitoring

Extensive European Community legislation has been adopted after contamination episodes and consequent increased awareness of food issues that followed. The 1999 fat contamination crisis of the Belgian food supply (Bernard et al., 1999) highlighted the absence of European Community legislation in this field. This absence greatly increased the difficulty of managing such problems at Community level. Although some national legislation had already been implemented in a few EU countries, such as Germany, there was a need for harmonized limits. The strategy consisted of a Community strategy for dioxins, furans and PCBs focused on two distinct aspects:

1. The environmental aspect, with implementation of measures to reduce the release of those contaminants into the environment.
2. The food and feed safety aspect, with actions to further decrease the presence of dioxin in foodstuffs and feedingstuffs.

The aim of this strategy is to bring the majority of the European population below the tolerable monthly intake of 70 pg/kg bw/month recommended by Joint FAO/WHO Expert Committee on Food Additives (JECFA) and the WHO (FAO/WHO, 2001; WHO, 2005).

3.1.4 The EU pillars strategy

The Belgian incident, and many others from the recent past, showed that the contamination of feeds is a major route of livestock exposure to dioxins. Also,

---

* Although the terms 'positive' and 'negative' are generally used when describing screening results, the EU legislation utilizes the term 'non-compliant' for 'positive' and 'compliant' for 'negative' samples

because dioxin-like compounds pass through various trophic levels of food chains into humans, the idea of controlling the early upstream food chain by setting strict limits in animal feedingstuffs was a logical proposal. For this purpose, valuable information on the occurrence of polychlorinated dibenzo-p-dioxins and polychlorinated dibenzofurans (PCDD/Fs) and dioxin-like polychlorinated biphenyls (DL-PCBs) in food collected from EU members' food contamination data was used to construct a database. At the early stage, only PCDD/F data were available for various matrices, with a dearth of information on DL-PCBs. Enough data were collected to establish initial regulatory levels for PCDD/Fs. This resulted in the so-called EU pillars strategy, and legislation requiring screening and confirmation.

Verstraete (2002) presented the legislative measures in feedingstuffs and foodstuffs as three pillars:

- Strict but feasible maximum levels in various food and feed matrices,
- Action levels, acting as a tool for 'early warning' of dioxin contaminations above background levels in food and feed,
- Target levels to be achieved in food and feed items in order to bring the population below the tolerable monthly intake.

Recently, it has been decided at EU level to remove the target levels and to limit the monitoring strategy with maximum and action levels.

Maximum levels became applicable from 1 July 2002 (Council Regulation (EC) No 2375/2001, 2001; Council Regulation 2001/102/EC, 2001). Then, the action levels were established by reducing the maximum levels by at least 30% with the aim of triggering actions on identification of sources and pathways of dioxin contamination (Commission Recommendation 2006/88/EC, 2006). Measures to reduce or to prevent such contamination should then be enforced. This entails permanent monitoring programs to detect the presence of those contaminants in food and feed inside the European market. Maximum and action levels are expressed in toxic equivalent quantities (TEQs) resulting from the sum of the seventeen 2,3,7,8 PCDD/F congeners using the WHO toxic equivalence factor (TEF) values (van den Berg et al., 1998). Later, the 12 DL-PCBs were incorporated in maximum levels, after more comprehensive data on background levels became available. This was completed by the end of 2006 (Commission Directive 2006/13/EC, 2006; Commission Regulation (EC) No 199/2006, 2006). Separate maximum levels for dioxins/furans on one side and DL-PCBs on the other side were applicable since 2006. In order to ensure a smooth switchover, the existing levels for the sum of the 17 PCDD/Fs continue to be applied, in addition to the newly set maximum levels for the sum of the 29 PCDD/Fs and DL-PCBs (Commission Regulation (EC) No 1881/2006, 2006). This point of view is indeed tenable, as often the sources of contamination are different, especially for action levels where separate measures are taken depending on whether PCDD/Fs or DL-PCBs exceed the action levels (Commission Recommendation 2011/516/EU, 2011). In addition, new limits for non-dioxin-like PCBs (NDL-PCBs) in food and feed are now proposed, together with the use of the WHO-2005 TEF values for PCDD/Fs and DL-PCBs (Commission Regulation 1259/2011, 2011; Commission Regulation 277/2012, 2012; van den Berg et al., 2006).

In order to facilitate free trade of goods, harmonization of acceptance criteria for dioxin analysis is needed (Malisch et al., 2001). Measures have been taken and analytical requirements for dioxin analysis in food and feed have been adopted (Commission Directive 2002/69/EC, 2002; Commission Directive 2002/70/EC, 2002) and have lately been revised in Commission Regulation (EC) No 1883/2006 and Commission Regulation (EC) No 152/2009 for food and feed, respectively (Commission Regulation (EC) No 1883/2006, 2006; Commission Regulation (EC) No 152/2009, 2009). It is an ongoing revision process, as both knowledge and analytical techniques progress.

The main consequences that followed from these legislative measures led to the set-up of large monitoring programs of the food chain (EFSA, 2010a; EFSA, 2010b). To cope with the great number of samples statistically required for monitoring, the strategy involves the use of cost-effective screening methods (capillary gas chromatography in combination with low resolution mass spectrometry, i.e. GC-LRMS, or AhR-based bioassays for TEQ determinations), characterized by a high sample throughput, together with the gold standard GC coupled to sector high resolution MS (GC-HRMS) reference method, used for confirmation. Food and feed samples generally show background levels and hence the monitoring strategy promotes the use of screening methods. They are used to sieve large numbers of samples for potential non-compliancy but, in addition, they should be characterized by false-compliant results (risk of β error) of less than 5%. The non-compliant samples need to be confirmed afterwards by the reference GC-HRMS method.

This chapter reports on the current situation of the screening-confirmatory strategy used for the detection of dioxins and PCBs in food and feed.

3.2 Biological versus physico-chemical screening for dioxins and PCBs in food and feed

Both biological and physico-chemical methods can be used for the screening of dioxins and PCBs in foods. On a total TEQ basis, both methods have to express false negative rates below 5%, a precision RSD below 30%, and obviously have a rate of false positives sufficiently low to make the use of a screening tool advantageous (Commission Regulation 278/2012, 2012). Additionally, all positive results have to be confirmed by the GC-HRMS method, and up to 10% of negative samples have also to be confirmed. Bioassays and GC-MS methods are, however, subjected to different levels of specific requirements and offer different information (Commission Regulation 252/2012, 2012; Commission Regulation 278/2012, 2012).

For bioassays, two different approaches can be performed: a pure screening approach and a semi-quantitative approach. For the pure screening approach, the response of samples is compared to that of a reference sample at the level of interest. The use of such matrix-specific reference samples requires taking into account variations in response factors and recovery rates (Hoogenboom et al., 2000). Furthermore, additional analyses of reference samples around the level of interest are often requested to demonstrate the good performance of the bioassay in the range concerned by the control value. The use of such matrix-specific reference samples further helps in reducing the potential risk of matrix effects due to endogenous or exogenous compounds present in samples (Van Wouwe et al., 2004a). Supplementary parameters concerning acceptable blank levels and precision must also be controlled. Samples with a response less than the reference are declared negatives (or compliant), those with a higher response are suspected positives (or non-compliant). For the semi-quantitative approach, the use of standard dilution series and replicate analyses is mandatory. Bioanalytical Equivalents (BEQs – analogous to TEQs) have to be calibrated versus 2,3,7,8-TCDD response factor, blank contributions from potential impurities have to be estimated, and recovery rates have to be calculated on the basis of reference samples. Additionally, other specific steps are required in the sample preparation, especially for the separation of the PCDD/F fraction from the PCB fraction, in order to allow proper normative use and possible comparison with the confirmatory GC-HRMS method (Van Wouwe et al., 2004b). Such an approach is therefore much more demanding in terms of resources, both analytical and economic. Further requirements (QC charts, replicate numbers, etc.) exist for bioassays, and vary depending on whether cell-based or kit-based bioassays are used (Commission Regulation 278/2012, 2012).

Unit resolution GC-MS methods can be used for semi-quantitative screening and require the use of specific standards, detailed calibration and the calculation of recovery rates. These allow reporting as TEQ results, but also provide additional, valuable congener-specific data. For GC-MS screening methods the required level of expertise is higher and is similar to that of the GC-HRMS methodology used for confirmatory purposes, the only differences being the type of mass analyzer to be used and the LOQs to be reached. Therefore, although originally foreseen as an economically viable alternative to GC-HRMS, the GC-LRMS screening approach is rarely implemented in practice. As a matter of fact, both general analytical criteria and sample preparation requirements being the same (preparing a sample for GC-LRMS or GC-HRMS requires the same efforts), the only remaining difference possibly consists in the investment related to the MS analyzer. Although higher than for most other alternative LRMS analyzers used in screening, the investment related HRMS sector instruments is not significant on a cost per sample basis. In reality, the real financial impact on a sample by sample basis based on a return of investment calculated on the lifetime of the analyzer is extremely low. Many analytical laboratories thus prefer to invest in the confirmatory physico-chemical tool rather than in its screening equivalent.

3.2.1 Biological screening tools

Various bioanalytical detection methods (BDM) exist and have been developed to measure dioxin-like activity in various sample matrices. *In vivo* assays are normally not used for food control as they are more complex in terms of the processes involved, can suffer from possible competitive inhibition, and are therefore more difficult to interpret as the number of biomarkers is important. Aryl hydrocarbon hydroxylase (AHH) and ethoxyresorufin-O-deethylase (EROD) assays utilized to determine cytochromes P4501A1 (CYP1A1) activity are the most reported *in vivo* examples (Safe, 1987). *In vitro* bioassays are also generally more popular because no euthanasia or invasive surgical techniques on animals are required, and they are normally less expensive and less time-consuming.

Cell-based assays

Next to enzymatic assays (AHH assay and EROD assay) (Safe, 1993), the most popular cell-based assay to screen for dioxin and dioxin-like compound in food-feed is the *in vitro* luciferase bioassay called CALUX (Chemically Activated LUciferase gene eXpression) (Murk *et al.*, 1996). It is a recombinant receptor/reporter gene assay system using several aspects of the AhR-dependent mechanism of action (Aarts *et al.*, 1995; Garrison *et al.*, 1996).

The CALUX assay is based on the use of eukaryotic cells, genetically modified to contain the firefly luciferase gene under the control of a promoter containing at least one DRE (Dioxin-Responsive Element). When these cells are exposed to dioxins, dioxins enter into the cells by easily crossing the phospholipidic membrane of the cells and bind to the cytoplasmic Ah receptor. The complex dioxin-AhR is then translocated into the nucleus of the cell and binds to DREs, inducing the expression of the luciferase gene, and subsequently the synthesis of the firefly luciferase protein. After substrate (ATP and luciferin) addition, one can measure the emission of light, which is correlated to the concentration of dioxin. The first CALUX assay was described by Aarts and co-workers in 1993 (Aarts *et al.*, 1993). Nowadays, at least two commercial systems exist, using either rat (dioxin-responsive (DR)-CALUX®, BDS (BDS, 2011)) or mouse hepatoma cells (CALUX®, XDS (XDS, 2011)). The relative effective potency (REP, biological activity in CALUX, relative to the TCDD) of CALUX based on rat and mouse cells is known to correlate well with reported TEF values (Garrison *et al.*, 1996; Scippo *et al.*, 2004), although some discrepancies can still contribute to variability between CALUX and GC-HRMS data (Carbonnelle *et al.*, 2004; Denison *et al.*, 2004; Vanderperren *et al.*, 2004). Other non-commercial assays have been developed with rat, mouse or human cells (Behnisch *et al.*, 2001). The application of the CALUX bioassay for the monitoring of dioxins in feed and food was among others reviewed by Windal et al. (2005a), Hoogenboom et al. (2006) and Scippo et al. (2008).

The chemical-activated fluorescent protein expression (CAFLUX) assay is another approach using the enhanced green fluorescent protein gene as a reporter gene for AhR activation instead of the firefly luciferase gene used in the CALUX. Despite potential advantages, such as simplicity of use and moderate cost compared to CALUX, CAFLUX may suffer from dynamic range limitations (Nagy et al., 2000).

Non-cell-based bioassays
A description of the non-cell-based assays capable to report on dioxin activity is available in a review by Behnisch et al. (2001). Among them, competitive immunoassays based on dioxin-specific antibodies produced by mammals are the most popular (Harrison and Carlson, 1997a). A major challenge is that because one has to report results in terms of TEQs for dioxin analysis, antibodies need to express binding affinities correlated to the TEF scale. Although it was evident in the case of bioassays, many studied antibodies did not respond to this attempt. In practice, 2,3,7,8-TCDD specific antibodies express various cross-reactivities towards other congeners (Stanker et al., 1987; Sugawara et al., 1998). Some of them exhibit cross-reactivities similar to the TEFs and, therefore, acceptable correlations are observed when compared to GC-MS TEQ data (Harrison and Carlson, 1997b). In addition to potential sensitivity issues, the high lipophilicity of PCDD/Fs gives compatibility problems with immunoassays which are most often required to be performed in aqueous media (Harrison and Eduljee, 1999). Solvent exchange between sample extraction (organic solvents) and final analysis is inevitable. The use of assay-compatible detergents and low concentration of water-miscible solvents such as methanol or DMSO can help to avoid larger variability in recovery rates.

Another commercially available receptor binding assay is the so-called 'Ah-Immunoassay'. It is based on AhR mediation. The additional use of antibodies specific to the AhR-xenobiotic complex allows colorimetric detection. The color development is thus proportional to the amount of transformed AhR. The detection limits achievable are as low as 1 pg per plate well (Kobayashi et al., 2002). This assay showed cross-reactivity (REP) values in accordance to the TEF scale (Hiraoka et al., 2001). One however needs to keep in mind that REP values are issued from a single 'affinity' measure, whereas TEFs are derived from multifold experiments and various additional assumptions.

Limitations of bioassays
These methods detect the overall exposure of many xenobiotic compounds via AhR mediation, and can therefore be subject to both false positive and negative responses. If it is a good tool for comparison of global dioxin-like

activity between groups, discrepancies might arise between absolute values and those reported from GC-HRMS experiments (Koppen et al., 2001). Even if proven to be robust enough to accommodate 'dirty' sample extracts (Murk et al., 1996), a dioxin-specific clean-up is required to overcome the lack of specificity, by removing non-dioxin compounds and reduce over estimation risks regarding GC-HRMS TEQ values.

Refinement of the methodology in order to separate activities due to various classes of chemicals is thus important. Most of the existing methodologies are, however, still tedious and require a lot of sample handling. Such excessive manipulations are definitely time-consuming, and are subject to poor recoveries and a high risk of cross-contamination. The bigger challenge is to establish simple clean-up dioxin-specific procedures that permit the use of the assay as a dioxin screening tool without a drastic increase in the analysis cost, which would remove the most interesting aspect of bioassay use in the field. It is a necessary condition to make bioanalytical methods not only final stage analytical tools, but a complete solution including sample extraction and specific clean-up. That is critical (1) to efficiently screen out negative samples from the entire set of samples to analyze without over production of false positives, and (2) to respond to major screening criteria.

3.2.2 Physico-chemical screening tools

It was clear over 30 years ago that GC-MS was the instrumental method of choice for persistent organic pollutant (POP) determinations and specially for PCDD/Fs. Mass spectrometry provides not only a very specific quantification, but also ensures the unambiguous identification of target compounds. However, not all GC-MS instrumentation can measure PCDD/Fs. As already mentioned, the complexity of the measurement of these compounds is also related to the low levels at which they occur in environmental matrices, but particularly in food and feed. The required sensitivity is achieved by a combination of high-resolution and high-mass accuracy using double focusing magnetic sector instruments, commonly called high-resolution mass spectrometers (HRMS). In the last few years, other MS-based detection techniques have been investigated as possible 'alternatives' to GC-HRMS for measurements of PCDD/Fs and related compounds in biological matrices (Focant et al., 2005a). Most of these, however, still suffer from a lack of sensitivity and can barely equal the current performance of the sector HRMS instruments. For this reason, the coupling of GC to alternative MS tools such as quadrupole ion-storage MS (GC-QISTMS) or triple quadrupole MS (GC-QQQMS), both instruments performing in tandem (MS/MS) mode, as well as comprehensive two-dimensional gas chromatography (GC×GC) coupled to time-of-flight MS (GC×GC-TOFMS), cannot directly be used as confirmatory tools but can potentially be used as screening tools.

GC-QISTMS/MS

Tandem in-time mass spectrometry based on ion trap MS has been developed and used to analyze PCDD/Fs and related compounds in various matrices over the last few years (Plomley et al., 2000). Basically, the lack of selectivity in full-scan mode with these benchtop instruments (i.e. mass unit resolution) is compensated by operating the instrument in MS/MS mode. The dioxin or furan precursor ion specifically loses a fragment of COCl•, which is characteristic of these molecules. A global overview of MS/MS scan functions occurring in-time is available in an early report (Plomley and March, 1996). The first step consists in isolating the most intense ion in the precursor ion cluster (e.g. m/z of 322 for TCDD), that is, the predominant transition $[M+2]^{+•}$ (Table 3.1). Then, collision induced dissociation (CID) is performed by multi-frequency irradiation (MFI), and specific daughter ions are produced. During the CID process, PCDD fragmentation is characterized by losses of Cl•, COCl•, 2COCl• and PCDF fragmentation by losses of Cl•, COCl•, $COCl_2$ and $COCl_3$•. The main fragment used for quantification by the ID technique is the loss of COCl• for both PCDDs and PCDFs, while the loss of Cl_2 characterizes the main fragment used for PCB quantification. This was reported earlier as an efficient screening and complementary method for the monitoring of dioxin levels in food and feed (Eppe et al., 2004).

As for the confirmatory GC-HRMS method, the quantification is carried out by isotope dilution (ID). The main difference is characterized here by the fact that instead of following the two most abundant molecular ions at 10 000 resolution in selected ion monitoring (SIM) mode, the two product ions $[M+2-CO^{35}Cl^•]$ and $[M+2-CO^{37}Cl^•]$ are monitored for both native and labelled molecular ions. This is called multiple reaction monitoring (MRM). Within a specified time window, the instrument alternately scans the native and the labelled congener. As for SIM GC-HRMS, the chromatogram is sliced into MRM time windows. As native $^{12}C_{12}$ congeners coelute with their corresponding labelled $^{13}C_{12}$ isomers, the native peak maxima have to fall within 3 s of their corresponding $^{13}C_{12}$ labelled analogues for identification. The specificity of MS/MS is achieved by monitoring two product ions and by checking their isotopic ratios. If the most abundant ion from the isotopic cluster is selected for MS/MS (i.e. [M+2] for TCDD), then the isotopic ratio for product ions does not follow the natural abundance of $N - 1$ Cl (i.e. 3 Cl). Indeed, there is one chance out of four of losing $CO^{37}Cl^•$ while there are three chances out of four of losing $CO^{35}Cl^•$. Thus, the ratio of product ions (i.e. the ratio 257/259) equals to 0.33 (Focant et al., 2001). The other ratios from tetra- to octa-chlorinated congeners can be calculated in the same way (Table 3.1). Obviously, to ensure the production of two different daughter ions, the precursor ions have to contain at least one ^{37}Cl. The analytical quality criterion for screening technique allows a broader range of isotopic ratios (i.e. ± 25%). Additionally, the labelled product ions are characterized by [M+11], as one carbon is lost during fragmentation.

Table 3.1 Main parameters optimized for GC–QISTMS/MS analysis of PCDD/Fs. The congener's classification corresponds to the elution order on Rtx5-MS 40 m column

Peak	Compounds	Window (min)	Molecular ions	CID (V)	Collision time (ms)	q value	Product ions	Isotopic ratios
1	2,3,7,8 TCDF	20–21.4	306 [M+2]	5.5	30	0.45	241/243	0.33
	2,3,7,8 TCDF $^{13}C_{12}$	—	318 [M+2]	5.5	30	0.45	252/254	0.33
2	2,3,7,8 TCDD	21.4–21.95	322 [M+2]	5	30	0.45	257/259	0.33
	2,3,7,8 TCDD $^{13}C_{12}$	—	334 [M+2]	5	30	0.45	268/270	0.33
3	1,2,3,7,8 PeCDF	21.95–25.7	340 [M+2]	6	30	0.45	275/277	0.25
	1,2,3,7,8 PeCDF $^{13}C_{12}$	—	352 [M+2]	6	30	0.45	286/288	0.25
4	2,3,4,7,8 PeCDF	—	340 [M+2]	6	30	0.45	275/277	0.25
	2,3,4,7,8 PeCDF $^{13}C_{12}$	—	352 [M+2]	6	30	0.45	286/288	0.25
5	1,2,3,7,8 PeCDD	25.7–29	356 [M+2]	6	30	0.45	291/293	0.25
	1,2,3,7,8 PeCDD $^{13}C_{12}$	—	368 [M+2]	6	30	0.45	302/304	0.25
6	1,2,3,4,7,8 HxCDF	29–33.5	374 [M+2]	6	30	0.45	309/311	0.20
	1,2,3,4,7,8 HxCDF $^{13}C_{12}$	—	386 [M+2]	6	30	0.45	320/322	0.20
7	1,2,3,6,7,8 HxCDF	—	374 [M+2]	6	30	0.45	309/311	0.20
	1,2,3,6,7,8 HxCDF $^{13}C_{12}$	—	386 [M+2]	6	30	0.45	320/322	0.20
8	2,3,4,6,7,8 HxCDF	—	374 [M+2]	6	30	0.45	309/311	0.20
	2,3,4,6,7,8 HxCDF $^{13}C_{12}$	—	386 [M+2]	6	30	0.45	320/322	0.20
9	1,2,3,4,7,8 HxCDD	—	390 [M+2]	6	30	0.45	325/327	0.20
	1,2,3,4,7,8 HxCDD $^{13}C_{12}$	—	402 [M+2]	6	30	0.45	336/338	0.20
10	1,2,3,6,7,8 HxCDD	—	390 [M+2]	6	30	0.45	325/327	0.20
	1,2,3,6,7,8 HxCDD $^{13}C_{12}$	—	402 [M+2]	6	30	0.45	336/338	0.20
11	1,2,3,7,8,9 HxCDD	—	390 [M+2]	6	30	0.45	325/327	0.20
	1,2,3,7,8,9 HxCDD $^{113}C_{12}$	—	402 [M+2]	6	30	0.45	336/338	0.20
12	1,2,3,7,8,9 HxCDF	—	374 [M+2]	6	30	0.45	309/311	0.20
	1,2,3,7,8,9 HxCDF $^{13}C_{12}$	—	386 [M+2]	6	30	0.45	320/322	0.20

(Continued)

Table 3.1 Continued

Peak	Compounds	Window (min)	Molecular ions	CID (V)	Collision time (ms)	q value	Product ions	Isotopic ratios
13	1,2,3,4,6,7,8 HpCDF	33.5–37.5	410 [M+4]	6	30	0.45	345/347	0.40
	1,2,3,4,6,7,8 HpCDF $^{13}C_{12}$	—	422 [M+4]	6	30	0.45	356/358	0.40
14	1,2,3,4,6,7,8 HpCDD	—	426 [M+4]	6	30	0.45	361/363	0.40
	1,2,3,4,6,7,8 HpCDD $^{13}C_{12}$	—	438 [M+4]	6	30	0.45	372/374	0.40
15	1,2,3,4,7,8,9 HpCDF	—	410 [M+4]	6	30	0.45	345/347	0.40
	1,2,3,4,7,8,9 HpCDF $^{13}C_{12}$	—	422 [M+4]	6	30	0.45	356/358	0.40
16	OCDD	37.5–43	460 [M+4]	6	30	0.45	395/397	0.33
	OCDD $^{13}C_{12}$	—	472 [M+4]	6	30	0.45	406/408	0.33
17	OCDF	—	444 [M+4]	6	30	0.45	379/381	0.33
	OCDF $^{13}C_{12}$	—	456 [M+4]	6	30	0.45	390/392	0.33

CID, collision-induced dissociation.

Native TCDD

$[C_{12}H_4{}^{35}Cl_4O_2]^{+\bullet}$ → $[C_{11}H_4{}^{35}Cl_3O]^+ + CO{}^{35}Cl^\bullet$
(m/z = 320) (m/z = 257)

$[C_{12}H_4{}^{35}Cl_3{}^{37}ClO_2]^{+\bullet}$ → $[C_{11}H_4{}^{35}Cl_3O]^+ + CO{}^{35}Cl^\bullet$
(m/z = 322) (m/z = 259)

Fig. 3.1 Selection of parent and product ions in MRM mode when using GC-QQQMS/MS for TCDD.

Table 3.2 Calculation of theoretical ratio of product ions of TCDD in MRM mode with GC-QQQMS/MS

Precursor ion (m/z)	Largest isotope precursor peak (%)	Relation $^{35}Cl/^{37}Cl$	Probability of ^{35}Cl loss in MS/MS	Product ions (m/z)	Relative abundance of product ions (%)	Theoretical ratio (qualifier/quantifier) of product ions
320	77.540	4/0	1.000	257 (quantifier)	77.540	0.969
322	100.000	3/1	0.750	259 (qualifier)	75.000	–

GC-QQQMS/MS

The basic principle of GC-QQQMS/MS (commonly referred to as GC-Triplequad MS) is similar to the principle GC-QISTMS/MS. Both operate in MRM mode but, instead of a sequential tandem-in-time events occurring inside the trap, leading to the characteristic loss of COCl$^\bullet$ and the ejection of product ions (MS/MS in-time), here the precursor ions are first selected in the first quadrupole, fragmented in the collision cell (second quadrupole), and finally the product ions are monitored in the third quadrupole (MS/MS in space). Two product ions are monitored using either QISTMS/MS or QQQMS/MS (e.g. product ions 257 and 259 m/z from TCDD). However, the fundamental difference between the two MS/MS approaches comes from the relationship between the product ions and their precursors. While the product ions 257 and 259 m/z arise directly from the same precursor ion when using an ion trap (e.g. 322 m/z, as mentioned in subsection GC-QISTMS/MS), the MRM acquisition mode (called product ions scan) in triple quadrupole consists of selecting two precursor ions from the isotope cluster giving the corresponding product ions 257 and 259 m/z. Obviously, the objective is to achieve the highest degree of selectivity and sensitivity. As a result, the appropriate selection of ions is driven by selecting the most abundant precursor ions that also give product ion ratio (257/259 m/z) as close as possible to the value of one. As an example, Fig. 3.1 shows the adequate selection of two precursor and two product ions in MRM mode when using GC-QQQMS/MS for TCDD. The calculation of the relative product ion abundance for TCDD is summarized in Table 3.2.

GC×GC-TOFMS

GC×GC has been developed to meet an increasing need for complex sample analysis and to address limitations such as peak capacity, dynamic range and restricted specificity of mono-dimensional (classical) GC systems (i.e. to improve the global efficiency of the separation) (Phillips and Beens, 1999). This technique is based on the fast sampling and transfer of the sample by a cryogenic modulator located between the first dimension (^{1}D) column and the second dimension (^{2}D) column connected in series. Series of high speed secondary chromatograms of a length equal to the modulation period P_M (3–10 s) are recorded one after another by a detector located at the end of ^{2}D. They are computerized by specific software and combined to describe the multidimensional elution pattern. In GC×GC the peak capacity is much larger than in classical GC, making GC×GC well suited to accommodating complex mixtures of compounds. Compared to classical GC, the analytical speed (number of compounds separated per unit of time) is greatly enhanced. GC×GC also improves specificity two retention times (t_R)), selectivity (phase combination) and sensitivity (peak compression resulting in signal enhancement because of mass conservation) (Dimandja, 2004).

GC×GC can be coupled to time-of-flight mass spectrometry (TOFMS) because it offers the fast acquisition rate required for accurate characterization of GC×GC peaks that have a width at the base of 100 to 250 ms. Sector and ion trap MS that are popular in the environmental field can barely be used because of the time it takes for those instruments to scan even a limited number of masses (SIM) and move to another mass cluster. TOFMS can acquire at rates up to 500 scans per s (Leclercq and Cramers, 1998), which corresponds to 50 data points per 100 ms peak. Contrary to sector, quadrupole, and traps operating in SIM mode, a full mass spectrum is collected during each TOFMS acquisition. Additionally, because all ions are collected virtually at the same time point of the chromatogram, accurate peak identification and quantification are thus possible based on deconvoluted ion currents (DIC). GC×GC coupled to TOFMS is therefore an analytical tool ideally suited for the measurement of families of contaminants in complex matrices such as food.

GC×GC-TOFMS can also be used with ID to quantify target compounds. For quantification of the selected PCBs and PCDD/Fs, two masses are summed up for both native and labelled compounds, and QA/QC measures similar to those used for GC-HRMS are applied (Table 3.3). GC×GC-TOFMS method LOQs are not as good as for GC-HRMS but are nevertheless at the level of 0.5 pg on-column for some congeners. Increasing sample sizes is not feasible in practice because the larger the sample size, the larger the quantities of solvents and sorbents, the higher the procedural blank control (BC) levels and the higher the LOQs (Focant et al., 2006). This single injection method correlates with the well-established and accepted reference GC-HRMS methods, which require at least two separate injections to report all the analytes. Most of the potentially interfering compounds are separated from the analytes of interest in the chromatographic GC×GC space, due to the increased peak capacity (Focant et al., 2005b).

Table 3.3 Principal chromatographic and mass spectrometric parameters for the GC×GC-IDTOFMS separation of the selected PCBs and PCDD/Fs

Peak number	Congener[a]	^1t_R (s)	^2t_R (s)	Quantification masses			Theoretical isotope ratios	Acceptable range (20%)
				$^{12}C_{12}$-natives	$^{13}C_{12}$-labels			
1	TriCB-28	727	1.91	256	258	268	270	0.98	0.78–1.18
2	TeCB-52	751	2.09	290	292	302	304	0.77	0.62–0.92
3	TeCB-80	895	2.11	—	—	302	304	0.77	0.62–0.92
4	PeCB-101	923	2.34	328	326	340	338	0.65	0.52–0.78
5	TeCB-81	1025	2.27	290	292	302	304	0.77	0.62–0.92
6	TeCB-77	1061	2.32	290	292	302	304	0.77	0.62–0.92
7	PeCB-123	1094	2.56	328	326	340	338	0.65	0.52–0.78
8	PeCB-118	1106	2.56	328	326	340	338	0.65	0.52–0.78
9	PeCB-114	1126	2.69	328	326	340	338	0.65	0.52–0.78
10	HxCB-153	1150	2.57	362	360	374	372	0.82	0.66–0.98
11	PeCB-105	1186	2.79	328	326	340	338	0.65	0.52–0.78
12	HxCB-138	1233	2.81	362	360	374	372	0.82	0.66–0.98
13	1,2,3,4-TCDD[b,c]	1252	2.56	—	—	328	—	—	—
14	2,3,7,8-TCDF	1264	2.56	304	306	316	318	0.77	0.62–0.92
15	2,3,7,8-TCDD	1292	2.46	320	322	332	334	0.77	0.62–0.92
16	PeCB-126	1340	2.46	328	326	340	338	0.65	0.52–0.78
17	HxCB-167	1381	2.69	362	360	374	372	0.82	0.66–0.98
18	HxCB-156	1476	2.84	362	360	374	372	0.82	0.66–0.98
19	HxCB-157	1496	2.89	362	360	374	372	0.82	0.66–0.98
20	HpCB-180	1512	2.81	396	394	408	406	0.98	0.78–1.18
21	1,2,3,7,8-PeCDF	1559	2.61	342	340	354	352	0.65	0.52–0.78
22	2,3,4,7,8-PeCDF	1675	2.61	342	340	354	352	0.65	0.52–0.78

(*Continued*)

Table 3.3 Continued

Peak number	Congener[a]	^1t_R (s)	^2t_R (s)	Quantification masses $^{12}C_{12}$-natives	Quantification masses $^{13}C_{12}$-labels	Theoretical isotope ratios	Acceptable range (20%)
23	1,2,3,7,8-PeCDD	1691	2.54	358 356	370 368	0.65	0.52–0.78
24	HxCB-169	1711	2.42	362 360	374 372	0.82	0.66–0.98
25	HpCB-189	1875	2.76	396 394	408 406	0.98	0.78–1.18
26	1,2,3,4,7,8-HxCDF	2025	2.52	376 374	388 386	0.82	0.66–0.98
27	1,2,3,6,7,8-HxCDF	2037	2.54	376 374	388 386	0.82	0.66–0.98
28	1,2,3,4,7,8-HxCDD	2185	2.42	392 390	404 402	0.82	0.66–0.98
29	1,2,3,6,7,8-HxCDD	2197	2.42	392 390	404 402	0.82	0.66–0.98
30	2,3,4,6,7,8-HxCDF	2209	2.46	376 374	388 386	0.82	0.66–0.98
31	1,2,3,7,8,9-HxCDD	2264	2.37	392 390	404 402	0.82	0.66–0.98
32	1,2,3,7,8,9-HxCDF	2296	2.54	376 374	388 386	0.82	0.66–0.98
33	1,2,3,4,6,7,8-HpCDF	2436	2.39	410 408	422 420	0.98	0.78–1.18
34	1,2,3,4,6,7,8-HpCDD	25119	2.46	426 424	438 436	0.98	0.78–1.18
35	1,2,3,4,7,8,9-HpCDF	2543	2.69	410 408	422 420	0.98	0.78–1.18
36	1,2,3,4,6,7,8,9-OCDD	2703	2.94	458 460	470 472	0.88	0.70–1.05
37	1,2,3,4,6,7,8,9-OCDF	2707	3.15	442 444	454 456	0.88	0.70–1.05

[a]Numbering of PCBs according to IUPAC.
[b]Congeners used for recovery calculation.
[c]This congener is $^{13}C_6$–1,2,3,4-TCDD only.

Comparison to the confirmatory method
GC-QISTMS/MS, GC-QQQMS/MS and GC×GC-TOFMS are capable of measuring low levels of PCBs and PCDD/Fs in foodstuffs. GC-QISTMS/MS and GC×GC-TOFMS cannot, however, currently reach the same sensitivity as GC-HRMS. GC×GC-TOFMS offers at least as much scope for QC as the other methods (isotopic dilution quantification, isotope ratio check, dual set of retention time check). The large ion volume in the TOFMS source makes it unlikely to be as influenced by sample extract quality as other classical small source types, a significant advantage for routine use. Furthermore, the TOFMS instrument showed itself (Focant *et al.*, 2005b) to be more robust than QISTMS, where the ion trap can easily be contaminated by matrix interfering ions, reducing sensitivity when used on a routine basis. Although the sensitivity of PCB measurement is adequate, further improvement in sensitivity at the sub-picogram level, together with reduced data handling and processing time requirements, are still needed for GC×GC-TOFMS. The integration of PCBs and other types of analytes into the EU regulation is making GC×GC-TOFMS a valuable comprehensive screening tool. One also has to mention that GC×GC coupled to micro electron capture detection (micro-ECD) has been reported for the measurement of selected POPs (Danielsson *et al.*, 2005). Valuable data were reported and, despite the lack of mass spectral data, this technique can also be considered as an EU standards screening tool.

The new generation of GC-QQQMS/MS instruments has shown a significant improvement in sensitivity. Kotz and co-workers have recently reported instrumental LOQ of 50 fg (absolute amount on-column) for TCCD (Kotz *et al.*, 2011). In addition, Fürst and co-workers achieved the same level of performance. The lowest calibration point for PCDD and PCDF congeners was 100 fg injected on-column with an excellent linearity from 0.1 to 10 pg injected on-column (Fürst *et al.*, 2010). This new generation of GC-QQQMS/MS instruments may partially fill the sensitivity gap with GC-HRMS sector instruments. Bernsmann *et al.* compared analytical results of food and feed extracts, injected by GC-HRMS and GC-QQQMS/MS. They demonstrated the suitability of the alternative MS/MS approach. They indicated that the results reported with GC-QQQMS/MS meet the requirements of EU legislation when performing analysis at the maximum and action levels (Bernsmann *et al.*, 2011).

3.3 Specific analytical requirements for biological and physico-chemical tools

This section neither provides an exhaustive list of analytical requirements to fulfil nor intends to offer instructions on how a dioxin measurement should be performed. We try to review some key criteria of the analytical methodologies of PCDD/F and PCB analyses in food and feed.

3.3.1 A performance-based measurement system (PBMS)

By setting regulatory limits in food and feed, analytical results reported by laboratories must be comparable in order to make correct decisions. At the time when no standardized or harmonized methods were available (i.e. at the beginning of 2000), the idea was to propose harmonized quality criteria for biological and GC-MS measurements (Behnisch *et al.*, 2001; Malisch *et al.*, 2001). The basic idea was to implement a kind of Performance-Based Measurement System (PBMS) rather than establish a standardized method for PCDD/Fs and PCBs in food and feed. In other words, every laboratory was allowed to use its own method, with the understanding that the laboratory had to demonstrate that the selected method was fit for purpose. The general recommendations described then can be considered as the foundation of the legislation that followed. Sets of practical criteria were proposed, and evolved to the ones that are currently used in EU regulation, such as those for the ID GC-HRMS method: the laboratory shall be accredited to ISO 17025, shall demonstrate its capability by obtaining satisfactory z-scores at relevant Proficiency Testing (PT), shall validate the performance of the method (accuracy and precision) in the range of tolerance, shall develop methods with the required sensitivity (LOQs should be at least one fifth of the level of interest), etc. In addition, based on the upper-, lower-, and medium-bound approach (Commission Directive 2002/69/EC, 2002), the difference between upperbound level and lowerbound level should not exceed 20% for foodstuffs with a dioxin contamination at the level of interest (for both maximum and action levels).

The focus of the general concept of harmonized quality criteria was not intended to offer specific or practical solutions to the analyst's daily challenges. It actually offered the necessary flexibility to develop state-of-the-art methods, to integrate new knowledge and technologies, but also took into account that laboratories may execute protocols in slightly different manners in daily routine or may need to modify their analytical procedures in order to control/eliminate interferences. The decisions taken at that time were clearly oriented to help analytical chemists to do and to use 'good science' with the aim of generating effective data that were fit for their intended use.

In parallel, quality criteria were also established for screening methods including GC-MS and bioassays. All the performance-based criteria were summarized in the publication of the Commission Regulation 1883/2006 and the Commission Regulation 152/2009 for food and feed, respectively (Commission Regulation (EC) No 1883/2006, 2006; Commission Regulation (EC) No 152/2009, 2009). The practical relevance of some criteria apparently not directly based on analytical logic is however questionable. For instance, statements on the acceptable recovery rate level ranges required to be able to report data (*the recoveries of the individual internal standards shall be in the range of 60% to 120%*) are probably not really that relevant. Does such a statement mean that you will report wrong data if your TCDD recovery is 59%? Probably not. If the ^{13}C-labelled 2,3,7,8-chlorine substituted internal PCDD/F standards are correctly spiked at the early stage of the analytical procedure, the principle of ID mass spectrometry technique automatically corrects the bias due to the loss of the analyte. Obviously, low recoveries could lead to a sensitivity issue but the sensitivity parameter, which is pivotal in ultra-trace analysis, is embedded in many other performance-based criteria such as the LOQs at one fifth of the level of interest or even the difference between lowerbound and upperbound that should not exceed 20% at the level of interest. All these criteria have to be fulfilled when generating daily data. Then it is easy to demonstrate that when a suitable amount of sample is taken, when an adequate concentration factor is applied (i.e. final volume in the GC vial) and when a sufficient volume of the final extract is injected into the GC column, accurate data can be reported even if recoveries are within a so-called unacceptable 10–60% range.

Over the last decade, the general trend in the amended EU documents related to the analysis for the official control of PCDD/Fs and DL-PCBs in food and feed have shown the addition of sometimes relevant, but often stringent specific criteria that could deviate from a PBMS approach. The underlying danger is that this could lead to a divergence from the first objective of the EU approach (i.e. acknowledging the flexibility required in this measurement process) by applying excessive guardrails on the analytical methods for PCDD/Fs and related compounds. We have to focus on the final objective of analytical methods – generating effective data in a sustainable manner. Because ultra-trace analysis of these compounds in complex matrices is a source for challenges, there is no choice other than implementing a performance-based measurement system if we desire to integrate new knowledge and technologies.

3.3.2 The need for cut-off values for screening approaches

Increasing scientific insight and experience from application of biological and physico-chemical tools in routine screening analysis of PCDD/Fs and related compounds have been gained over the past 10 years. The primary objective of screening methods is to select samples that are suspected to be non-compliant from the compliant samples. A sample declared 'compliant' in screening may prove to be non-compliant after confirmation with GC-HRMS, thus the result from screening was a false-compliant result. A sample declared 'non-compliant' in screening may prove to be compliant in GC-HRMS, turning out to be a false-non-compliant result. The fractions of those results should be termed false-compliant rates (β-error) and false-non-compliant rates (α-error). The fitness for purpose of screening methods is to keep α- and β-errors at or below an acceptable level (e.g. 5%). To this end, cut-off values for the screening approach have been recently introduced. The determination of cut-off values is extensively described in the new EU legislation (Commission Regulation 252/2012, 2012; Commission Regulation 278/2012, 2012). As an example, Fig. 3.2 shows how a bioassay cut-off value should be set for decision over compliance with the respective maximum limits (ML) or transfer of the sample to a the confirmatory GC-HRMS with a risk of false-compliant result $\beta < 5\%$.

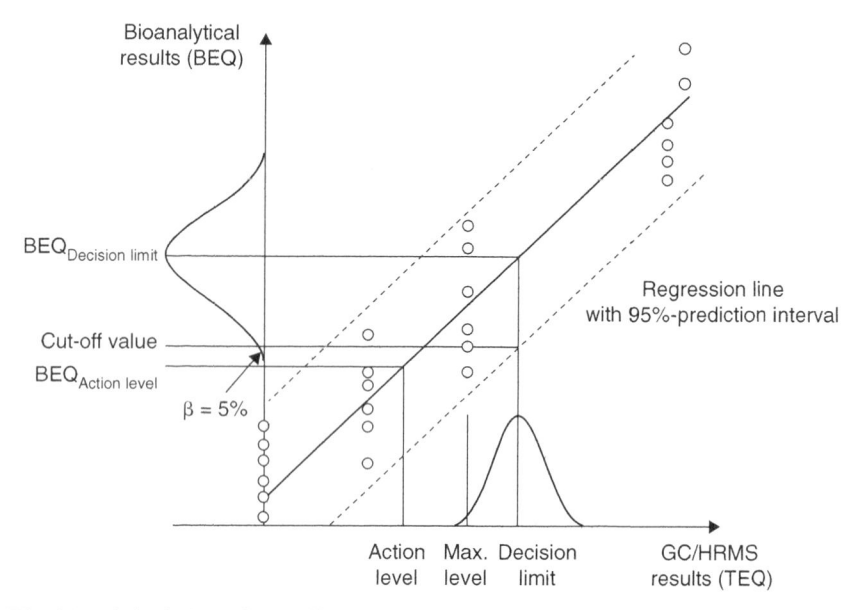

Fig. 3.2 Calculation of cut-off values based on a 95% level of confidence implying a false-compliant rate < 5% and a RSD_R < 25%: 1. from the *lower* band of the 95% prediction interval at the GC/HRMS decision limit, 2. from multiple analysis of samples ($n \geq 6$) contaminated at the GC/HRMS decision limit as the *lower* endpoint of the data distribution (represented in the figure by a bell-shaped curve) at the corresponding mean BEQ value.

3.4 Quantitative versus semi-quantitative approach

In analytical chemistry, the distinction between qualitative and quantitative methods is usually reported. Qualitative methods, also called screening methods, may have the potential for providing semi-quantitative results but are solely used for reporting a yes/no decision as indication of levels above or below certain ranges, e.g. limit of detection, limit of quantification, cut-off values or above maximum levels. For control of maximum and action levels for PCDD/Fs and dioxin-like compounds in feed and food, screening methods (bioanalytical and physico-chemical methods) can be applied. These are based on comparison of the analytical result with a cut-off value, as described in Section 3.3.2.

Semi-quantitative methods give an approximate indication of the concentration of the analyte (no measurement uncertainty associated with the reported result). While the numerical result does not meet the requirements for quantitative methods, it may be useful as information on the range of the analyte concentration and helpful for the analyst in deciding the calibration range for the confirmatory test subsequently to be performed and for QC purposes.

Quantitative methods meet the requirements for accuracy, trueness and precision, and other relevant validation information as detailed in Commission

Regulation (EC) No 1883/2006 and No 152/2009 (Commission Regulation (EC) No 1883/2006, 2006; Commission Regulation (EC) No 152/2009, 2009).

3.5 Validation QA/QC

As will later also be highlighted in the framework of the confirmatory GC-HRMS method (Section 3.6.2), a number of key performance parameters have to be met for bioanalytical or chemical screening approaches used specifically for compliance assessment with legal limits. In this context, to cope with the primary purpose of a screening tool, consisting of systematically eliminating all compliant samples while highlighting those samples which require further investigation by confirmatory GC-HRMS, strict but adapted QC criteria have also to be precisely designed for bioassays and GC-MS methods. They are summarized in the publication of the Commission Regulation 252/2012 and the Commission Regulation 278/2012 for food and feed, respectively (Commission Regulation 252/2012, 2012; Commission Regulation (EC) No 278/2012, 2012). Among them, setting the cut-off concentration to detect potentially non-compliant samples has already been stated, and is a fundamental performance characteristic (Section 3.3.1). In addition, Windal and co-workers proposed a series of specific quality control setting measures for the CALUX bioassay, to estimate the dioxin-like activity in marine biological samples (Windal et al., 2005b). Briefly, the proposal encompasses the assessment of the quality of the measurement plates, the verification of procedural blank levels, the introduction of a quality control sample analyzed simultaneously with each series of samples, the set-up of corresponding quality control charts with adapted limits including interpretation rules and decision making, the recommendations when establishing the dose–response curves of TCDD, and proposals for correcting results with the percentage of recovery. All the performance criteria mentioned here are minimum needed to statistically maintain the rates of false-compliancy and false-non-compliancy at or under the acceptable level of 5%.

3.6 Confirmatory methods for dioxins and PCBs in food and feed

After screening methods have been used to sieve large numbers of samples, non-compliant samples have to be confirmed by the reference GC-HRMS method.

3.6.1 State-of-the-art GC-HRMS

Once extracted and purified from matrix interferences, the chromatographic separation of the target compounds has to take place before analysis. Due to the range of volatilities of PCDD/Fs and related compounds, GC is used to separate the different congeners and to allow non-ambiguous identification. Because of chromatographic peak capacity and ion recording issues, a fractionation step is required during the sample preparation procedure to isolate the planar species (PCDD/Fs and non-ortho-PCBs) from the non-planar species (Mono-ortho-PCBs and NDL-PCBs). Those two fractions are injected separately in GC-HRMS.

Capillary GC columns ensure the required selectivity, especially for the separation of congeners of the same chlorination level. The separation characteristics and elution profiles of PCDD/Fs and PCBs on GC columns are well documented (Ryan et al., 1991). An appropriate combination of column length, internal diameter and stationary phase polarity is needed. For the analysis of food-feed samples, non-polar GC columns are usually used. They allow separation between homologue groups and can separate 2,3,7,8-substituted congeners from each other (Fraisse et al., 1994), showing the efficiency of a non-polar (methyl polysiloxane with 5% phenyl) column to separate the 2,3,7,8-toxic congeners from the rest, especially for 2,3,7,8-TCDF. The apolar GC column phase is the most widely used column for 2,3,7,8 PCDD/Fs and non-ortho PCBs analysis in foodstuffs, feedingstuffs and human samples. Usually 40–60m columns with 0.18–0.25 mm internal diameter and 0.15–0.25 μm film thickness are selected.

For the PCB fraction, apart from apolar columns, the 8% phenyl polycarborane-siloxane is also often used. The carborane group has a high affinity for PCBs with a low degree of ortho-substitution. Although this phase does not allow the separation of all the 209 PCB congeners, it separates some critical pairs of co-elutions present with other phases (Larsen et al., 1995). For example, indicator trichlorinated PCB-28 and PCB-31 (not followed), pentachlorinated mono-ortho PCB-123 and PCB-118, as well as hexachlorinated PCB-163 (not followed) and the indicator hexachlorinated PCB-138, are separated. A 25 m × 0.25 mm × 0.2 μm 8% phenyl polycarborane-siloxane stationary phase seems to be a good compromise between the required resolving power and the GC run time of roughly 30 mins. It should be mentioned that, to date, none of the existing stationary phases is capable of the separation of all the PCB congeners. Even emerging hyphenated methods such as the GC×GC TOFMS mentioned earlier can at the most separate 192 congeners (Focant et al., 2004).

Basic information on the reference GC-HRMS method is available in recent reviews (Eppe et al., 2006; Reiner et al., 2006). The extreme sensitivity of the HRMS instrument is gained by the ability to monitor specific characteristic ions in the mass spectrum of the target compound. It is called the selected ion monitoring mode or SIM. By this scan mode, a few femtograms of 2,3,7,8-TCDD injected in GC-HRMS can be detected. This performance is unmatched by any other techniques and it makes GC-HMRS the reference or the 'gold standard' method for PCDD/Fs. Even after an extensive clean-up and a high-resolution chromatographic separation, the risk of interference is still high. For that reason, the resolution ($M/\Delta M$) of the mass spectrometer should be set at least at 10 000 (10% valley definition). This allows mass discrimination at the 0.03–0.05 mass unit (dalton) level in the tetra- to octa-substituted congeners mass range. The number of ions which can be measured at any one time is generally limited, because at least ten sampling points for each GC peak are needed in order to get a Gaussian peak for accurate integration and quantification. Selected ions are therefore grouped in various segments. In addition, to control the mass accuracy, a lock mass (e.g. perfluorokerosene (PFK)) is measured in each segment cycle and a lock mass check is often included. The lock mass is ideally located within the measured mass range of each segment.

As mentioned, within each segment, the number of compounds (i.e. the number of specific ions) is limited. A quantification mass (the most abundant peak from the ions cluster) and a confirmatory mass (based on relative isotopic ratio of naturally occurring chlorinated isotopes) are required for native as well as for internal standard (Table 3.4). Thus, four masses are needed for one target congener. A maximum of three to four compounds of differing masses can therefore be measured. In order to overcome this limitation, the chromatogram is sliced in time windows by grouping target compounds based on their retention time. The chromatographic challenge is then to bring the compound by groups (chromatographic windows) with no overlap, which would result in loss of congener measurement. Hence, one of the major drawbacks of SIM mode is the necessity to redefine windows any time the chromatographic parameters are modified (i.e. cutting or changing the column).

The high resolution gas chromatography (HRGC) serves the essential requirement for isomeric separation, which is required to separate occurring congeners that show different, or no, TEF values. The retention time of native and labelled standard peaks must be within a range of 2 sec. To control the chlorination level and therefore the identity and the absence of interfering compounds, the measurement of the isotopic composition of the two most intense ions of both native and ^{13}C-labelled ion clusters must be ± 15% about the theoretical ion abundance ratio (Table 3.4). Any deviation outside this range will cause rejection of the congener's result.

The power of mass spectrometry in quantitative analysis is, in fact, further enhanced by the ID technique. This technique consists of spiking samples with an ideal internal standard (surrogate), which is the isotopically labelled standard (e.g. $^{13}C_{12}$ 2,3,7,8-TCDD), showing almost identical characteristics to the compound of interest (e.g. $^{12}C_{12}$ 2,3,7,8-TCDD). The small mass difference (e.g. 12 m/z) enables the discrimination between the compound of interest and its internal standard (Table 3.4). A calibration performed for all the PCDD/Fs and DL-PCBs with known amounts of native and internal standard congeners allows the calculation of the relative response ractor (RRF). The RRF takes into account the discrepancy that can be observed during MS ionization between natives and internal labelled standards. Thus, the RRF

Table 3.4 Target masses for PCDD/Fs and non-*ortho* PCBs in SIM mode for HRMS

Congeners	Window (min)	Monitored ions		Ion dwell time (ms)	Interscan time (ms)	Theoretical isotopic ratios	15% for isotopic ratios
		Quantitation ion	Confirmation ion				
TCB	20–26	291.9194 [M+2]	289.9224 [M]	110	10	0.77	0.65–0.88
TCB $^{13}C_{12}$		303.9597 [M+2]	301.9626 [M]	40		0.77	0.65–0.88
Lock mass		316.9824 [I]	316.9824 [I]	50			
TCDF	26–30	305.8987 [M+2]	303.9016 [M]	100	10	0.77	0.65–0.88
TCDF $^{13}C_{12}$		317.9389 [M+2]	315.9419 [M]	15		0.77	0.65–0.88
TCDD		321.8936 [M+2]	319.8965 [M]	100		0.77	0.65–0.88
TCDD $^{13}C_{12}$		333.9339 [M+2]	331.9368 [M]	15		0.77	0.65–0.88
TCDD $^{13}C_6$*		331.9078 [M+6]		85			
PeCB		325.8804 [M+2]	327.8775 [M+4]	100		0.64	0.56–0.75
PeCB $^{13}C_{12}$		337.9207 [M+2]	339.9177 [M+4]	15		0.64	0.56–0.75
Lock mass		330.9792 [I]	330.9792 [I]	50			
PeCDF	30–35	339.8597 [M+2]	337.8627 [M]	120	10	0.61	0.53–0.71
PeCDF $^{13}C_{12}$		351.9000 [M+2]	349.9029 [M]	15		0.61	0.53–0.71
PeCDD		355.8546 [M+2]	353.8576 [M]	150		0.61	0.53–0.71
PeCDD $^{13}C_{12}$		367.8949 [M+2]	365.8978 [M]	15		0.61	0.53–0.71
HxCB		359.8415 [M+2]	361.8385 [M+4]	100		0.81	0.69–0.94
HxCB $^{13}C_{12}$		371.8817 [M+2]	373.8788 [M+4]	15		0.81	0.69–0.94
Lock mass		380.9760 [I]	380.9760 [I]	50			

HxCDF	35–42	373.8207 [M+2]	375.8178 [M+4]	150		0.69–0.94
HxCDF $^{13}C_{12}$		385.8610 [M+2]	387.8580 [M+4]	15		0.69–0.94
HxCDD		389.8156 [M+2]	391.8127 [M+4]	150		0.69–0.94
HxCDD $^{13}C_{12}$		401.8559 [M+2]	403.8530 [M+4]	15		0.69–0.94
Lock mass		380.9760 [I]	380.9760 [I]	50	10	0.81
						0.81
						0.81
						0.81
HpCDF	42–47	407.7818 [M+2]	409.7788 [M+4]	150		0.88–1.20
HpCDF $^{13}C_{12}$		419.8220 [M+2]	421.8190 [M+4]	15		0.88–1.20
HpCDD		423.7767 [M+2]	425.7737 [M+4]	150		0.88–1.20
HpCDD $^{13}C_{12}$		435.8169 [M+2]	437.8140 [M+4]	15		0.88–1.20
Lock mass		430.9728 [I]	430.9728 [I]	50	10	1.04
						1.04
						1.04
						1.04
OCDD	47–52	459.7348 [M+4]	457.7377 [M+2]	150		0.75–1.01
OCDD $^{13}C_{12}$		471.7750 [M+4]	469.7780 [M+2]	15		0.75–1.01
OCDF		443.7398 [M+4]	441.7428 [M+2]	150		0.75–1.01
OCDF $^{13}C_{12}$		455.7801 [M+4]	453.7830 [M+2]	15		0.75–1.01
Lock mass		466.9728 [I]	466.9728 [I]	50	10	0.89
						0.89
						0.89
						0.89

* Syringe standard added prior to GC-HRMS analysis and used for recovery.

value directly affects the congener quantification as indicated in the following equation:

$$[\text{congener}]_i = \frac{(A^1\text{native},i + A^2\text{native},i) \times Q_i}{(A^1_{\text{labelled},i} + A^2_{\text{labelled},i}) \times RRF_i \times m}$$

where $[\text{congener}]_i$ is the concentration of the congener i (e.g. ng/kg); areas $A^1_{\text{native},i}$ and $A^2_{\text{native},i}$ are the areas of the quantitation and confirmation ions for the native congener i; $A^1_{\text{labelled},i}$ and $A^2_{\text{labelled},i}$ are the areas of the quantitation and confirmation ions for its corresponding labelled compound i; Q_i is the amount of the corresponding recovery standard i spiked (e.g. ng) in the sample; RRF_i is the relative response factor of the congener i; and m is the weight of the sample (e.g. kg).

3.6.2 Validation quality assurance/quality control (QA/QC)

The objective of QA/QC programmes is to control analytical measurements errors at levels acceptable to the end-user of the data and to assure that analytical results have a high probability of acceptable quality. It is generally evaluated on the basis of its measurement uncertainty associated with the final result which should match the end-user requirements. Accreditation according to the Standard ISO/CEN 17025 is an adequate way to implement QA/QC programmes in a routine laboratory and is now routinely applied to dioxin analysis.

One of the most important features of a QA/QC programme is the use of procedural blank and internal quality control (IQC) samples. They have to be included with the series of real samples in daily routine work. Both have to follow as much as possible the entire analytical protocol. IQCs have to be characterized by sufficient homogeneity and long-term stability to ensure that the analytical system remains under control. When available, they should as far as possible match analyte, level and matrix of real samples tested. Control charts (plots of the data from procedural blanks or IQC plotted in a chronological way) provide the most effective mechanism for interpreting data as shown in Fig. 3.3.

These graphs provide pictures of the performances of the analytical system. Limits and decision rules have to be implemented in order to use the charts efficiently and to maintain the system stabilized and under control. The strategy of the decision process, the corrective and preventive actions to be taken when a lack of control is observed, should be clearly identified and followed. The use of certified reference materials (CRMs or SRMs) and the participation in relevant interlaboratory studies complete the requirements for assessing the accuracy of the analytical method. For instance, participation in proficiency testing (PT) exercises is a useful way for assessing the quality of a laboratory's results. In chemical analysis, the statistical evaluation of PT

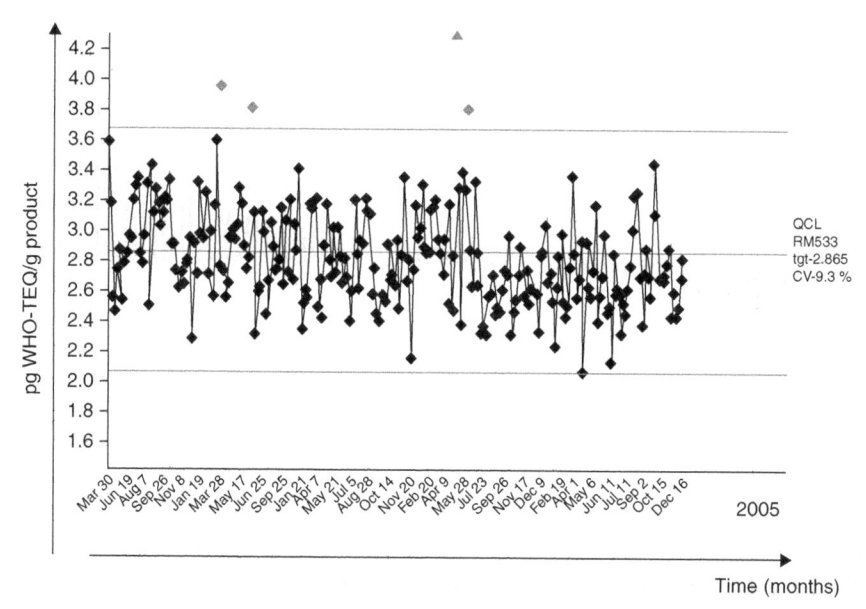

Fig. 3.3 Quality control chart of PCDD/Fs in milk powder.

results is based on the scoring system as recommended in the International Harmonized Protocol (Thompson *et al.*, 2006). In this system, the participant's result is converted into a z-score that gives a valuable indication of the performance of the laboratory. The z-scores are calculated as follows:

$$z = \frac{x - X}{\sigma_p}$$

where x is the lab result; X is the assigned value; and σ_p is the target value for the standard deviation.

In fact, a z-value is nothing more than an estimate of the error in the result scaled in standard deviation units (σ_p). Figure 3.4 gives an overview of the z-scores distribution for the sum of the PCDD/Fs in TEQ for egg yolk. Data are from the PT on dioxins in FOOD 2006 (Smastuen Haug and Becher, 2006). Statistics and chemometrics are tools that are more and more involved in the extraction of the information from QA/QC data with the ultimate aim of improving the quality of the generated data.

3.6.3 Potential alternative confirmatory methods

As described in Section Comparison to the confirmatory method, new technical developments of sensitive GC-QQQMS/MS for the determination of PCDD/Fs and PCBs in food and feed have been recently presented

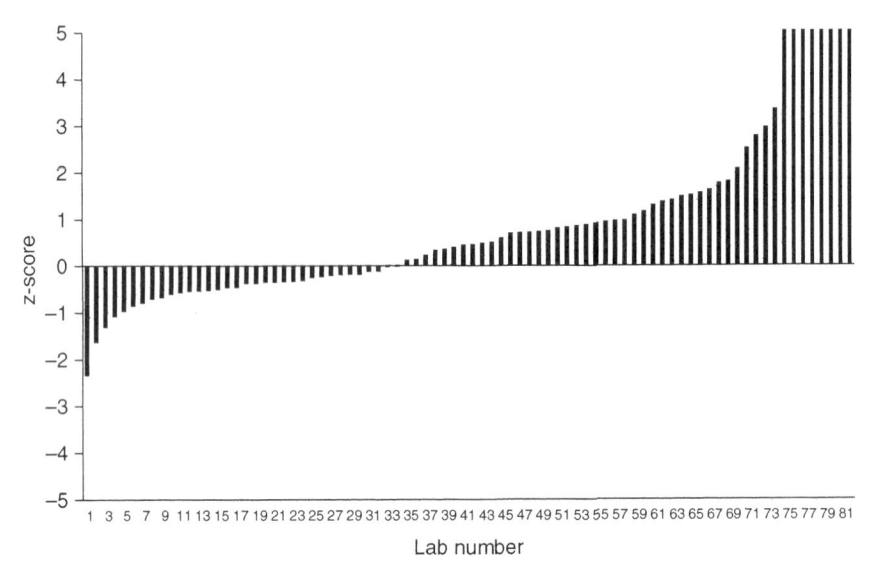

Fig. 3.4 Distribution of z-scores: PCDD/Fs TEQ in egg yolk.

(Bernsmann et al., 2011; Kotz et al., 2011). Both papers concluded that the new generation of GC-QQQMS/MSs are suitable for the measurement of PCDD/Fs and dioxin-like PCBs in food and feed at the level of interest. Those results triggered interest at the European level to investigate the possible use of new MS techniques or instruments for the determination of PCDD/Fs and PCBs in food and feed as a confirmatory method in addition to the gold standard GC-HRMS method. It means that the new candidate instrumentation should provide full information enabling the PCDD/Fs and DL-PCBs to be identified and quantified unequivocally at the level of interest. It is therefore assumed that the new instrumentation should meet the same requirements set for confirmatory method as already described in 252/2012 and 278/2012 (Commission Regulation 252/2012, 2012; Commission Regulation 278/2012, 2012). However, additional specific analytical criteria for GC-MS/MS are required. One should mention that the highest degree of specificity should be met by monitoring at least two specific precursor ions, giving each one specific corresponding product ion. Such an approach is reported for residue analyses in Commission Decision 2002/657/EC setting criteria for the performance of analytical methods and the interpretation of results, based on the earning of identification points (Commission Decision 2002/657/EC, 2002). In addition, product ion ratio (as calculated in Table 3.2 Section GC-QQQMS/MS for TCDD) should fall within an acceptable range of tolerance as already stated for ions monitored in SIM mode for GC-HRMS method. Although any new candidate MS technique for confirmatory use should meet the GC-HRMS performance criteria, the use of potentially less sensitive alternatives could require the differentiation between different categories of confirmatory methods, based on the purpose of analysis that could be either to check for sample compliance or to determine low background levels for re-evaluation of maximum or action levels (European Union reference laboratory, 2011). So far, only the GC-HRMS method can offer both types of information reliably. Future developments of GC(×GC)-HRTOFMS technology could potentially bring interesting insight in this area.

3.7 Future trends

This chapter focuses mainly on the description of analytical requirements that a screening method or a confirmatory method should successfully meet to generate effective data for the monitoring of PCDD/Fs and DL-PCBs in food and feed. Regulations, and specifically EU regulations, specify method performance criteria rather than adherence to a particular method (e.g. such as EPA 1613). The basic idea is appropriate because it is based on the fact that the best estimate of the true value (i.e. the accuracy) can be obtained by several equivalent analytical methods. The applicability of the selected methodology is established through the expression of a number of measurement quality objectives, which encompass accuracy, bias, precision, sensitivity, selectivity and robustness. However, the new trend in the amended EU documents recently published shows the addition of excessive guardrails on the analytical methods for PCDD/Fs and related compounds, which may hinder progress with the consequence of returning to standardized methods. Because ultra-trace analysis of these compounds in complex matrices is a source of challenge, there is no other choice than implementing a PBMS if we desire to integrate new knowledge and technologies (e.g. GC-MS/MS as a candidate confirmatory method).

Future trends will probably include study and consideration of emerging techniques, such as GC×GC coupled to either sector HRMS or HRTOFMS, and possibly ion mobility mass spectrometry. For GC×GC-HRsectorMS, although the scanning rate limitation of sector instruments does not make it as the detector of choice for GC×GC, they can be used with cryogenic modulation to use the modulator as a signal enhancer for selected congeners exhibiting extremely low levels (Patterson *et al.*, 2011). This would ease the measurement of real samples and limit the number of non-detected compounds to a minimum. GC×GC-HRTOFMS would be much easier to use, as full-scan data acquisition could be performed instead of SIM. Such a comprehensive instrument, offering the required acquisition speed, sensitivity, and resolution is unfortunately not yet available at the moment. No doubt it will come in the near future.

We have already emphasized the benefit of this additional dimension of chromatographic separation as a valuable comprehensive screening tool for POPs. However, the dimension of separation is not only limited to chromatographic based principle. One can imagine separating isobaric congeners with ion mobility spectrometry (IMS). A research group from Waters recently published preliminary data on the analysis of PCDD/F and PCB congeners by gas chromatography-electron impact ionization and ion mobility mass spectrometry (GC-EI-Q-IMS-TOF) (Rhys Jones *et al.*, 2011). The analytical approach could be of particular interest as it opens new perspectives to separate compounds which are isobaric (i.e. congeners) or near-isobaric in mass (mix of halogenated compounds) and co-eluting on a GC column.

Because of all the effort made by the EU on both analytical and food-feed continuous control aspects, it is frequently affirmed that Europe has one of the highest levels of food safety in the world. So far, however, only a very limited set of analytes is included in those monitoring programmes. What about all the other potential harmful molecules present in our food that we do not look for? For example, the new POPs recently ratified and added in the annex A of the Stockholm Convention such as hexobromodiphenyl, pentabromodiphenyl ether, perfluorooctanesulfonic acid (PFOS) and more recently endosulfan (Stockholm convention, 2001). What about potential synergic effects of mixture of untargeted toxicants?

Next to the continuous discussion regarding the potential use of such or such alternative tools for specific PCB and dioxin monitoring, it would perhaps be more beneficial for our health to focus effort on the development of analytical approaches that would allow enlargement of the list of target compounds to more 'exotic' (un)suspected persistent molecules. A more proactive and exhaustive approach is probably needed to more appropriately ensure higher levels of food quality.

3.8 Sources of further information and advice

Further information on how the EC screening-confirmatory approach is regulated are available in the Commission Regulation, Council Regulation, Commission Recommendation and Commission Directive that are given in the reference list.

3.9 References

J.M. AARTS, M.S. DENISON, L.H.J. HAAN, A.C. SCHALK, M.A. COX and A. BROUWER, *Organohal. Compds.* **13** (1993) 361.

J.M. AARTS, M.S. DENISON, M.A. COX, A.C. SCHALK, P.M. GARRISON, K. TULLIS, L.H.J. HAAN and A. BROUWER, *Eur. J. Pharmacol. Environ.'. Toxicol.* **293** (1995) 463.

BDS (BioDetection Systems). Available at http://www.biodetectionsystems.com/, accessed in June 2011.

P.A. BEHNISCH, R. ALLEN, J. ANDERSON, A. BROUWER, D.J. BROWN, T.C. CAMPBELL, L. GOEYENS, R.O. HARRISON, R. HOOGENBOOM, I. VAN OVERMEIRE, W. TRAAG and R. MALISCH, *Organohal. Compds.* **50** (2001) 59.

P. BEHNISCH, K. HOSOE and S.-I SAKAI, *Environ. Int.* **27** (2001) 413.
A. BERNARD, C. HERMANS, F. BROECKAERT, G. DE POORTER, A. DE COCK and G. HOUINS, *Nature* **401** (1999) 231.
T. BERNSMANN, D. BAUMEISTER, P. FÜRST and C SANDY, *Organohal. Compds.* **73** (2011).
S. CARBONNELLE, J. VAN LOCO, I. VAN OVERMEIRE, I. WINDAL, N. VAN WOUWE, S. VAN LEEUWEN and L. GOEYENS, *Talanta* **63** (2004) 1255.
Commission Decision 2002/657/EC of 12 August 2002 implementing Council Directive 96/23/EC concerning the performance of analytical methods and the interpretation of results. *Off. J. Eur. Union*, L221 (2002) 8.
Commission Directive 2002/69/EC of 26 July 2002 laying down the sampling methods and the methods of analysis for the official control of dioxins and the determination of dioxin-like PCBs in foodstuffs. *Off. J. Eur. Union*, L209 (2002) 5.
Commission Directive 2002/70/EC of 26 July 2002 establishing requirements for the determination of levels of dioxins and dioxin-like PCBs in feedingstuffs. *Off. J. Eur. Union*, L209 (2002) 15.
Commission Directive 2006/13/EC of 3 February 2006 amending Annexes I and II to Directive 2002/32/EC of the European Parliament and of the Council on undesirable substances in animal feed as regards dioxins and dioxin-like PCBs. *Off. J. Eur. Union*, L32 (2006) 44.
Commission Recommendation 2006/88/EC of 6 February 2006 on the reduction of the presence of dioxins, furans and PCBs in feedingstuffs and foodstuffs. *Off. J. Eur. Union*, L42 (2006) 26.
Commission Recommendation 2011/516/EU of 23 August 2011 on the reduction of the presence of dioxins, furans and PCBs in feed and food. *Off. J. Eur. Union*, L218 (2011) 23.
Commission Regulation (EC) No 199/2006 of 3 February 2006 amending Regulation (EC) No 466/2001 setting maximum levels for certain contaminants in foodstuffs as regards dioxins and dioxin-like PCBs. *Off. J. Eur. Union*, L32 (2006) 34.
Commission Regulation (EC) No 1881/2006 of 19 December 2006 setting maximum levels for certain contaminants in foodstuffs. *Off. J. Eur. Union*, L364 (2006) 5.
Commission Regulation (EC) No 1883/2006 of 19 December 2006 laying down methods of sampling and analysis for the official control of levels of dioxins and dioxin-like PCBs in certain foodstuffs. *Off. J. Eur. Union*, L364 (2006) 32.
Council Regulation 2001/102/EC of 27 November 2001 amending Directive 1999/29/EC on the undesirable substances and products in animal nutrition. *Off. J. Eur. Union*, L6 (2001) 45.
Commission Regulation 1259/2011 of 2 December 2011 amending Regulation (EC) No 1881/2006 as regards maximum levels for dioxins, dioxin-like PCBs and non dioxin-like PCBs in foodstuffs.
Commission Regulation 277/2012 amending Annexes I and II to Directive 2002/32/EC of the European Parliament and of the Council as regards maximum levels and action thresholds for dioxins and polychlorinated biphenyls.
Commission Regulation 278/2012 amending Regulation (EC) No 152/2009 as regards the determination of the levels of dioxins and polychlorinated biphenyls.
Commission Regulation 252/2012 laying down methods of sampling and analysis for the official control of levels of dioxins, dioxin-like PCBs and non dioxin-like PCBs in certain foodstuffs and repealing Commission Regulation (EC) 1883/2006.
Commission Regulation (EC) No 152/2009 of 27 January 2009 laying down the methods of sampling and analysis for the official control of feed. *Off. J. Eur. Union*, L54 (2009) 1.
Council Regulation (EC) No 2375/2001 of 29 November 2001 amending Commission Regulation (EC) No 466/2001 setting maximum levels for certain contaminants in foodstuffs. *Off. J. Eur. Union*, L321 (2001) 1.

M.S. DENISON, B. ZHAO, D.S. BASTON, G.C. CLARK, H. MURATA and D.-H. HAN, *Talanta* **63** (2004) 1123.

J.M.D. DIMANDJA, *Anal. Chem.* **76** (2004) 167A.

EFSA, 2010a. Scientific report, results of the monitoring of non dioxin-like PCBs in food and feed. *EFSA J.* **8**, 1701–1736. 594

EFSA, 2010b. Scientific report, results of the monitoring of dioxin levels in food and feed. *EFSA J.* **8**, 1385–1421.

G. EPPE, J.-F. FOCANT, C. PIRARD and E. DE PAUW, *Talanta* **63** (2004) 1135.

G. EPPE, J.-F. FOCANT and E. DE PAUW, *The Encyclopedia of Mass Spectrometry, Hyphenated Methods.* In NIESSEN, W.M.A. (Ed.), Elsevier, Amsterdam, The Netherlands, pp. 531–541, 2006.

European Union reference laboratory, Core working group on possible use of new MS techniques / instruments for determination of dioxin and PCBs in feed and food, ongoing discussion, EU Working Expert Group, 2011.

FAO/WHO (2001). 57th JECFA Meeting, Rome, 5–14 June 2001: Summary and Conclusions, *Annex* **4**, pp. 24–40

J.-F. FOCANT, G. EPPE and E. DE PAUW, *Chemosphere* **43** (2001) 417.

J.F. FOCANT, A. SJÖDIN and D.J. PATTERSON, *J. Chromatogr. A* **1040** (2004) 227.

J.-F. FOCANT, C. PIRARD, G. EPPE and E. DE PAUW, *J. Chromatogr. A* **1067** (2005a) 265.

J.F. FOCANT, G. EPPE, M.-L. SCIPPO, A.C. MASSART, C. PIRARD, G. MAGHUIN-ROGISTER and E. DE PAUW, *J. Chromatogr. A* **1086** (2005b) 45.

J.-F. FOCANT, A.-C. MASSART, G. SCHOLL, G. EPPE, C. PIRARD and E. DE PAUW, *J. Chromatogr. A* **1130** (2006) 97.

D. FRAISSE, O. PAISSE, L. NGUYEN HONG and M.F. GONNORD, *Fresenius J. Anal. Chem.* **348** (1994) 154.

P. FÜRST, T. BERNSMANN, D. BAUMEISTER and C. SANDY, Agilent Technologies publication 5990–6594EN (2010); http://www.chem.agilent.com/Library/applications/5990-6594EN.pdf

P.M. GARRISON, K. TULLIS, J.M. AARTS, A. BROUWER, J.P. GIESY and M.S. DENISON, *Fundam. Appl. Toxicol.* **30** (1996) 194.

R.O. HARRISON and R.E. CARLSON, *Chemosphere* **34** (1997a) 915.

R.O. HARRISON and R.E. CARLSON, *Organohal. Compds.* **31** (1997b) 139.

R.O. HARRISON and G.H. EDULJEE, *Sci. Tot. Environ.* **239** (1999) 1.

M. HIRAOKA, Y. KOBAYASHI, K. OGIWARA, T. NAKANISHI, C. CAMPBELL and G.D. WHEELOCK, *Organohal. Compds.* **53** (2001) 401.

R. HOOGENBOOM, L. PORTIER, C. ONSTENK, T. POLMAN, A. HAMERS and W. TRAAG, *Organohal. Compds.* **45** (2000) 180.

R. HOOGENBOOM, L. GOEYENS, S. CARBONNELLE, J. VAN LOCO, H. BEERNAERT, W. BAEYENS, W. TRAAG, T. BOVEE, G. JACOBS and G. SCHOETERS, *Trends Anal. Chem.* **25** (2006) 410.

Y. KOBAYASHI, A. LUNDQUIST, T. UECHI, K. ASHIEDA, K. SASAKI, B. HUGUES and T. KAISE, *Organohal. Compds.* **58** (2002) 337.

G. KOPPEN, A. COVACI, R. VAN CLEUVENBERGEN, P. SCHEPENS, G. WINNEKE, V. NELEN and G. SCHOETERS, *Toxicol. Lett.* **123** (2001) 59.

A. KOTZ, R. MALISCH, K. WAHL, N. BITOMSKY, K. ADAMOVIC, I. GERSTEISEN, S. LESWAL, J. SCHÄCHTELE, R. TRITSCHLER and H. WINTERHALTER, *Organohal. Compds.* **73** (2011) 688.

B. LARSEN, M. CONT, L. MONTANARELLA and N. PLATZNER, *J. Chromatogr. A*, **708** (1995) 115.

P.A. LECLERCQ and C.A. CRAMERS, *Mass Spectrom. Rev.* **17** (1998) 37.

R. MALISCH, B. BAUMANN, P. BEHNISCH, R. CANADY, D. FRAISSE, P. FÜRST, D. HAYWARD, R. HOOGENBOOM, R. HOOGERBRUGGE, D LIEM, O. PAPKE, W. TRAAG and T. WIESMULLER, *Organohal. Compds.* **50** (2001) 53.

A.J. MURK, J. LEGLER, M.S. DENISON, J.P. GIESY, C. VAN DE GUCHTE and A. BROUWER, *Fund. Appl. Toxicol.* **33** (1996) 149.
A.J. MURK, A. JONAS, A. BROUWER, P.E.G. LEONARDS and M.S. DENISON, *Organohal. Compds.* **27** (1996) 291.
S.R. NAGY, G. LUI, K. LAM and M.S. DENISON, *Organohal. Compds* **45** (2000) 232.
D.G. PATTERSON JR., S.M. WELCH, W.E. TURNER, A. SJÖDIN and J.-F. FOCANT, *J. Chromatogr. A* **1218** (2011) 3274.
J.B. PHILLIPS and J. BEENS, *J. Chromatogr. A* **856** (1999) 331.
J.B. PLOMBEY and R.E. MARCH, *Anal. Chem.* **68** (1996) 2345.
J.B. PLOMLEY, M. LAUSEVIC and R.E. MARCH, *Mass Spectrom. Rev.* **19** (2000) 305.
E.J. REINER, R.E. CLEMENT, A.B. OKEY and C.H. MARVIN. *Anal. Bioanal. Chem.* **386** (2006) 791.
G. RHYS JONES, K. RICHARDSON and M. GREEN, poster presentation, American Society for Mass Spectrometry (ASMS proceedings), (2011).
J.J. RYAN, H.B.S. CONACHER, L. G. PANOPIO, B.P.Y. LAU, J.A. HARDY and Y. MASUDA, *J. Chromatogr.* **541** (1991) 131.
S. SAFE, *Chemosphere* **16** (1987) 791.
S. SAFE, *Environ. Health Perspect.* **101** (1993) 317.
M.-L. SCIPPO, G. EPPE, E. DE PAUW and G. MAGHUIN-ROGISTER, *Talanta* **63** (2004) 1193.
M.-L. SCIPPO, G. EPPE, C. SAEGERMAN, G. SCHOLL, E. DE PAUW, G. MAGHUIN-ROGISTER and J.-F. FOCANT, 2008. Chapter 14 Persistent Organochlorine Pollutants, Dioxins and Polychlorinated Biphenyls. In Y. PICO (Ed.), *Comprehensive Analytical Chemistry* **51**. Elsevier, Amsterdam, pp. 457–506.
L. SMASTUEN HAUG and G. BECHER, *Seventh Round of an International Study*, Norwegian Institute of Public health, Report Number 7 (2006).
Y. SUGAWARA, S.J. GEE, J.R. SANBORN, S.D. GILMAN and B.D. HAMMOCK, *Anal. Chem.* **70** (1998) 1092.
L.H. STANKER, B. WATKINS, N. ROGERS and M. VANDERLAAN, *Toxicology* **45** (1987), 229.
Stockholm Convention on Persistent Organic Pollutants (POPs), Stockholm, 2001. Available at www.pops.int
M. THOMPSON, S. ELLISON and R. WOOD, *Pure Appl. Chem.* **78** (2006) 145.
M. VALCARCEL, S. CARDENAS and M. GALLEGO, *Trend Anal. Chem.* **18** (1999) 685.
H. VANDERPERREN, N. VAN WOUWE, S. BEHETS, I. WINDAL, I. VAN OVERMEIRE and A FONTAINE, *Talanta* **63** (2004) 1277.
M. VAN DEN BERG, L. BIRNBAUM, A. BOSVELD, B. BRUNSTROM, P. COOK, M. FEELEY, J.P. GIESY, A. HANBERG, R. HASEGAWA, S. KENNEDY, T. KUBIAK, J.C. LARSEN, F.V. LEEUWEN, K. LIEM, C. NOLT, R. PETERSON, L. POELLINGER, S. SAFE, D. SCHRENK, D. TILLITT, M. TYSKLIND, M. YOUNES, F. WAERN and T. ZACHAREWSKI, *Environ. Health Perspect.* **106** (1998) 775.
M. VAN DEN BERG, L.S. BIRNBAUM, M. DENISON, M.D. VITO, W. FARLAND, M. FEELEY, H. FIEDLER, H. HAKANSSON, A. HANBERG, L. HAWS, M. ROSE, S. SAFE, D. SCHRENK, C. TOHYAMA, A. TRITSCHER, J. TUOMISTO, M. TYSKLIND, N. WALKER and R.E. PETERSON, *Toxicol. Sci.* **93** (2006) 223.
N. VAN WOUWE, I. WINDAL, H. VANDERPERREN, G. EPPE, C. XHROUET, A.-C. MASSART, N. DEBACKER, A. SASSE, W. BAYENS, E. DE PAUW, F. SARTOR, H. VAN OYEN and L. GOEYENS, *Talanta* **63** (2004a) 1157.
N. VAN WOUWE, I. WINDAL, H. VANDERPERREN, G. EPPE, C. XHROUET, E. DE PAUW, L. GOEYENS and W. BAYENS, *Talanta* **63** (2004b) 1269.
F. VERSTRAETE, *Organohal. Compds.* **55** (2002) 1.
WHO (World Health Organization) *PCBs and dioxins in salmon*. WHO, Geneva, 2005. Available at http://www.who.int/foodsafety/chem/pcbsalmon/en/print.html. Accessed on February 15, 2007.

I. WINDAL, M.S. DENISON, L.S. BIRNBAUM, N. VAN WOUWE, W. BAEYENS and L. GOEYENS, *Environ. Science Technol.* **39** (2005a) 7357.

I. WINDAL, N. VAN WOUWE, G. EPPE, C. XHROUET, V. DEBACKER, W. BAYENS, E. DE PAUW and L. GOEYENS, *Environ. Sci. Technol.* **39** (2005b) 1741.

XDS (Xenobiotic Detection System). Available at http://www.dioxins.com/, accessed in June 2011.

4

Screening and confirmatory methods for the detection of heavy metals in foods

F. Arduini and G. Palleschi, Università di Roma Tor Vergata, Italy

DOI: 10.1533/9780857098917.1.81

Abstract: This chapter highlights the importance of both screening methods and confirmatory methods for heavy metal analysis in food. The analysis focuses on the most important heavy metals in terms of adverse health effects, such as lead, cadmium, zinc and arsenic. The chapter covers the use of electrochemical, optical and bioassay techniques as screening methods for heavy metal detection in food. Regarding confirmatory methods, atomic absorption spectrometry, atomic fluorescence spectrometry, inductively coupled plasma mass spectroscopy and inductively coupled plasma optical emission spectrometry are discussed. A detailed description of screening applications and confirmatory methods in food safety and control is given. Sample treatments are also reported, highlighting the practical problems of potential heavy metal contamination encountered during sample treatment and measurement. The final section deals with quality assurance and validation for heavy metals in food. Future trends for both screening procedures and confirmatory methods are highlighted.

Key words: heavy metal detection, screening analytical methods, confirmatory analytical methods, food.

4.1 Introduction

The presence of heavy metals such as mercury, lead, cadmium and arsenic in food, at levels of μg/L or higher, can have an adverse impact on consumer health. These heavy metals may occur as a result of environmental and industrial contamination. Such heavy metals tend to be stable and to persist in the environment, thus bioaccumulating in the food chain. Additionally, the production process, packaging materials and food storage may be sources of heavy metal contamination. The food sector therefore needs analytical techniques to control the safety and quality of food products and also to exercise control over the manufacturing process itself.

L. S. Jackson from the National Center for Food Safety and Technology in the U.S. Food and Drug Administration[1] observed that, before the seventeenth century, scientists were only able to detect gross adulteration in foods, and it was not until later in the century that British chemist Robert Boyle began developing new assays for food adulteration. In the 1850s Arthur Hill Hassall, a British physician, started his investigation into food adulteration through chemical tests. In the latter half of the twentieth century, analytical chemistry advanced to allow the detection and quantification of food chemicals at far lower levels than previously possible. If in the 1940s only the balance and simple spectrophotometers were available for quantitative analysis, the 1950s brought gas chromatography, the 1960s high-performance liquid chromatography, the 1970s mass spectrometry and afterwards the development and optimization of analytical techniques for food components and contaminants allowing improvement in quality and safety control of food.

In the safety control sector, the detection of highly toxic heavy metals, such as mercury, arsenic, lead and cadmium, is important. In order to analyse the heavy metal content of food samples, both confirmatory and screening methods are reported in the literature. Confirmatory methods are characterised by low limits of detection (LOD) and high accuracy, but require large instruments, laboratory set-ups and highly specialised personnel. They are thus characterised by a high cost of analysis. A welcome improvement would be simpler instruments to render on-site application possible, and thus to reduce costs. These requirements should be taken into consideration because the increasingly complex food chain requires effective traceability systems that guarantee security and efficiency in all parts of the chain. The priority here is to install control systems characterised by high sensitivity, short analysis time, and the possibility of automated 'on-line' systems that minimise the need for skilled personnel. In this context, screening methods seem an attractive choice. They have the advantage of direct on-site application, they are cost-effective, and relatively easy to use. This approach is particularly useful in the food industry, where long analysis time leads to food spoilage with significant cost implications. Screening methods can control the food chain, and in the case of a positive response, confirmatory methods can be used to quantify accurately the heavy metal content. In this way screening methods can serve to reduce analysis costs.

For the detection of heavy metals in foods using screening methods, the issue of sample treatment should be considered. The aim should be for sample treatment to be performed *in situ* without a complex procedure, particularly for simple matrices such as potable water, which require only a dilution step. In the case of complex matrices such as milk, a simple sample treatment procedure is not always possible for high protein content. In addition, special attention should be given to sample, storage and treatment, because of possible environmental contamination during preparation and/or analysis. Practical information on how to avoid contamination should be followed, as highlighted in our recent paper[2]. The European Commission, for example, in directive CE N333/2007 sets out the rules for sampling and analysis for the official control of lead, cadmium, mercury, inorganic tin, 3-MCPD, and benzo(a)pyrene levels in foodstuffs. We highlight the rules for sample preparation for heavy metals. In paragraph C.2.2.2.1, the Regulation states:

> The analyst shall ensure that samples do not become contaminated during sample preparation. Wherever possible, apparatus and equipment coming into contact with the sample shall not contain those metals to be determined and be made of inert materials e.g. plastics such as polypropylene, polytetrafluoroethylene (PTFE) etc. These should be cleaned with acids to minimise the risk of contamination. High quality stainless steel may be used for cutting edges [CE N333/2007].

In the best practice laboratory glassware should be avoided, and instead PTFE beakers, round-bottomed flasks in polypropylene, polypropylene centrifuge tubes, etc., should be used. In addition, before each measurement, all containers and pipette tips should be washed several times with ultrapure 65% nitric acid diluted 1:10, followed by washing three times with double-distilled water. Another key point is the selection of appropriate ultrapure trace analysis reagents. Also, for measurements in flow systems, it is important to have all tubing connecting components of the flow system made of PTFE.[3] In a paper published in the *Journal of Agricultural and Food Chemistry* S. P. Dolan *et al.* from the Center for Food Safety and Applied Nutrition, U.S. Food and Drug Administration, report that tacky floor mats were placed outside and inside each laboratory entrance. The vessels used for microwave digestion were cleaned with micro-90 liquid laboratory grade detergent and warm water and with trace metal grade nitric acid using a microwave cleaning program, and the vessel was then rinsed with water and dried under class 100 HEPA-filtered laminar flow air in order to avoid environmental contamination.[4] This procedure highlights how important it is to work carefully in order to avoid contamination.

Finally, measurements must be carried out in an appropriate location. If the sample contains concentrations at low ppb or ppt levels, it is highly recommended that preparation and analysis of the samples be carried out in clean-room laboratories, to avoid environmental contamination and subsequent overestimation of lead content. In addition it would be useful to confirm daily, before measurement, the purity of the reagents by measuring the solvent blank. If the amount of lead obtained is in the range of the usual blanks,[5] the sample analysis can continue. Otherwise, an additional clean is necessary, or new portions of non-contaminated trace reagents should be used.

We have highlighted the importance of not only a sensitive and accurate method of analysis, but also sampling and treatment characterised by high accuracy. For the analytical methods described, screening and confirmatory methods reported in literature from 2000 to the present will be described for lead, cadmium, mercury and arsenic ion detection, heavy metals characterised by salient toxicity and presence in food samples.

4.2 Screening methods for heavy metal detection in foods

Screening methods may be defined as analytical methods, characterised by short analysis time, cost-effectiveness, and simple measurement. The several methods reported in the literature can be classified as electrochemical, optical and bioassay. These are described below.

4.2.1 Electrochemical methods

Electrochemical sensors are suitable for use in screening due to cost-effectiveness, ease of use and potential for miniaturization. The electrochemical-based glucose biosensor used every day by diabetic patients for monitoring glycaemia is relevant in order to demonstrate that the electrochemical transducer has all the features suitable for use in screening. Among several electrochemical techniques, such as potentiometric, amperometric, conductimetric and impedance, electrochemical stripping analysis is widely recognised as a powerful technique for detecting some heavy metals, including mercury, lead, cadmium and arsenic. This is due to its ability to detect trace and ultra-trace levels. Among electrochemical stripping analyses, anodic stripping is the most widely adopted technique. This consists of two steps: in the first a negative potential to reduce the heavy metal at the electrode surface is applied; in the case of a mercury electrode the metal ions reach the electrode by diffusion and convection where, in the presence of negative potential, they are reduced as follows:

$$M^{n+} + ne^- \rightarrow M(Hg)$$

to form with the mercury amalgam. After the time of deposition, stirring is stopped and the second step consists of the potential scanned anodically, so the amalgamated metals are reoxidised and stripped out of the electrode as follows:

$$M(Hg) \rightarrow M^{n+} + ne^- + Hg$$

This type of analysis is a multianalyte analysis, in which it is possible simultaneously to measure several heavy metals such as Pb^{2+}, Cd^{2+}, Zn^{2+} and Cu^{2+}. In detail, the position of the peak is specific for each heavy metal and the height of the peak is proportional to the heavy metal concentration.

In the case of potentiometric stripping analysis, the reoxidation of metal is carried out by an oxidizing agent present in solution, such as O_2.

Since the invention of polarography by Heyrovsky, in which a liquid working electrode (hanging drop mercury electrode) is used,[6] mercury has become the preferred electrode for electrochemical stripping analysis, because of the high overpotential of hydrogen that allows its use as a useful negative potential. Even if there is a growing interest in reducing the use of toxic compounds, this type of electrode still appears in literature from 2000.

Locatelli and Torsi have developed an analytical procedure using differential pulse anodic stripping voltammetry (DPASV) and hanging mercury drop electrode (HDME) as a working electrode to analyse metals, such as lead and cadmium, in several matrices (such as wholemeal, wheat and maize meal). The sample was digested using concentrated HCl suprapure at 130 °C for 3 h, filtered on Whatman No 541 filter paper and diluted to obtain a final acid concentration equal to 0.5 M, which is the working solution.[7] The same treatment procedure was used with the square wave anodic stripping voltammetry (SWASV) technique.[8] This analytical procedure was applied to detect copper, zinc, antimony, lead and cadmium. The detection limits were equal to 0.031 and 0.029 µg/L for lead and cadmium, respectively, in standard solution. For commercial meals 0.05 ± 0.01 µg/g, 0.04 ± 0.01 µg/g and 0.05 ± 0.01 µg/g of cadmium in wholemeal, wheat meal and maize meal, were found respectively. For lead 0.66 ± 0.06 µg/g, 0.49 ± 0.04 µg/g and 0.47 ± 0.05 µg/g, were found respectively. In order to evaluate the analytical procedure, reference materials were tested. Good precision and accuracy were found (standard deviation and relative error of between 3% and 6%).[8] The same authors, in another paper, used the same electrode HDME with SWASV to detect chromium, lead, tin, antimony, copper and zinc in the same matrices, but using digestion with a concentrated $HCl-HNO_3-H_2SO_4$ acid attack mixture. The LOD for lead measured in standard solutions was 0.019 µg/L and in wholemeal BCR-CRM 1890.037 µg/g in agreement with the certified value.[9] Derivate potentiometric stripping analysis was applied to detect cadmium, copper, lead and zinc in various citrus essential oils. For heavy metal extraction the authors used two procedures involving: i) concentrated hydrochloric acid treatment, and ii) acid–alcoholic dissolution, demonstrating the suitability of both. The first treatment requires more steps but the total analysis time is shorter than for the second. Detection limits obtained for cadmium and lead in four matrices (lemon, mandarin, sweet orange and bergamot) were between 0.31 and 0.61 ng/g. Samples of various citrus oil were analysed. Cadmium concentrations lower than 2 ng/g were found in lemon and mandarin oil samples while cadmium concentrations equal to 8.25 ± 0.21 ng/g, and 23.31 ± 1.00 ng/g were found in sweet orange and bergamot oils, respectively. For lead, sweet orange and bergamot oils showed very similar lead content, (75.94 ± 2.06 ng/g and 77.92 ± 1.02 ng/g, respectively) while a greater amount was found in lemon and mandarin oils 107.87 ± 3.32 ng/g and 180.28 ± 6.15 ng/g.[10] The same group of metals was also measured using this technique and a mercury film electrode in olive oil. The mercury film electrode was prepared electrochemically on the surface of a glassy carbon electrode applying a potential of –950 mV for 1 min in a solution of 1000 μg/mL of Hg. The use of this electrode reduces Hg usage and can thus be considered more environmentally friendly than HDME. The sample pretreatment consisted of mixing oil with chloridric acid heated at 50°C for 45 min, followed by separation by means of a separation funnel and clean-up using a carbon column.[11] In the case of more complex matrices, such as margarine and butter, the sample was digested using nitric acid and hydrogen peroxide heated to 150–160°C. In these cases, the measurement was carried out using galvanostatic stripping chronopotentiometry with a mercury film electrode. The detection limit found for both cadmium and lead was 0.02 μg/L. The mean concentrations of cadmium and lead found in different types of margarine and butter varied from 9.2 to 26.2 μg/kg and 9.1 to 16.4 μg/kg.[12] In the cases of fruit juice and skimmed milk powder, acid digestion was carried out for several hours, even at temperatures of up to 500°C. The long sample pretreatment was justified for slow dry-ashing to prevent volatilization losses. Figure 4.1 shows the stripping chronopotentiometric curves for cadmium and lead in a sample treated before and after heavy metal additions in order to calculate the amount of cadmium and lead using the standard addition method.[13] Anodic stripping analysis can also be coupled with a flow cell, as reported by Munoz and Palmero, that renders the analysis automatic after the sample treatment has been carried out by hand (milk, in this case).[14] The HDME was also used as an electrode for adsorptive stripping analysis, in which a complexant agent was used for the adsorptive collection of complexes with heavy metals, to be analysed on the HDME. Luminol was used as a chelating agent to detect copper and cadmium accurately (obtaining LOD for cadmium equal to 0.02 ng/mL) in rice, tomato, tea, spinach, tap and mineral water.[15] The same approach was used by Abbasi *et al.* using 2-mercaptobenzothiazole as a chelating agent on the HDME for lead and cadmium measurement, obtaining LODs of 0.017 ng/mL and 0.01 ng/mL, respectively. The system was applied in rice, soya and sugar, with recovery values between 94.2% and 101.6%.[16]

Chemistry and analytical chemistry have recently begun reducing or eliminating the use and generation of hazardous substances.[17] Regulation (EC) No 1102/2008 of the European Parliament has banned exports of metallic mercury and certain mercury compounds. Alternative electrodes must be found. To this purpose, several approaches have been investigated in order to obtain a sensitive, reproducible and accurate sensor for heavy metal detection in food. Chemically modified electrodes may be one alternative.

A glassy carbon electrode was coated with Lagmuir–Blodgett film of p-allylcalix[4]arene (AllylCA-GCE) to detect thallium and cadmium in water samples at μg/L levels.[18] The principle of this method involves the chelation of metal ions on the surface of AllylCA-GCE, the reduction of accumulated metal ions and the electrochemical stripping of the reductive material back into solution.[18] The lead was also measured using glassy carbon modified with overoxidised polypyrrole doped with 2(2-pyridylazo)chromotropic acid anion ($PACh^{2-}$). In this case $PACh^{2-}$ acts as a chelating agent and as a counter

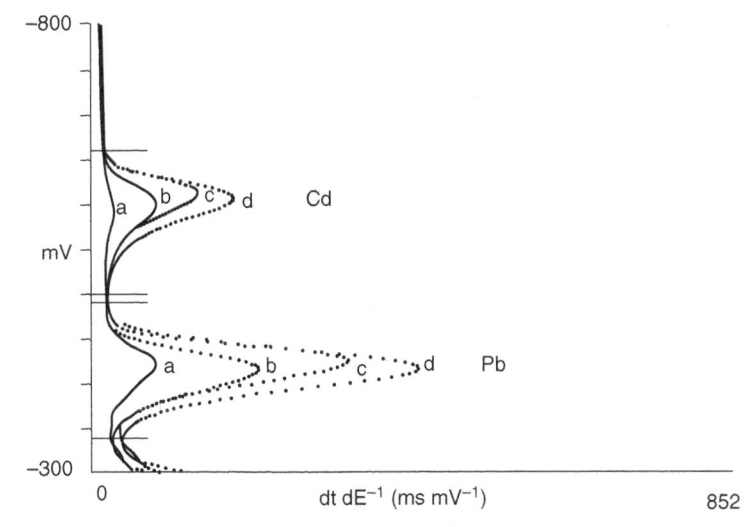

Fig. 4.1 Voltammograms of cadmium and lead in a sample of fruit juice (a) before and (b–d) after one, two and three additions of both 100 µL of a solution containing 0.5 ng/µL of cadmium and 100 µL of a solution containing 1.0 ng/ µL of lead (b–d, respectively) (Reprinted from reference 13 with permission of Elsevier).

anion within the polypyrrole matrix. The lead was then accumulated on the modified GCE via the formation of Pb-PACh complex at open circuit. The electrode with the Pb^{2+} $PACh^{2-}$ was then transferred in an acetate buffer for the reduction of the accumulated Pb^{2+} to Pb^0 followed by the stripping analysis. The system was tested for the determination of lead in potable water.[19]

In some cases reported, the measurements were carried out in solution deaerated by bubbling nitrogen. This requirement limits the process for on-site application, thus research was focused on the stripping measurement that does not require absence of oxygen, as reported in the following examples. The silver electrode was used to detect lead and cadmium at nanomolar concentration in potable water (detection limit for lead 0.05 nM and 1 nM for cadmium) using subtractive-anodic stripping voltammetry without oxygen removal. In this case, in order to destroy organic complexants and surfactants that negatively affect sensor response, the water sample was acidified to pH 2 with nitric acid immediately after collection, followed by wet-ashing.[20] A gold electrode was used for the stripping analysis of heavy metals. Simultaneous detection of As^{3+}, Hg^{2+} and Cu^{2+} was performed using a highly sensitive platform based on gold nanoelectrode ensembles. The gold nanoelectrode was prepared by a colloidal chemical approach on a thiol-functionalised sol-gel derived three dimensional silicate network, assembled on a gold electrode (Fig. 4.2). The system showed high sensitivity towards mercury and arsenic (0.02 µg/L) and was applied to the arsenic monitoring of potable water of West Bengal in India, finding 51.4 ± 1 µg/L of arsenic, higher than the

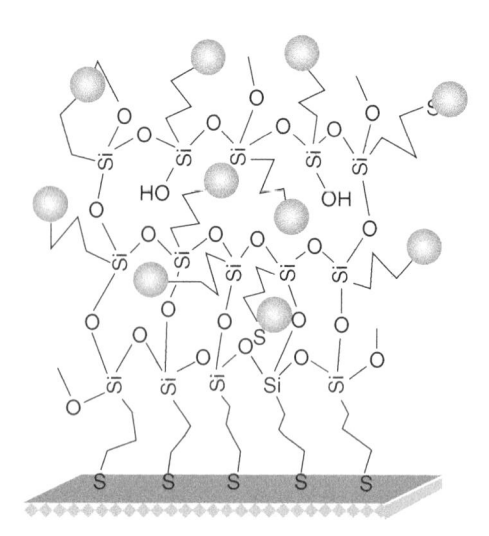

Fig. 4.2 Scheme of the gold nanoelectrode ensembles grown by colloidal chemical approach on thiol-functionalised sol-gel derived three dimensional silicate network, assembled on a gold electrode for arsenic detection (Reprinted with permission from reference 21, Copyright (2009) American Chemical Society).

10 μg/L set by WHO as a guideline value.[21] The As^{3+} together with As^{5+} was also measured using anodic stripping voltammetry with a gold-modified boron doped diamond electrode, reaching a detection limit equal to 5 μg/L and 100 μg/L for As^{3+} and As^{5+}, respectively. This was applied to spiked samples from tap water in Yokohama, Japan.[22] Platinum nanoparticles were used to modify screen-printed electrodes to detect As^{3+}, at μg/L levels, and in this case the application was confined to water samples.[23] However, we emphasise the advantage of the screen-printed electrodes (SPE) over other types usually adopted (i.e. glassy carbon electrodes), because SPEs in modern electroanalytical chemistry are recognised as successful sensors with low current background, wide potential window and ease of surface modification. In addition, these types of sensors are small, easy to use and cost-effective. Their mass production (Fig. 4.3) also renders them suitable for single use.

In the search for a non-mercury electrode for heavy metal stripping analysis, the bismuth electrode was also demonstrated as a valid alternative. In 2000, for the first time, a bismuth-coated carbon electrode was used as a sensor for stripping voltammetric analysis of lead, cadmium and zinc.[24] The detection of lead, cadmium and zinc at the surface of a bismuth electrode is due to the capacity of bismuth to form a 'fused alloy' with some heavy metals. It has the advantage of not requiring the removal of dissolved oxygen during stripping analysis, and is also characterised by a wide negative potential window, an advantage reported in several papers. It is one of the less toxic heavy metals.[25,26] In the case of the Bi electrode, the M^{n+} at negative potential is reduced to form a heavy metal bismuth alloy M(Bi),

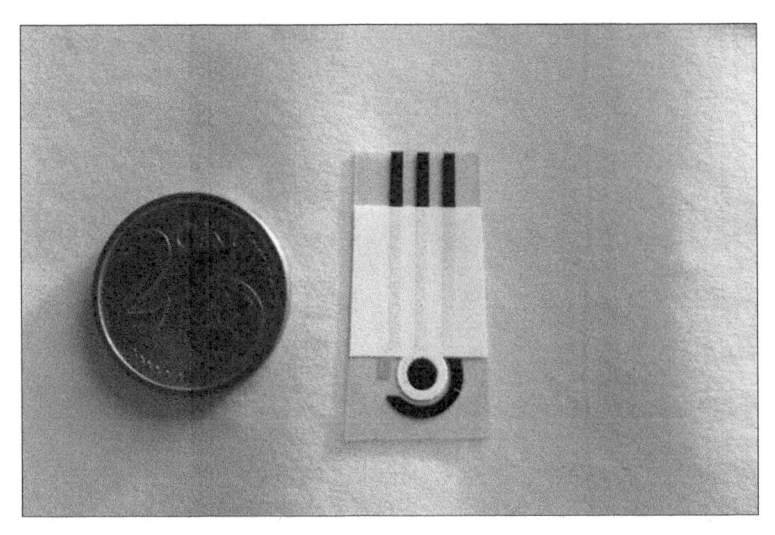

Fig. 4.3 Photograph of a screen-printed electrode.

followed, as in the case of Hg electrode, by M reoxidation (M(Bi) →M^{n+} + ne$^-$+ Bi). The Bi electrode was applied for heavy metal detection in potable water,[27] tea,[28] wine,[29] cabbage lettuce, celery and spinach[30] with sufficiently accurate results.

Recently, within the framework of a European project known as Biocop[98], a sensitive procedure for lead detection in milk was developed, keeping in mind that the European regulation set the legal limit to 20 μg/kg. Briefly, a milk treatment procedure was developed using a wet digestion with HCl, HClO$_4$ and H$_2$O$_2$ combined with an ultrasonic treatment, and lead was detected by means of SPEs modified with the polymer Nafion® and a Bi film (Fig. 4.4). Other electrochemical methods are present in the literature, using amperometric detection. A microchip based on capillary electrophoresis was developed, able to separate the different metal species with a final amperometric detector using an SPE. The system was applied to detect Cd^{2+}, Pb^{2+} and Cu^{2+} in juice samples, obtaining a recovery range from 91.93% to 107.9%.[31] Our research group developed an electrochemical method for mercury detection. A sensor based on SPE, modified with a stable dispersion of cost-effective nanostructured carbon black (CB) N220, was developed for the detection of several thiol-containing compounds. The very high sensitivity obtained towards thiols was then used as a basis for developing an analytical method for mercury ion detection, a non-electroactive complex (thiol–Hg) having formed in the presence of the metal (Fig. 4.5).[32] By selecting an appropriate concentration of thiol, a concentration of mercury as low as 5×10^{-9} M (1 ppb) was detected. Satisfactory recovery was obtained when the system was tested on potable water samples.

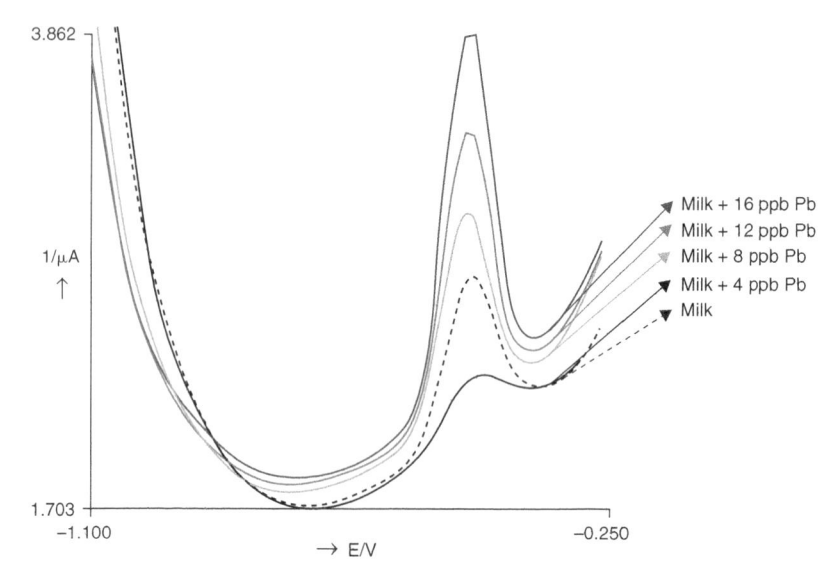

Fig. 4.4 Voltammograms obtained using anodic stripping analysis and SPE modified with Nafion® and bismuth film in treated milk and after several additions of lead to apply the standard addition method.

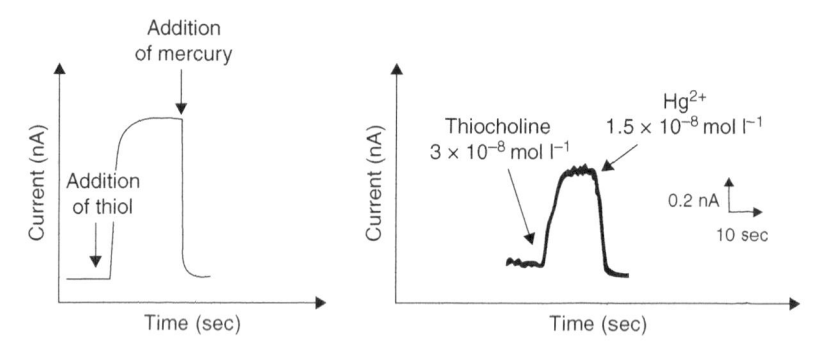

Fig. 4.5 Typical current–time curve obtained before and after adding mercury ions: applied potential +300 mV vs Ag/AgCl, phosphate buffer 0.05 M + KCl 0.1 M, pH 7.4, thiol (thiocholine) 3×10^{-8} M (Reprinted from reference 32 with permission of Elsevier).

4.2.2 Optical methods

Optical methods for heavy metal detection in food analysis are based principally on simple colorimetric or even chemiluminescent methods. Chemiluminescence is defined as the generation of light through a chemical reaction producing radical species. The species can react to produce unstable intermediates that decompose with the formation of excited species. Their application is limited to Cr^{3+}, Co^{2+}, Ni^{2+}, Cu^{2+}. For example in the case of

Cr^{3+}, the method is based on the measurement of light emitted from catalysed oxidation of luminol by H_2O_2. This system was applied to Cr^{3+} in food samples.[33] However, optical detection methods for heavy metals, such as As^{3+}, Pb^{2+}, Cd^{2+}, Hg^{2+} on which we have focused, mainly consist of spectrophotometry. The presence of heavy metals is related to the variation of absorbance, and concentration is correlated with absorbance by means of Beer's law. Lead may be measured using a dithizone that will form a coloured complex with it.[6] Lead can be detected using several chromogenic agents that form a coloured complex such as di-bromo-*p*-methyl-bromosulfonazo (DBMBSA)[34] or dibromohydroxyphenylporphyrin.[35] In the case of DBMBSA, LOD was found equal to 2.45 μg/L, the system was applied to detect lead in soy, vinegar, red and white grape wine, with good recovery values.

In this case, the samples were dried in an electric furnace at 550–600°C and ashed in a muffle furnace, treated with acids and extracted with a separating funnel.[34] In the case of dibromohydroxyphenylporphyrin, the LOD was 1.2 μg/L. Lead was detected in several types of food, such as salted egg and pickled vegetables. In this case the sample treatment consisted of a digestion with concentrated HNO_3 for 12 h at RT, then at 120°C for 1 h with successive additions of H_2O_2. The solution was then boiled to dryness and the residue was dissolved in 3 mL of 0.1% (v/v) HNO_3.[35] In order to increase sensitivity, the liquid–liquid microextraction was coupled with UV-Vis spectrometry for the detection of cadmium and copper using dithizone and diethyldithiocarbamate as chromogenic agents. Liquid–liquid microextraction preconcentration using tetrachloromethane as extraction solvent (schematised in Fig. 4.6) allows for very low detection limits of 0.01 ng/L and 0.5 μg/L, for cadmium and copper. The system was used to detect amounts of these heavy metals in tap and mineral water, tea, rice and defatted milk powder with satisfactory accuracy.[36] In the cases of tea, rice and defatted milk powder, closed-vessel microwave digestion was chosen.

An interesting example for heavy metal detection using the optical method is based on a recent trend in which properly functionalised gold nanoparticles are used. Gold nanoparticles can be functionalised with compounds such as gallic acid[37] or thio-compounds[38,39] able to react with heavy metals. The reaction between heavy metal and the compound present on the surface of gold nanoparticles allows the gold nanoparticles to aggregate as shown by Kalluri *et al.* (Fig. 4.7), and a consequent change of absorption profile (Fig. 4.8).[39] In this example the authors used glutathione (GSH), dithiothreitol (DTT) and cysteine (Cys) modified gold nanoparticles in the presence of 2,6-pyridinedicarboxylic acid (PDCA) to selectively detect arsenic. The system allows the detection of As^{3+} as low as 1 μg/L in standard solution. The colorimetric method was tested to detect the amount of arsenic in potable water. To exactly quantify the amount of As, the authors have used another spectroscopy technique known as dynamic light scattering. This technique allows small changes in particle size to be determined. In this case LOD is as low as 10 ppt for As^{3+} and good results in terms of accuracy were found

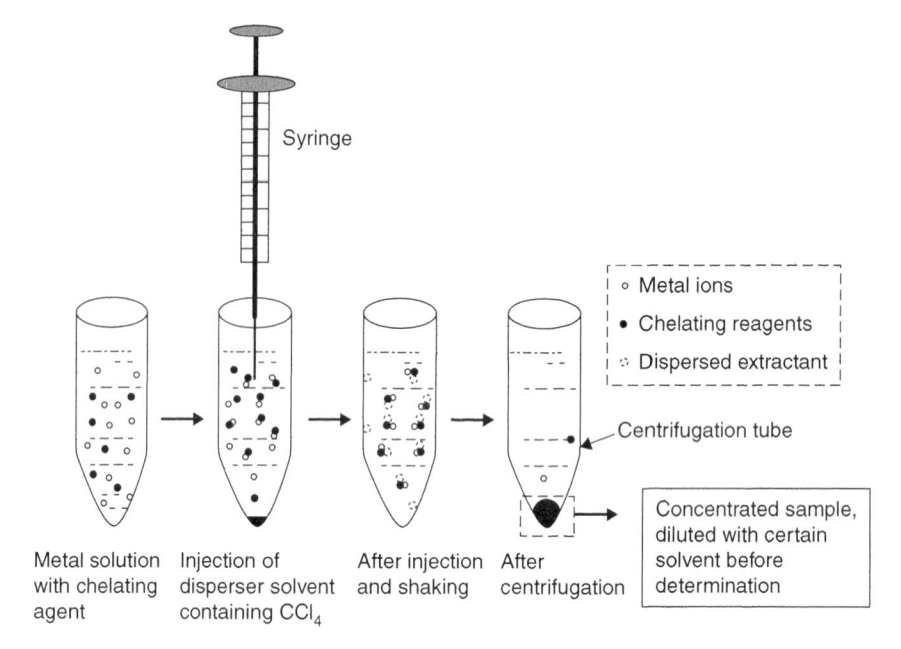

Fig. 4.6 Scheme of liquid–liquid microextraction preconcentration (Reprinted from reference 36 with permission of Elsevier).

when tested with potable water samples.[39] Using this novel and interesting approach we have found several papers focused on heavy metal detection in standard solution, but with infrequent application to real samples and, when reported, limited only to potable water samples.

4.2.3 Bioassay methods

Bioassay methods are characterised by the presence of a biocomponent (antibody, enzyme, microorganism, tissue, nucleic acid) that renders a high level of specificity, reacting with a given analyte or substrate. For example, bacteria-based bioassays for arsenic detection in natural water were reviewed in 2009 by Diesel *et al.* in the *Analytical and Bioanalytical Chemistry* journal.[40] A biosensor for As^{3+} detection was recently reported.[41] The biosensor developed is an enzyme inhibition biosensor. The enzyme was used as a biocomponent and the measurement of the target analyte was carried out by measuring enzyme activity with the target analyte both present and absent. The percentage of inhibition can be calculated using the following equation:

$$I\% = \frac{A_0 - A_i}{A_0}$$

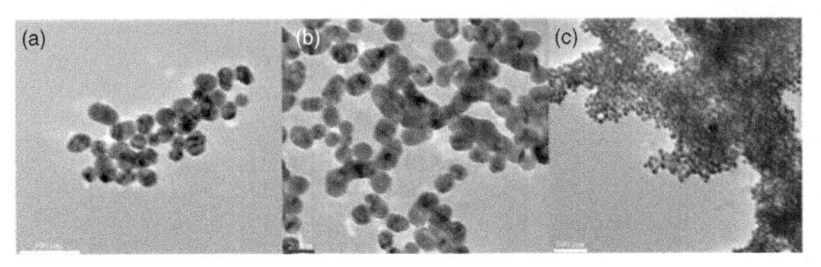

Fig. 4.7 TEM image showing thiol-modified gold nanoparticles (a) before, (b) after addition of As^{3+} 80 ppb and (c) 250 ppt (Reprinted from reference 39 with permission of John Wiley and Sons 2009).

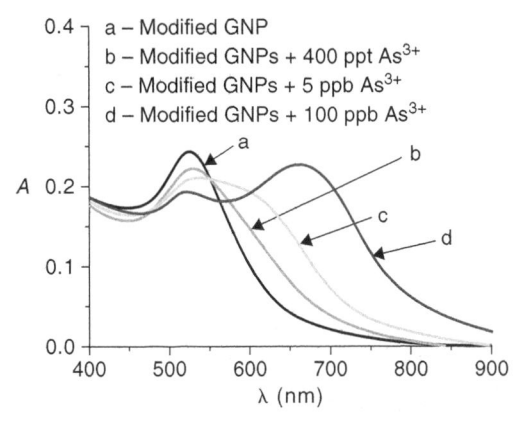

Fig. 4.8 Absorption profile of thiol-modified gold nanoparticles (a) before and (b–d) after addition of As^{3+} (b–d) (Reprinted from reference 39 with permission of John Wiley and Sons 2009).

where:

$I\%$ = percentage of inhibition
A_0 = the enzyme activity in absence of analyte
A_i = the enzyme activity in presence of analyte

The percentage of inhibition correlates with the concentration of target analyte, so it is possible to evaluate the unknown concentration of the analyte. The As^{3+} was detected by measuring the acetylcholinesterase (AChE) activity in both the absence and presence of As^{3+}. The enzymatic activity was evaluated measuring the enzymatic product electrochemically by means of SPE. Thus with As^{3+} present, a decreased signal was registered (Fig. 4.9). In addition, the working electrode was used as support to immobilise by covalent bonds the AChE. The As^{3+}, as reported by the authors, is a more powerful inhibitor than other metal ion, such as mercury, nickel and copper. The biosensor was tested in spiked tap water samples with good recovery

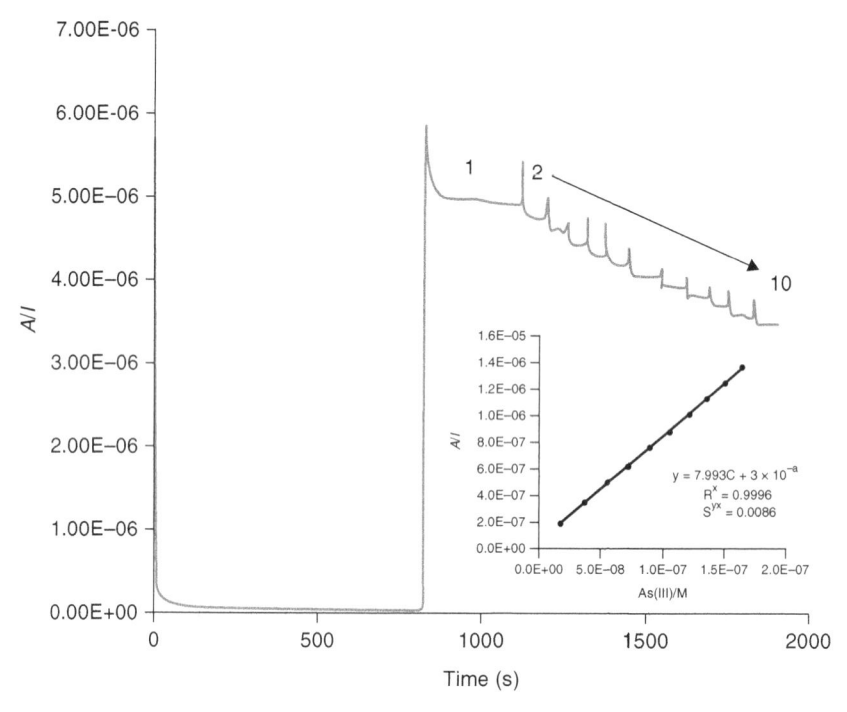

Fig. 4.9 Typical amperometric recording for an acetylthiocholine iodide concentration: (1) 3.64×10^{-4} M and consecutive addition of aliquots of As^{3+} solution into the cell to give an overall concentration of: (2) 1.90×10^{-8} M, (3) 5.60×10^{-8} M, (4) 7.40×10^{-8} M, (5) 9.0×10^{-8} M, (6) 1.07×10^{-7} M, (7) 1.22×10^{-7} M, (8) 1.37×10^{-7} M, (9) 1.52×10^{-7} M and (10) 1.66×10^{-7} M. The inset shows the relative calibration plot (Reprinted with the permission of the authors from reference 41).

values.[41] Another type of bioassay is the immunoassay, in which the biocomponent is an antibody able to bind the antigen (analyte). An interesting example was reported by Sasaki et al.[42] who developed an immunochromatography system for cadmium analysis using anti-Cd-EDTA antibody, highly specific to Cd-EDTA. This allowed cadmium detection at a range 0.01 mg/L to 0.1 mg/L with a 20% mean coefficient of variance. The system is schematised in Fig. 4.10 and, once developed, can be used 'on-site'. However, it is important to stress that for cadmium detection in food samples, such as milled rice grain, rice leaves, tomato, tomato leaves and lettuce leaves, a treatment that includes a step at 130°C for 5–6 hours is required. In the case of tomato, a treatment is required that includes a step at 130°C for several hours. Thus, even if the screening analytical system allows for on-site measurement, in the case of a complex matrix, sample treatment restricts measurement to a laboratory although using miniaturised and cost-effective instrumentation.[42]

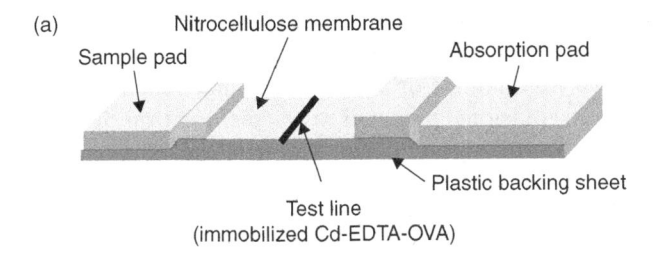

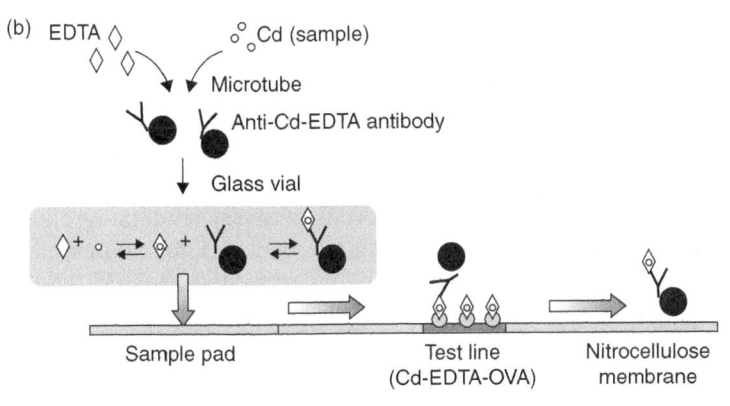

Fig. 4.10 (a) Cadmium immunochromatography system: scheme of the system and (b) scheme of the analysis procedure (Reprinted with permission from reference 42, Copyright (2009) American Chemical Society).

Direct competitive and indirect competitive enzyme-linked immunosorbent (ELISA) assays for Cd^{2+} detection in farm produce, apple juice, rice flour, wheat flour, tea and spinach samples have been reported.[43,44] In the case of direct competitive ELISA, an LOD of 0.20 μg/L was obtained with recovery values in farm produce ranging from 100.47% to 103.86%,[43] while in the case of indirect competitive ELISA an LOD equal to 1.95 μg/L was found with a recovery value[44] in apple juice, rice flour, wheat flour, tea and spinach samples ranging between 97.67% and 107.08%.

4.3 Confirmatory methods for heavy metal detection in foods

Confirmatory methods may be characterised as having high accuracy and low detection limits. However, they need a laboratory set-up, large instrumentation and skilled personnel. These methods are classified in this section by the function of the technique used: atomic absorption spectrometry, atomic fluorescence spectrometry, inductively coupled plasma mass spectroscopy and inductively coupled plasma optical emission spectrometry.

4.3.1 Atomic absorption spectrometry

In atomic absorption spectrometry (AAS), atoms in the form of atomic vapour are generated (using flame or an electrothermal system), then a portion are thermally and collisionally excited to a higher electronic energy level, before being returned to their ground energy state by emitting photons. The amount of the analyte is proportional to the amount of radiation absorbed by ground-state atoms regulated by Beer's law.[6]

The flame AAS was the first type of AAS used, due to its inherent simplicity and low operational cost. However, it is characterised by a low sensitivity of 0.01 mg/L,[45] mainly relies on the nebulizer system characterised by only 5–10% efficiency, and has a very short residence time for analysis in the absorption volume.[46] Several methods are reported in the literature to preconcentrate the elements prior to introducing the sample to flame, such as the beam injection flame furnace AAS. In this case the system was coupled with ultrasound solid–liquid extraction for sample treatment consisting of leaching metals from powdered materials in slurry containing diluted acid solution.[46] The LOD found was 0.65 µg/L and 32 µg/L for cadmium and lead, respectively in standard solutions, which correspond to 0.033 µg/g and 1.6 µg/g, respectively in sample for 2% (m/v) slurry (200 mg/10 mL). Using this analytical system several foods were analysed, such as lettuce, mushroom, aubergine, fish and mussels, with the highest content of cadmium found in mushroom (10.15 ± 0.49 mg/kg) and of lead in straight lettuce (20.04 ± 0.15 mg/kg).[46] Another way to increase the sensitivity of flame AAS is Donnan dialysis, which involves the migration of the sample across a membrane barrier into a receiver volume which, in order to have a preconcentration factor, is smaller.

Antonia and Allen have developed a process involving flow-injection Donnan dialysis with flame AAS for lead detection in sweeteners obtaining a detection limit of 350 ng/g and recovery values between 89% and 105%.[47] A different system to increase the sensitivity of flame AAS used coprecipitation (i.e. with aluminium hydroxide[48] followed by centrifugation and washing of the precipitate); this system allows for the detection of cadmium and lead at µg/L levels in mineral water.[48] This preconcentration can also be carried out by liquid–liquid extraction, using dithizone as a complexant, as reported by Carasek et al.[49] or Amorin and Ferreira.[50] However, the most widely used method to preconcentrate the sample is solid-phase extraction using various supports such as carbon, silica gel, cellulose and amberlite resin.[45,51] For example, Karve and Rajor used an octadecyl bonded silica membrane disk modified with Cyanex302 for lead detection with an LOD of 1 µg/L. The tea, pepper and wine samples were analysed after an acid digestion, obtaining a lead content of 48.8 mg/kg, 22.7 mg/kg and 13.4 mg/L.[45] Melek et al. detected cadmium and lead with an LOD of 0.43 µg/L and 0.65 µg/L using dibenzyldithiocarbamate chelates on Dowex Optipore V-493, and tested on tea and potable water certified materials with satisfactory accuracy.[52] Yildiz et al. have detected lead in food samples (cheese, bread, baby food, honey, milk, wine) after microwave digestion using minichromatographic column with 1-phenylthiosemicarbazide as ligand and Dowex Optipore L-493 resin as adsorbent with an LOD of 0.55 µg/L.[53] Yebra-Biurrun et al. have used Chelite P as chelating resin, with aminomethylphosphoric acid group that allows an LOD of 0.011 µg/g and 0.25 µg/g for cadmium and lead, respectively, to be attained. The analytical system was put to the test in mussel samples, obtaining a good agreement with reference methods and certified values.[54] Parham et al. have used sulphur as a ligand packed in a glass column for lead and cadmium detection in water samples, obtaining a LOD of 3.2 µg/L and 0.2 µg/L, respectively.[55] Portugal et al. have used 2-(2-thiazolylazo)-p-cresol as ligand for determination of cadmium and lead and cloud point extraction. This procedure requires a micellar phase obtained by using Triton X-114. The method allowed the LOD of 0.077 µg/L and 1.05 µg/L for cadmium and lead, respectively and was applied to samples of potable water.[56] The cloud point extraction was even used for arsenic detection in potable water samples.[57] Arsenic detection was also carried out by precipitation with aluminium hydroxide followed by hydride generation AAS allowing an LOD of 0.012 µg/L, and was applied with certified potable water samples with a relative error of −4.0%.[58]

Electrothermal AAS is a good alternative for the detection of heavy metals at trace levels. It uses a modifier that stabilises the heavy metals,[59] is more sensitive than flame AAS, and allows the analysis of solid and semi-solid samples with minimal manipulation. Total arsenic was, for example, detected by means of electrothermal AAS with Pd $(NO_3)_2$ as a modifier and by precipitating arsenic from digested samples using a weakly acidic silver solution, avoiding spectral interference from cations because they do not precipitate with silver cations. The LOD achieved was 0.3 µg/L and the system was tested in wheat flour from different countries. The highest value was found in the sample from Norway (26 ± 2 ng/g).[60] Bruhn et al. have studied different modifiers selecting the Rh 0.5 µg and have obtained an LOD for As equal to 42 pg. The analytical system coupled with UV-photooxidation sample treatment (dogfish muscle) allows good recovery of 102%.[61] The electrothermal AAS was also used for the study of As speciation in baby foods.[62]

Electrothermal analysis was also applied for lead detection in human milk using 75 µg of Pd as a modifier, obtaining an LOD of 5.0 µg/L and a recovery percentage of (109.8 ± 5.4)%.[63] The lead with cadmium was detected with a dispersive liquid–liquid microextraction reaching a very low LOD for lead and cadmium, 10 ng/L and 4 ng/L, respectively, and when applied in bottled and tap water it attained good recovery values.[64] Cadmium was detected in a sample of Orujo spirit (a traditional distillate produced in Galicia from vinasse after alcoholic fermentation). In this case, several modifiers for cadmium stabilization were investigated (Pd-$Mg(NO_3)_2$, $(NH_4)H_2PO_4$-$Mg(NO_3)_2$, Ir, W-Ir, W-Ru and W, obtaining with the last the highest sensitivity and lowest LOD (0.01 µg/L)). Among the 18 samples tested, only two have shown cadmium higher than 1.00 µg/L.[65] Testing for mercury was carried out in food colourants using electrothermal AAS with introduced slurry.[66] This procedure

98   Persistent organic pollutants and toxic metals in foods

involved the direct introduction of the samples suspension[67] and a fast program in order to avoid mercury volatilization, obtaining an LOD of 59 pg. A mercury content lower than 0.22 µg/g was found by analysing the colour additives.[66] However, in the case of mercury, the usual method is AAS using a cold vapour method. It was, for example, applied for mercury detection in mussels, clams[68] and canned fish samples[69] obtaining very low detection limits (i.e. 10 ng/L).[68]

4.3.2 Atomic fluorescence spectrometry

Atomic fluorescence is a spectroscopic process based on the absorption of a certain wavelength of radiation by an atomic vapour and subsequent radiational deactivation of the excited atoms toward the detection device. Both the absorption and subsequent atomic emission processes occur at wavelengths characteristic of the atomic species present.[70] The main advantage of fluorescence detection compared to absorption measurement is greater sensitivity, because the fluorescence signal has a very low background. The resonant excitation provides selective excitation of the analyte, avoiding interference.

Using this technique and tungsten coil electrothermal vapourization, cadmium was detected in rice and water samples. In this case, the sample treatment was carried out by cloud point extraction using Triton X-114 and dithizone allowing a very low LOD (0.01 µg/L).[71] The mercury speciation analysis was performed in several papers[72–74] by using cold vapour atomic fluorescence spectrometry coupled with chromatography, reaching a very low LOD (i.e. 6 ng/L).[74] This technique was also used for arsenic determination using hydride generation, which makes use of volatile arsines formation with $NaBH_4$ for arsenic detection in milk samples,[75] water samples,[76] seafood[77] vegetables and cereals.[78] In this case, as highlighted by the De la Guardia research group, when it is necessary to generate covalent hydrides the residue organic matter could interfere with hydride generation, thus the use of a high dilution factor with consequent decreased sensitivity is required. The researchers highlighted the advantages of using dry-ashing that completely destroys organic samples and allows a preconcentration of the samples with a very low LOD; however, it is important to stress that it is necessary to ensure that no elements are lost during the ashing process.[78]

4.3.3 Inductively coupled plasma mass spectroscopy or inductively coupled plasma optical emission spectrometry

Inductively coupled plasma mass spectroscopy (ICP-MS) is based on the same principle as atomic emission spectrometry. In this case the atomization and ionization of the analyte is carried out using high temperature argon plasma (ICP) followed by analysis based on its mass to charge ratio (MS). The detection can be also carried out by means of optical emission (ICP-OES).

The latter approach was used for the detection of mercury and methylmercury in seafood. The tissue of sea food samples was treated by acid leaching using an ultrasonic bath followed by on-line microcolumn separation/preconcentration. The LOD found was 72 ng/L for mercury. This method was applied for mercury determination in spiked and certified material, obtaining, in the case of spiked samples, recovery values in the range of 87.0%–104.6%.[79] The As content can be detected using ICP-OES and hydride generation. It is also possible to detect As^{5+}, reducing it by using KI solution. By means of ICP-OES with hydride generation an LOD of 3.6 µg/kg was found in seafood samples[80]. The ICP-OES was also applied to heavy metal detection such as cadmium and lead in table olives[81].

Arsenic, cadmium and mercury were analysed using ICP-MS in cereals. In this case a slurry flow-injection chemical generation was used. Briefly, a cereal slurry was prepared using cereal with HCl followed by heating at 85 °C for 10 min. After cooling, thiourea and Co was added, and the slurry was analysed by ICP-MS. In this case, vapour generation as a sample introduction technique was used, better than a conventional pneumatic nebulizer, obtaining LODs of 0.10 ng/g, 0.16 ng/g and 0.07 ng/g for arsenic (As^{3+}), cadmium and lead, respectively.[82] For the detection of cadmium, mercury and lead in rice flour, Li and Jiang have developed an ultrasonic slurry sampling-electrothermal vapourization as sample introduction, reaching LODs of 0.4 ng/g, 0.53 ng/g and 0.69 ng/g, respectively.[83] ICP-MS was also used for the detection of arsenic, cadmium, mercury and lead in edible marine species,[84] arsenic in wheat flour,[85] mercury in seafood[86] and lead in wines.[87] For sample treatment, a digestion at high temperature for several hours can be adopted[84] or leaching in acid solution by using an ultrasound-assisted extraction procedure.[86] However, the best treatment for avoiding contamination and the loss of heavy metals is microwave processing[88,89] using closed vessels that requires little time, for example no more than 20 min for wheat flour.[85]

4.4 Quality assurance and method validation

Quality in analytical laboratories is essential. The results of chemical analyses are important in many areas, such as control of food quality and safety.

There is frequent confusion over the meaning of *quality control*, *quality assurance* and *quality system*, as highlighted by Quevauviller *et al.*[90] *Quality control* is designed to provide a quality product, *quality assurance* is designed to ensure that the quality control activities are being implemented properly, and a *quality system* is a set of procedures involving the organization of the structure, responsibilities, processes and resources for implementing quality management.

Although these concepts and procedures are well understood by modern laboratories (whether commercial or with national responsibilities), there is an enormous gap in the transfer of knowledge to smaller laboratories.[90] In research papers, this approach is rarely mentioned. Quality assurance was discussed by Leufroy *et al.* in a paper published in 2011 on Talanta in the case of arsenic detection in sea food.[91] In this case, the authors detected arsenic species by ion exchange chromatography (IEC) coupled to ICP-MS, following microwave assisted extraction. Quality assurance was controlled as follows:

- the sample solutions were analysed in batches, including internal quality control such as the 5 point calibration standards to monitor linearity with $R^2 \geq 0.995$;
- a reagent blank was used to monitor possible cross-reaction or memory effect;
- certified reference materials were tested to check the trueness;
- standard solutions were analysed every six samples and at the end of the sequence to monitor instrument drift;
- control charts were produced (Fig. 4.11) indicating that the concentrations found were well within the confidence interval.

The following rules should be followed when implementing a quality system, as reported by Benoliel[92]:

- Planning: the establishment of a plan, an activity schedule and a deadline for formalizing the accreditation application will ensure the accreditation process is conducted more efficiently and consequently with greater staff support;
- Responsibility of the management: the most senior management body in the laboratory is responsible for defining quality policy, laying down a set of objectives aimed at satisfying the customer and the internal requirements of the company;
- Training of key staff: professional training on the subject of quality and particularly on accreditation is of fundamental importance;
- Preparation of the documentation system;
- Preparation of the quality manual (QM) on the basis of the criteria laid down in Standard EN 45001:

    45001: the QM will define the laboratory's quality policy and objectives and outline the quality system framework. In the QM, reference must be made to all quality system documentation and care must be taken to ensure all of the information is correctly linked; the QM and its attachments, should be easy to consult;

- Preparation and implementation of procedures, i.e. administrative, general and calibration techniques, and operation of the equipment;
- Testing methods: whenever these differ from standardised methods;
- Validation of analytical methods and implementation of the analytical quality control system: the purpose of a testing laboratory is to measure,

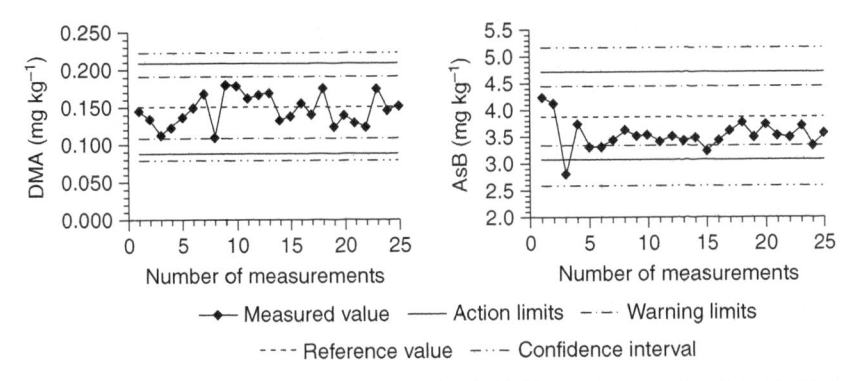

Fig. 4.11 Control charts for dimethylarsine (DMA) and arsenobetaine (AsB) in BCR 627 from 25 measurements (Reprinted from reference 91 with permission of Elsevier).

examine, test, calibrate or generally determine the characteristics or behaviour of materials or products;
• Conducting internal audits: to assess whether the quality guarantee system complies with Standard EN 45001.

As reported, validation is a point of the quality system. In the research papers, validation is often reported only as a linear range: LOD, LOQ (limit of quantification) and trueness (agreement between the value found with the proposed method and certified materials). In our opinion, a very good validation was performed in the case of arsenic, cadmium, chromium and lead in milk by means of dynamic reaction cell inductively coupled plasma mass spectrometry by D'Ilio et al. from the Istituto Superiore di Sanità.[93] For in-house validation several parameters were investigated by the authors such as:

• selectivity: to evaluate the presence of interference ions;
• trueness by recovery at three levels of concentration: in this case the authors have used the recovery due to the lack of certified reference materials in liquid form for trace elements;
• repeatability and within-laboratory reproducibility at three levels of concentration: repeatability and within-laboratory reproducibility were calculated as percentage variation coefficient by analysing three independent sets of samples, six spiked and three unspiked, on three different days;
• LOD and LOQ calculation;
• range of linearity;
• measurement of uncertainty: as reported by the authors and also Tavernies et al.[94] Its calculation is important to demonstrate the quality of the results in a laboratory. In fact a measurement of the uncertainty should be associated with results. Uncertainty is the accumulation of all relevant unknown factors.

Quality assurance was also evaluated in the paper using an internal quality control procedure to check the blank level and drift of the instrumentation.

In our opinion, particularly in the case of trace level detection of heavy metals, quality assurance should be carried out both in laboratory analysis and in research work, in order to avoid contamination and therefore an overestimation of heavy metal content.

4.5 Future trends

Future trends can be summarised in three main points. In the case of screening analysis, the research will be focused towards rendering a more sensitive system. In this way an interesting, recently developed approach is based on the use of DNA sensors that are very sensitive. For example, a DNA-based evanescent wave optical biosensor for rapid detection of mercury ions was recently developed.[95] This thymine–thymine DNA probe is able to bind mercury to form T-Hg^{2+}-T complex and, as structured, the high concentrations of mercury lead to a lower fluorescence signal. The system developed allows for a detection limit of 2.1 nM and was applied to potable water with satisfactory results. In future, other approaches will render the measurement automatic[96] and the instrumentation miniaturised in order to be simpler to use and more cost-effective.[97]

In the case of the confirmatory method, the aims are: i) to further improve sensitivity, and ii) to reduce interference. For example, in the case of ICP-MS, quadruple-based spectrometers are currently the most used instruments, owing to their low cost. However, formation of spectral interference, originating from monoatomic and polyatomic ions produced in plasma, makes the quantification of several trace elements difficult using low-resolution quadruple mass discrimination. To overcome this drawback, improvement in this instrumentation may be a possible resolution, for example using magnetic sector high resolution spectrometer.[89]

The last point relates to the sample treatment. In our opinion, in order to avoid environmental contamination and to have a complete digestion of complex matrices such as milk, microwave assisted digestion is highly recommended. However, the challenge is to have a miniaturised system able to treat the sample on-site coupled to screening methods.

4.6 References

1. JACKSON, LS (2009), "Chemical food safety issues in the United States: past, present and future", *Journal of Agricultural and Food Chemistry*, **57**, 8161–8170.
2. ARDUINI F, CALVO, J, AMINE, A, PALLESCHI, G and MOSCONE, D (2010), "Bismuth-modified electrodes for lead detection", *Trends in Analytical Chemistry*, **29**, 1295–1304.

3. VIEIRA DOS SANTOS AC and MASINI, JC (2006), "Development of a sequential injection anodic stripping voltammetry (SI-ASV) method for determination of Cd(II), Pb(II) and Cu(II) in wastewater samples from coatings industry", *Analytical and Bioanalytical Chemistry*, **385**, 1538–1544.
4. DOLAN, SP, NORTRUP, DA, BOLGER PM and CAPAR, SG (2003), "Analysis of dietary supplements for arsenic, cadmium, mercury and lead using inductively coupled plasma mass spectrometry", *Journal of Agricultural and Food Chemistry*, **51**, 1307–1312.
5. ANNIBALDI A, TRUZZI C, ILLUMINATI S, BASSOTTI E and SCARPONI G (2007), "Determination of water-soluble and insoluble (dilute-HCl-extractable) fractions of Cd, Pb and Cu in Antarctic aerosol by square wave anodic stripping voltammetry: distribution and summer seasonal evolution at Terra Nova Bay (Victoria Land)", *Analytical and Bioanalytical Chemistry*, **387**, 977–998.
6. CHRISTIAN, DC (2003), Analytical Chemistry 6th edition, Wiley international edition, JOHN WILEY and SONS, New York, USA.
7. LOCATELLI, C and TORSI, G (2003), "Analytical procedures for the simultaneous voltammetric determination of heavy metals in meals", *Microchemical Journal*, **75**, 233–240.
8. LOCATELLI, C (2004), "Heavy metals in matrices of food interest: sequential voltammetric determination at trace and ultratrace level of copper, lead, cadmium, zinc, arsenic, selenium, manganese and iron in meals", *Electroanalysis*, **16**, 1478–1486.
9. LOCATELLI, C and TORSI, G (2004), "Simultaneous square wave anodic stripping voltammetric determination of Cr, Pb, Sn, Sb, Cu, Zn in presence of reciprocal interference: application to meal matrices", *Microchemical Journal*, **78**, 175–180.
10. LA PERA, L, SAITTA, M, DI BELLA, G and DUGO, G (2003), "Simultaneous determination of Cd (II), Pb (II), and Zn (II) in citrus essential oils by derivative potentiometric stripping analysis", *Journal of Agricultural and Food Chemistry*, **51**, 1125–1129.
11. LA PERA, L, LO CURTO, S, VISCO, A, LA TORRE, L and DUGO, G (2002), "Derivative potentiometric stripping analysis (dPSA) used in the determination of cadmium, copper, lead and zinc in Sicilian olive oils", *Journal of Agricultural and Food Chemistry*, **50**, 3090–3093
12. SZLYK, E and SZYDLOWSKA-CZERNIAK, A (2004), "Determination of cadmium, lead and copper in margarines and butters by galvanostatic stripping chronopotentiometry", *Journal of Agricultural and Food Chemistry*, **52**, 4064–4071
13. LO COCO, F, MONOTTI, P, COZZI, F and ADAMI, G (2006), "Determination of cadmium and lead in fruit juice by stripping chronopotentiometry and comparison of two sample pretreatment procedures", *Food Control*, **17**, 966–970.
14. MUNOZ, E and PALMERO, S (2004), "Determination of heavy metals in milk by potentiometric stripping analysis using a home-made flow cell", *Food Control*, **15**, 635–641.
15. ABBASI, S, BAHIRAEI, A and ABBASAI, F (2011), "A highly sensitive method for simultaneous determination of ultra trace levels of copper and cadmium in food and water samples with luminol as a chelating agent by adsorptive stripping voltammetry", *Food Chemistry*, **129**, 1274–1280.
16. ABBASI, S, KHODARAHMIYAN, K and ABBASAI, F (2011), "Simultaneous determination of ultra trace amounts of lead and cadmium in food samples by adsorptive stripping voltammetry", *Food Chemistry*, **128**, 254–257.
17. ARMENTA, S, GARRIGUES, S and DE LA GUARDIA, M (2006), "Green analytical chemistry", *Trends in Analytical Chemistry*, **27**, 497–511.

18. DONG, H, ZHENG, H, LIN, L and YE, B (2006), "Determination of thallium and cadmium on a chemically modified electrode with Langmuir-Blodgett film of p-allylcalix[4]arene", *Sensors and Actuators B*, **115**, 303–308.
19. WANEKAYA, A and SADIK, OA (2002), "Electrochemical detection of lead using polypyrrole films", *Journal of Electroanalytical Chemistry*, **537**, 135–143.
20. BONFIL, Y and KIROWA-EISNER, E (2002), "Determination of nanomolar concentrations of lead and cadmium by anodic-stripping voltammetry at silver electrode", *Analytica Chimica Acta*, **457**, 285–296.
21. JENA, BK and RAJ, CR (2008), "Gold nanoelectrode ensembles for the simultaneous electrochemical detection of ultratrace arsenic, mercury and copper", *Analytical Chemistry*, **80**, 4836–4844.
22. YAMADA, D, IVANDINI, TA, KOMATSU, M, FUJISHIMA, A and EINAGA, Y (2008), "Anodic stripping voltammetry of inorganic species of As^{3+} and As^{5+} at gold-modified boron doped diamond electrodes", *Journal of Electroanalytical Chemistry*, **615**, 145–153.
23. SANLLORENTE-MENDEZ, S, DOMINGUEZ-RENEDO, O and ARCOS-MARTINEZ, MJ (2009), "Determination of arsenic (III) using platinum nanoparticle-modified screen-printed carbon-based electrodes", *Electroanalysis*, **21**, 635–639.
24. WANG, J, LU, J, HOCEVAR, SM, FARIAS, PAM and OGOREVC, B (2000), "Bismuth-coated carbon electrodes for anodic stripping voltammetry", *Analytical Chemistry*, **72**, 3218–3222.
25. WANG, J (2005), "Stripping analysis at bismuth electrodes: A review", *Electroanalysis*, **17**, 1341–1346.
26. KOKKINOS, C and ECONOMOU, A (2008), "Stripping analysis at bismuth-based electrodes", *Current Analytical Chemistry*, **4**, 183–190.
27. WANG, J, LU, J, HOCEVAR, SB and OGOREVC, B (2001), "Bismuth-coated screen-printed electrodes for stripping voltammetric measurements of trace lead", *Electroanalysis*, **13**, 13–16.
28. IDRAOUI, I, RHAZI, ME and AMINE, A (2007) "Fibrinogen coated bismuth film electrodes for voltammetric analysis of lead and cadmium using the batch injection analysis", *Analytical Letters*, **40**, 349–368.
29. OLIVEIRA SALLES, M, RUAS DE SOUZA, AP, NAOZUKA, J, VITORIANO DE OLIVEIRA, P and BERTOTTI, M (2009) "Bismuth modified gold microelectrode for Pb(II) determination in wine using alkaline medium", *Electroanalysis*, **21**, 1349–1442.
30. XU, H, ZENG, L, HUANG, D, XIAN, Y and JIN, L (2008), "A Nafion-coated bismuth film electrode for the determination of heavy metals in vegetable using differential pulse anodic stripping voltammetry: An alternative to mercury-based electrodes", *Food Chemistry*, **109**, 834–839.
31. CHAILAPAKUL, O, KORSRISAKUL, S, SIANGPROH and GRUDPAN, K (2008), "Fast and simultaneous detection of heavy metals using a simple and reliable microchip-electrochemistry route: An alternative approach to food analysis", *Talanta*, **74**, 683–689.
32. ARDUINI, F, MAJORANI, C, AMINE, A, MOSCONE, D and PALLESCHI, G (2011), "Hg^{2+} detection by measuring thiol groups with a highly sensitive screen-printed electrode modified with a nanostructured carbon black film", *Electrochimica Acta*, **56**, 4209–4215.
33. LIU, M, LIN, Z and LIN, JM (2010), "A review on applications of chemiluminescence detection in food analysis", *Analytica Chimica Acta*, **670**, 1–10.
34. FANG, G, MENG, S, ZHANG, G and PAN, J (2001), "Spectrophotometric determination of lead in foods with di-bromo-*p*-methyl-bromosulfonazo", *Talanta*, **54**, 585–589.
35. LI, Z, TANG, J and PAN, J (2004), "The determination of lead in preserved food by spectrophotometry with dibromohydroxyphenylporphyrin", *Food Control*, **15**, 565–570.

36. WEN, X, YANG, Q, YAN, Z and DENG, Q (2011), "Determination of cadmium and copper in water and food samples by dispersive liquid-liquid microextraction combined with UV-VIS spetcrophotometry", *Microchemical Journal*, **97**, 249–254.
37. HUANG, KW, YU, CJ and TSENG, WL (2010), "Sensitivity enhancement in the colorimetric detection of lead (II) ion using gallic acid-capped gold nanoparticles: Improving size distribution and minimizing interparticle repulsion", *Biosensors and Bioelectronics*, **25**, 984–989.
38. HUNG, YL, HSIUNG, TM, CHEN, YY and HUANG, CC (2010), "A label-free colorimetric detection of lead ions by controlling the ligand shells of gold nanoparticles", *Talanta*, **82**, 516–522.
39. KALLURI, JR, ARBNESHI, T, KHAN, SA, NEELY, A, CANDICE, P, VARISLI, B, WASHINGTON, M, MCAFEE, S, ROBINSON, B, BANERJEE, S, SINGH, AK, SENAPATI, D and RAY, PC (2009), "Use of gold nanoparticles in a simple colorimetric and ultrasensitive dynamic light scattering assay: selective detection of arsenic in groundwater", Angewandte Chemie-International Edition, **48**, 9668–9671.
40. DIESEL, E, SCHREIBER, M and VAN DER MEER, JR (2009), "Development of bacteria-based bioassays for arsenic detection in natural waters", *Analytical and Bioanalytical Chemistry*, **394**, 687–693.
41. SANLLORENTE-MENDEZ, S, DOMINGUEZ-RENEDO, O and ARCOS-MARTINEZ, MJ (2010), "Immobilization of acetylcholinesterase on screen-printed electrodes. Application to the determination of arsenic (III)", *Sensors*, **10**, 2119–2128.
42. SASAKI, K, YONGVONGSOONTORN, N, TAWARADA, K, OHNISHI, Y, ARAKANE, T, KAYAMA, F, ABE, K and OGUMA, N (2009), "Cadmium purification and quantification using immunochromatography", *Journal of Agricultural and Food Chemistry*, **57**, 4514–4519.
43. LIU, GL, WANG, JF, LI, ZY, LIANG, SZ, LIU, J and WANG, XN (2009), "Development of direct competitive enzyme-linked immunosorbent assay for the determination cadmium residue in farm produce", *Applied Biochemistry and Biotechnology*, **159**, 708–717.
44. LIU, GL, WANG, JF, LI, ZY, LIANG, SZ and WANG, XN (2009), "Immunoassay for cadmium detection and quantification", *Biomedical and Environmental Sciences*, **22**, 188–193.
45. KARVE, M and RAJGOR, RV (2007), "Solid phase extraction of lead on octadecyl bonded silica membrane disk modified with Cyanex302 and determination by flame atomic absorption spectrometry", *Journal of Hazardous Materials*, **141**, 607–613.
46. ALEIXO, PC, JUNIOR, DS, TOMAZELLI, AC, RUFINI, IA, BERNDT, H and KRUG, FJ (2004), "Cadmium and lead determination in foods by beam injection flame furnace atomic absorption spectrometry after ultrasound-assisted sample preparation", *Analytica Chimica Acta*, **512**, 329–337.
47. ANTONIA, A and ALLEN, LB (2001), "Extraction and analysis of lead in sweeteners by flame-injection donnan dialysis with flame atomic absorption spectroscopy", *Journal of Agricultural and Food Chemistry*, **49**, 4615–4618.
48. DONER, G and EGE, A (2005), "Determination of copper, cadmium and lead in seawater and mineral water by flame atomic absorption spectrometry after coprecipitation with aluminum hydroxide", *Analytica Chimica Acta*, **547**, 14–17.
49. CARASEK, E, TONJES, JW and SCHARF, M (2002), "A new method of microvolume back-extraction procedure for enrichment of Pb and Cd and determination by flame atomic absorption spectrometry", *Talanta*, **56**, 185–191.
50. AMORIM, FAC and FERREIRA, SLC (2005), "Determination of cadmium and lead in table salt by sequential multi-element flame atomic absorption spectrometry", *Talanta*, **65**, 960–964.
51. ALVES, VN, BORGES, SSO, NETO, WB and COELHO, NMM (2001), "Determination of low levels of lead in beer using solid-phase extraction and detection by flame atomic

absorption spectrometry", *Journal of Automated Methods and Management in Chemistry*, Article ID 464102, 6 pages.
52. MELEK, E, TUZEN, M and SOYLAK, M (2006), "Flame atomic absorption spectrometric determination of cadmium (II) and lead (II) after their solid phase extraction as dibenzyldithiocarbamate chelates on Dowex Optipore V-493", *Analytica Chimica Acta*, **578**, 213–219.
53. YILDIZ, O, CITAK, D, TUZEN, M and SOYLAK, M (2011), "Determination of copper, lead and iron in water and food samples after column solid phase extraction using 1-phenylthiosemicarbazide on Dowex Optipore L-493 resin", *Food and Chemical Toxicology*, **49**, 458–463.
54. YEBRA-BIURRUN, MC, CANCELA-PEREZ, S and MORENO-CID-BARINAGA, A (2005), "Coupling continuous ultrasound-assisted extraction, preconcentration and flame atomic absorption spectrometric detection for the determination of cadmium and lead in mussel samples", *Analytica Chimica Acta*, **533**, 51–56.
55. PARHAM, H, POURREZA, N and RAHBAR, N (2009), "Solid phase extraction of lead and cadmium using solid sulfur as a new metal extractor prior to determination by flame atomic absorption spectrometry", *Journal of Hazardous Materials*, **163**, 588–592.
56. PORTUGAL, LA, FERREIRA, HS, DOS SANTOS, WNL and FERREIRA, SLC (2007), "Simultaneous pre-concentration procedure for the determination of cadmium and lead in potable water employing sequential multi-element flame atomic absorption spectrometry", *Microchemical Journal*, **87**, 77–80.
57. ULUSOY, HI, AKCAY, M and GURKAN, R (2011), "Development of an inexpensive and sensitive method for the determination of low quantity of arsenic species in water samples by CPE-FAAS", *Talanta*, **85**, 1585–1591.
58. TUZEN, M, CITAK, D, MENDIL, D and SOYLAK, M (2009), "Arsenic speciation in natural water samples by coprecipitation-hydride generation atomic absorption spectrometry combination", *Talanta*, **78**, 52–56.
59. KORN, MGA, DE ANDRADE, JB, DE JESUS, DS, LEMOS, VA, BANDEIRA, MLSF, DOS SANTOS, WNL, BEZERRA, MA, AMORIM, FAC, SOUZA, AS and FERREIRA, SLC (2006), "Separation and preconcentration procedures for the determination of lead using spectrometric techniques: A review", *Talanta*, **69**, 16–24.
60. GONZALEZ, MM, GALLEGO, M and VALCARCEL, M (2001), "Determination of arsenic in wheat flour by electrothermal atomic absorption spectrometry using a continuous precipitation-dissolution flow system", *Talanta*, **55**, 135–142.
61. BRUHN, CG, BUSTOS, CJ, SAEZ, KL, NEIRA, JY and ALVAREZ, SE (2007), "A comparative study of chemical modifiers in the determination of total arsenic in marine food by tungsten coil electrothermal atomic absorption spectrometry", *Talanta*, **71**, 81–89.
62. LOPEZ-GARCIA, I, BRICENO, M and HERNANDEZ-CORDOBA, M (2011), "Non-chromatographic screening procedure for arsenic speciation analysis in fish-baby foods by using electrothermal atomic absorption spectrometry", *Analytica Chimica Acta*, **699**, 11–17.
63. FALOMIR, P, ALEGRIA, A, BARBERA, A, FARRE, R and LAGARDA, MJ (1999), "Direct determination of lead in human milk by electrothermal atomic absorption spectrometry", *Food Chemistry*, **64**, 111–113.
64. RIVAS, RE and LOPEZ-GARCIA, I (2009), "Determination of trace of lead and cadmium using dispersive liquid-liquid microextraction followed by electrothermal atomic absorption spectrometry", *Microchimica Acta*, **166**, 355–361.
65. FARINAS, MV, GARCIA, JB, GARCIA MARTIN, S, PENA CRECENTE, R and HERRERO LATORRE, C (2007), "Direct determination of cadmium in *Orujo* spirit samples by electrothermal atomic absorption spectrometry: comparative study of different chemical modifiers", *Analytica Chimica Acta*, **591**, 231–238.

66. VINAS, P, PARDO-MARTINEZ, M, LOPEZ-GARCIA, I and HERNANDEZ-CORDOBA, M (2002), "Rapid determination of mercury in food colorants using electrothermal atomic absorption spectrometry with slurry sample introduction", *Journal of Agricultural and Food Chemistry*, **50**, 949–954.
67. VINAS, P, PARDO-MARTINEZ, M and HERNANDEZ-CORDOBA, M (2000), "Rapid determination of selenium, lead and cadmium in baby food samples using electrothermal atomic absorption spectrometry and slurry atomization", *Analytica Chimica Acta*, **412**, 121–130.
68. VEREDA ALONSO, E, SILES CORDERO, MT, GARCIA DE TORRES, A, CANADA RUDNER, P and CANO PAVON, JM (2008), "Mercury speciation in sea food by flow injection cold vapor atomic absorption spectrometry using selective solid phase extraction", *Talanta*, **77**, 53–59.
69. VOEGBORLO, RB and ADIMADO, AA (2010), "A simple classical wet digestion technique for the determination of total mercury in fish tissue by cold-vapor atomic absorption spectrometry in a low technology environment", *Food Chemistry*, **123**, 936–940.
70. ENCYCLOPEDIA OF ANALYTICAL CHEMISTRY (2000), R.A. Meyers (Ed.), John Wiley & Sons Ltd, Chichester.
71. WEN, X, WU, P, CHEN, L and HOU, X (2009), "Determination of cadmium in rice and water by tungsten coil electrothermal vaporization-atomic fluorescence spectroscopy and tungsten coil electrothermal atomic absorption spectrometry after cloud point extraction", *Analytica Chimica Acta*, **650**, 33–38.
72. LI, Y, YAN, XP, DONG, LM, WANG, SW, JIANG, Y and JIANG, DQ (2005), "Development of an ambient temperature post-column oxidation system for high-performance liquid chromatography on-line coupled with cold vapor atomic fluorescence spectrometry for mercury speciation in seafood", *Journal of Analytical Atomic Spectrometry*, **20**, 467–472.
73. LIANG, LN, JIANG, GB, LIU, JF and HU, JT (2003), "Speciation analysis of mercury in seafood by using high-performance liquid chromatography on-line coupled with cold-vapor atomic fluorescence spectrometry via a post column microwave digestion", *Analytica Chimica Acta*, **477**, 131–137.
74. EBDON, L, FOULKES, ME, LE ROUX, S and MUNOZ-OLIVAS, R (2002), "Cold vapor atomic fluorescence spectrometry and gas chromatography-pyrolysis-atomic fluorescence spectrometry for routine determination of total and organometallic mercury in food samples", *Analyst*, **127**, 1108–1114.
75. CAVA-MONTESINOS, P, DE LA GUARDIA, A, TEUTSCH, C, CERVERA, ML and DE LA GUARDIA, M (2003), "Non-chromatographic speciation analysis of arsenic and antimony in milk generation atomic fluorescence spectrometry", *Analytica Chimica Acta*, **493**, 195–203.
76. LI, N, FANG, G and ZHU, H (2009), "Determination of As (III) and As (V) in water samples by flow injection online sorption preconcentration coupled to hydride generation atomic fluorescence spectrometry", *Microchimica Acta*, **165**, 135–141.
77. JESUS, JP, SUAREZ, CA, FERREIRA, JR and GINÈ, MF (2011), "Sequential injection analysis implementing multiple standard additions for As speciation by liquid chromatography and atomic fluorescence spectrometry (SIA-HPLC-AFS)", *Talanta*, **85**, 1364–1368.
78. MATOS-REYES, MN, CERVERA, ML, CAMPOS, RC and DE LA GUARDIA, M (2010), "Total content of As, Sb, Se, Te and Bi in Spanish vegetables, cereals and pulses and estimation of the contribution of these foods to the Mediterranean daily intake of trace elements", *Food Chemistry*, **122**, 188–194.
79. XIONG, C and HU, B (2007), "Online YPA4 resin microcolumn separation/preconcentration coupled with inductively coupled plasma optical emission

108 Persistent organic pollutants and toxic metals in foods

spectrometry (ICP-OES) for the speciation analysis of mercury in seafood", *Journal of Agricultural and Food Chemistry*, **55**, 10129–10134.

80. BOUTAKHRIT, K, CLAUS, R, BOLLE, F, DEGROODT, JM and GOEYENES, L (2005), "Open digestion under reflux for the determination of total arsenic in seafood by inductively coupled plasma atomic emission spectrometry with hydride generation", *Talanta*, **66**, 1042–1047.
81. LOPEZ-LOPEZ, A, LOPEZ, R, MADRID, F and GARRIDO-FERNANDEZ, A (2008), "Heavy metals and mineral elements not included on the nutritional labels in table olives", *Journal of Agricultural and Food Chemistry*, **56**, 9475–9483.
82. CHEN, FY and JIANG, SJ (2009), "Slurry sampling flow injection chemical vapor generation inductively coupled plasma mass spectrometry for the determination of As, Cd, and Hg in cereals", *Journal of Agricultural and Food Chemistry*, **57**, 6564–6569.
83. LI, PC and JIANG, SJ (2003), "Electrothermal vaporization inductively coupled plasma-mass spectrometry for the determination of Cr, Cu, Cd, Hg and Pb in rice flour", *Analytica Chimica Acta*, **495**, 143–150.
84. FALCO, G, LLOBET, JM, BOCIO, A and DOMINGO, JL (2006), "Daily intake of arsenic, cadmium, mercury, and lead by consumption of edible marine species", *Journal of Agricultural and Food Chemistry*, **54**, 6106–6112.
85. SAHAYAM, AC, CHAURASIA, SC and VENKATESWARLU, G (2010), "Dry ashing of organic rich matrices with palladium for the determination of arsenic using inductively coupled plasma-mass spectrometry", *Analytica Chimica Acta*, **661**, 17–19.
86. BATISTA, BL, RODRIGUES, JL, DE SOUZA, SS and OLIVEIRA SOUZA, VC (2011), "Mercury speciation in seafood samples by LC_ICP_MS with a rapid ultrasound-assisted extraction procedure: Application to the determination of mercury in Brazilian seafood samples", *Food Chemistry*, **126**, 2000–2004.
87. LARCHER, R, NICOLINI, G and PANGRAZZI, P (2003), "Isotope ratios of lead in Italian wines by inductively coupled plasma mass spectrometry", *Journal of Agricultural and Food Chemistry*, **51**, 5956–5961.
88. NOEL, L, DUFAILLY, V, LEMAHIEU, N, VASTEL, C and GUERIN, T (2005), "Simultaneous analysis of cadmium, lead, mercury and arsenic content in foodstuffs of animal origin by inductively coupled plasma/mass spectrometry after closed vessel microwave digestion: method validation", *Journal of AOAC International*, **88**, 1811–1821.
89. HUSAKOVA, L, URBANOVA, I, SRAMKOVA, J, CERNOHORSKY, T, KREJCOVA, A, BEDNARIKOVA, M, FRYDOVA, E, NEDELKOVA, I and PILAROVA, L (2011), "Analytical capabilities of inductively coupled plasma orthogonal acceleration time-of-flight mass spectrometry (ICP-oa-TOF-MS) for multi-element analysis of food and beverage", *Food Chemistry*, **129**, 1287–1296.
90. QUEVAUVILLER, P, CAÈMARA, C, KRAMER AND KJM (1999), "EC initiatives for quality assurance training in analytical chemistry", *Trends in Analytical Chemistry*, **18**, 644–649.
91. LEUFROY, A, NOEL, L, DUFAILLY, V, BEAUCHEMIN, D and GUERIN, T (2011), "Determination of seven arsenic species in seafood by ion exchange chromatography coupled to inductively coupled plasma-mass spectrometry following microwave assisted extraction: Method validation and occurrence data", *Talanta*, **83**, 770–779.
92. BENOLIEL, MJ (1999), "Step-by-step implementation of a quality system in the laboratory", *Trends in Analytical Chemistry*, **18**, 632–638.
93. D'ILIO, S, PETRUCCI, F, D'AMATO, M, DI GREGORIO, M, SENOFORTE, O and VIOLANTE, N (2008), "Method validation for determination of arsenic, cadmium, chromium

and lead in milk by means of dynamic reaction cell inductively coupled plasma mass spectrometry", *Analytica Chimica Acta*, **624**, 59–67.
94. TAVERNIES, I, VAN BOCKSTAELE, E and DE LOOSE, M (2004), "Trends in quality in the analytical laboratory. I. Traceability and measurement uncertainty of analytical results", *Trends in Analytical Chemistry*, **18**, 644–649.
95. LONG, F, GAO, C, SHI, HC, HE, M, ZHU, AN, KLIBANOV, AM and GU, AZ (2011), "Reusable evanescent wave DNA biosensor for rapid, highly sensitive, and selective detection of mercury ions", *Biosensors and Bioelectronics*, **26**, 4018–4023.
96. MONTICELLI, D, CICERI, E and DOSSI, C (2007) "Optimization and validation of an automated voltammetric stripping technique for ultratrace metal analysis", *Analytica Chimica Acta*, **594**, 192–198.
97. PALM INSTRUMENTS COMPANY, www.palmsens.com [Accessed 2 November 2012]
98. CALVO QUINTANA, J, ARDUINI, F, AMINE, A, VAN VELZEN, K, PALLESCHI, G and MOSCONE D (2012), "Part two: Analytical optimisation of a procedure for lead detection in milk by means of bismuth-modified screen-printed electrodes", *Analytica Chimica Acta*, **736**, 992–99.

5

Responding to food contamination incidents: principles and examples from cases involving dioxins

C. Tlustos, W. Anderson and R. Evans,
Food Safety Authority of Ireland, Ireland

DOI: 10.1533/9780857098917.1.110

Abstract: The consequences of a food contamination incident can be severe, not only affecting consumers' health but also causing disruption to trade and having substantial impact on the economy. Therefore, to minimise the impact of an incident, preparedness, coordination and consistency in management, as well as transparency in actions taken, are imperative. The basic elements of a crisis response, with a particular focus on traceability and management of a product withdrawal or recall, as well as risk communication strategies, are presented.

Key words: food incident, food traceability, food withdrawal, food recall, risk communication.

5.1 Introduction

Despite best efforts and increased awareness of risks associated with the food chain, there have been numerous incidents involving chemical contaminants. These can largely be categorised into incidents related to production and/or handling of food and feed, or stem from incidents due to accidental pollution issues. Whatever the cause, consequences can be severe, not only affecting consumers' health but also causing disruption to trade and having substantial impact on the economy. Therefore, to minimise the impact of an incident, preparedness, coordination and consistency in management, as well as transparency in actions taken, are imperative.

This chapter discusses general principles that may be used in response to food incidents, and uses dioxins as a case study to examine responses to contamination incidents from a regulator's point of view.

5.1.1 Recent major incidents concerning dioxins

The major source of human exposure to dioxins is through food (greater than 90%), with food of animal origin being the predominant source (EFSA, 2010). As such, food contamination incidents can lead to a significant increase in exposure to these substances. Over recent years numerous such contamination incidents have occurred, some of which are listed in Table 5.1.

Table 5.1 Overview of recent dioxin and polychlorinated biphenyls (PCBs) contamination incidents

Year	Dioxin contamination incident	Reference
1998	Contamination of milk, butter and meat due to the use of contaminated citrus pulp in feedstuffs in Germany	Carvalhaes et al., 1999; Malisch, 2000
1999	Contamination of feed fat with PCB-containing oil in Belgium	Bernhard et al., 1999
	Contamination of feed due to the use of contaminated kaolin as anti-caking agent in Switzerland	Schmid et al., 2000
2000	Contamination of animal feed due to contaminated sawdust used as carrier in pre-mixed choline chloride in Germany	Llerene et al., 2001
2003	Contamination of bakery waste due to use of waste wood in the drying process in Germany	Hogenboom et al., 2003
2004	Contamination of poultry, meat and eggs due to pentachlorophenol (PCP)-contaminated wood shavings as litter in Italy	Diletti et al., 2005
	Contamination of pig feed due to inclusion of waste fat contaminated by a malfunction in gelatine processing in the Netherlands	Hoggenboom et al., 2006
	Contamination of milk by use of potato by-products contaminated by kaolinic clay during sorting in animal feed in the Netherlands	Hoogenboom et al., 2005
2008	Contamination of pork due to contaminated zinc oxide used as feed ingredient in Chile	Kim et al., 2009
	Contamination of buffalo milk due to illegal waste burning in Italy	Borello et al., 2008; Scortichini et al., 2008; CRL, 2009
	Contamination of pork and beef due to contamination of bakery waste via burning of contaminated oil in Ireland	Tlustos, 2009
	Contamination of Guar Gum with dioxins related to the presence of pentachlorophenol	Wahl et al., 2008
2010	Contamination of pigs and poultry via dioxin-contaminated oil in animal feed in Germany	BMELV, 2011

5.1.2 Basic elements of a crisis response

In 2009, the World Health Organization (WHO, 2009) defined the 'disaster management cycle' as a continuous process by which governments, businesses and civil society plan for and reduce the impact of incidents by acting at different stages of an incident's life cycle.

The life cycle comprises the following stages: prevention – preparation – detection and alert – response – recovery.

In the case of a food incident, it is vital that all actions taken are coordinated and communicated appropriately. To that effect, many countries have prepared strategies, plans or Codes of Practice (FSAI, 2004), to be implemented in the event of an incident. These plans usually set out the roles and responsibilities of the various collaborating bodies and identify a designated control point responsible for the coordination of the response. Often, the key roles, i.e. risk assessment, risk management, risk communication, etc., are separate from, but dependent on each other, which underlines the importance of one coordinating central post, to ensure consistency and transparency in the management of the incident. The recognition of the latter from experiences and lessons learnt over the last ~ 25 years has led to the development of the so-called risk analysis paradigm.

5.2 The risk analysis paradigm

Risk analysis encompasses a conceptual strategy that was developed in the 1980s and 1990s with the aim of introducing standardisation and transparency to the approaches taken to evaluate health risks arising from potential hazards. The risk analysis paradigm includes three interlinked elements—risk assessment, risk management and risk communication. The three elements combine science-based, socio-economic, political and cultural considerations in assessing, managing and communicating risk.

5.2.1 Risk assessment

The evaluation of safety and suitability of food for human consumption is often not straight forward and requires further consideration. In the European Union, Regulation (EC) No. 178/2002 lays down the general principles and requirements of food law, and provides for a legal framework encompassing the identification of unsafe food. Risk assessment is an essential element in identifying unsafe food. The purpose of risk assessment is to characterise the nature and probability of adverse effects to human health arising from exposure to a hazard. It involves four steps, namely hazard identification, hazard characterisation, exposure assessment and risk characterisation. Depending on the physicochemical properties of the substance and its eventual environmental fate, different types of mechanism may have to be applied in each step.

1. Hazard Identification: the identification of known or potential health effects associated with a particular agent
2. Exposure Assessment: the qualitative or quantitative evaluation of the degree of intake likely to occur
3. Hazard Characterisation: the qualitative or quantitative evaluation of the nature of the adverse effect associated with the hazard
4. Risk Characterisation: the integration of hazard identification, hazard characterisation and exposure

The above steps should be undertaken based on reliable scientific data and logical reasoning, leading to a conclusion which expresses the possibility of occurrence and the severity of a hazard's impact on the environment, or health of a given population including the extent of possible damage, persistency, reversibility and delayed effect. However, it is not possible in all cases to complete a comprehensive assessment of risk, for example in emergency situations, where timing is critical and data are limited. A number of conservative assumptions may be necessary and frequently a 'worst case scenario' approach is used in determining the character of the risk. To fully aid risk managers in making a decision based on the risk assessment performed, information on the reliability of the data used and inherent uncertainties of the assessment should always be provided.

For more in-depth information on risk assessment please see Chapter 8, by D. Benford.

5.2.2 Risk management

Risk management is defined within Codex[1] as the process of weighing policy alternatives in the light of the results of risk assessment and, if required, selecting and implementing appropriate control options, including regulatory measures (WHO/FAO 1997). During a food contamination incident this practically translates into deciding on what to do with the implicated food, based on the outcome of the risk assessment and other legitimate factors. Depending on the magnitude of the risk and numerous other factors, a decision has to be made whether to withdraw the food from circulation and also, if necessary, whether to also recall it from the consumer. The consequences of such action also need to be taken into consideration, and processes implemented to handle them. For example, the 2008 Irish dioxin contamination incident involved live animals, which had to be identified using a positive

---

[1] The Codex Alimentarius Commission was created in 1963 by FAO and WHO to develop food standards, guidelines and related texts, such as codes of practice under the Joint FAO/WHO Food Standards Programme. The main purposes of this Programme are protecting health of the consumers, ensuring fair trade practices in the food trade, and promoting coordination of all food standards work undertaken by international governmental and non-governmental organisations.

release system and if found contaminated, destroyed following a predefined protocol. Management of the disposal of the considerable amount of carcasses created a significant challenge to the Irish authorities.

As discussed earlier, risk assessments in crisis situations are rarely complete, due to the limited availability of necessary information. Risk assessors attempt to address uncertainty in their assessments using worst case scenarios and conservative assumptions. Therefore, risk managers may decide that the uncertainty is such that a precautionary approach is necessary. In Europe, this has led to the elaboration of the precautionary principle in Regulation (EC) No. 178/2002. The precautionary principle may be invoked when the potentially dangerous effects of a phenomenon, product or process have been identified by a scientific and objective evaluation, and this evaluation does not allow the risk to be determined with sufficient certainty. However, the precautionary principle may not be used to justify arbitrary decisions (EC, 2000) and should start with a scientific evaluation that is as complete as possible, and should ideally identify at each stage the degree of scientific uncertainty. Where action is deemed necessary, measures based on the precautionary principle should be, inter alia:

- *proportional* to the chosen level of protection,
- *non-discriminatory* in their application,
- *consistent* with similar measures already taken,
- *based on an examination of the potential benefits and costs* of action or lack of action (including, where appropriate and feasible, an economic cost/benefit analysis),
- *subject to review,* in the light of new scientific data, and
- *capable of assigning responsibility for producing the scientific evidence* necessary for a more comprehensive risk assessment.

Section 5.3 of this chapter reviews the above mentioned management options and actions required to achieve them.

5.2.3 Risk communication

In the industrialised world, food safety has become a global issue influencing political careers and policy, selling television air time, newspapers and magazines and frightening the majority of consumers (Anderson 2005).

Risk communication was defined in 1995 by a WHO/FAO expert consultation as 'an interactive process of exchange of information and opinion on risk among risk assessors, risk managers, and other interested parties', involving all aspects of communication among risk assessors, risk managers and the public (WHO/FAO 1998).

Whilst in theory this seems a straight forward task, in practice it encompasses a fine balancing act, which needs to take into consideration a multitude of factors, many of which are emotive.

In the midst of a food incident it is therefore vital that all stakeholders keep an open dialogue and that information is communicated in a comprehensible, transparent and open manner. It is also imperative that release of conflicting messages by different parties be avoided at all cost, as it undermines the credibility and trustworthiness of the information relayed. Coordination of risk communication between different bodies involved in a crisis is an essential element of crisis management. Section 5.5 of this chapter examines the relevance of these elements during a food incident in more detail.

5.3 Food traceability

Being able to trace food through the production and distribution chain is a fundamental requirement of incident management. Traceability systems also bring other business advantages, such as better control of logistics and efficient stock management. However, these considerations are not the subject of this chapter. When food is contaminated through whatever means, it is essential that a food business is able to trace where that food has gone as a prelude to recall or withdrawal as necessary. It is important to be able to trace the inputs to a process, such as animals, feeds, food or packaging, as well as to track the final product up through the food chain to the customer. At the most basic level, a working traceability system can focus on suppliers and customers; however, most food businesses find added advantage in being able to link incoming material from suppliers to finished goods being sent to customers. For this purpose, a system of internal or process traceability is necessary. Often food businesses have to meet a minimum legal requirement, but increasingly customer requirements seek systems that go beyond the legal minimum. For example, in Europe a food business has a compliant traceability system if it can identify the suppliers of the food and its immediate customers who received the food (not including the consumer) – the so-called 'one up one down' approach to traceability. However, retail multiples via private standards like the British Retail Consortium standard (BRC) require a level of internal process traceability as well.

Traceability systems can tend towards complexity, particularly in large multi-product food businesses. As complexity increases, it is necessary to employ technological solutions to maintain a manageable traceability process. In a small business with a single product, a manual record production and maintenance system is a workable solution, but in a larger business there is a need to increasingly turn to information technology to maintain the key requirements of any traceability system. These key requirements are simply accuracy and efficient recall of information.

There are generally two main objectives of a traceability system. Firstly, to identify uniquely a batch of food or feed and the raw material batches used in its production, and track the physical flow of that batch forward through the food and feed chain to the customer. Secondly, to create and maintain accurate traceability records about that batch of food/feed, and to retrieve that information on demand. Best practice demands the documentation of the traceability system for clear communication both to internal and external customers as well as the competent authorities. Food businesses should follow a clear step by step approach to establishing such a traceability system.

5.3.1 Creation and maintenance of a traceability system

The Food Safety Authority of Ireland has developed guidance on developing traceability systems (FSAI, 2010) that forms the basis of the information detailed below. Other guides also exist and cover similar information (e.g. CAC, 2006; EC, 2010). Although not written specifically for feed, this document could also be used by feed businesses as a basis of their traceability systems.

The scope of the system should determine if one or more of the three key areas need to be covered. These are supplier traceability, customer traceability and process (internal) traceability. Once established, it is essential to decide which products need to be covered and which raw materials are necessary to track. The next step is to decide on the batch size. The smaller the batch size, the greater the amount of traceability information generated and the more complex the system, and vice versa. However, the bigger the batch size, the more product that may need to be recalled in the event of a contamination incident. Continuous production poses particular unique problems regarding the definition of a true batch. In fact, a true batch may not exist, and therefore many businesses tend to batch during continuous production by setting a start and end production date or time, in the knowledge that each batch has an inseparable link to the batches that have come before or after. A food business must find the appropriate balance between a small batch size to limit the extent of recalls and a batch size that is so small that the traceability system becomes too complex to achieve the principle goals of accuracy and efficiency.

Having established a batch definition and the scope of the system, the food/feed business must decide what information it is necessary to generate and keep. A possible minimum dataset could be the name and contact details of suppliers and customers, as well as a list of items supplied or delivered, dates of delivery and quantities. Other information, such as date of durability of ingredients or finished goods, may also be captured by some businesses. It is not so easy to stipulate a minimum list of information necessary for process traceability because this is process specific. However, most good businesses will be able to generate process sheets that capture traceability information of ingredients for each batch of food/feed produced, and in addition this will also be associated with essential quality assurance information for the production of the batch.

Record keeping is a challenge for any food/feed business. The European Union (EU) have produced guidance in this regard (EC, 2010). According to these guidelines, the traceability information for food product with short shelf life should be kept for at least 6 months beyond the use-by date. For less perishable product (e.g. with best-before dates) 5 years is the suggested time span for record keeping. For products with shelf life longer than 5 years, they suggest that records are kept for the duration of the product's shelf life plus 6 months. Of course, this is only a guideline but food businesses must be aware that products are in circulation beyond their shelf life, and also that health issues associated with a product can occur after the shelf life of the product is over.

It is important to periodically test and review the traceability system to ensure it is fit for purpose and that staff are familiar with its workings. Such reviews are essential if products change formulation or if new products are added. Otherwise, testing and review is best achieved on an annual cycle. An audit approach to the system is recommended, consisting of a horizontal and vertical check of accuracy and efficiency. A horizontal audit will look at the information accuracy for several batches at the same point in the process. A vertical audit will follow several batches from the customer to the supplier, looking at the accuracy of the information in the system as well as the system's overall ability to provide the necessary information within a reasonable time frame. The audit should be documented, and action should be taken to resolve any problems. The solutions implemented should also be documented and the traceability documentation updated if necessary.

5.4 Food recall and withdrawal

Food recall and withdrawal are risk management procedures laid down in general food law and refer to removal of unsafe food from the market. Article 19 of the Regulation (EC) No.178/2002 on the general principles of food law specifies the requirements of food business operators regarding the withdrawal and recall of unsafe foods.

5.4.1 Food recall and withdrawal systems

If unsafe food has been placed on the market then it may need to be recalled or withdrawn. Recall is generally viewed as a public process in which unsafe food must be removed from the market whilst informing consumers. Withdrawal, on the other hand, is viewed as a business to business communication process to remove unsafe food from the distribution chain before it has reached the consumer.

Efficient and effective recall systems reduce the exposure of the public to harmful food, and therefore protect public health. However, during a recall the reputation of the food business is exposed and public scrutiny of how the recall is handled can affect that reputation and consequently affect future trade. The ideal situation is to ensure that the recall is completed quickly and accurately to limit public scrutiny and the impact on public health. Adverse public comment on a recall, or escalating public health impact, can quickly lead to a crisis of trust in the food business, not only by the public but also by other trading partners in the industry. Quick, decisive action has been shown to limit damage to business reputation (Casey *et al.*, 2010; Jacob *et al.*, 2010).

5.4.2 Creation and maintenance of a food recall/withdrawal system

It is unlikely that a recall of food can be either efficiently or effectively executed unless a food business has taken measures beforehand to develop a recall system. Traceability and recall are interlinked, since food recall must be based on good traceability information which if available, can be used to limit the recall and the impact on the food business. Comprehensive guidance on developing recall systems and managing food recalls has been provided by the Food Safety Authority of Ireland (FSAI, 2010), on which the following information has been based. Other guidance documents are also available (e.g. CFIA, 2007; FSANZ, 2008)

There are three main stages in developing a food recall/withdrawal system: establishing a documented policy, establishing a documented plan and review, and testing the plan. A food recall/withdrawal policy demonstrates the food business's commitment to the protection of public health and compliance with legal requirements. The policy should clearly convey the objective of a recall/withdrawal system, and also make a statement about resource allocation to support the process.

The recall/withdrawal plan consists of a number of related documents. Importantly, it is necessary to identify a food incident team by name, job title and contact details (both in and out of normal business hours). The team should be led by a food incident coordinator with authority to take the necessary decisions during an incident. The incident team should establish and maintain a key contacts list, a decision tree, notification procedures, an incident log template, and a review and testing procedure. The contact list must be updated regularly and consist of 24 hour contacts for suppliers, customers, competent authorities and internal staff. The decision tree should be designed to clarify the actions and thought processes leading to a decision to recall or withdraw a food. The notification procedures should cover business customers, the competent authorities and the public. Templates of notices for each cohort should be developed and maintained. It is also important that the notification procedures consist of a check list of information about the unsafe food that needs to be communicated, such as the name and details of the food, the traceability information and the nature of the problem. During an incident communications and decisions will happen in very quick succession and it is important, for review and also for legal reasons, that a log of the actions taken in managing the incident is completed during the recall/withdrawal process. The log should record all correspondence, including phone calls, and corresponding decisions with links to original documentation. Staff should be trained on the incident plan, taking particular care to outline the initial information management process that leads to activation of the plan. Staff should be empowered to report any information that suggests unsafe food has been placed on the market to management and the incident coordinator to ensure that the plan is activated appropriately.

5.4.3 Managing a food recall/withdrawal

During a food incident, the coordinator leads the incident team through the recall/withdrawal process based on the documented plan. The incident coordinator is responsible for ensuring that the incident log is maintained up to date and available to the incident team and senior managers for review. The key part of the process is the risk assessment and subsequent identification of whether a food is unsafe or not. To help food businesses make the right decisions, it is wise to adopt an approach that considers the food to be unsafe unless it can be proved otherwise. Underestimation of the risk has historically led to public health issues and irrevocable damage to a company's reputation. Early communication with the competent authorities in such circumstances may help food and feed businesses with the accuracy of their risk assessment. Another pitfall for food businesses is delayed action caused by a desire to have all the necessary information available. Information is always scarce at the start of a food incident, and food businesses need to avoid undue delay in recalls whilst they wait for 'further information'.

The food incident team should compile all the traceability information on the affected food and communicate this accurately to customers and consumers so that unsafe food can be removed from the market. There are two key points to remember with public recall notices or communications. Firstly, avoid turning the recall information into a marketing exercise that emphasises the virtues of the company. Secondly, be accurate in communicating the risk rather than using vague statements about quality. Both of these potential problems can delay actions taken by consumers to protect their health by avoiding the unsafe food.

If food is returned it must be reconciled, quarantined and, if necessary, destroyed. This information should be used to decide when to close the food incident. It should be understood that it is rare for all affected food placed on the market to be accounted for during the recall. When a recall/withdrawal is complete, it is important that a food business review its performance during the incident based on the food incident log. Any improvements in the food recall/withdrawal system should be undertaken immediately.

5.4.4 Use of rapid detection methods

Once a food incident has been discovered, further analysis is often required to identify contaminated batches and/or further investigate the source or spread of the contamination. Depending on the type of chemical, rapid detection methods may be available. Suitability of the latter very much depends on the chemical in question, methodology applied and information sought. Some rapid detection methods do not provide the same amount of detail or sensitivity as conventional methodologies, information which is often required for determination of the potential source or nature of contamination (i.e. congener fingerprints), but may be sufficient in providing information on presence or absence of a chemical, which can form the basis of a positive release system. Rapid detection methods may as such aid in increased sample throughput once set up, and may present great benefit in prevention and subsequent early detection of contamination issues. The reader is referred to Part 1 of this book, which is dedicated to chemical analysis and which provides details on available rapid and conventional detection methods.

5.5 Risk communication strategies

Risk communication involves various aspects of information exchange between risk assessors, risk managers and the public (WHO/FAO 1998). Depending on the information recipient and the type of information that needs to be provided, different methods and systems for distributing the information are available. This chapter provides an overview of communication strategies typically employed during a food incident.

5.5.1 International risk communication between risk managers

When an incident involving unsafe food has been identified, rapid communication of information on the nature of the incident, the products involved and measures being taken to protect consumers (e.g. withdrawal or recall of product) needs to be made to consumers, trade customers, national food safety control bodies alike, and in turn to food safety bodies in other countries, particularly neighbouring countries or for example within the EU, with other Member States. The information that needs to be provided, and the method for distributing the information, will of course depend on the intended audience. At the international level, the two main systems in operation for distribution of detailed information between government bodies are the International Food Safety Authorities Network (INFOSAN) system operated by the WHO, and within Europe the Rapid Alert System for Feed and Food (RASFF) system, which is managed by the European Commission. Both of these systems comprise a section containing detailed, commercially sensitive information on products that may be affected by a particular incident, access to which is restricted to national governments, and also a section where more general information on the nature of food safety related incidents is publicly available. Both systems are described in more detail below, but both are tools which enable the rapid dissemination of detailed information on food safety related incidents to a network of relevant government bodies responsible for food safety.

International Food Safety Authorities Network (INFOSAN)
The INFOSAN network was established in 2004 following the adoption by the Codex Alimentarius Commission of the 'Principles and Guidelines for the Exchange of Information in Food Control Emergency Situations' (Codex, 2004). The network is managed by the WHO in collaboration with the Food and Agriculture Organization of the United Nations (FAO). Further information on the structure and operation of INFOSAN is available (WHO, 2007), but briefly the international network comprises 177 Member States with a dedicated emergency contact point identified in each country, through which information can be exchanged and further disseminated on food safety related issues. When certain food safety related incidents with potentially international public health significance occur within a member country of the Network, the International Health Regulations 2005 (WHO, 2005) require that the designated contact point communicates information on the incident to the INFOSAN Secretariat for further dissemination, thus providing a rapid exchange of relevant information on the identification, assessment and management of food safety related incidents (WHO, 2007). In addition, member countries have also identified other relevant organisations to receive information on food incidents disseminated by the INFOSAN network. As well as coordinating information exchange during food safety related incidents, the network also provides more general information notes on food safety related issues, and these are available in the different official languages of the WHO. Given the breadth of membership of the network, member countries are able to obtain detailed information on food incidents that are developing outside their usual sphere of information exchange but which, given the global nature of food supply chains, could well have direct implications for their domestic markets. One recent example of an incident where the INFOSAN network was able to provide information on a chemical contamination incident involved melamine adulteration of milk in China which resulted in many secondary products being contaminated. One such product, a milk-based confectionary identified in the communications from INFOSAN as being potentially contaminated, was discovered to be on sale in Ireland and on analysis was found to contain elevated levels of melamine and was removed from sale.

Rapid Alert System for Food and Feed (RASFF)
A system analogous to INFOSAN exists for the exchange of information on food safety related incidents between the European institutions (European Commission and the European Food Safety Authority) and the competent authorities responsible for food and feed safety in the 27 EU Member States along with those in the EEA and EFTA Member Countries. The RASFF has been in operation (in one form or another) for the rapid exchange of

information for over 30 years; however, in 2002 the introduction of Regulation 178/2002 laying down the general principles and requirements of food law (European Commission, 2002) formalised the structure and operation of the system that is in place today. Each Member State of the EU is required to provide information to the Commission and to other Member States when they identify a food contamination incident that constitutes a serious risk to human health. Information must be provided, via completion of a standardised template, on the nature of the incident, the risk associated with the incident, the products implicated, the distribution of those products within the Member State concerned, to other Member States and to Third Countries outside the EU, and also information on the risk management measures being taken to address the contamination incident. Since very detailed, often commercially sensitive, information on the identity of contaminated products and their distribution to trade customers is contained within the notifications, Member States are required to respect the confidentiality of the information provided unless publication is required to protect consumers. Detailed information on the products and their distribution is essential to enable competent authorities to be able to trace and possibly withdraw or recall product; however, it is possible that if routinely made public it could be used for commercial gain by competitors. However, the requirement to routinely respect the confidentiality of the information does not preclude governments providing detailed information on affected products to consumers where this is considered necessary in order to ensure protection of public health.

In addition to the detailed information disseminated via the RASFF system, less detailed information is publicly available via the RASFF Portal, a fully searchable on-line database provided by the EC which contains summary information on all notifications issued via the RASFF system since 2006 (the web address of the Portal is provided below under Section 5.7 'Sources of further information'). The Portal allows searches to be undertaken on the basis of a number of different criteria including date of notification, product type, country of origin and/or destination of product, and nature of hazard. More than 300 individual notifications related to contamination of foods and feed with so-called Industrial Contaminants (including dioxins and PCBs) or heavy metals were communicated via the RASFF system in 2010. This system is particularly useful for the identification of trends, such as types of contaminant or countries where there might be a problem with production.

International communication between risk assessors
Whilst the formalised information exchange networks described under Sections 'International Food Safety Authorities Network (INFOSAN)' and 'Rapid Alert System for Food and Feed (RASFF)' are fundamental to the rapid identification and management of food safety incidents, these are only able to function effectively in the time of crisis if other strong 'informal' information exchange networks have been established under normal working conditions. Close cooperation and collaboration between various organisations

with wide ranging expertise is essential for the effective management of food safety related incidents. Whilst good working relationships can and have been established between organisations and individuals on an ad hoc basis through participation or cooperation in other forums, arrangements have increasingly been put in place in recent years by international bodies to help foster closer cooperation and collaboration in a more structured way between individual organisations. A very good example of such a network is the EFSA Advisory Forum, which regularly brings together members of all of the food safety authorities from the European Member States for discussion and information exchange on a range of topical issues related to food safety. Not only does the exchange of information at such meetings help to progress consideration of particular issues, the contacts, understanding and trust developed at such meetings are invaluable when effective management of incidents in a crisis situation requires immediate access to information and expertise that may only be available in another Member State or international organisation.

5.5.2 National communication between risk managers

Enforcement agencies
It is fair to say that no two countries operate identical food safety enforcement systems, and some of the arrangements can be rather complex with national, regional and local government bodies all involved in some way in the control of food safety. It is therefore vital that close cooperation is maintained throughout the national food safety network, with roles and responsibilities clearly defined and their effective application being assessed on a regular basis.

For example, in Ireland, the competent authority with responsibility for food safety is the Food Safety Authority of Ireland. Before the Authority was set up in the late 1990s there were 48 different agencies and authorities in Ireland with some form of responsibility for official controls on food safety, this in a country with a population of approximately 4 million people. The legislation that established the Authority makes provision for the Authority to enter into a series of service contracts with the official agencies in Ireland to undertake the necessary controls to ensure food safety on behalf of the Authority. These service contracts, along with additional Memorandums of Understanding with certain other official agencies, clearly set out the roles and responsibilities of the Authority and the various different bodies, and clarify reporting arrangements to ensure that the most (cost-) effective food safety control system is in operation.

Obviously, the main focus of enforcement agencies during any food incident is to ensure consumers are protected from contaminated product. Therefore, to ensure this protection is as effective as possible, systems should be in place that ensure dissemination of accurate and comprehensive information on the risk associated with the product and the risk management actions considered necessary, e.g. withdrawal or recall of product from the market. Whilst in theory it is often relatively straightforward for the role of each agency to be identified, as noted in Section 5.1.2 above, it is beneficial to operate a system where one central authority is responsible for ensuring that various functions are taken forward without any unnecessary degree of overlap, but also ensuring, (probably more importantly) that there are no gaps in the official control system that could mean product or processes remain outside official control.

Industry
When disseminating information on a food-related incident, a fine balance needs to be struck between, on one hand, maintaining the confidentiality of commercially sensitive information that needs to be available for the effective management of an incident by food safety authorities, and on the other, providing sufficient information to industry to allow suppliers to take action where necessary to withdraw or recall specific batches of contaminated product, to ensure consumers are protected but without compromising a supplier's other existing or future markets. It is possible to do this effectively whilst maintaining the required balance, although an assessment of the likely impact of releasing detailed information needs to be made on a case-by-case basis.

Early identification and assessment of the implications associated with instances of contamination of the food chain are often vital in the subsequent effective management of any resulting incident. It is therefore vital that close contact and a certain degree of trust is developed between regulatory authorities and the food industry, so that food (or feed) businesses become more inclined to provide information without delay to the competent authority on issues related to the contamination of the food chain that may have been identified through the course of their own controls. Such early identification of problems with specific products can help to mitigate the extent of any withdrawal or recall that may become necessary, thus ultimately benefitting the food producer specifically, also helping to maintain consumer trust in the food safety control system.

5.5.3 Risk communication with the public
As noted in Section 5.2.3 of this chapter, the concept of risk communication is well defined, and the impact of different approaches has been studied in great depth in other publications and will not be repeated here. In terms of communication of risk to consumers, whether in the context of a specific food contamination incident, or more generally on the issue of food safety, government bodies are often just one provider of information in a very crowded media environment. Where, in the past, most government departments communicated with consumers and stakeholders via press releases, and official reports aimed at 'educating people' (Anderson, 2000), in recent years the principle ways people communicate and obtain information are increasingly via electronic and social media, in addition to the more traditional forms of

broadcast and print media. It is therefore important that during a food safety crisis, when an incident is evolving rapidly and information is constantly being updated, communication via all of these routes is used effectively, and information provided via the respective different media is coordinated and tailored to the expected audience. It is sometimes difficult for the message from an official agency to compete with other information in the public domain that may be based on opinion and perception; however, it is essential that provision of up-to-date, accurate, open and user-friendly information is maintained by food safety agencies, thereby building the trust and confidence of consumers, and ultimately allowing them to make fully informed choices.

For example, during the 2008 Irish dioxin incident, in addition to the more formal route of press conferences, the Food Safety Authority constantly provided up-to-date information on the incident via the website and the existing advice line, including information on the products being recalled and on the risk associated with the contamination of pork products, as well as more general information on disposal of product, etc., and other information in the form of Question and Answers (Q + As). The website quickly became a principle source of reliable information for consumers and food businesses alike, and saw the number of visits to the site increase by over 4000% on the weekend when the recall was announced, with some 19 000 unique visits to the site. In addition, the advice line, which typically deals with some 900 enquiries a month, received 2660 calls on the day after the announcement (FSAI, 2009). Staff from the Authority were also made available for interview on a range of broadcast media, both domestic and international, so that consistently accurate and clear information could be made available to consumers through as many channels as possible. It is also important, when communicating information on an incident that the link between the risk assessment and subsequent risk management actions is explained, so that consumers can understand why certain measures are being taken. For example, during the Irish dioxin crisis, the FSAI needed to explain to consumers that, whilst a total recall of all pork products was being undertaken and how this protected the consumer, it would not have been the case if product had been allowed to remain on the market.

5.6 Future trends

Numerous contamination incidents that have occurred in the food chain in recent years have highlighted the need for improvement and/or review of systems in place. In particular, traceability systems, prevention strategies and risk communication have been found lacking, and are discussed in more detail.

5.6.1 Improved traceability systems

The dioxin incident in Ireland in 2008 (Tlustos, 2009) highlighted the shortcomings of the traceability system in the pork-meat sector. Due to co-mingling of produce, and the inability to separate contaminated from uncontaminated meat, a total recall of pork and pork products, of which only about 10% were actually contaminated, had to be issued to ensure a fast response to protect consumer health. This resulted in enormous economic losses, which might have been prevented if a better traceability system were in place. That said, the traceability systems used by the Irish pork processors were compliant with all legal requirements, and were similar to systems used by industry throughout the world.

5.6.2 Improved prevention strategies

Experiences from recent incidents have highlighted the need for improved prevention strategies. In particular, a call for increased control of the feed chain has been issued numerous times. As Section 5.1.1 shows, almost all of the recent food incidents relating to dioxins were due to contamination of feed. The most recent incident in 2010 in Germany, prompted the German authorities to propose the introduction of more stringent controls in the feed chain on a European-wide basis. In particular, over recent years it has been suggested that a uniform registration and licensing protocol be established for feed producers, that feed production should be kept separate from other commercial activities, and possibly even that a positive list be established to guarantee transparency and uniformity in the feed production area.

5.6.3 Improved risk communication

Whilst the management of the 2008 dioxin incident in Ireland has generally received favourable reviews, the public was still left somewhat confused about the necessity to remove all pork when evidently there was no risk to consumers' health. The action was perceived at the time in some quarters to be 'a waste of money' and 'over the top' as a direct result of disseminated information not having been conveyed in a fully comprehensible manner. Despite the fact that the actions of the FSAI have been vindicated subsequently (Casey *et al.*, 2010; Jacob *et al.*, 2010), this again highlights the importance of ensuring that all information released during a crisis provides total transparency and clarity on the actions taken.

5.7 Sources of further information

Community Reference Laboratory for Dioxins and PCBs in Feed and Food (CRL) (2009), Analytical capacities of National Reference Laboratories (NRLs) and Official Laboratories (OFLs) in case of dioxin incidents in the feed and food chain and conclusions for management in crisis situations. Available at:

- http://ec.europa.eu/food/food/chemicalsafety/contaminants/emergency_analyses_en.pdf

- http://www.who.int/foodsafety/micro/riskanalysis/en/
- http://www.who.int/water_sanitation_health/dwq/gdwq3_4.pdf
- http://europa.eu/legislation_summaries/consumers/consumer_safety/l32042_en.htm
- http://www.who.int/foodsafety/fs_management/infosan/en/
- http://ec.europa.eu/food/food/rapidalert/rasff_portal_database_en.htm

5.8 References

ANDERSON, W. A. (2000) The future relationship between the media, the food industry and the consumer. In David I Thurnham and Terry A Roberts (Eds.), Health and the food-chain, British Medical Bulletin, **56** (1): 254–268.

BERNARD, A., HERMANS, C., BROECKAERT, F., de Poorter, G., de Cock, A. and Houins, G. Food contamination by PCBs and dioxins. *Nature*, 16 September 1999, **401**: 231–232.

BORRELLO, S., BRAMBILLA, G., CANDELA, L., DILETTI, G., GALLO, P., IACOVELLA, G., IOVANE, LIMONE, A., MIGLIORATI, G., PINTO, O., SARNELLI, P., SERPE, L., SCORTICHINI, G. and DI DOMENICO (2008) Management of the 2008 "Buffalo milk crisis" in the campania region under the perspective of consumer protection. *Organohalogen Compounds*, **70**: 891–893.

BUNDESMINISTERIUM FÜR ERNÄHRUNG, LANDWIRTSCHAFT UND VERBRAUCHERSCHUTZ (BMELV) (2011) Background information: Dioxins in feed fats Available at http://www.bmelv.de/SharedDocs/Standardartikel/EN/Food/DioxinSummaryReport.html [Accessed 25/5/2011]

CAC (CODEX ALIMENTARIUS COMMISSION) (2004) Principles and guidelines for the exchange of information in food safety emergency situations *CAC/GL 19–1995, Rev. 1–2004* http://www.fao.org/docrep/009/y6396e/Y6396E07.htm

CAC (CODEX ALIMENTARIUS COMMISSION) (2006) Principles for traceability/product tracing as a tool within a food inspection and certification system *CAC/GL 60–2006* http://www.codexalimentarius.net/download/standards/10603/CXG_060e.pdf

CARVALHAES, G., BROOKS, P. and KRAUSS, T. (1999) Lime as the source of PCDD/F contamination in citrus pulp pellets from Brazil. *Organohalogen Compounds*, **41**: 137–140.

CASEY, D. K., LAWLESS, J. S. and WALL, P. G. (2010) A tale of two crises: the Belgian and Irish dioxin contamination Incidents. *British Food Journal*, **112** (10) :1077–1091.

CFIA (CANADIAN FOOD INSPECTION AGENCY) (2007) Food Recalls: Make a Plan and action it (Manufacturers guide) http://www.inspection.gc.ca/english/fssa/recarapp/rap/mgguide.shtml

DILETTI, G., CECI, R., DE MASSIS, M.R., SCORTICHINI, G. and MIGLIORATI, G. (2005) A case of eggs contamination by PCDD/Fs in Italy: analytical levels and contamination source identification. *Organohalogen Compounds*, **67**: 1460–1461.

EC (EUROPEAN COMMISSION) (2010) Guidance on the implementation of articles 11,12,14,17,18,19 and 20 of Regulation (EC) No. 178/2002 on general food law. http://ec.europa.eu/food/food/foodlaw/guidance/docs/guidance_rev_8_en.pdf

EFSA (2010) Monitoring of dioxins in food and feed. *EFSA Journal 2010*, **8**(3): 1385 (36 pages).

EUROPEAN COMMISSION (2000) Communication from the Commission on the precautionary principle. COM (2000) 1 final. Brussels, 2 February 2000 http://eurlex.europa.eu/LexUriServ/LexUriServ.do?uri=COM:2000:0001:FIN:EN:PDF

EUROPEAN COMMISSION (2002) Regulation (EC) No. 178/2002 laying down the general principles and requirements of food law, establishing the European Food Safety Authority and laying down procedures in matters of food safety. *Official Journal of the European Communities*, **L31** 1 February 2002 http://eur-lex.europa.eu/LexUriServ/LexUriServ.do?uri=OJ:L:2002:031:0001:0024:EN:PDF

FSAI (2004) Code of Practice No 5 Food Incidents and Food Alerts. http://www.fsai.ie/WorkArea/DownloadAsset.aspx?id=8576

FSAI (2009) FSAI News, Volume 11, Issue 1 – Irish Dioxin Crisis Supplement http://www.fsai.ie/uploadedFiles/News_Centre/Newsletters/Newsletters_Listing/FINAL(8).pdf

FSAI (2010) Recall and Traceability. Guidance Note 10 (rev 2) http://www.fsai.ie/gn10productrecallandtraceabilityrevision2.html

FSANZ (FOOD STANDARDS AUSTRALIA NEW ZEALAND) (2008) Food industry recall protocol: a guide to conducting a food recall and writing a food recall plan. 6th Ed. http://www.foodstandards.gov.au/_srcfiles/Food%20Recall_WEB.pdf

HOOGENBOOM, L. A., VAN EIJKEREN, J. C., ZEILMAKER, M. J., MENGELERS, M. J., HERBES, R. and TRAAG, W. A. (2006) A novel source for dioxins present in waste fat from gelatine production. *Organohalogen Compounds*, **68**: 193–196.

HOOGENBOOM, R. and TRAAG, W. (2003) The German bakery waste incident. *Organohalogen Compounds*, **64**: 13–16.

HOOGENBOOM, R. L., ZEILMAKER, M. J., KAN, K. A., MENGELERS, M. B., VAN EIJKEREN, J. and TRAAG, W. A. (2005) Kaolinic clay derived dioxins in potato by-products. *Organohalogen Compounds*, **67**: 1470–1473.

JACOB, C. J., LOK, C., MORLEY, K. and POWELL, D. A. (2010) Government management of two media-facilitated crises involving dioxin contamination of food. *Public Understanding of Science*, **20**: 261–269.

KIM, M. K., CHOI, S. W., PARK, J. Y., KIM, D. G., BONG, Y. H., JANG, J. H., SONG, S. O., CHUNG, G. S. and GUERRERO, P. (2009) Dioxin contamination of Chilean pork from zinc oxide in feed. *Organohalogen Compounds*, **71**: 173–176.

LLERENA, J. J., ABAD, E., CAIXACH, J. and RIVERA, J. (2001) A new episode of PCDDs/PCDFs feed contamination in Europe: The choline chloride. *Organohalogen Compounds*, **51**: 283–286.

MALISCH, R. (2000) Increase of the PCDD/F-contamination of milk, butter and meat samples by use of contaminated citrus pulp. *Chemosphere*, **40**: 1041–1053.

SCHMID, P. and UTHRICH, W. (2000) Dioxin contamination of kaolins (bolus alba): Monitoring of PCDDS and PCDFS in kaolin, feed and foodstuffs of animal origin. *Organohalogen Compounds*, **47**: 173–176.

WAHL, K., KOTZ, A., MALISCH, R., HAEDRICH, J., ANASTASSIADES, M. and SIGALOVA, I. (2008) The guar gum case: contamination with PCP and dioxins and analytical problems. *Organohalogen Compounds*, **70**. http://www.dioxin20xx.org/pdfs/2008/08-301.pdf

WHO (2005) International Health Regulations (2nd Edition) http://www.who.int/ihr/9789241596664/en/index.html

WHO (2007) International Food Safety Authorities Network (INFOSAN) http://www.who.int/foodsafety/fs_management/infosan_1007_en.pdf

WHO (2009) Manual for the Public Health Management of Chemical Incidents. Printed by the WHO Document Production Services, Geneva, Switzerland. http://whqlibdoc.who.int/publications/2009/9789241598149_eng.pdf

WHO/FAO (1997) Risk Management and Food Safety. Report of a Joint FAO/WHO Consultation Rome, Italy, 27–31 January 1997. FAO Food and Nutrition Paper 65. ftp://ftp.fao.org/docrep/fao/w4982e/w4982e00.pdf

WHO/FAO (1998) The Application of Risk Communication to Food Standards and Safety Matters. Report of a Joint FAO/WHO Expert Consultation. Rome, Italy, 2–6 February 1998. FAO Food and Nutrition Paper 70. http://www.fao.org/docrep/005/x1271e/x1271e00.htm

6
Uptake of organic pollutants and potentially toxic elements (PTEs) by crops

C. Collins, University of Reading, UK

DOI: 10.1533/9780857098917.1.129

Abstract: The controlling factors on the uptake of contaminants by crops are outlined. These are presented using two categories of contaminants: potentially toxic elements (PTEs) and organic pollutants. For each of these categories, the processes described are soil-to-root transfer and the subsequent transfer to other plant parts. The modelling approaches used to predict these processes are also presented, again separating the PTEs and organic pollutants. Equations from the less complex models are provided. However, models cannot cope with the large variability frequently seen between data sets. It is therefore proposed that simple measurement techniques which can predict the potential for contamination using the available fraction are one potential way to address this. Finally, the chapter discusses how polluted soils can be treated to reduce the uptake of pollutants by plants.

Key words: potentially toxic elements (PTEs), organic pollutants, crop contamination, modelling, monitoring.

6.1 Introduction

As discussed in the previous chapters, there is a whole suite of contaminants that may potentially contaminate food. Guidance levels for these are produced by the European Economic Commission (EEC) regulations (e.g. EEC 1881/2006) and the European Food Safety Authority.[1-4] In some cases this is a consequence of pollutants entering the feedstock, as in the case of the Belgian chicken crisis in January 1999 where high PCBs were recovered in the meat,[5] and the high dioxin levels reported in German pigs in 2011.[6] Alternatively, contamination of the food chain may occur from the direct contamination of crop products as a consequence of plant uptake of pollutants from the soil and, to a lesser extent, the atmosphere, which is the subject of this chapter. Two scenarios dominate this pathway: (i) application of sewage sludge to agricultural land, and (ii) crop production on historically contaminated sites, including the development of brownfield sites for homes with gardens. The former has resulted in legislation, such as the EEC Council Directive No. 86/278/EEC.[7] This Directive prohibits the sludge from sewage treatment plants from being used in agriculture unless specified requirements are fulfilled, including the testing of the sludge and the soil. Toxic elements subject to the legislation are copper (Cu), zinc (Zn), cadmium (Cd), lead (Pb), mercury (Hg) and chromium (Cr). Point sources of pollution from previous activities have resulted in a number of crop monitoring studies, particularly in urban areas[8-10] and at former mine sites.[11-14] From such studies, potential health risks have been determined – these frequently involve using crop-uptake models within broader exposure-assessment models; these are described later in the chapter.

6.1.1 Separation of metals and metalloids from organic pollutants

Although the root is the most important site for pollutant entry from soil, this is not the edible fraction for many plants. It is only if a chemical accumulates in the edible parts that there is a potential risk to human health. Water and solutes are transported under hydrostatic pressure, with the water potential increasing in the following order – soil solution < root cells < xylem sap < leaf cells << atmosphere. This gradient is created during transpiration, where water is drawn in through the root system to replace evaporative loss from the leaves. Chemicals must cross the Casparian strip (a suberised cell layer) to enter the xylem. This configuration requires a suite of protein channels, carriers and pumps for ionised compounds, so it is a zone of selectivity. Most anions require active uptake to maintain plant requirements, whereas a number of cations e.g. Ca^{2+} require active exclusion to prevent accumulation when their concentration is high in the soil solution.[15] By contrast, the transfer of organic pollutants* occurs by passive diffusion from the soil water into the root with subsequent transfer to the shoot driven by the movement of water in the transpiration stream.[16] For the majority of the pollutants of concern, their transfer is governed by the hydrophobicity of the pollutants and the interaction at the point where they cross a cell membrane such as the Casparian strip. Differences in uptake of PTEs and organic pollutants also occur in the transfer from the soil to the pore water. The fraction of organic carbon is the prime influence for organic pollutants. For PTEs, organic carbon is again important but this time, because of its contribution to the Cation Exchange Capacity of the

---

* This excludes the ionisable pesticides, which are not considered further here, as they are not routinely detected in crops or in quality standards used for food products.

soil, which is influenced by the type of clay, the binding to these exchange sites is heavily influenced by the pH.

6.2 Uptake of organic pollutants by plants

The transport of organic pollutants into crops can be divided into a number of processes: the transfer from the soil to the root, root to shoot transfer, vapour uptake from the atmosphere and particle deposition. The dominant pathway from these is determined by the physico-chemical characteristic of the compound concerned. These processes and how they can be modelled are described below.

6.2.1 Soil–root transport

Experiments on the uptake of non-ionised chemicals from a hydroponic solution have shown that uptake consists of two stages.[17,18] Firstly, equilibration between the chemical concentration in the aqueous phase within the plant root and the external solution, and secondly, chemical sorption on to lipophilic root solids. These solids include lipids in membranes and cell walls.[18] Uptake from the external solution is often expressed as a root concentration factor (RCF), which is the ratio of chemical concentration in the root to the concentration found in an external solution.[17,19]

Lipophilic organic chemicals possess a greater tendency to partition into plant root lipids than hydrophilic chemicals. Polycyclic aromatic hydrocarbons (PAH), chlorobenzenes, polychlorinated biphenyl (PCB), and polychlorinated dibenzo-p-dioxins/dibenzofurans (PCDD/F) have all been found at elevated levels in plant roots.[20,21] Briggs et al.[17] found a linear relationship between the octanol–water partition coefficient (K_{OW}) of non-ionised chemicals and the observed RCF. Environmental scientists often use K_{OW} as a surrogate for chemical lipophilic tendency. For example, Wild et al. reported that non-ionised organic chemicals with log K_{OW} > 4 have a high potential for retention in plant roots;[22] while Cousins and Mackay[18] suggested that, for organic chemicals with log K_{OW} < 2 and a dimensionless Henry's Law constant (H) of less than 100, the plant aqueous phase was the most important storage compartment. Although chemical properties are important predictors of uptake potential, the physiology and composition of the plant root itself is also a significant influence, with differences in uptake potential explained by the varying types and amounts of lipids in root cells.[23] However, the available data across a range of plant species is limited.[24]

Chemical transfers from the soil into the root are primarily mediated by the uptake of soil pore water during plant transpiration. Therefore, the factors that influence the chemical concentration in pore water also exert control over the passive-uptake process. Organic chemicals can be sorbed or bound to several components in soil, including clays, iron oxides and organic matter, although it is the last that usually exerts the strongest influence on the pore water concentration. As the organic matter content of a soil increases (typically measured using the weight fraction of organic carbon present), the proportion of the chemical in the pore water decreases and consequently the total amount of chemical taken up by the root decreases. Several researchers, including Karickhoff[25] and Sablijc et al.,[26] have found empirical relationships between a chemical's lipophilicity and its affinity to sorb to soil organic matter. These QSAR relationships have been described as over-simplistic.[27]

6.2.2 Transfer from roots to other plant parts

Briggs et al.[17] derived an empirical relationship for predicting chemical concentration in the xylem transpiration stream from its concentration in external aqueous solution, based on experiments investigating the uptake of non-ionised chemicals into barley plants. They adopted the earlier definition of Shone and Wood[19] in defining the transpiration stream concentration factor (TSCF) as the ratio of chemical concentration in the transpiration stream to the concentration found in an external solution. The TSCF could be predicted from knowledge of a chemical's lipophilicity, with maximum uptake at a log K_{ow} of about 1.8; further studies by Hsu et al.[28] and Burken and Schnoor[29] reported similar relationships. More recent work with a significantly wider range and number of compounds, particularly where log K_{ow} < 1, has proposed a sigmoidal relationship with a maximum TSCF at log K_{ow} of 1.[30] These relationships between the TSCF and chemical lipophilicity are summarised in Fig. 6.1.

After transport in the stem, water and solutes diffuse laterally into adjacent tissues and may become concentrated in plant shoots, tubers and

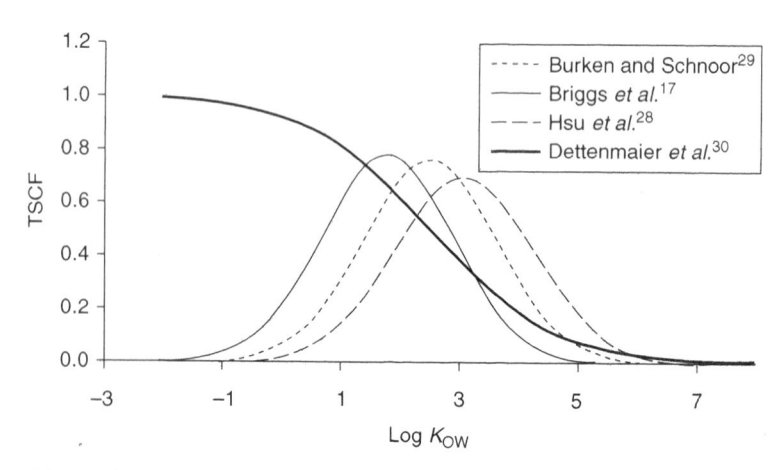

Fig. 6.1 Variation in predicted TSCF with chemical octanol–water partition coefficient.[17, 28–30]

fruits.[31] This is a two-step process, beginning with equilibrium partitioning between water in the vascular system and the aqueous solution in cell tissues, followed by sorption to cell walls. Briggs *et al.*[17] and Barak *et al.*[32] showed that partitioning to plant stems is linearly proportional to chemical lipophilicity for non-ionised organic chemicals. Therefore, the lipid composition of above-ground plant tissues is likely to be an important factor determining chemical retention and accumulation. Fractionation along the stem was observed in a laboratory,[33] but was not observed in the field, probably because of a significant aerial source.[34]

6.2.3 Vapour or gas uptake from ambient air

In addition to the root system, another potential pathway for plant uptake is the absorption of chemical vapour from ambient air by shoots. This pathway differs from root uptake because it is mediated via gaseous exchange rather than through aqueous solution. As a result, this pathway is likely to be important not just for highly volatile pollutants but for those with a strong preference to partition to air rather than water.[35] This has been shown to be the main uptake pathway into above-ground plant parts for a variety of organic chemicals, including PAHs,[36, 37] PCBs[38] and tetra- and hexa-chlorinated PCDD/Fs.[39]

Air-to-plant concentration factors for a variety of organic chemicals have been estimated by empirical studies.[40–42] Various studies have reported a good correlation between shoot uptake and chemical properties including the K_{OW}, Henry's Law constant, molecular weight and the octanol–air partition coefficient.[43–47] McLachlan[48] used knowledge of the partitioning behaviour of organic chemicals to build an interpretative framework to identify those compounds where plant contamination was most likely to be *dominated by* foliar uptake. Gaseous uptake is the primary pathway for chemicals with an octanol–air partition coefficient (log K_{OA}) less than 11, although between log K_{OA} of 8.5 to 11 this process will be kinetically limited as the plant will not come into contact with enough air for it to become saturated with the chemical and will not approach equilibrium.

6.2.4 Particulate deposition on plant surfaces

Organic chemicals bound to soil particles may be deposited on above-ground leaves and shoots as a result of wind re-suspension or rain splash. This is a significant pathway for dioxins, local soils accounted for 70–90% and 20–40% of total PCDD/F deposition in urban and rural areas, respectively.[49] The dry deposition of suspended particles, with subsequent permeation into the cuticle, represents the major pathway of contamination for PAH[36] and PCCD/Fs.[37] Wet deposition is the process of gravitational coagulation of solid particles with water droplets. It is thought to be the dominant deposition mechanism for organic chemicals with Henry's Law constant of less than 1×10^{-6},

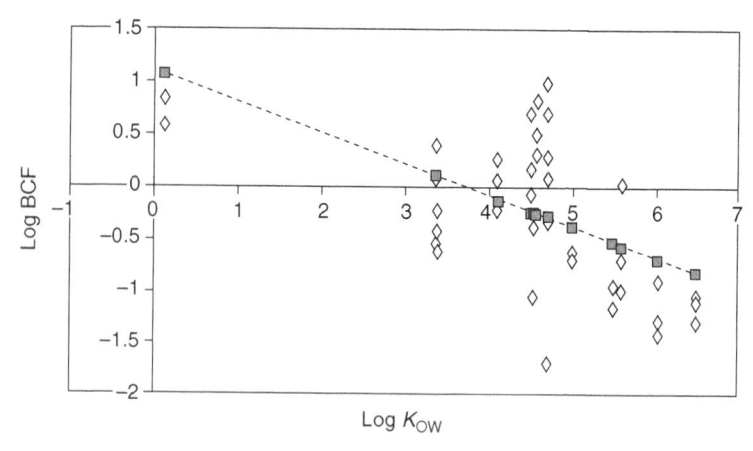

Fig. 6.2 Variation in bioconcentration factor (BCF) with log K_{OW} from a range of studies (taken from reference 50).

although the majority of particles will not be intercepted by vegetation and will return directly to the soil.[18]

6.2.5 Modelling

The models for predicting plant uptake of organic pollutants have been reviewed, but the variability in the data is such that it is difficult to propose a robust approach.[50] Note in Fig. 6.2. that there is a three order magnitude variation in the BCF at a log K_{OW} of 4.5.

The approach of the majority of workers is to use the physico-chemical characteristics of the pollutants to determine the accumulation of the pollutant. Two of the most basic models are those of Trapp and Matthies[51] and Travis and Arms[52] expanded below.

Travis and Arms:

$$\log C_v = 1.588 - 0.578 \log K_{OW}$$

where:

C_v = vegetation concentration
K_{OW} = octanol–water partition coefficient

Trapp and Matthies:

$$\frac{dC_{leaf}}{dt} = -\alpha_{soil} C_{leaf} + \beta_{soil}$$

where:

$$\alpha_{soil} = K_{plant\ growth}$$
$$\beta_{soil} = Q \cdot TSCF \cdot (C_{water}/V_{leaf})$$

where:

Q = transpiration stream flux (m³ s⁻¹) (1.15 × 10⁻⁸)
TSCF = the higher value from $0.784 \exp[-(\log K_{ow} - 1.78)^2/2.44]$
or $0.7 \exp[-(\log K_{ow} - 3.07)^2/2.78]$
C_{water} = concentration in soil water (mg m⁻³)
V_{leaf} = leaf volume (m³) (0.002)
$K_{plant\ growth}$ = pseudo-first order rate constant for dilution growth plant by growth (per day) (0.035)
C_{water} is calculated from the bulk soil concentration (C_{soil})
$$C_{water} = C_{soil}/K_d$$

where:

$$K_d = 10^{(0.72 \log K_{ow} + 0.49)} \cdot f_{OC}$$
f_{OC} = fraction of organic carbon in soil

6.3 Uptake of PTEs by plants

Accumulation of PTEs, like the organics, can be described by a number of discrete processes where the properties of the element and the soil. Simple models can also be constructed to describe the fate of PTEs in the soil plant system.

6.3.1 Soil–root transport

In general, finer textured soils contain more total PTEs than those which are coarse, as they have higher clay contents and hence more binding sites. The redox status of the soil is also of considerable importance when determining if a metal ion is likely to be in solution and therefore available for uptake i.e. bioavailable. For example, the chalcophile ('sulphur loving') elements (e.g., Cu, Pb, Cd) form insoluble sulphides in reducing conditions. The oxidation state will also affect the mobility of metals in solution e.g. Cr (II) and Cr (III) cations sorb to exchange sites on the soil (usually clays and humic materials), whereas the Cr (VI) anion is more mobile as it does not readily sorb to these fractions. The humic materials are often contained in the dissolved organic carbon fraction in the soil pore water; this increases the solubility of some metals by forming complexes with them e.g. Cu. In general, metal ions have a higher uptake at neutral pH and lower pH (≤ 7) than at higher pH (> 7) (Fig.

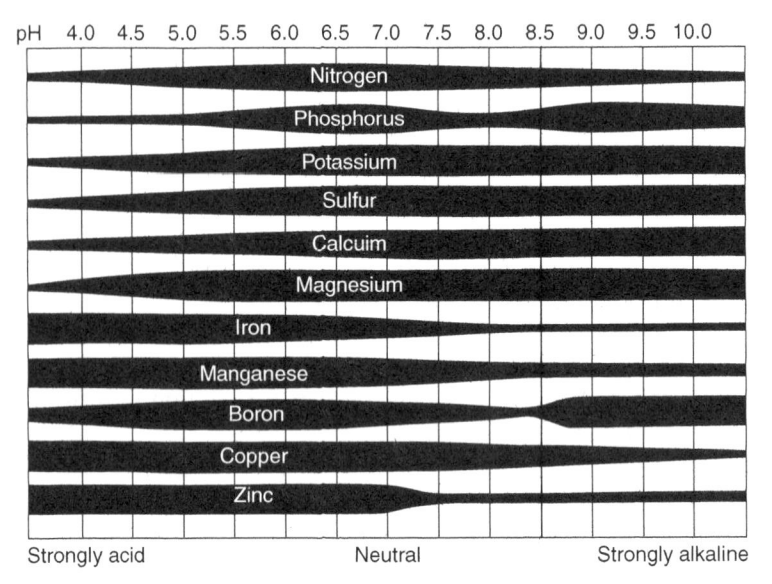

Fig. 6.3 Plant uptake of nutrients dependent on soil pH. The general relation of pH to the availability of plant nutrients in the soil: the thicker the bar, the more available the nutrient.

6.3). The transfer between the soil and the soil solution is governed by the partitioning coefficient (K_d), which is used in a number of models to predict the potential for crop uptake.

6.3.2 Accumulation in the shoot

Cadmium is readily taken up by plants.[53–57] There is a range in the transfer rates to plants (Table 6.1) and this can be seen at the intra- and inter-species level.[58] Hyperaccumulators of metals are a subject of much interest in the phytoremediation community, but they have relevance to crops too, with certain crop families e.g. brassicas, known to accumulate higher levels than others.[59] Rice has been described as a natural arsenic accumulator.[60] Under normal circumstances rice plants actively take up large amounts of silicon from the soil, unlike its close relatives wheat, barley and oats. They use silicon to strengthen the stem and the husk that protects the grain against pest attack. Arsenic and silicon are chemically very similar under the anaerobic soil conditions found in flooded rice paddies, and as a result arsenic literally fits into the silicon transporters and is transferred to the grain. The concentrations of PTEs in plant material decline from root to grain, as was observed for organics. In rice, Pb, As and Cd concentrations declined in the following order root >> shoot > husk > whole grain.[61] Specific transporter proteins (phyotochelatins) have been reported for a number of metals,[62, 63] and it maybe that variation in these within-crop varieties will reduce crop contamination.[64]

Table 6.1 Range of K_d and BCFs reported for PTEs

Element	K_d value[a] (m³ kg⁻¹)			BCF[b] (mg kg⁻¹ f.wt. plant/mg kg⁻¹ d.wt. soil)		
	Soil type			Percentile		
	Sand	Loam	Clay	5	50	95
As	0.005–50	0.005–50	0.005–50	0.0014	0.025	0.15
Cd	0.0037–1.5	0.0016–0.99	0.09–3.3	0.09	0.55	4.6
Hg	0.05–22	0.05–22	0.05–22	0.02	0.33	3
Pb	0.0027–27	0.99–270	0.0055–54	0.002	0.015	0.083

[a] From reference 65.
[b] From reference 66.

However, while these details are essential for developing our continued understanding of the uptake process, they are too detailed for the purposes of modelling. This is highlighted by the two data sets in Table 6.1. The differences in the reported values for the K_d and BCF for the same PTE in the same soil type can be several orders of magnitude. Clearly then modelling at the process level where we consider complexation is not required.

6.3.3 Particulate deposition on plant surfaces

Dry deposition from the atmosphere is a significant source of PTEs to soils, for Cu, Ni, Pb and Zn.[67] Point sources, such as smelters and roads, can significantly affect the local dust composition.[68,69] The direct deposition of dusts is a source of plant contamination.[69] The dust particles maybe so fine that they can penetrate the leaf, or be weathered and subsequently penetrate the leaf.[70] They will therefore not be removed by the washing involved in culinary practice. Following atmospheric deposition, As and Cd may be transported to the root and other underground edible organs, while Pb was not transported.[71]

6.3.4 Modelling

Following a wide-ranging review, the Dutch National Institute for Health and Environment (RIVM) developed a regression model based on field data:[72]

$$\log[\text{C-plant}] = \text{constant} + b*\log(Q) + c*\text{pH} + d*\log(\%OC) + e*\log(\%\text{clay})$$

where:

C-plant = metal concentration in edible plant part (mg kg⁻¹ d.m.)
Q = total metal concentration in the soil (mg kg⁻¹ d.m.)
pH = $-^{10}\log[H^+]$

%OC = organic carbon content of soil
%clay = clay content of soil
b,c,d,e = constants

The constants were crop and soil type specific, the correlation coefficients with the field data were not always high, and in fact the authors state '*A high correlation coefficient may be caused by data from too limited a number of contamination situations*'.

In England and Wales, modelling for contaminated land is undertaken using the CLEA model, and the authors of the documentation underpinning the model also noted a wide range of the results found in the literature.[73] They propose the following approach, which they acknowledge is cautious:

$$CR = \frac{\delta}{\theta_w + \rho_s K_d}$$

where:

CR = soil-to-root concentration factor (mg g^{-1} f.wt plant per mg g^{-1} d.wt. soil)
δ = soil–plant availability factor (dimensionless)
θ_w = water filled soil porosity (cm^3 cm^{-3})
ρ_s = is the dry soil bulk density (g cm^{-3})
K_d = soil–water partition coefficient (cm^3 g^{-1})

The soil–plant availability factor is set to 5 for most PTEs e.g. arsenic, cadmium, lead, mercury and nickel. It should be noted that a very high range in K_d values have been reported for these elements (Table 6.1).

6.4 *In situ* monitoring of plant available pollutants

One potential solution to the wide variation seen in the transfer of pollutants from soils to plants (Fig. 6.2, Table 6.1) is to develop simple chemical mimics of this process so that it can be measured easily, either *in situ* or by using a soil sample in the laboratory. These techniques fall into two broad categories – mild extractions that can predict the plant available fraction in the soil, such as EDTA for metals,[74] and butanol for organics.[75] These have not proved to be very robust and have been superseded by diffusion gradient films techniques (DGFT), and these exist from both metals[76] and organics.[77] These consist of an outer film or gel through which the contaminant diffuses, and an internal sink where it accumulates. They have provided good predictions for a suite of metals, but operate most effectively when diffusive uptake is the limiting factor on plant uptake.[76] DGFTs are at an earlier stage of development for organics.

6.4.1 Strategies for the prevention of crop contamination

In order to reduce the contamination of crops, a number of soil amendments have been proposed. These usually alter the chemical composition of the soil to reduce the level of contamination in the soil pore water, thus reducing plant uptake. Liming to increase the pH in order to reduce the concentration of cations in solution (Fig. 6.3) has been successful.[78] There is also considerable interest in using wastes from other streams in order to reduce crop contamination to achieve a win–win situation. Examples of these include: ochres from the mining industry for arsenic,[79] municipal composts which have been used for both metals and organics,[80,81] biochar[82,83] and animal manures.[84,85] These treatments are very dependent on the feedstock. For example, after the addition of compost, an increase in Cu mobility was observed as a consequence of changes in pH and dissolved carbon, but Cd mobility decreased,[86] while in another study Pb mobility decreased and As increased.[87]

6.5 Conclusions

The basic principles governing the uptake by crops of organic pollutants and PTEs are known. However, there are large differences in the concentrations in crops and these are dependent on a suite of factors such as soil type, crop type and the chemical form of the contaminant. These wide variations make it difficult to develop non-conservative models to protect human health because a high degree of conservatism is required to cover all situations. Rather than modelling, simple chemo-mimetic extractions may provide a better tool to predict crop contamination. When the potential for contamination is known, it may be possible to use secondary products as cheap and effective amendments to reduce pollutant uptake by crops.

6.6 References

1. EFSA, 2008. Scientific Opinion of the Panel on Contaminants in the Food Chain on a request from the European Commission on Polycyclic Aromatic Hydrocarbons in Food. *EFSA Journal*, **724**: p. 1–114.
2. EFSA, 2009. Scientific Opinion of the Panel on Contaminants in the Food Chain on a request from the European Commission on Cadmium in Food. *EFSA Journal*, **980**: p. 1–139.
3. EFSA, 2010. EFSA Panel on Contaminants in the Food Chain (CONTAM); Scientific Opinion on Lead in Food. *EFSA Journal*, **8**: p. 1570–1717.
4. EFSA, 2005. EFSA Panel on Contaminants in the Food Chain (CONTAM); Scientific Opinion on Arsenic in Food. *EFSA Journal*, **209**(7): p. 1351–1550.
5. COVACI, A., S. VOORSPOELS, P. SCHEPENS, P. JORENS, R. BLUST and H. NEELS, 2008. The Belgian PCB/dioxin crisis – 8 years later – An overview. *Environmental Toxicology and Pharmacology*, **25**(2): p. 164–170.
6. EUROPA, E.C., 2011. *Food Contamination – Dioxin in Gemany.* http://ec.europa.eu/food/food/chemicalsafety/contaminants/dioxin_germany_en.htm, Accessed 2 October 2012

7. EEC, (1986). Council Directive 86/278/EEC on the protection of the environment, and in particular of the soil, when sewage sludge is used in agriculture. Accessible from http://eur-lex.europa.eu/LexUriServ/LexUriServ.do?uri=CELEX:31986L0278:EN:NOT.
8. KULHANEKA, A., S. TRAPP, M. SISMILICH, J. JANKU and M. ZIMOVA, 2005. Crop-specific human exposure assessment for polycyclic aromatic hydrocarbons in Czech soils. *Science of the Total Environment*, **339**(1–3): p. 71–80.
9. NADAL, M., M. SCHUHMACHER and J.L. DOMINGO, 2004. Levels of PAHs in soil and vegetation samples from Tarragona County, Spain. *Environmental Pollution*, **132**(1): p. 1–11.
10. VOUTSA, D. and C. SAMARA, 1998. Dietary intake of trace elements and polycyclic aromatic hydrocarbons via vegetables grown in an industrial Greek area. *Science of the Total Environment*, **218**(2–3): p. 203–216.
11. BROWN, G., 1995. The effects of lead and zinc on the distribution of plant-species at former mining areas of western-europe. *Flora*, **190**(3): p. 243–249.
12. CHOPIN, E.I.B. and B.J. ALLOWAY, 2007. Distribution and mobility of trace elements in soils and vegetation around the mining and smelting areas of Tharsis, Riotinto and Huelva, Iberian Pyrite Belt, SW Spain. *Water Air and Soil Pollution*, **182**(1–4): p. 245–261.
13. COBB, G.P., K. SANDS, M. WATERS, B.G. WIXSON and E. DORWARD-KING, 2000. Accumulation of heavy metals by vegetables grown in mine wastes. *Environmental Toxicology and Chemistry*, **19**(3): p. 600–607.
14. KIEN, C.N., N.V. NOI, L.T. SON, H.M. NGOC, S. TANAKA, T. NISHINA and K. IWASAKI, 2010. Heavy metal contamination of agricultural soils around a chromite mine in Vietnam. *Soil Science and Plant Nutrition*, **56**(2): p. 344–356.
15. TAIZ, L. and E. ZEIGER, *Plant Physiology*. 5th ed. 2010: Sinauer Associates. 782.
16. COLLINS, C., M. FRYER and A. GROSSO, 2006. Plant uptake of non-ionic organic chemicals. *Environmental Science & Technology*, **40**(1): p. 45–52.
17. BRIGGS, G.G., R.H. BROMILOW, A.A. EVANS and M. WILLIAMS, 1983. Relationships between lipophilicity and the distribution of non-ionized chemicals in barley shoots following uptake by the roots. *Pesticide Science*, **14**(5): p. 492–500.
18. COUSINS, I.T. and D. MACKAY, 2001. Strategies for including vegetation compartments in multimedia models. *Chemosphere*, **44**(4): p. 643–654.
19. SHONE, M.G.T. and A.V. WOOD, 1977. Longitudinal movement and loss of nutrients, pesticides, and water in barley roots. *Journal of Experimental Botany*, **28**(105): p. 872–885.
20. DUARTE DAVIDSON, R. and K.C. JONES, 1996. Screening the environmental fate of organic contaminants in sewage sludge applied to agricultural soils. 2. The potential for transfers to plants and grazing animals. *Science of the Total Environment*, **185**(1–3): p. 59–70.
21. WILD, S.R., M.L. BERROW, S.P. MCGRATH and K.C. JONES, 1992. Polynuclear aromatic-hydrocarbons in crops from long-term field experiments amended with sewage-sludge. *Environmental Pollution*, **76**(1): p. 25–32.
22. WILD, S.R. and K.C. JONES, 1992. Polynuclear aromatic hydrocarbon uptake by carrots grown in sludge-amended soil. *Journal of Environmental Quality*, **21**(2): p. 217–225.
23. BRIMILOW, R.H. and K. CHAMBERLAIN, 1995. Principles governing uptake and transport of chemicals, in S. TRAPP and J.C. MCFARLANE (Eds.), *Plant Contamination: Modelling and Simulation of Organic Chemical Processes*, Lewis Publishers, London. p. 38–64.
24. TRAPP, S. and L. PUSSEMEIR, 1991. Model calculations and measurements of uptake and translocation of carbamates by bean plants. *Chemosphere*, **22**: p. 327–345.

25. KARICKHOFF, S.W., 1981. Semiempirical estimation of sorption of hydrophobic pollutants on natural sediments and soils. *Chemosphere*, **10**(8): p. 833–846.
26. SABLJIĆ, A., H. GÜSTEN, H. VERHAAR and J.L.M. HERMENS, 1995. QSAR modelling of soil sorption. Improvements and systematics of log Koc vs. log Kow correlations. *Chemosphere*, **31**(11–12): p. 4489–4514.
27. DOUCETTE, W.J., 2003. Quantitative structure-activity relationships for predicting soil-sediment sorption coefficients for organic chemicals. *Environmental Toxicology and Chemistry*, **22**(8): p. 1771–1788.
28. HSU, F.C., R.L. MARXMILLER and A.Y.S. YANG, 1990. Study of root uptake and xylem translocation of cinmethylin and related-compounds in detopped soybean roots using a pressure chamber technique. *Plant Physiology*, **93**(4): p. 1573–1578.
29. BURKEN, J.G. and J.L. SCHNOOR, 1998. Predictive relationships for uptake of organic contaminants by hybrid poplar trees. *Environmental Science & Technology*, **32**(21): p. 3379–3385.
30. DETTENMAIER, E.M., W.J. DOUCETTE and B. BUGBEE, 2009. Chemical hydrophobicity and uptake by plant roots. *Environmental Science & Technology*, **43**(2): p. 324–329.
31. MCFARLANE, J.C., 1995. Plant transport of organic chemicals, in S. TRAPP and J.C. MCFARLANE (Eds.), *Plant Contamination – Modelling and Simulation of Organic Chemical Processes*, Lewis Pub., Boca Raton, FL.
32. BARAK, E., B. JACOBY and A. DINOOR, 1983. Adsorption of systemic pesticides on ground stems and in the apoplastic pathway of stems, as related to lignification and lipophilicity of the pesticides. *Pesticide Biochemistry and Physiology*, **20**(2): p. 194–202.
33. MCCRADY, J.K., C. MCFARLANE and F.T. LINDSTROM, 1987. The transport and affinity of substituted benzenes in soybean stems. *Journal of Experimental Botany*, **38**(196): p. 1875–1890.
34. TAO, S., X.C. JIAO, S.H. CHEN, W.X. LIU, R.M. COVENEY, L.Z. ZHU and Y.M. LUO, 2006. Accumulation and distribution of polycyclic aromatic hydrocarbons in rice (Oryza sativa). *Environmental Pollution*, **140**(3): p. 406–415.
35. BELL, R.M., 1992. Higher Plant Accumulation of Organic Pollutants from Soils, US EPA. Report no. EPA/600/SR-92/138.
36. NAKAJIMA, D., Y. YOSHIDA, J. SUZUKI and S. SUZUKI, 1995. Seasonal-changes in the concentration of polycyclic aromatic- hydrocarbons in azalea leaves and relationship to atmospheric concentration. *Chemosphere*, **30**(3): p. 409–418.
37. WELSCH-PAUSCH, K., M.S. MCLACHLAN and G. UMLAUF, 1995. Determination of the principal pathways of polychlorinated dibenzo-p-dioxins and dibenzofurans to Lolium-multiflorum (Welsh Ray Grass). *Environmental Science & Technology*, **29**(4): p. 1090–1098.
38. BOHME, F., K. WELSCH-PAUSCH and M.S. MCLACHLAN, 1999. Uptake of airborne semi-volatile organic compounds in agricultural plants: Field measurements of interspecies variability. *Environmental Science & Technology*, **33**(11): p. 1805–1813.
39. MENESES, M., M. SCHUHMACHER and J.L. DOMINGO, 2002. A design of two simple models to predict PCDD/F concentrations in vegetation and soils. *Chemosphere*, **46**(9–10): p. 1393–1402.
40. BACCI, E., M.J. CEREJEIRA, C. GAGGI, G. CHEMELLO, D. CALAMARI and M. VIGHI, 1990. Bioconcentration of organic-chemical vapors in plant-leaves – the azalea model. *Chemosphere*, **21**(4–5): p. 525–535.
41. BACCI, E. and C. GAGGI, 1987. Chlorinated-hydrocarbon vapors and plant foliage – kinetics and applications. *Chemosphere*, **16**(10–12): p. 2515–2522.
42. BACCI, E. and C. GAGGI, 1986. Chlorinated pesticides and plant foliage – translocation experiments. *Bulletin of Environmental Contamination and Toxicology*, **37**(6): p. 850–857.

43. TOPP, E., I. SCHEUNERT, A. ATTAR and F. KORTE, 1986. Factors affecting the uptake of c-14-labeled organic-chemicals by plants from soil. *Ecotoxicology and Environmental Safety*, **11**(2): p. 219–228.
44. TOLLS, J. and M.S. MCLACHLAN, 1994. Partitioning of semivolatile organic-compounds between air and Lolium-multiflorum (Welsh Ray Grass). *Environmental Science & Technology*, **28**(1): p. 159–166.
45. PATERSON, S., D. MACKAY and A. GLADMAN, 1991. A fugacity model of chemical uptake by plants from soil and air. *Chemosphere*, **23**(4): p. 539–565.
46. RYAN, J.A., R.M. BELL, J.M. DAVIDSON and G.A. OCONNOR, 1988. Plant uptake of non-ionic organic-chemicals from soils. *Chemosphere*, **17**(12): p. 2299–2323.
47. KOMP, P. and M.S. MCLACHLAN, 1997. Octanol/air partitioning of polychlorinated biphenyls. *Environmental Toxicology and Chemistry*, **16**(12): p. 2433–2437.
48. MCLACHLAN, M.S., 1999. Framework for the interpretation of measurements of SOCs in plants. *Environmental Science & Technology*, **33**(11): p. 1799–1804.
49. KAO, A.S. and C. VENKATARAMAN, 1995. Estimating the contribution of reentrainment to the atmospheric deposition of dioxin. *Chemosphere*, **31**(10): p. 4317–4331.
50. COLLINS, C.D., I. MARTIN and M.E. FRYER, 2006. Evaluation of models for predicting plant uptake of chemicals from soil Science Report – SC050021/SR, Environment Agency: Bristol.
51. TRAPP, S. and M. MATTIES, 1995. Generic one compartment model for the uptake of organic chemicals by foliar vegetation. *Environmental Science & Technology*, **29**: p. 2333–8.
52. TRAVIS, C.C. and A.D. ARMS, 1988. Bioconcentration of organics in beef, milk, and vegetation. *Environmental Science & Technology*, **22**(3): p. 271–274.
53. DIJKSHOORN, W., L.W. VANBROEKHOVEN and J.E.M. LAMPE, 1979. Phytotoxicity of zinc, nickel, cadmium, lead, copper and chromium in 3 pasture plant-species supplied with graduated amounts from the soil. *Netherlands Journal of Agricultural Science*, **27**(3): p. 241–253.
54. DIJKSHOORN, W., J.E.M. LAMPE and L.W. VANBROEKHOVEN, 1981. Influence of soil-PH on heavy-metals in ryegrass from sludge-amended soil. *Plant and Soil*, **61**(1–2): p. 277–284.
55. CHANG, A.C., A.L. PAGE, K.W. FOSTER and T.E. JONES, 1982. A comparison of cadmium and zinc accumulation by 4 cultivars of barley grown in sludge-amended soils. *Journal of Environmental Quality*, **11**(3): p. 409–412.
56. PERALTA-VIDEA, J.R., J.L. GARDEA-TORRESDEY, E. GOMEZ, K.J. TIEMANN, J.G. PARSONS and G. CARRILLO, 2002. Effect of mixed cadmium, copper, nickel and zinc at different pHs upon alfalfa growth and heavy metal uptake. *Environmental Pollution*, **119**(3): p. 291–301.
57. BHOGAL, A., F.A. NICHOLSON, B.J. CHAMBERS and M.A. SHEPHERD, 2003. Effects of past sewage sludge additions on heavy metal availability in light textured soils: implications for crop yields and metal uptakes. *Environmental Pollution*, **121**(3): p. 413–423.
58. ENGQVIST, G. and A. MARTENSSON, 2005. Cadmium uptake in field pea cultivars grown under French and Swedish conditions. *Acta Agriculturae Scandinavica Section B-Soil and Plant Science*, **55**(1): p. 64–67.
59. BROADLEY, M.R., N.J. WILLEY, J.C. WILKINS, A.J.M. BAKER, A. MEAD and P.J. WHITE, 2001. Phylogenetic variation in heavy metal accumulation in angiosperms. *New Phytologist*, **152**(1): p. 9–27.
60. HEIKENS, A., G.M. PANAULLAH and A.A. MEHARG, 2007. Arsenic behaviour from groundwater and soil to crops: Impacts on agriculture and food safety, in G.W. Ware (Ed.), *Reviews of Environmental Contamination and Toxicology*, **189**: p. 43–87.

61. LEI, M., B.Q. TIE, P.N. WILLIAMS, Y.M. ZHENG and Y.Z. HUANG, 2011. Arsenic, cadmium, and lead pollution and uptake by rice (Oryza sativa L.) grown in greenhouse. *Journal of Soils and Sediments*, **11**(1): p. 115–123.
62. DE LA ROSA, G., J.R. PERALTA-VIDEA, M. MONTES, J.G. PARSONS, I. CANO-AGUILERA and J.L. GARDEA-TORRESDEY, 2004. Cadmium uptake and translocation in tumbleweed (Salsola kali), a potential Cd-hyperaccumulator desert plant species: ICP/OES and XAS studies. *Chemosphere*, **55**(9): p. 1159–1168.
63. FARINATI, S., G. DALCORSO, S. VAROTTO and A. FURINI, 2010. The Brassica juncea BjCdR15, an ortholog of Arabidopsis TGA3, is a regulator of cadmium uptake, transport and accumulation in shoots and confers cadmium tolerance in transgenic plants. *New Phytologist*, **185**(4): p. 964–978.
64. DHANKHER, O.P., Y.J. LI, B.P. ROSEN, J. SHI, D. SALT, J.F. SENECOFF, N.A. SASHTI and R.B. MEAGHER, 2002. Engineering tolerance and hyperaccumulation of arsenic in plants by combining arsenate reductase and gamma-glutamylcysteine synthetase expression. *Nature Biotechnology*, **20**(11): p. 1140–1145.
65. THORNE, M., R. WALKE and P. MAUL, 2005. *The PRISM foodchain modelling software: parameter values for the soil plant model*. Report of the UK Food Standards Agency QRS-1198A-3, version 1.1.
66. VERSLUIJS, C.W. and P.F. OTTE, 2001. *Accumulation of Metals in Plants as a Function of Soil Type* (in Dutch), RIVM: Bilthoven. Report No. 711701024.
67. AZIMI, S., P. CAMBIER, I. LECUYER and D. THEVENOT, 2004. Heavy metal determination in atmospheric deposition and other fluxes in northern France agrosystems. *Water Air and Soil Pollution*, **157**(1–4): p. 295–313.
68. WANG, Q.R., Y.S. CUI, X.M. LIU, Y.T. DONG and P. CHRISTIE, 2003. Soil contamination and plant uptake of heavy metals at polluted sites in China. *Journal of Environmental Science and Health Part a-Toxic/Hazardous Substances & Environmental Engineering*, **38**(5): p. 823–838.
69. DIETL, C., M. WABER, L. PEICHL and O. VIERLE, 1996. Monitoring of airborne metals in grass and depositions. *Chemosphere*, **33**(11): p. 2101–2111.
70. UZU, G., S. SOBANSKA, G. SARRET, M. MUNOZ and C. DUMAT, 2010. Foliar lead uptake by lettuce exposed to atmospheric fallouts. *Environmental Science & Technology*, **44**(3): p. 1036–1042.
71. DE TEMMERMAN, L., A. RUTTENS and N. WAEGENEERS, 2012. Impact of atmospheric deposition of As, Cd and Pb on their concentration in carrot and celeriac. *Environmental Pollution*, **166**: p. 187–195.
72. BRAND, E., P.F. OTTE and J.P.A. LIJZEN, 2007. CSOIL 2000 an exposure model for human risk assessment of soil contamination. *A model description*. Accessible at: http://rivm.openrepository.com/rivm/handle/10029/13385/.
73. JEFFRIES, J. and I. MARTIN, 2008. Updated technical background to the CLEA model. *Science Report*, **SC050021/SR3**.
74. SHAHID, M., E. PINELLI and C. DUMAT, 2012. Review of Pb availability and toxicity to plants in relation with metal speciation; role of synthetic and natural organic ligands. *Journal of Hazardous Materials*, **219**: p. 1–12.
75. GOMEZ-EYLES, J.L., M.T.O. JONKER, M.E. HODSON and C.D. COLLINS, 2012. Passive samplers provide a better prediction of PAH bioaccumulation in earthworms and plant roots than exhaustive, mild solvent, and cyclodextrin extractions. *Environmental Science & Technology*, **46**(2): p. 962–969.
76. DEGRYSE, F., E. SMOLDERS, H. ZHANG and W. DAVISON, 2009. Predicting availability of mineral elements to plants with the DGT technique: a review of experimental data and interpretation by modelling. *Environmental Chemistry*, **6**(3): p. 198–218.
77. TAO, Y.Q., S.Z. ZHANG, Z.J. WANG and P. CHRISTIE, 2008. Predicting bioavailability of PAHs in soils to wheat roots with triolein-embedded cellulose acetate mem-

branes and comparison with chemical extraction. *Journal of Agricultural and Food Chemistry*, **56**(22): p. 10817–10823.
78. HOODA, P.S. and B.J. ALLOWAY, 1996. The effect of liming on heavy metal concentrations in wheat, carrots and spinach grown on previously sludge-applied soils. *Journal of Agricultural Science*, **127**: p. 289–294.
79. DOI, M., G. WARREN and M.E. HODSON, 2005. A preliminary investigation into the use of ochre as a remedial amendment in arsenic-contaminated soils. *Applied Geochemistry*, **20**(12): p. 2207–2216.
80. FARRELL, M., W.T. PERKINS, P.J. HOBBS, G.W. GRIFFITH and D.L. JONES, 2010. Migration of heavy metals in soil as influenced by compost amendments. *Environmental Pollution*, **158**(1): p. 55–64.
81. ANTIZAR-LADISLAO, B., J. LOPEZ-REAL and A.J. BECK, 2006. Bioremediation of polycyclic aromatic hydrocarbons (PAH) in an aged coal-tar-contaminated soil using different in-vessel composting approaches. *Journal of Hazardous Materials*, **137**(3): p. 1583–1588.
82. GOMEZ-EYLES, J.L., T. SIZMUR, C.D. COLLINS and M.E. HODSON, 2011. Effects of biochar and the earthworm Eisenia fetida on the bioavailability of polycyclic aromatic hydrocarbons and potentially toxic elements. *Environmental Pollution*, **159**(2): p. 616–622.
83. XU, T., L.P. LOU, L. LUO, R.K. CAO, D.C. DUAN and Y.X. CHEN, 2012. Effect of bamboo biochar on pentachlorophenol leachability and bioavailability in agricultural soil. *Science of the Total Environment*, **414**: p. 727–731.
84. ATAGANA, H.I., 2004. Co-composting of PAH-contaminated soil with poultry manure. *Letters in Applied Microbiology*, **39**(2): p. 163–168.
85. BROWN, S., R. CHANEY, J. HALLFRISCH, J.A. RYAN and W.R. BERTI, 2004. *In situ* soil treatments to reduce the phyto- and bioavailability of lead, zinc, and cadmium. *Journal of Environmental Quality*, **33**(2): p. 522–531.
86. BEESLEY, L., E. MORENO-JIMENEZ and J.L. GOMEZ-EYLES, 2010. Effects of biochar and greenwaste compost amendments on mobility, bioavailability and toxicity of inorganic and organic contaminants in a multi-element polluted soil. *Environmental Pollution*, **158**(6): p. 2282–2287.
87. UDOVIC, M. and M.B. MCBRIDE, 2012. Influence of compost addition on lead and arsenic bioavailability in reclaimed orchard soil assessed using Porcellio scaber bioaccumulation test. *Journal of Hazardous Materials*, **205**: p. 144–149.

7
Transfer and uptake of dioxins and polychlorinated biphenyls (PCBs) into sheep: a case study

S. W. Panton, F. Smith and A. Fernandes, Food and Environment Research Agency (FERA), UK and C. Foxall, University of East Anglia, UK

DOI: 10.1533/9780857098917.1.145

Abstract: This chapter discusses the uptake of dioxins and polychlorinated biphenyls (PCBs) present at background levels into farm animals, specifically sheep, reared in accordance with common commercial practices. Residues measured in samples of grass, soil, feed and milk were used to determine input flux. Residues in samples of lamb meat, kidney and liver were used in conjunction with calculated input fluxes to determine biotransfer factors for selected compounds. Residues found were in agreement with those reported in previous studies. Biotransfer factors presented here can be used to assess the possibility of lamb liver exceeding EU maximum toxic equivalents (TEQ) values.

Key words: dioxins, polychlorinated biphenyls, sheep, biotransfer factors.

7.1 Introduction

One of the basic requirements for understanding and controlling human exposure to environmental contaminants is to understand their transfer and uptake into animals used for food. Most existing studies to date have involved the deliberate addition of contaminants to animal diets. Extrapolating the data from such studies to the real world situation, where contaminants are usually present at much lower levels (typically at parts per billion or less) and in various combinations, is fraught with numerous uncertainties, for example differences in bioavailabilty and metabolism.

Polychlorinated biphenyls (PCB) are a group of organochlorine compounds that were made by catalysed batch chlorination of biphenyl. Depending on

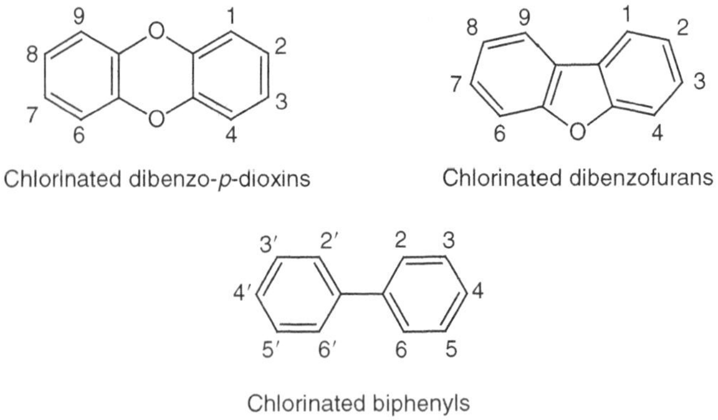

Fig. 7.1 Basic structures and labelling of dioxins and PCBs.

the number of chlorine atoms (1–10) and their position on the two rings, 209 different compounds or 'congeners' are possible. PCBs had widespread use in numerous industrial applications, generally in the form of complex technical mixtures. These applications included heat transfer fluids, hydraulic fluids, plasticisers, and capacitor and transformer dielectrics. Commercial production of PCBs was ended in the early 1980s due to toxicological and environmental concerns, and now are only synthesised for research purposes.

Polychlorinated dibenzo-*p*-dioxins (PCDD) and polychlorinated dibenzofurans (PCDF), commonly referred to as dioxins, are tricyclic aromatic compounds consisting of two benzene rings joined by one (furans) or two (*p*-dioxins) oxygen atoms. Depending on the number of chlorines (1–8) and their distribution between the benzene rings, there are 75 possible PCDD and 135 possible PCDF congeners. They are formed as trace by-products of waste incineration, chemical production, metal processing, paper bleaching, and in automotive exhausts. Various legislations have been enacted throughout the world over the last 30 years to minimise the release of dioxins into the environment because of their toxicity and stability. Figure 7.1 shows the structure of PCBs, PCDD/Fs and the numbering of the carbon atoms in the two rings.

PCBs and PCDD/Fs display a wide range of toxic action, which includes immunotoxicity and endocrine disruption. However, their most important and therefore most studied toxic effects are due to their binding with the aryl hydrocarbon receptor (AhR). This is often referred to as 'dioxin-like' toxicity, and is displayed by PCDD/Fs with 2,3,7,8 substitution and other similar compounds such as non- or mono-ortho-PCBs with 3,3',4,4' substitution. On this basis, PCBs can be divided into two groups: planar or dioxin-like PCBs (DL-PCBs) and non-dioxin-like PCBs (NDL-PCBs). Dioxin-like toxic effects include tumour development, birth defects, reproductive abnormalities, dermatological disorders and a plethora of other adverse effects (National Research Council of the National Academies, 2006).

As dioxins and DL-PCBs generally occur in the environment as complex mixtures with varying composition dependent on the source, an estimation of the toxic potential cannot be performed by simply summing the concentrations of the determined congeners. In order to compare the toxicity of a mixture of congeners, the concept of toxic equivalents (TEQ) based on different toxic equivalency factors (TEF) for the congeners has been introduced. It is assumed that the individual dioxins and dioxin-like compounds bind to AhR with different affinities. It is also assumed that the effects are additive. By definition, the most toxic congener, 2,3,7,8-tetrachlorodibenzo-p-dioxin (2,3,7,8-TCDD) is assigned a value of 1 and the TEFs for the other 16 toxic PCDD/Fs and 12 toxic DL-PCBs are between 0.00003 and 1. Thus, a TEF indicates an order of magnitude estimate of the toxicity of a dioxin-like compound relative to 2,3,7,8-TCDD. To calculate the total TEQ value of a sample, the concentration of each congener is multiplied with its TEF and they are then added together. TEF values for the 16 toxic PCDD/Fs and 12 toxic DL-PCBs are given in Table 7.1.

The physiochemical properties of PCBs and PCDD/Fs, such as high temperature stability, low water solubility, resistance to metabolism and high lipophilicity, have led to their environmental persistence and potential for bioaccumulation. As a result, PCBs and PCDD/Fs are widespread throughout the ecosystem. They occur in water, soil, sediments and biota throughout rural and urban environments (Zook and Rappe, 1994). The food chain is widely recognised as the primary route of human exposure to PCBs and PCDD/Fs (Startin, 1994; Travis and Nixon, 1996). Studies on the occurrence of these contaminants have shown that food of animal and marine origin poses the highest potential for human exposure (Liem and Theelen, 1997; Focant *et al.*, 2002; Fernandes *et al.*, 2010). Dairy products, fish, eggs and poultry are thought to provide as much as 90% of such exposure in adults (USEPA, 1994).

Detailed information on the history, properties, sources and toxicology of PCBs and PCDD/Fs are beyond the scope of this chapter. Readers requiring more background information should refer to the following: USEPA (1994), WHO (2002), European Commission (2002), National Research Council of the National Academies (2006), Fiedler (2008), EFSA (2010).

7.2 Uptake pathways and sources

The major pathway by which PCBs and PCDD/Fs can be transferred into grazing animals such as sheep is ingestion. Due to the properties of PCBs and PCDD/Fs coupled with data from studies on other farm animals, inhalation and dermal routes of uptake are of minimal significance. The importance of the ingestion pathway for grazing animals is highlighted by McLachlan (1997) in a mass balance study of PCBs on a lactating cow under normal animal husbandry conditions. The report showed that, for all the congeners

Table 7.1 WHO TEFs

Compound	TEF$_{WHO98}$	TEF$_{WHO05}$
PCDDs		
2,3,7,8-TCDD	1	1
1,2,3,7,8-PeCDD	1	1
1,2,3,4,7,8-HxCDD	0.1	0.1
1,2,3,6,7,8-HxCDD	0.1	0.1
1,2,3,7,8,9-HxCDD	0.1	0.1
1,2,3,4,6,7,8-HpCDD	0.01	0.01
OCDD	0.0001	0.0003
PCDFs		
2,3,7,8-TCDF	0.1	0.1
1,2,3,7,8-PeCDF	0.05	0.03
2,3,4,7,8-PeCDF	0.5	0.3
1,2,3,4,7,8-HxCDF	0.1	0.1
1,2,3,6,7,8-HxCDF	0.1	0.1
1,2,3,7,8,9-HxCDF	0.1	0.1
2,3,4,6,7,8-HxCDF	0.1	0.1
1,2,3,4,6,7,8-HpCDF	0.01	0.01
1,2,3,4,7,8,9-HpCDF	0.01	0.01
OCDF	0.0001	0.0003
Non-ortho-PCBs		
PCB-77	0.0001	0.0001
PCB-81	0.0001	0.0003
PCB-126	0.1	0.1
PCB-169	0.01	0.03
Mono-ortho-PCBs		
PCB-105	0.0001	0.00003
PCB-114	0.0005	0.00003
PCB-118	0.0001	0.00003
PCB-123	0.0001	0.00003
PCB-156	0.0005	0.00003
PCB-157	0.0005	0.00003
PCB-167	0.00001	0.00003
PCB-189	0.0001	0.00003

From *EFSA Journal* 8(3) 2010.

studied, over 99% of exposure occurred through feeding. Inhalation of air and ingestion of water each contributed less than 1%.

PCB and PCDD/F accumulation in vegetation can occur by the following pathways: (i) atmospheric deposition, both wet and dry, of contaminated particulates onto exposed plant surfaces, (ii) uptake of airborne vapours by aerial plant parts, and (iii) root uptake and translocation to upper plant parts. That the dominant contribution to PCDD/F and PCB levels in vegetation arises from atmospheric deposition is now well established (Furst *et al.*, 1993; McLachlan, 1997). The adsorption or absorption of PCDD/Fs and PCBs into plant roots and their subsequent translocation into other parts of the plant is minimal (Muller et al., 1993).

A high proportion of PCBs in ambient air occurs in the vapour phase (Manchester-Neesvlg and Andren, 1989), with relatively little of the volatilised material being absorbed by airborne particulate matter (Eisenreich et al., 1981). In contrast, PCDD/Fs are largely associated with particulate matter (Harrad and Jones, 1992).

In view of the dominance of the aerial deposition pathway, it might be expected that the levels of PCDD/Fs and PCBs in vegetation would reflect the seasonal trends in ambient air concentrations. However, the evidence for such variations in levels is not clear cut. Although seasonality in air–grass transfer of PCBs has been reported (Thomas et al., 1998a), a survey of pasture PCB concentrations has shown them to be remarkably consistent throughout England (Thomas et al., 1998b). A study of PCB concentrations in silage produced at different times of year (Thomas et al., 1999) also failed to reveal any significant seasonal variations.

Soil ingestion by sheep has been measured under a variety of conditions, mainly as an integral part of studies to examine the effect of soil ingestion on teeth wear in farm animals (Healy and Ludwig, 1965; Healy et al., 1967; Healy, 1968; Healy and Drew, 1970). These studies demonstrated that soil ingestion was inversely related to the availability of forage when pasture was the sole animal feed source and indicated that the lowest soil intake values (as little as 1–2% of dry matter intake) tended to occur in the spring when forage availability was at its peak.

When vegetation was sparse during the autumn or winter months, soil intake was estimated to be as great as 9% of the diet (Healy et al., 1967). Based on the work of Healey and his co-workers, Fries (1996) calculated that soil intake by sheep was approximately 4.5% of the dry matter intake when the animals were grazing for 365 days per year and pasture was the only feed source. Soil ingestion by a 50 kg animal was found to average 45 g/day when dry feed matter intake was 1 kg/day.

A study of the levels of soil ingestion by sheep in Ireland (McGrath et al., 1982) estimated that the animals ingested up to 400g soil/kg of body weight during four month rearing period. This equates to around 110 g/day for a 50 kg animal. The authors found that rainfall and stocking density were the principal factors affecting soil ingestion. The impact of vegetation availability was less evident.

Although the daily intake of soil by a sheep is likely to be less than 5% of dry feed intake, it is possible that owing to the known persistence of PCDD/Fs and PCBs in the upper soil layers, soil may still represent a significant source of such dietary contaminants. This source may be of enhanced significance if animal manure or sewage sludge has been applied to the grazing pastures on a regular basis.

Another path is the direct transfer of PCBs and PCDD/Fs from the ewe to her offspring, since they can be transferred through the placenta and milk to newborn and suckling animals. It has been shown that some PCBs, notably PCB-118 and PCB-153, can cross the placenta from ewes into the foetus (Berg *et al.*, 2010); in this study relatively high doses of PCB-118 and PCB-153 were administered. As mentioned earlier, extrapolating from such high levels to those usually experienced in real world situations should be done with caution. Prior to the present study, there appears to have been few detailed investigation into the levels of PCBs and PCDD/Fs in ewe's milk. A recent survey into PCDD/Fs and DL-PCBs in sheep and goats milk (Esposito *et al.* 2010) found mean total TEQ of 1.7 pg /g fat for samples compliant with the EU maximum residue limit (EU-MRL). PCDD/Fs contributed 50% to mean TEQ. Milk is the only source of sustenance for lambs during the first 5–6 weeks of their lives. Lambs continue to consume milk, albeit in reducing importance, throughout the weaning period, which is usually around three months for lowland-reared sheep. Because of sparser vegetation available on hill farms, highland sheep are not fully weaned until they are approximately five months old.

Compared to cattle, sheep usually receive relatively little supplementary feed. However, ewes are often provided with supplementary feed during winter conditions and in the period prior to lambing. Consumption levels of these supplements are usually minor compared with the intake of grass. It is therefore possible that the relative contribution of grass to the total dietary PCB and PCDD/F intake of sheep may exceed that observed in cattle.

Consideration of the potential pathways for the uptake of PCBs and PCDD/Fs in sheep suggests that the ingestion of feed, vegetation, soil and mother's milk are the main contributors to body burdens in these animals.

Considerable uncertainty exists over the relative contribution of the various dietary components such as vegetation, commercial feed, soil, etc., to the overall intake of PCBs and PCDD/Fs by livestock (Harrad and Smith, 1997). The same authors also point to the lack of data on the 'relative bioavailability' of these contaminants in different dietary components as an additional source of uncertainty.

A comparison of typical PCB and PCDD/F concentrations in sheep with those for other farm animals is restricted by the lack of data for sheep tissue. However, two studies have provided some useful insights into species variations. In the first of these (Liem and Theelen, 1997), chicken, horse, mutton, goat and pork samples (animal fat and liver) were collected from slaughterhouses across the Netherlands. The samples were then combined to provide nationally representative composites for each animal species. The results indicated that the total WHO TEQ concentrations (PCDD/Fs and DL-PCBs) in the mutton samples were similar to those in beef and chicken, but significantly higher than the levels in pork meat. In contrast, levels in liver samples from sheep were much higher than those in liver samples from cattle, chickens or pigs. The liver:meat concentration ratio was highest for sheep. Significant shifts in congener pattern between liver and meat samples from sheep were noted, with the former being characterised by the dominance of hexachlorinated furans.

In a similar survey but in Belgium (Focant et al., 2002), in which a range of samples from slaughterhouses or retail outlets was analysed, total WHO TEQ concentrations (PCDD/Fs and DL-PCBs) in mutton were also found to be similar to those in beef, and much higher than the concentrations in pork. In view of the small number of mutton samples analysed, the authors advised that the results must be treated with some caution. The results suggested that the uptake of PCBs and PCDD/Fs into meat and other tissue varies substantially from species to species.

7.3 Transfer of PCBs and polychlorinated dibenzo-*p*-dioxins and dibenzofurans (PCDD/Fs) into animal tissues

A variety of approaches have been adopted to quantitatively characterise the transfer of contaminants from the diets of livestock into milk and other animal products. The terms used most frequently are bioconcentration factors (BCF), carry-over rates (COR) and biotransfer factors (BTF) (McLachlan, 1997; Fries et al., 1999; Thomas et al., 1999). These three approaches are defined by the following equations:

$$\text{Bioconcentration factor (BCF)} = \frac{C_{product}}{C_{feed}} \qquad [7.1]$$

$$\text{Carry-over rate (COR)} = \frac{F_{milk}}{F_{feed}} \qquad [7.2]$$

$$\text{Biotransfer factor (BTF)} = \frac{C_{product}}{F_{total}} \qquad [7.3]$$

where C = concentration in ng per g fat; F = input flux in ng per day and $F_{total} = F_{feed} + F_{veg} + F_{milk} + F_{soil}$

Each calculation method has its advantages and disadvantages and the procedure adopted usually depends upon the circumstances of the study and the type of data available. For instance, it is appropriate to use BCFs where contaminant inputs are known to originate from a single feed source. Where the PCB and PCDD/Fs found in a particular foodstuff originate from multiple sources, BCFs are less useful.

CORs are more useful than BCFs when multiple sources are involved as they take account of the total intake flux from all dietary sources. Such an approach is particularly suited to studies involving dairy cattle, where it is relatively easy to measure the PCB and PCDD/F fluxes in milk on a regular basis.

152 Persistent organic pollutants and toxic metals in foods

The BTF approach is rather more versatile than that of BCFs or CORs, in that the total intake of PCBs and PCDD/Fs from all dietary sources can be calculated as a daily input flux and related to the contaminant concentration in the various animal products, such as milk or meat, under study. Using this approach, it is possible to estimate the relative importance of sources such as milk, feed, soil, grass, etc., at different stages of an animal rearing programme.

A number of studies have attempted to measure the transfer of PCBs and PCDD/Fs into livestock animals. Most of these investigations have focused on dairy cattle. Only two previous research projects have attempted detailed investigations of transfers to farm animals reared using typical husbandry methods and exposed to background levels of PCBs and PCDD/Fs (McLachlan, 1997; Thomas *et al.*, 1999). Both of these studies were concerned with uptake into cattle. A study by Costera *et al.* (2006) investigated CORs from contaminated hay to goat milk. In general, CORs were higher for DL-PCBs compared to PCDD/Fs, with ranges of 5% to 90% and 1% to 40%, respectively. There was a general trend of decreasing COR with increasing chlorination for PCDDs. For PCDFs, the trend was with a few exceptions, PeCDF >= HxCDF > TCDF > HpCDF > OCDF. Such a general relationship between chlorination level and CORs for PCBs could not be established.

In the remainder of this chapter we will look at data produced as part of a study into the transfer and uptake of PCDD/Fs and PCBs in sheep (Foxall *et al.*, 2004). Lowland and highland sheep were reared to market readiness using typical commercial husbandry practices. During the respective rearing periods, closely matched sets of feed, soil, grass and animal products were collected and analysed.

7.4 Experimental rearing, sampling and analysis

A brief outline of the experimental details are given below. Further details are recorded in Foxall *et al.* (2004).

7.4.1 Selection of farms

Lowland sheep were reared at Easton College, near Norwich, United Kingdom as part of the normal commercial activities of the farm. Highland sheep were reared at Carlcroft, a College farm in Northumberland, United Kingdom.

For both lowland and highland programmes, the pastures used during the study received no inputs of fertilisers or other agrochemicals during the five years prior to the start of the rearing programme. The pastures were in rural areas with no history of industrial use. Highland ewes were lambed outdoors and therefore had no contact with bedding or buildings. The lowland ewes were kept indoors in a lambing shed of conventional design for a period of two months before the commencement of lambing.

No instances of ill health amongst the ewes or lambs were recorded in either programme.

7.4.2 Samples collected

In order to facilitate comparisons between the two sheep rearing programmes, the slaughter and sampling schedules adopted were designed to be as similar as possible.

Four pregnant lowland ewes and four highland ewes were selected after ultrasonic scanning confirmed that they were carrying twins. Dietary intakes were measured over a period of 28 days prior to slaughter of each animal.

Two of the lambs were slaughtered aged 48 days with two more at 87 days (lowland) or 91 days (highland). The two remaining pairs of lambs were slaughtered at around 130 days (lowland) or 150 days (highland). The mothers of each pair of lambs were slaughtered on the same day as their offspring. Samples of meat from the same cuts were taken from all the animals and additionally samples of liver and kidney were taken from the oldest lambs. Samples of milk from each ewe were collected at intervals throughout the period that the lambs were suckling. Samples of grass, soil and feeds were also taken. A total of 16 and 18 samples were analysed for the lowland and highland sheep programmes, respectively.

Animal identities together with the codes of samples analysed from the ewes and their respective lambs are detailed in Tables 7.2 and 7.3.

7.4.3 Sampling procedures

All samples were weighed, logged then stored at −15°C prior to analysis. Details of the sampling procedures for soil, grass and feed were as described by Foxall *et al.* (2004).

Table 7.2 Lowland sheep programme – identities, sample codes of ewes and their respective offspring

Ewe					Offspring					Age[1] (days)
ID code	Sample codes				ID Code	Sex	Sample codes			
	Milk	Meat	Liver	Kidney			Meat	Liver	Kidney	
Y179		EBM			Y179/03	F	EBP			48
					Y179/04	F	EBQ			48
Y174	EBE				Y174/05	M	ECC			87
					Y174/06	F	ECD			87
Y168					Y168/32	F			EDA	134
					Y168/33	F				134
Y195	EBG	ECL			Y195/22	F	ECQ	ECU		127
	EBX				Y195/23	M	ECR			127

[1] Age of lamb on day of slaughter.

154 Persistent organic pollutants and toxic metals in foods

Table 7.3 Highland sheep programme – identities, sample codes of ewes and their respective offspring

Ewe					Offspring					
ID code	Sample codes				ID Code	Sex	Sample codes			Age[1] (days)
	Milk	Meat	Liver	Kidney			Meat	Liver	Kidney	
A		GBM			A1	M	GBP			48
					A2	M	GBQ			48
B	GBE				B1	F	GCD			91
					B2	M	GCC			91
C					C1[2]	F	–	–	–	–
					C2	M	GCQ	GCU	GCY	153
D	GBG	GCL			D1	F	GCR			152
	GBX				D2	F				

[1] Age of lamb on day of slaughter.
[2] Lamb C1 died when 3 days old.

Milk
A minimum of 150 mL was collected wherever possible. The milk collected was expressed directly into pre-cleaned 1 L Duran bottles.

Meat
Samples of muscle tissue were taken from the hind legs of sheep carcass. Each sample was wrapped in a double layer of hexane-washed aluminium foil and placed in a sealed polypropylene box.

Liver and kidney
Liver and kidney samples were removed from each sheep carcass. Each sample was weighed, logged and stored under the same conditions as the meat samples.

7.4.4 Determination of intake

The direct measurement of the quantities of vegetation consumed by sheep is clearly impractical without seriously interfering with normal animal husbandry practices. The vegetation intakes were therefore calculated on the basis of the measured body weights of the lambs and the estimated body weights of the ewes using the relationships developed by Abbot (2003) as shown in Table 7.4.

Sheep regulate their food intake on a dry matter basis, and consequently vegetation intakes are typically expressed in the agricultural literature in dry matter terms. However, the vegetation samples collected during this work were analysed on a 'whole weight' basis, and so the calculated dry matter intakes needed to be converted to 'whole weight' equivalents. Although the

Table 7.4 Daily vegetation consumption estimates

Animal	Lowland sheep	Highland sheep
Lactating ewe (single lamb)	139g dm/kg bw$^{0.75}$	118gdm/kg bw$^{0.75}$
Lactating ewe (twins)	152g dm/kg bw$^{0.75}$	129gdm/kg bw$^{0.75}$
Dry ewe	100g dm/kg bw$^{0.75}$	85gdm/kg bw$^{0.75}$
Lamb	100g dm/kg bw$^{0.75}$	85 gdm/kg bw$^{0.75}$

dm = dry matter.
bw = body weight.
bw$^{0.75}$ known as metabolic weight.

dry matter content of sheep pasture would be expected to range from 14–30% depending on conditions, stage of growth, etc., the mean value for pasture grazed according to typical animal husbandry practices is generally taken to be 20% (Nix, 1999) and this figure was used here.

To avoid stress to the animals, the ewes were not weighed. However, the weight of a lactating lowland ewe is typically 70 kg and that of highland ewes 60 kg, so these figures were used in the calculation of the vegetation intakes.

As is normal practice in the industry, lowland ewes were provided with supplementary feed and hay in the period prior to lambing.

For highland sheep, the ewes were given access to supplementary feed blocks in the form of licks during the winter months. Hay was also provided to the ewes when falls of snow prevented normal grazing. This is normal practice in the husbandry of such animals.

It is possible, owing to the sparser vegetation generally available, that highland sheep could ingest more soil during grazing than the lowland sheep. However, much of the rearing period was characterised by ample rainfall and, as a consequence, the availability of vegetation was in fact little different from that for lowland sheep. It was assumed that the soil intakes for both highland and lowland sheep were similar, i.e. 2% of dry matter intake.

Ewe's milk production depends upon a number of factors, including weight of ewe, age, number of offspring, level of nutrition and stage of lactation (Cardellino and Benson, 2002). Milk production typically starts at around 0.6 L/day following the birth of the lambs and rises sharply to around 1 L/day within the following few days. It peaks at around 2.5 L/day three to four weeks after the lambs are born (Abbott, 2003; Morton, 2003). For healthy ewes that are feeding twins, this profile is unlikely to vary significantly from animal to animal and was therefore used to calculate the milk intakes of the individual lambs reared during the study.

Milk production in a highland ewe generally follows a different profile from that of its lowland counterpart. For the highland sheep breed used in this study, a lactating ewe typically weighing 60 kg produces around 0.5 L/day at the birth of the lambs and increases to around 1 L/day within the first seven days. Production subsequently peaks at around 1.5 L/day three to four weeks after the lambs are born (Abbott, 2003).

It was assumed that all the milk produced by the ewe was actually consumed by their lambs (rather than being reabsorbed) and that it was equally divided between the two lambs. As lambs became older and vegetation was an increasingly significant part of their diet, daily milk intakes diminished.

7.4.5 Materials

Reference standards of PCDDs, PCDFs, PCBs and $^{13}C_{12}$ materials for use as internal standards were sourced from Cambridge Isotope Laboratories (Andover, MA, USA) as solutions in n-nonane, iso-octane or hexane with a specified 10% tolerance on concentration.

Dichloromethane, methanol, toluene, hexane and n-nonane were purchased as doubly glass distilled (Rathburn, Scotland) and checked for contamination before use. Alumina (Sigma Chemical Company, USA) was activated by heating overnight in a muffle furnace at 450°C. All other chemicals employed were Analytical Reagent grade materials. Reagents were checked in-house prior to use.

All equipment was scrupulously cleaned and thoroughly rinsed with dichloromethane prior to use. Care was taken to avoid airborne contamination of containers by keeping vials capped even when empty, and covering flasks and concentration tubes with cleaned aluminium foil.

7.4.6 Extraction, purification and analysis

Samples were analysed for the seventeen 2,3,7,8-chlorine substituted PCDD/Fs and a range of PCB congeners (IUPAC Nos. 18, 28, 31, 52, 77, 81, 99, 101, 105, 114, 118, 123, 126, 128, 138, 153, 156, 157, 167, 169, 180, 189). The extraction and purification of samples were carried out as previously reported (Fernandes et al., 2004). In brief, portions of the samples were fortified with internal standard solutions prior to solvent extraction. The crude extract obtained was quantitatively transferred and chromatographed using an apparatus containing modified silica followed by activated carbon on glass fibres where the analytes were fractionated on the basis of their planarity.

The two fractions containing (i) ortho-PCBs, and (ii) non-ortho-PCBs and PCDD/Fs were purified using acid hydrolysis and activated alumina. Where required, fractions were further purified using alumina. The extracts were concentrated and the appropriate sensitivity standards were added to each fraction prior to instrumental analysis.

Instrumental analysis using GC-HRMS for PCDDs, PCDFs and non-ortho-PCBs, and GC-LRMS for ortho-PCBs was carried out as described by Fernandes et al. (2004).

7.4.7 Quality control

All analytical data were assessed for compliance with published acceptance criteria (Ambidge *et al.*, 1990).

The quality control samples were reference materials prepared by the BCR (Maier *et al.*, 1995): 'RM534, PCDDs and PCDFs in spiked milk powder-higher level' and 'CRM 350, PCBs in mackerel oil' (Griepink *et al.*, 1988). Results obtained for certified congeners in the samples were compared with the concentrations assigned by BCR studies and all fell within acceptable ranges. The reporting limit quoted for all analytes was the limit of determination that prevailed in that instance.

7.4.8 TEQs

WHO TEQ values for PCDD/Fs and PCBs reported in this chapter are based on the 1998 TEF (TEF_{WHO98}) values (van den Berg *et al.*, 1998) for the relevant analytes, to allow comparison with data from previous studies and at the time of studies were the values used under current European legislation. New TEF_{WHO05} values have since come into effect in 2012. Both TEF_{WHO98} and TEF_{WHO05} are detailed in Table 7.1.

TEQ_{WHO98} values were calculated as follows: the rounded result for each congener was multiplied by the appropriate TEQ_{WHO98} and summed to give total TEQ_{WHO98}. In cases where congeners were reported as a '<' value, lower bound TEQ_{WHO98} were calculated by treating the result as if zero and upper bound TEQ_{WHO98} by treating the result as if present at the reporting limit. TEQs were rounded to two decimal places.

7.5 Results and discussion for PCDD/Fs, dioxin-like PCBs (DL-PCBs) and ICES6 PCBs

The next sections discuss results for PCDD/Fs, DL-PCBs and ICES6 PCBs (PCB-28, 52, 101, 138, 153 and 180). Full details of all the congeners analysed in the study are given in Foxall *et al.* (2004). All results are given on a fat weight basis except those for soil and grass which are on a whole weight basis.

7.5.1 Lowland sheep – PCB and PCDD/F concentrations

Tables 7.5 and 7.6 present the PCDD/F and DL-PCB levels respectively in the feed, milk, meat, liver, kidney, soil and grass samples analysed. The corresponding total WHO TEQ values are summarised in Tables 7.7, while values for ICES6 PCBs are presented in Table 7.10.

In general, the most prominent PCDD/F congeners were, in order of concentration, OCDD, 2,3,4,7,8-PeCDF, 1,2,3,4,6,7,8-HpCDD, 1,2,3,6,7,8-HxCDD

Table 7.5 Lowland sheep PCDD/F residue data

Sample type Code	Feed EAF	Milk EBE	Milk EBG	Milk EBX	Meat EBM	Meat ECL	Meat EBP	Meat EBQ	Meat ECC	Meat ECD	Meat ECQ	Meat ECR	Liver ECU	Kidney EDA	Soil EAK	Grass EAJ
Fat %	5.70	8.12	8.54	4.98	20.56	6.79	3.56	5.77	1.92	9.04	7.52	6.03	4.88	27.94	#	#
2,3,7,8-TCDD	<0.45	0.23	0.28	0.17	0.3	0.12	0.41	0.3	<0.23	0.17	0.2	0.2	0.4	0.15	<0.18	<0.01
1,2,3,7,8-PeCDD	<0.63	0.67	0.67	0.46	0.57	0.31	0.92	0.76	0.44	0.41	0.52	0.41	1.96	0.51	0.28	<0.01
1,2,3,4,7,8-HxCDD	0.41	0.28	0.29	0.24	0.33	0.2	0.42	0.44	0.3	0.17	0.31	0.23	1.81	0.24	0.34	<0.01
1,2,3,6,7,8-HxCDD	0.56	0.54	0.54	0.61	0.57	0.35	0.89	0.84	0.45	0.32	0.48	0.49	2.29	0.56	0.57	<0.01
1,2,3,7,8,9-HxCDD	0.6	0.21	0.19	0.2	0.18	0.11	0.18	0.16	<0.19	0.09	0.11	0.12	0.83	0.1	0.56	<0.01
1,2,3,4,6,7,8-HpCDD	9.57	1.02	0.9	0.52	0.89	0.56	0.97	1.17	1.06	0.39	0.62	0.43	8	0.41	8.34	0.07
OCDD	245.1	2.34	2	1.13	0.99	0.65	1.43	2.29	2.71	0.66	1.12	0.82	18.33	0.39	39.89	0.21
2,3,7,8-TCDF	0.66	0.05	0.07	0.09	0.08	0.08	<0.08	<0.08	0.19	0.05	<0.08	<0.08	0.11	<0.04	0.63	<0.01
1,2,3,7,8-PeCDF	<0.26	0.04	0.05	0.05	<0.04	<0.05	<0.04	<0.04	<0.13	<0.04	0.04	0.05	<0.07	0.04	0.55	<0.01
2,3,4,7,8-PeCDF	<0.49	1.17	0.89	0.72	0.83	0.42	0.96	0.88	0.46	0.48	0.61	0.46	18.06	0.69	0.85	<0.01
1,2,3,4,7,8-HxCDF	<0.38	0.5	0.37	0.34	0.32	0.21	0.52	0.45	0.35	0.21	0.28	0.29	6.74	0.29	0.78	<0.01
1,2,3,6,7,8-HxCDF	<0.49	0.34	0.24	0.22	0.28	0.16	0.32	0.26	<0.26	0.13	0.21	0.18	6.01	0.19	0.55	<0.01
1,2,3,7,8,9-HxCDF	<0.33	<0.02	0.02	0.04	<0.01	<0.06	<0.01	<0.01	<0.17	<0.05	<0.01	<0.01	<0.09	<0.01	0.05	<0.01
2,3,4,6,7,8-HxCDF	<0.40	0.34	0.22	0.25	0.23	0.13	0.27	0.23	0.22	0.11	0.16	0.19	6.81	0.15	0.74	<0.01
1,2,3,4,6,7,8-HpCDF	1.62	0.38	0.29	0.22	0.23	0.14	0.41	0.64	0.39	0.2	0.31	0.21	9.87	0.17	5.19	0.03
1,2,3,4,7,8,9-HpCDF	<0.31	0.05	<0.03	0.06	<0.05	<0.05	0.09	<0.05	<0.16	<0.04	0.08	0.08	0.71	<0.02	0.29	<0.01
OCDF	1.94	0.08	0.06	0.15	0.06	<0.08	0.27	0.09	0.35	0.08	0.23	0.18	1.12	0.13	4.2	0.02
WHO TEQ (ng/kg fat) lower	0.36	1.73	1.6	1.2	1.5	0.77	2.08	1.76	0.84	0.93	1.19	1	14.04	1.17	1.3	<0.01
WHO TEQ (ng/kg fat) upper	1.86	1.73	1.6	1.2	1.5	0.78	2.1	1.77	1.14	0.94	1.2	1.01	14.05	1.17	1.48	0.03

Note: # all results in ng/kg fat weight except soil and grass which are on a whole weight basis.

Table 7.6 DL-PCB residue data for lowland sheep samples

Sample type Code	Feed EAF	Milk EBE	Milk EBG	Milk EBX	Meat EBM	Meat ECL	Meat EBP	Meat EBQ	Meat ECC	Meat ECD	Meat ECQ	Meat ECR	Liver ECU	Kidney EDA	Soil EAK	Grass EAJ
Fat %	5.70	8.12	8.54	4.98	20.56	6.79	3.56	5.77	1.92	9.04	7.52	6.03	4.88	27.94	#	#
Non-ortho-PCBs (ng/kg)																
PCB-77	22.02	2.39	2.39	2.82	1.74	6.02	1.46	1.43	14.73	2.38	1.6	2.44	3.51	1.14	2.5	0.33
PCB-81	1.46	0.42	0.45	0.66	<1.01	0.67	<1.01	<1.02	1	0.28	<1.02	<1.04	1.4	0.34	<0.2	0.02
PCB-126	<2.49	4.72	5.25	5.06	5.54	5.52	6.13	4.36	4.45	4.01	5.39	5.84	51.59	5.4	0.97	0.05
PCB-169	<0.42	2.04	2.21	2.23	2.8	2.5	4.22	4.65	2.81	2.42	3.8	4.32	4.13	4.39	0.34	<0.01
WHO TEQ (ng/kg fat) lower	<0.01	0.49	0.55	0.53	0.58	0.58	0.66	0.48	0.47	0.43	0.58	0.63	5.2	0.58	0.1	0.01
WHO TEQ (ng/kg fat) upper	0.26	0.49	0.55	0.53	0.58	0.58	0.66	0.48	0.47	0.43	0.58	0.63	5.2	0.58	0.1	0.01
Mono-ortho-PCBs (µg/kg)																
PCB-105	<0.59	0.16	0.2	0.14	0.18	0.14	0.29	0.26	<0.30	0.14	0.21	0.23	<0.15	0.15	<0.04	<0.01
PCB-114	<0.59	<0.10	<0.09	<0.10	<0.10	<0.10	<0.10	<0.10	<0.30	<0.09	<0.10	<0.10	<0.15	<0.04	<0.04	<0.01
PCB-118	<0.59	0.41	0.53	0.35	0.52	0.34	0.82	0.72	0.4	0.36	0.49	0.61	0.27	0.35	0.05	<0.01
PCB-123	<0.59	<0.10	<0.09	<0.10	<0.10	<0.10	<0.10	<0.10	<0.30	<0.09	<0.10	<0.10	<0.15	<0.04	<0.04	<0.01
PCB-156	<0.59	<0.10	0.11	0.11	0.14	0.17	0.21	0.25	<0.30	0.15	0.27	0.24	0.15	0.21	<0.04	<0.01
PCB-157	<0.59	<0.10	<0.09	<0.10	<0.10	0.11	<0.10	<0.10	0.41	<0.09	<0.10	<0.10	<0.15	0.04	<0.04	<0.01
PCB-167	<0.59	<0.10	<0.09	<0.10	<0.10	<0.10	<0.10	<0.10	<0.30	<0.09	<0.10	<0.10	<0.15	<0.04	<0.04	<0.01
PCB-189	<0.59	<0.10	<0.09	<0.10	<0.10	<0.10	<0.10	<0.10	<0.30	<0.09	<0.10	<0.10	<0.15	<0.04	<0.04	<0.01
WHO TEQ (ng/kg fat) lower	<0.01	0.06	0.13	0.1	0.14	0.19	0.22	0.22	0.25	0.13	0.21	0.2	0.1	0.18	0.01	<0.01
WHO TEQ (ng/kg fat) upper	1.13	0.23	0.24	0.23	0.26	0.26	0.34	0.34	0.64	0.23	0.33	0.33	0.3	0.2	0.08	0.02

Note: # all results given on a fat weight basis except soil and grass which are on a whole weight basis.

Table 7.7 Summary of TEQ values for lowland sheep

Sample type Code	Feed EAF	Milk EBE	Milk EBG	Milk EBX	Meat EBM	Meat ECL	Meat EBP	Meat EBQ	Meat ECC	Meat ECD	Meat ECQ	Meat ECR	Liver ECU	Kidney EDA	Soil EAK	Grass EAJ
Fat %	5.70	8.12	8.54	4.98	20.56	6.79	3.56	5.77	1.92	9.04	7.52	6.03	4.88	27.94	#	#
Dioxin	1.86	1.73	1.6	1.2	1.5	0.78	2.1	1.77	1.14	0.94	1.2	1.01	14.05	1.17	1.48	0.03
Non-ortho-PCB	0.26	0.49	0.55	0.53	0.58	0.58	0.66	0.48	0.47	0.43	0.58	0.63	5.2	0.58	0.1	0.01
Ortho-PCB	1.13	0.23	0.24	0.23	0.26	0.26	0.34	0.34	0.64	0.23	0.33	0.33	0.3	0.2	0.08	0.02
Sum of WHO TEQs (upper)	**3.25**	**2.45**	**2.39**	**1.96**	**2.34**	**1.62**	**3.10**	**2.59**	**2.25**	**1.60**	**2.11**	**1.97**	**19.55**	**1.95**	**1.66**	**0.06**

Note: # soil and grass results are in ng/g whole weight. All other results are in ng/g fat weight.

Table 7.8 Summary of TEQ values for highland sheep

Sample type Code	Feed GAF	Feed GAN	Feed GAO	Milk GBE	Milk GBG	Milk GBX	Meat GBM	Meat GCL	Meat GBP	Meat GBQ	Meat GCC	Meat GCD	Meat GCR	Meat GCQ	Liver GCU	Kidney GCY	Soil GAH	Grass GAI
Fat %	2.16	1.73	11.28	15.57	8.69	4.44	2.19	1.76	2.47	3.54	5.74	5.35	3.54	8.95	7.57	6.51	#	#
Dioxin	1.44	1.74	0.71	2.1	3.02	0.29	0.42	0.18	1.87	1.9	0.49	0.62	0.39	0.43	6.4	0.67	2.83	0.04
Non-ortho-PCB	0.16	0.22	0.14	0.74	1.34	0.2	0.2	0.14	0.91	1.00	0.21	0.31	0.26	0.2	1.61	0.15	0.17	0.01
Ortho-PCB	2.67	3.25	0.48	0.28	0.26	0.17	0.21	0.19	0.36	0.31	0.19	0.25	0.22	0.2	0.2	0.33	0.06	0.04
Sum of WHO TEQs (upper)	**4.27**	**5.21**	**1.33**	**3.12**	**4.62**	**0.66**	**0.83**	**0.51**	**3.14**	**3.21**	**0.89**	**1.18**	**0.87**	**0.83**	**8.21**	**1.15**	**3.06**	**0.09**

Note: # soil and grass results are in ng/g whole weight. All other results are in ng/g fat weight.

Table 7.9 ICES6 PCBs residue data for highland sheep

Sample type Code	Feed GAF	Feed GAN	Feed GAO	Milk GBE	Milk GBG	Milk GBX	Meat GBM	Meat GCL	Meat GBP	Meat GBQ	Meat GCC	Meat GCD	Meat GCR	Meat GCQ	Liver GCU	Kidney GCY	Soil GAH	Grass GAI
Fat %	2.16	1.73	11.28	15.57	8.69	4.44	2.19	1.76	2.47	3.54	5.74	5.35	3.54	8.95	7.57	6.51	#	#
PCB-28	<1.4	<1.7	3.66	0.14	0.14	<0.09	<0.11	<0.1	0.24	0.25	<0.08	<0.11	<0.1	<0.1	<0.1	<0.17	<0.03	<0.02
PCB-52	<1.4	<1.7	1.35	<0.1	<0.1	<0.09	<0.11	<0.1	<0.11	<0.11	<0.08	<0.11	<0.1	<0.1	<0.1	<0.17	<0.03	<0.02
PCB-101	<1.4	<1.7	<0.25	<0.1	<0.1	<0.09	<0.11	<0.1	<0.11	<0.11	<0.08	<0.11	<0.1	<0.1	<0.1	<0.17	0.03	<0.02
PCB-138	<1.4	<1.7	<0.25	1.15	1.07	0.17	0.32	0.15	1.81	1.32	0.71	0.7	0.6	0.57	0.75	0.57	0.15	<0.02
PCB-153	<1.4	<1.7	<0.25	2.85	2.82	0.35	0.99	0.36	4.83	3.62	2.16	2.28	2.25	1.57	2.33	1.53	0.13	<0.02
PCB-180	<1.4	<1.7	<0.25	0.97	0.92	0.12	0.78	0.1	1.63	1.23	0.77	0.82	0.85	0.51	0.45	0.51	0.07	<0.02

Note: # all results are µg/kg fat weight basis except for soils and grass which are on a whole weight basis.

162 Persistent organic pollutants and toxic metals in foods

and 1,2,3,7,8-PeCDD in most of the analysed samples. Considering DL-PCBs, the prominent congeners were PCB-118 and 156, while PCB-153, 138, and 180 dominated the ICES6.

OCDD was the prominent congener detected in feed and soil samples accounting for around 90% and 65% respectively of total PCDD/F residues detected. Milk TEQs were highest during the early stages of lactation and then show a decline over time. Similar trends have been reported in dairy cattle (Thomas *et al.*, 1999). The behaviour of individual congeners over time varied. Levels of most PCDD/F and DL-PCB congeners in milk were constant, while OCDD, 1,2,3,7,8-PeCDD, 1,2,3,4,6,7,8-HpCDD, 2,3,4,7,8-PeCDF, PCB-105 and PCB-118 showed a significant decline.

The decline in the TEQ in ewe's meat over time was due to partial excretion of residues in its milk. A low but significant proportion would also have been excreted via faeces, which was not analysed during this study. The observed declined in residues in ewe's meat was not uniform across all the congeners; average decline for PCDD/Fs was 50%, for DL- PCBs there was a small or no decline in residues while there were significant increases in the levels of PCB-138, PCB-153 and PCB-180 over the same time.

Lamb meat total (sum of PCDD/F and PCB) TEQs declined with age before showing a slight increase at the end of study period. Overall, the meat TEQs for the lowland sheep samples appeared to be slightly higher than those reported for pig and chicken meat (Fernandes *et al.*, 2011).

TEQ levels in kidneys were similar to those found in meat. The TEQ for sheep's liver was much higher than that for meat from the same animal (compare ECQ and ECU). The liver sample was also characterised by a much lower PCB contribution to the total TEQ and a predominance of many of the hepta and octa substituted PCDD/Fs. Total PCDD/F and PCB TEQ exceeded the current EU limits for liver.

The subtle differences in the actual dioxin and PCB profiles between sample types and over time (e.g. compare EBG, EBM, ECL) may be due to their differential metabolism. However, not enough samples were analysed to draw any unambiguous conclusions.

7.5.2 Highland sheep – PCB and PCDD/F concentrations

The concentrations for ICES6 PCBs are given in Table 7.9. Tables 7.11 and 7.12 list the PCDD/F and DL-PCB levels and TEQs, respectively in the feed, milk, meat, liver, kidney, soil and grass samples analysed. The corresponding total WHO TEQ values are shown in Table 7.8.

Considering TEQs, trends for highland sheep were generally similar to those for lowland sheep. These include trends for the milk TEQs to decrease as lactation proceeds, for ewe meat TEQs to decline with age, and for the TEQ of the liver sample to be much higher than in meat from the same animal. However, there were notable differences. TEQ levels in highland lamb meat decreased throughout the study. The TEQs for meat from

Table 7.10 ICES6 PCBs residue data for sheep

Sample type Code	Feed EAF	Milk EBE	Milk EBG	Milk EBX	Meat EBM	Meat ECL	Meat EBP	Meat EBQ	Meat ECC	Meat ECD	Meat ECQ	Meat ECR	Liver ECU	Kidney EDA	Soil EAK	Grass EAJ
Fat %	5.70	8.12	8.54	4.98	20.56	6.79	3.56	5.77	1.92	9.04	7.52	6.03	4.88	27.94	#	#
PCB-28	<0.59	0.14	0.12	0.15	0.13	<0.10	0.12	<0.10	<0.30	<0.09	0.1	0.16	<0.15	0.05	0.04	<0.01
PCB-52	<0.59	<0.10	<0.09	<0.10	<0.10	<0.10	<0.10	<0.10	<0.30	<0.09	<0.10	<0.10	<0.15	<0.04	<0.04	<0.01
PCB-101	<0.59	<0.10	0.1	<0.10	<0.10	0.2	<0.10	<0.10	0.5	<0.09	<0.10	<0.10	<0.15	<0.04	0.06	<0.01
PCB-138	<0.59	0.61	0.79	0.67	1	1.88	1.6	1.6	4.84	0.81	1.26	1.4	1.26	1.06	0.19	<0.01
PCB-153	<0.59	1.17	1.39	1.46	1.99	3.79	2.93	3.07	8.27	1.72	2.93	3.35	4.89	2.85	0.14	<0.01
PCB-180	<0.59	0.38	0.44	0.58	0.94	4.29	1.03	1.07	16.31	0.65	0.95	1.12	1.01	1.13	0.05	<0.01

Note: # All results are in µg/kg fat weight except soil and grass which are on a whole weight basis.

Table 7.11 PCDD/F residue data for highland sheep samples

Sample type Code	Feed GAF	Feed GAN	Feed GAO	Milk GBE	Milk GBG	Milk GBX	Meat GBM	Meat GCL	Meat GBP	Meat GBQ	Meat GCC	Meat GCD	Meat GCR	Meat GCQ	Liver GCU	Kidney GCY	Soil GAH	Grass GAI
Fat %	2.16	1.73	11.28	15.57	8.69	4.44	2.19	1.76	2.47	3.54	5.74	5.35	3.54	8.95	7.57	6.51	#	#
2,3,7,8-TCDD	<0.44	<0.53	<0.08	0.2	0.23	<0.04	<0.02	<0.04	0.18	0.22	0.04	<0.02	<0.04	0.05	0.1	<0.15	<0.12	<0.01
1,2,3,7,8-PeCDD	<0.33	<0.4	0.21	0.74	1.1	0.08	0.13	<0.05	0.77	0.77	0.16	0.29	0.14	0.18	0.71	0.26	0.41	<0.01
1,2,3,4,7,8-HxCDD	<0.39	<0.47	0.14	0.6	0.71	0.03	0.14	0.03	0.43	0.39	0.13	0.16	0.12	0.12	0.89	0.14	0.62	<0.01
1,2,3,6,7,8-HxCDD	<0.39	<0.47	0.21	1.23	2.25	0.07	0.31	0.06	1.11	0.92	0.29	0.33	0.38	0.24	1.29	0.33	1.07	<0.01
1,2,3,7,8,9-HxCDD	<0.33	<0.4	0.67	0.28	0.89	0.03	0.04	0.02	0.19	0.19	0.05	0.06	<0.01	0.04	0.38	0.06	0.9	0.01
1,2,3,4,6,7,8-HpCDD	<2.66	<3.19	2.94	1.15	2.56	0.14	0.89	0.19	1.03	0.89	0.66	0.43	0.28	0.2	4.34	0.37	17.65	0.11
OCDD	10.12	16.62	70.8	1.11	0.83	<0.75	1.27	0.85	1.33	1.08	2.31	0.57	0.66	0.42	3.26	2.06	78.11	0.42
2,3,7,8-TCDF	<1.77	<2.13	0.4	0.09	<0.08	<0.07	<0.12	<0.09	<0.13	<0.13	<0.09	<0.12	<0.09	<0.09	<0.09	<0.09	1.22	0.02
1,2,3,7,8-PeCDF	<0.28	<0.33	0.28	0.04	<0.03	<0.03	<0.02	<0.06	<0.02	<0.02	<0.02	<0.02	<0.06	<0.06	<0.06	<0.07	1.27	<0.01
2,3,4,7,8-PeCDF	<0.28	<0.33	0.31	1.47	1.75	0.22	0.28	<0.09	1.13	1.17	0.32	0.36	0.2	0.23	8.3	0.28	1.75	0.01
1,2,3,4,7,8-HxCDF	<0.44	<0.53	0.2	0.84	1.42	0.12	0.25	0.06	0.75	0.67	0.3	0.27	0.24	0.18	4.66	0.23	2.65	0.01
1,2,3,6,7,8-HxCDF	<0.33	<0.4	0.18	0.56	1.03	0.09	0.16	0.04	0.39	0.43	0.13	0.15	0.08	0.08	3.07	0.11	1.67	0.01
1,2,3,7,8,9-HxCDF	<0.44	<0.53	<0.08	<0.01	<0.01	<0.01	<0.02	<0.01	<0.02	<0.02	<0.02	<0.02	<0.01	<0.01	<0.01	<0.04	0.06	<0.01
2,3,4,6,7,8-HxCDF	<0.55	<0.66	0.17	0.42	1.35	0.17	0.14	<0.04	0.31	0.32	0.13	0.11	0.08	0.07	3.05	0.09	1.85	0.02
1,2,3,4,6,7,8-HpCDF	<2.11	3.36	0.63	0.38	0.96	0.11	0.28	0.1	0.6	0.51	0.38	0.23	0.14	0.13	4.44	0.31	15.43	0.07
1,2,3,4,7,8,9-HpCDF	<0.28	<0.33	<0.05	<0.02	0.06	<0.01	<0.02	<0.02	<0.02	<0.02	<0.01	<0.02	<0.02	<0.02	0.35	<0.05	1.24	<0.01
OCDF	1.75	1.63	0.38	0.07	0.09	<0.06	0.12	0.18	0.15	0.14	0.19	<0.07	<0.07	<0.07	0.55	0.16	27.6	0.06
WHO TEQ (ng/kg fat) lower	<0.01	0.04	0.62	2.09	3.01	0.24	0.39	0.02	1.85	1.88	0.47	0.58	0.33	0.42	6.39	0.5	2.71	0.01
WHO TEQ (ng/kg fat) upper	1.44	1.74	0.71	2.1	3.02	0.29	0.42	0.18	1.87	1.9	0.49	0.62	0.39	0.43	6.4	0.67	2.83	0.04

Note: # all results in ng/kg fat weight except soil and grass which are on a whole weight basis.

market ready animals are slightly lower in highland compared to lowland sheep, and this contrast was also reflected in more highland sheep samples having larger numbers of congeners below the limit of detection. In fact, even though TEQs were initially (day 48) higher in highland lamb, they declined at a much faster rate. Another difference was the decline in highland milk TEQ – over 80% compared with 40% for lowland sheep over a comparable time.

For soil, the TEQ was double that seen in the lowland soil sample. If it is assumed that the soil consumption rate for both lowland and highland sheep were the same, we would have expected to see higher TEQ values for highland sheep. Since this was not the case, we can assume that PCDD/Fs in soil may have low bioavailability in sheep.

7.5.3 Comparisons with previous data

Meat
Relatively few previous studies (Liem and Theelen, 1997; Focant *et al.*, 2002; Rose *et al.*, 2005) have reported on PCB and PCDD/F concentrations in sheep, but they all lie within a relatively narrow concentration range of between 3.1 and 5.0 ng TEQ/kg fat with a mean value of 4.0 ng TEQ/kg fat. The contribution of dioxin-like PCBs (DL-PCBs) to the total TEQ ranges from 37–53% (mean 47%). The corresponding values from the present survey (range 44–57%; mean 50%) thus closely agree with the published data.

The values from the present study (range 0.67–2.0 ng TEQ/kg fat; mean 1.4 ng TEQ/kg fat) are somewhat lower than previously reported.

Liver and kidney
TEQ levels observed in both lowland and highland sheep livers exceeded the EU-MRL. The livers were obtained from 130–150 day old lambs, the typical age at which 'spring lamb' for consumption in the UK is slaughtered. Liver data from UK surveys conducted by MAFF (1997), FSA (2000) and Rose *et al.* (2005) were in agreement with values observed in this study.

Bruns-Weller *et al.* (2010) analysed 77 lamb liver samples and found that 71 samples exceeded the combined PCDD/F and PCB EU-MRL of 12 ng TEQ/g fat. Although TEQs from our present study were much lower than that from Bruns-Weller *et al.* (2010), similar congener profiles were observed. The same authors did not find any correlation between TEQ levels and age.

It is well established that PCDD/Fs concentrate in the liver, due to the lipophilic nature of these contaminants and the role the liver plays in xenobiotic and lipid metabolism. Binding to proteins within liver cells as well as to AhR may lead to the preferential partition of PCDD/Fs into the liver.

The results from a control kidney sample collected during the study of foot and mouth disease pyres (FSA, 2002) had a TEQ of 1.3 ng TEQ/kg fat. Total TEQ concentrations (range 1.2–2.0; mean 1.6 ng TEQ/kg fat) from the

Table 7.12 DL-PCB residue data for highland sheep samples

Sample type Code	Feed GAF	Feed GAN	Feed GAO	Milk GBE	Milk GBG	Milk GBX	Meat GBM	Meat GCL	Meat GBP	Meat GBQ	Meat GCC	Meat GCD	Meat GCR	Meat GCQ	Liver GCU	Kidney GCY	Soil GAH	Grass GAI
Fat %	2.16	1.73	11.28	15.57	8.69	4.44	2.19	1.76	2.47	3.54	5.74	5.35	3.54	8.95	7.57	6.51	#	#
Non-ortho-PCBs (ng/kg)																		
PCB-77	38.84	31.45	161.48	1.91	2.03	1.68	1.37	1.34	4.07	3.11	1.25	1.12	1.12	1.02	<1.04	2.77	3.25	0.48
PCB-81	2.5	2.09	8.68	0.51	0.71	0.31	0.2	0.16	0.37	0.34	0.12	0.11	0.16	0.12	0.28	0.25	0.24	0.03
PCB-126	1.5	2.13	1.25	6.83	12.76	1.92	1.84	1.39	8.19	9.27	1.66	2.65	1.96	1.65	15.66	1.28	1.62	0.08
PCB-169	<0.22	<0.27	0.11	5.41	5.91	0.45	2.01	0.37	9.16	6.89	3.99	4.73	6.48	3.1	4.5	2.53	1.14	0.02
WHO TEQ (ng/kg fat) lower	0.15	0.22	0.14	0.74	1.34	0.2	0.2	0.14	0.91	1.00	0.21	0.31	0.26	0.2	1.61	0.15	0.17	0.01
WHO TEQ (ng/kg fat) upper	0.16	0.22	0.14	0.74	1.34	0.2	0.2	0.14	0.91	1.00	0.21	0.31	0.26	0.2	1.61	0.15	0.17	0.01
Mono-ortho-PCBs (µg/kg)																		
PCB-105	<1.4	<1.7	<0.25	0.2	0.16	<0.09	<0.11	<0.1	0.25	0.18	<0.08	<0.11	<0.1	<0.1	<0.1	<0.17	<0.03	<0.02
PCB-114	<1.4	<1.7	<0.25	<0.1	<0.1	<0.09	<0.11	<0.1	<0.11	<0.11	<0.08	<0.11	<0.1	<0.1	<0.1	<0.17	<0.03	<0.02
PCB-118	<1.4	<1.7	<0.25	0.49	0.37	<0.09	0.11	<0.1	0.63	0.52	0.21	0.2	<0.1	0.18	0.16	0.2	0.04	<0.02
PCB-123	<1.4	<1.7	<0.25	<0.1	<0.1	<0.09	<0.11	<0.1	<0.11	<0.11	<0.08	<0.11	<0.1	<0.1	<0.1	<0.17	<0.03	<0.02
PCB-156	<1.4	<1.7	<0.25	0.18	0.18	<0.09	<0.11	<0.1	0.28	0.22	0.13	0.17	0.15	<0.1	<0.1	<0.17	<0.03	<0.02
PCB-157	<1.4	<1.7	<0.25	<0.1	<0.1	<0.09	<0.11	<0.1	<0.11	<0.11	<0.08	<0.11	<0.1	<0.1	<0.1	<0.17	<0.03	<0.02
PCB-167	<1.4	<1.7	<0.25	<0.1	<0.1	<0.09	<0.11	<0.1	<0.11	<0.11	<0.08	<0.11	<0.1	<0.1	<0.1	<0.17	<0.03	<0.02
PCB-189	<1.4	<1.7	<0.25	<0.1	<0.1	<0.09	<0.11	<0.1	<0.11	<0.11	<0.08	<0.11	<0.1	<0.1	<0.10	<0.17	<0.03	<0.02
WHO TEQ (ng/kg fat) lower	<0.01	<0.01	<0.01	0.16	0.14	<0.01	0.01	<0.01	0.23	0.18	0.09	0.11	0.08	0.02	0.02	0.02	<0.01	<0.01
WHO TEQ (ng/kg fat) upper	2.67	3.25	0.48	0.28	0.26	0.17	0.21	0.19	0.36	0.31	0.19	0.25	0.22	0.2	0.2	0.33	0.06	0.04

Note: # Congener residues expressed on a fat basis except for soil and grass where results are on a whole weight basis.

present study are entirely consistent with this value. The mean percentage PCB contribution to the total TEQ (41%) also agrees well with 46% found in the Rose *et al.* (2005) study.

More recently, Fernandes *et al.* (2010) investigated PCDD/Fs and PCBs in offal and found that nearly half the lamb liver samples analysed exceeded the EU-MRL for PCDD/Fs of 6 ng TEQ/g fat. The mean total TEQ for kidneys was 1.1 ng TEQ/g fat, which is in line with results from the present study.

7.5.4 Biotransfer factors (BTFs)

BTFs were calculated for each of the 29 PCDD/F and PCB congeners. However, only a representative set of five compounds have been selected for discussion.

- 2,3,7,8-TCDD
- 2,3,4,7,8-PeCDF
- 3,3',4,4',5-pentachlorobiphenyl (PCB-126)
- 3,3',4,4',5,5'-hexachlorobiphenyl (PCB-169)
- 2,2',4,4',5,5'-hexachlorobiphenyl (PCB-153)

This selection reflects a number of considerations. Toxicological significance was an important consideration (2,3,7,8-TCDD and 2,3,4,7,8-PeCDF have TEF_{WHO98} of 1 and 0.5, respectively), as was the occurrence of as few concentrations below detection limits as possible. Previous studies (e.g. Fries *et al.*, 1999; Thomas *et al.*, 1999) have reported that transfer rates can vary appreciably according to differences in chemical and physical properties, so it is important to reflect such contrasts in the chosen congeners. Finally, another objective was to highlight differences between types of samples. Following scrutiny of all the BTF results, PCB-169 was selected as a representative of the non-ortho-PCBs partly on the latter grounds. Table 7.13 summarises BTFs in lowland and highland sheep for the above mentioned congeners.

General trends observed were that BTFs for lowland sheep were higher than highland sheep; the highest BTFs were for liver, while values for meat and kidneys were similar, for both low and highland sheep. However, it can be seen that BTFs for PCB-169 did not obey the general trend; transfer factors for lamb meat, liver and kidney were roughly equal in lowland sheep, while in highland sheep PCB-169 BTFs for lamb meat and liver were similar, but much higher than kidney. Transfer factors for PCB-169 and PCB-153 were higher than those for 2,3,78-TCDD and 2,3,4,7,8,-PeCDF for all sample types. Ewes had lower BTFs than their lambs.

The highest observed BTFs were for 2,3,4,7,8-PeCDF and PCB-126 in liver, which were at least an order of magnitude greater than those for meat.

Table 7.13 Biotransfer factors

Sample code	Sample type	Age at slaughter (days)	Input flux (ng/day)	Fat content (%)	BTF (day/kg fat)				
					2,3,7,8-TCDD	2,3,4,7,8-PeCDF	PCB-126	PCB-169	PCB-153
Lowland sheep									
EBP	Meat	48	554	3.56	12.9	8.98	11.98	20.1	22.68
EBQ	Meat	48	550	5.77	9.51	8.26	8.54	22.17	23.8
ECC	Meat	87	1128	1.92	3.68	5.1	10.73	24.39	89.45
ECD	Meat	87	1110	9.04	2.77	5.4	9.8	21.21	18.82
ECQ	Meat	127	1358	7.52	2.59	5.94	12.25	35.42	31.81
ECR	Meat	127	1395	6.03	2.52	4.36	12.95	39.43	35.56
ECU	Liver	127	1358	4.88	5.18	175.74	117.21	38.5	53.1
EDA	Kidney	134	1395	27.94	1.89	6.54	11.98	40.07	30.25
EBM	Meat	Adult	3407	20.56	1.52	3.37	5.59	13.4	10.24
ECL	Meat	Adult	3407	6.79	0.61	1.7	5.57	11.96	19.51
Highland sheep									
GBP	Meat	48	880	2.47	6.92	7.64	9.8	18.33	18.58
GBQ	Meat	48	869	3.54	8.56	7.94	11.12	13.81	13.96
GCC	Meat	91	1629	5.74	0.86	4.15	4.14	35.35	22.97
GCD	Meat	91	1592	5.35	0.44	4.77	6.75	42.84	24.79
GCR	Meat	152	1965	3.54	0.71	2.17	4.14	48.43	20.15
GCQ	Meat	153	2148	8.95	0.81	2.29	3.2	21.22	12.88
GCU	Liver	153	2148	7.57	1.63	82.49	30.34	30.81	19.11
GCY	Kidney	153	2148	6.51	2.44	2.78	2.48	17.32	12.55
GBM	Meat	Adult	5078	2.19	0.14	1.18	1.53	5.89	3.47
GCL	Meat	Adult	5078	1.76	0.27	0.38	1.16	1.08	1.26

Significant differences in meat BTFs for some congeners (i.e. PCB-153) were observed between lambs from the same ewe slaughtered at the same time. This difference may have been related to fat content. Higher fat content in lambs can be correlated with greater milk intake (Cañeque et al., 2001). In the calculation of input fluxes it was assumed that both lambs from a single ewe had the same milk intake, but it may have been the case that one lamb was dominant with respect to the other, and therefore would consume a greater proportion of the milk produced.

7.6 Conclusions and future trends

Several broad trends are apparent from the results discussed in this chapter. Firstly, the analytical data obtained are generally in good agreement with the results of other studies and appear consistent with the PCDD/F and PCB levels to be expected in rural locations.

Another feature was that the total TEQs for meat declined with age from birth and were due to a decrease in the PCDD/F contribution to the TEQ.

The analytical data was suitable for the purpose of calculating congener specific PCDD/F and PCB biotransfer factors for sheep reared under typical commercial husbandry conditions.

There were congener specific trends, particularly with changes in the relative importance of inputs such as milk, commercial feed, grass or soil. Lower detection limits, especially for PCBs in soil and feed, as well as data on the bioavailability of residues in the different dietary components, would have allowed for more accurate determination on their contribution to BTFs. It would have been ideal to have collected more samples of milk, meat, kidney and liver at more time points over a longer duration, but this was not possible within the time and cost constraints of the study.

Biotransfer factors for PCDD/Fs in sheep liver were very high, such that a sheep that has consumed grass, feed and milk with PCDD/Fs levels below the EU limits yielded liver samples which exceeded the EU-MRL. Using BTFs calculated here it would be possible to estimate the likelihood of lambs liver exceeding the EU limits from residues measured in feed, grass and milk consumed by lambs.

7.7 Acknowledgements

The authors would like to thank Paul Dunning, Andrew Farley, David Norton and Chris Nix from Easton College, Norwich and Rachel Donkin from Northumberland College of Further Education for their assistance. The authors are grateful to the UK Food Standards Agency for funding this work.

7.8 References

ABBOTT, K. (2003). Royal Veterinary College, London – personal communication.
AMBIDGE, P. F., COX, E. A., CREASER, C. S., GREENBERG, M., DE M. GEM, M. G., GILBERT, J., JONES, P. W., KIBBLEWHITE, M. G., LEVEY, J., LISSETER, S. G., MEREDITH, T. J., SMITH, L., SMITH, P., STARTIN, J. R., STENHOUSE, I. and WHITWORTH, M. (1990). Acceptance criteria for analytical data on polychlorinated dibenzo-p-dioxins and polychlorinated dibenzofurans. *Chemosphere* **21**, 999–1006.
BERG, V., LYCHE, J. L., GUTLEB, A. C., LIE, E., UTNE SKAARE, J., ALEKSANDERSEN, M., ROPSTAD, E. (2010) Distribution of PCB118 and PCB153 and hydroxylated PCB metabolites (OH-CBs) in maternal, fetal and lamb tissues of sheep exposed during gestation and lactation. *Chemosphere* **80**, 1144–1150.
BRUNS-WELLER, E., KNOLL, A. and HEBERER, T. (2010). High levels of polychlorinated dibenzo-dioxins/ furans and dioxin like-PCBs found in monitoring investigations of sheep liver samples from Lower Saxony, Germany. *Chemosphere* **78**, 653–658.
CAÑEQUE, V., VELASCO, S., DÍAZ, M., PÉREZ, C., HUIDOBRO, F., LAUZURICA, S, MANZANARESAND, C. and GONZÁLEZ, J. (2001). Effect of weaning age and slaughter weight on carcass and meat quality of Talaverana breed lambs raised at pasture. *Anim. Sci.* **73**(1), 85–95.
CARDELLINO, R. A. and BENSON, M. E. (2002). Lactation curves of commercial ewes rearing lambs. *J. Anim. Sci.* **80**, 23–27.
COSTERA, A., FIELD, C., MARCHAND, B., LE BIZEC, B. and RYVHEN, G. (2006). PCDD/F and PCB transfer to milk in goats exposed to a long term intake of contaminated hay. *Chemosphere* **64**(4), 650–657.
EFSA (2010). Results of the monitoring of dioxin levels in food and feed. *EFSA Journal* **8**(3), 1385 pp36].
EISENREICH, S. J., LOONEY, B. B. and THORNTON, J. D. (1981). Airborne organic contaminants in the Great Lakes ecosystem. *Environ. Sci. Technol.* **15**, 30–38.
ESPOSITO, M., SERPE, F. P., CAVALLO S., PELLICANÒ, R., GALLO, P., COLARUSSO, G., D'AMBROSIO, R., BALDI, L., IOVANE, G. and SERPE, L. (2010). A survey of dioxins (PCDDs and PCDFs) and dioxin-like PCBs in sheep and goat milk from Campania, Italy. *Food Addit. Contam. B.* **3**(1), 58–63.
EUROPEAN COMMISSION (2002). *Preparatory actions in the field of dioxins and PCBs*, Brussels, European commission.
FERNANDES, A., MORTIMER, D., GEM, M. and ROSE, M. (2010). Dioxins and PCBs in offal: Occurrence and dietary exposure. *Chemosphere* **81**, 536–540.
FERNANDES, A., WHITE, S., D'SILVA, K. and ROSE, M. (2004). Simultaneous determination of PCDDs, PCDFs, PCBs and PBDEs in food. *Talanta* **63**, 1147–1155.
FERNANDES, A. R., FOXALL, C., LOVETT, A., ROSE, M. and DOWDING, A. (2011). The assimilation of dioxins and PCBs in conventionally reared farm animals: Occurrence and Biotransfer factors. *Chemosphere* **83**, 815–822.
FIEDLER (2008). Chapter1 Part1 in E. MEHMETLI and B. KOUMANOVA (Eds.), *The Fate of Persistent Organic Pollutants in the Environment*, 3–12. New York, Springer.
FOCANT, J.-F., EPPE, G., PIRARD, C., MASSART, A.-C., ANDRE, J.-E. and DE PAUW, E. (2002). Levels and congener distributions of PCDDs, PCDFs and non-ortho PCBs in Belgian foodstuffs. Assessment of dietary intake. *Chemosphere* **48**, 167–179.
FOXALL, C., LOVETT, A., NIX, C., SHIELS, A., DONKIN, R., FERNANADES, A., WHITE, S. and ROSE, M. (2004). Transfer and uptake of organic contaminants into meat and eggs of chickens, sheep, and pigs. *FSA project report C01020*. London, Food Standards Agency.
FRIES, G. F. (1996). Ingestion of sludge applied organic chemicals by animals. *Sci. Total. Environ.* **185**, 93–108.
FRIES, G. F., PANSTENBACH, D. J., MATHER, D. B. and LUKSEMBERG, W. J. (1999). A congener specific evaluation of transfer of chlorinated dibenzo-p-dioxins and dibenzofurans

to milk of cows following ingestion of pentachlorophenol-treated wood. *Environ. Sci. Technol.* **33**, 1165–1170.

FSA (2000). Dioxins and PCBs in the UK Diet: 1997. *Total Diet Study (Number 04/00)*. London, Food Standards Agency, UK.

FSA (2002). *Report on dioxins and dioxin-like PCBs in foods from farms close to a foot and mouth disease pyres*. London, Foods Standards Agency.

FURST, P., KRAUSE, G. H. M., HEIN, D., DELSCHEN, T. and WILMERS, K. (1993). PCDD/PCDF in cow's milk in relation to their levels in grass and soil. *Chemosphere* **27**, 1349–1357.

GRIEPINK, B., WELLS, D. E. and FERREIRA, M. F. (1988). The certification of the contents (mass fraction) of chlorobiphenyls (IUPAC Nos 28, 52, 101, 118, 138, 153 and 180) in two fish oils: cod-liver oil CRM No 349; mackerel oil CRM No 350, Report EUR11520EN, Commission of the European Communities, Community Bureau of Reference.

HARRAD, S. J. and JONES, K. C. (1992). A source inventory and budget for chlorinated dioxins and furans in the United Kingdom environment. *Sci. Total Environ.* **126**, 89–107.

HARRAD, S. J. and SMITH, D. J. T. (1997). Evaluation of a terrestrial food chain model for estimating foodstuff concentrations of PCDD/Fs. *Chemosphere* **34**, 1723–1737.

HEALEY, W. B. and LUDWIG, T. G. (1965). Wear of sheep's teeth I. The role of ingested soil. *New Zealand, J Agric. Res.* **8**, 737–752.

HEALEY, W. B., CUTRESS, T. W. and MICHIE, C. (1967). Wear of sheep's teeth IV. Production of soil ingestion and tooth wear by supplementary feeding. *New Zealand, J. Agric. Res.* **10**, 201–209.

HEALEY, W. B. (1968). Ingestion of soil by dairy cows. *New Zealand, J. Agric. Res.* **11**, 487–499.

HEALEY, W. B. (1967). Ingestion of soil by sheep. *Proc. N. Z. Soc. Anim. Prod.* **27**, 109–120.

HEALEY W. B. and DREW, K R (1970). Ingestion of soil by hoggets grazing swedes. *New Zealand J. Agric. Res.* **13**, 940–944.

LIEM, A. K. D. and THEELEN, R.M.C. (1997). Dioxins: Chemical analysis, exposure and risk assessment. Thesis, National Institute of Public Health and the Environment, Bilthoven, The Netherlands. ISBN 90-393-2012-8.

MAFF (1997). *Dioxins and polychlorinated biphenyls in foods and human milk*. Food Surveillance Information Sheet, 105. Ministry of Agriculture, Fisheries and Food.

MAIER, E. A., VAN CLEUVENBERGEN, R., KRAMER, G. N., TUINSTRA, L. G. M. TH. and PAUWELS, J. (1995). BCR (non-certified) reference materials for dioxins and furans in milk powder, *Fresen. J. Anal. Chem.* **352**, 179–183.

MCGRATH, D., POOLE, D. B. R., FLEMING, G. A. and SINNOTT, J. (1982). Soil ingestion by grazing sheep. *Ir. J. Agric. Res.* **21**, 135–145.

MCLACHLAN, M. S. (1997). A simple model to predict accumulation of PCDD/Fs in an agricultural food chain. *Chemosphere* **34**, 1263–1276.

MANCHESTER-NEESVLG, J.B. and ANDREN, A.W. (1989). Seasonal variation in the atmospheric concentration of polychlorinated biphenyl congeners. *Environ. Sci. Technol.* **23**, 1138–1146.

MORTON, A. (2003). Easton College – personal communication.

MULLER, J. F., HULSTER, A., PAPKE, O., BULL, M. and MARSCHNER, H. (1993). Transfer pathways of PCDD/Fs to fruits. *Chemosphere* **27**, 195–201.

NIX, J. (1999). *Farm management pocketbook* (29th Edition). Wye College Press, Kent. ISBN 0 86266 059 9.

NATIONAL RESEARCH COUNCIL OF THE NATIONAL ACADEMIES (2006). *Health risks from dioxins and related compounds. Evaluation of EPA reassessment*. Washington D.C., National Academies Press.

ROSE M., HARRISON N., GEM M., FERNANDES A., WHITE S., DUFF M., COSTLEY C., LEON I., PETCH R.S., HOLLAND J., CHAPMAN A. (2005). Dioxins and polychlorinated biphenyls (PCDD/Fs and PCBs) in food from farms close to foot and mouth disease animal pyres. *J. of Environ. Monit.* **7**, 378–383.

STARTIN, J.R. (1994). Dioxins in food. In SCHECTER, A (Ed.), *Dioxins and health*, Plenum, New York, 115–137.

THOMAS, G. O.,. SWEETMAN, A. J., OCKENDEN, W. A., MACKAY, D and JONES, K. C. (1998a). Air-pasture transfer of PCBs. *Environ. Sci. Technol.* **32**, 936–942.

THOMAS, G. O., SMITH, K. E. C., SWEETMAN, A. J. and JONES, K. C. (1998b). Further studies of the air-pasture transfer of PCBs. *Environ. Pollut.* **102**, 119–128.

THOMAS, G. O., SWEETMAN, A. J. and JONES, K. C. (1999). Input-output balance of polychlorinated biphenyls in a long-term study of lactating daily cows. *Environ. Sci. Technol.* **33**, 104–112.

TRAVIS, C. C. and NIXON, A. G. (1996). Exposure to dioxin. In HESTER, R. E. and HARRISON, R. M. (Eds.), *Chlorinated organic micropollutants*. Royal Society of Chemistry, Cambridge, 17–30. ISBN: 978-1-84755-049-1.

USEPA (1994). *Estimating exposure to dioxin-like compounds*. Report No. EPA/600/6-88/005 Ca -c, Office of Research and Development, Washington DC, USA.

VAN DEN BERG, M., BIRNBAUM, L. S., BOSVELD, A. T. C., BRUNSTROM, B., COOK, P., FEELEY, M., GIESY, J. P., HANBERG, A., HASEGAWA, R., KENNEDY, S. W., KUBIAK, T., LARSEN, J. C., VAN LEEUWEN, F. X. R., LIEM, A. K. D., NOLT, C., PETERSON, R. E., POELLINGER, L., SAFE, S. H., SCHRENK, D., TILLIT, D., TYSKLIND, M., YOUNES, M., WAERN, F., ZACHAREWSKI, T. (1998). Toxic equivalency factors (TEFs) for PCBs, PCDDs, PCDFs for humans and wildlife. *Environ. Health Persp.* **106**, 775–792.

WHO (2002). Section 5.2 in: *Fifty-seventh report of the Joint FAO/WHO Expert Committee on Food Additives. WHO Technical Report Series 909*, pp. 121–149, Geneva, FAO.

ZOOK, D. R. and RAPPE, C. (1994). Environmental sources, distribution and fate of polychlorinated dibenzodioxins, dibenzofurans and related chlorines. In SCHECHTER, A. (Ed.), *Dioxins and Health*, John Wiley and Sons, New Jersey, 79–113.

8

Risk assessment of chemical contaminants and residues in foods

D. J. Benford, Food Standards Agency, UK

DOI: 10.1533/9780857098917.1.173

Abstract: This chapter describes the approaches used in assessment of risks associated with chemical contaminants and residues of plant protection products and veterinary drugs in food. Risk assessment consists of the interlinked stages of hazard identification, hazard characterisation, exposure assessment and risk characterisation. Depending on the completeness of the toxicological database and the properties of a specific chemical, the aim is to set health-based guidance values, representing intakes judged to be without appreciable risk, or to identify margins of exposure between reference points associated with the dose–response curve and estimated human dietary exposure. The impact of chemical risk on the process of risk management varies depending on the regulatory context.

Key words: risk assessment, chemical contaminants, residues, pesticides, veterinary drugs.

Note: This chapter was first published as Chapter 1 'Risk assessment of chemical contaminants and residues in food' by D. J. Benford, in *Chemical contaminants and residues in food*, edited by D. Schrenk, Woodhead Publishing Ltd, 2012, ISBN: 978-0-85709-058-4.

8.1 Introduction

Chemical risk assessment provides the scientific basis for decisions aimed at ensuring, maintaining and improving the safety of human exposure to chemicals. This chapter describes the approaches used in assessment of risks associated with chemical contaminants and residues of plant protection products and veterinary drugs in food. The risk assessments that underpin development of regulatory measures for these chemicals in food are generally conducted by authoritative independent committees of scientific experts, such as the scientific panels of the European Food Safety Authority (EFSA), and the bodies that advise the Food and Agriculture Organization (FAO) and World Health Organization (WHO) of the United Nations, i.e. the Joint FAO/WHO Expert Committee on Food Additives (JECFA) and the Joint FAO/WHO Meeting on Pesticide Residues (JMPR).

8.1.1 Risk assessment paradigm and definitions

Risk assessment is defined as 'a process intended to calculate or estimate the risk to a given target organism, system or (sub)population, including the identification of attendant uncertainties, following exposure to a particular agent, taking into account the inherent characteristics of the agent of concern as well as the characteristics of the specific target system' (IPCS 2009a). In the context of chemicals in food, the term safety assessment is also sometimes used, wherein safety is the 'practical certainty that adverse effects will not result from exposure to an agent under defined circumstances' (IPCS 2009a). Risk is defined as the 'probability of an adverse effect in an organism, system or (sub)population caused under specified circumstances by exposure to an agent' (IPCS 2009a).

The risk assessment is a well-established independent scientific process, which, together with risk management and risk communication, constitutes risk analysis (see Fig. 8.1). It is considered important to separate the activities of risk assessment from those of risk management in order to ensure the scientific independence of the assessment, since risk management and communication are also influenced by political and socio-economic considerations. However, in order to ensure that the outputs of the risk assessment are useful, it is beneficial for risk managers to communicate and interact with risk assessors during the process, particularly during the initial problem formulation (also known as framing the question). Thus, the relationship between risk assessment and risk management is an interactive, often iterative, process.

As shown in Fig. 8.1, the risk assessment process consists of four linking steps, relating to hazard, exposure and risk. Risk is determined by both the hazard and the exposure. If there is no exposure, then there will be no risk. The higher the exposure, the more likely it becomes that there will be a risk. In chemical risk assessment, hazard is defined as an 'inherent property of an agent or situation having the potential to cause adverse effects when an organism, system or (sub)population is exposed to that agent'. This differs from microbiological risk assessment, wherein the hazard is generally considered to be the biological agent, rather than its properties. There can be a number of different hazards associated with an individual chemical, influenced by the route, magnitude and duration of exposure, and the exposed population (e.g. different life stages).

Risk assessment of chemical contaminants and residues 175

Fig. 8.1 The risk analysis paradigm.

8.1.2 Chemical contaminants and residues in food

Chemical contaminants in food can be defined as environmental contaminants, which transfer from the environment into the food chain, or as process contaminants, which are generated in food as a result of chemical reactions occurring during cooking and processing. Food contact materials can also be a source of chemicals (e.g. formaldehyde, melamine, phthalates and primary aromatic amines) with the potential to leach into food. Environmental contaminants include ubiquitous pollutants such as dioxins and heavy metals. To some extent these may be naturally present in the environment, but they can also be increased by anthropogenic activity. Contaminants can also arise from toxins produced by fungi (e.g. aflatoxins, fumonisins, ochratoxin A), plants (e.g. pyrrolizidine alkaloids) and algae (e.g. saxitoxins, okadeic acid).

Process contaminants generated during cooking include acrylamide, furan and heterocyclic amines. Other processes leading to formation of contaminants include fermentation (e.g. ethyl carbamate, 3-monochloropropanediol) and disinfection (e.g. trihalomethanes).

Some chemical contaminants are not readily categorised: for example, polycyclic aromatic hydrocarbons can be generated during cooking and drying but also present from the environment. Aluminium can be present in food naturally, from environmental contamination, from leaching from food contact materials and also due to the use of approved food additives. Similarly, nitrate is produced naturally in plants, but can also be a contaminant and an approved food additive. Contaminants do not have a function in food or food production, and their presence may be considered undesirable. However,

176 Persistent organic pollutants and toxic metals in foods

they are often unavoidable, and found in wide ranges of foods as a result of increasingly sensitive methods of analytical detection.

The term 'residues' is applied to plant protection products (e.g. agricultural pesticides) or veterinary drugs in food products. Clearly these have a purpose in food production, and there is much greater potential for controls on their conditions of use, and hence their presence in food.

8.2   Risk assessment

The generic risk assessment approach is appropriate for all types of contaminants and residues. The key difference is the availability of data. Plant protection products and veterinary drugs are subject to approval processes requiring the manufacturers to provide a dossier that includes toxicological studies conducted to approved guidelines and standards. For contaminants, there is generally no sponsor to provide a complete toxicological dataset, and studies published in the scientific literature have often been conducted using protocols that limit their applicability for risk assessment purposes (e.g. lack of dose–response data). However, epidemiological data are sometimes available, allowing the risk assessment to be based on human data.

8.2.1   Hazard identification

The aim of hazard identification is to establish the type and nature of adverse effects that an agent has an inherent capacity to cause in an organism, system or (sub)population.

An adverse effect is 'a change in the morphology, physiology, growth, development, reproduction or lifespan of an organism, system or (sub)population that results in an impairment of functional capacity, an impairment of the capacity to compensate for additional stress or an increase in susceptibility to other influences' (IPCS 2009a).

Hazard identification generally involves a wide range of toxicological tests to define the potential for harm to arise (irrespective of dose) at different stages of the life cycle. These tests involve single and repeat dose exposure and aim to identify adverse effects such as general systemic toxicity, effects on the reproductive, immune, endocrine and nervous systems, and tumorigenicity. Particular importance is frequently attributed to the results of tests for mutagenicity or genotoxicity. Mutagenicity is the potential to induce mutation, i.e. a permanent change in the amount or structure of the genetic material of an organism. Mutations may involve individual genes, blocks of genes, or the structure or number of whole chromosomes. Genotoxicity is a broader term, which also includes endpoints associated with the potential to result in mutation whilst not necessarily reflecting a permanent change in the genetic material, such as damage to DNA, production of DNA adducts, unscheduled DNA synthesis subsequent to DNA damage and sister chromatid exchange.

Mutations of somatic cells are passed to descendent daughter cells, which can lead to cancer when associated with the activation and expression of oncogenes, or the loss or inactivation of tumour suppressor genes. Whilst potential carcinogenicity is normally the main focus of mutagenicity testing, mutations in the germ cells may be transferred to the offspring, which may lead to inherited disorders.

There is broad consensus that, unless there is evidence to the contrary, chemical substances that are genotoxic and carcinogenic have the potential to cause DNA damage at any level of exposure and that such damage may lead to tumour development (EFSA 2005a; FAO/WHO 2006a). Furthermore, this mode of action is generally assumed to be relevant to humans (Boobis *et al.*, 2006). Such substances are not authorised for use in food production, although they can be unavoidably present as contaminants in food. Examples include acrylamide, aflatoxins, ethyl carbamate and polycyclic aromatic hydrocarbons.

In contrast, for carcinogens that act by a non-genotoxic mode of action the tumours generally arise subsequent to other effects, such as cytotoxicity, cell proliferation or hormonal effects, for which it may be possible to assume a threshold and/or which may be based on a biological mechanism not relevant to humans (Boobis *et al.*, 2006).

For some food contaminants, human data from epidemiological studies, case studies or outbreaks of ill health provide support for the human relevance of adverse effects observed in animal studies, or even provide the primary basis for hazard identification. For example, the recent EFSA and JECFA evaluations of arsenic, cadmium and lead have noted that the data from experimental animals provide evidence for the plausibility of the observations in the epidemiological studies, and the human data were considered sufficient to use as the basis of the evaluations (EFSA 2009a; 2009b; 2010; FAO/WHO 2011a; 2011b). For contaminants with chronic effects, it is generally not possible to establish a causal link with human illness in the absence of supporting data from experimental animals. In contrast, direct associations can be made for chemicals with acute effects. Outbreaks of human illness led to the discovery of marine biotoxins responsible for diarrhoeic, paralytic and amnesic shellfish poisoning, and the human data have been used as the basis for establishing acute reference doses (ARfD) (EFSA 2008a; 2008b; 2009c; 2009d).

8.2.2 Hazard characterisation

Hazard characterisation is closely linked to hazard identification. This is the qualitative and, wherever possible, quantitative description of the inherent properties of an agent or situation having the potential to cause adverse effects. It should, where possible, include a dose–response assessment and its attendant uncertainties. This supports identification of the most important adverse effect(s), i.e. those occurring at the lowest doses and considered also likely to occur in humans. The dose–response relationship(s) for these effects are analysed in order to define a level that either had no effect in the critical study (e.g. no observed adverse effect level (NOAEL) or induced predetermined level of effect or response, to be used in risk characterisation. It also includes consideration of interspecies differences and human variability in the absorption, distribution, metabolism and excretion (toxicokinetics) and in the biological response (toxicodynamics), and the completeness of the database.

Health-based guidance values
For chemicals that are not genotoxic, and certain classes of chemicals with a genotoxic mechanism with a demonstrated threshold effect, the aim of hazard characterisation is generally to set a health-based guidance value. This is a level of exposure that is without appreciable risk to health over a defined period. The term 'no appreciable risk' is used because absolute safety, or zero risk, cannot be guaranteed unless it is possible to guarantee zero exposure. For plant protection products and veterinary medicines, the common health-based guidance value is the acceptable daily intake (ADI), which is an estimate of the amount of a chemical in food or drinking water, expressed on a body weight basis, that can be ingested daily over a lifetime without appreciable health risk to the consumer, derived on the basis of all the known facts at the time of the evaluation. The term 'acceptable' is used because plant protection products and veterinary medicines are subject to an approval process. In addition to, or instead of, the health-based guidance values referring to lifetime exposure, an ARfD is sometimes set for chemicals with the potential to cause effects following short-term exposure. The ARfD relates to the amount of a substance in food or drinking water, expressed on a body weight basis, that can be ingested in a period of 24 h or less without appreciable health risk, and can be numerically equal to or greater than the ADI or tolerable daily intake (TDI). If estimated exposure (see Section 8.2.3) is below the relevant health-based guidance value(s) then the product can be approved and is considered acceptable. For similar reasons ADIs are also established for food additives. In contrast, the term 'tolerable' is used for contaminants, since they are not deliberately used in food, but may be unavoidable. There are some differences in terminology used by different authorities: for example, the JECFA uses the term 'provisional' for contaminants in food, but the TDI and provisional maximum tolerable daily intake (PMTDI) are essentially equivalent to the ADI. For contaminants with cumulative properties, a longer reference period is sometimes used, reflecting the need to average exposure over a long period of time. This is generally the (provisional) tolerable weekly intake (PTWI or TWI), but the JECFA has also established provisional tolerable monthly intakes (PTMIs) for dioxins and cadmium, which have very long half-lives in the human body.

Uncertainty factors
The approach to setting health-based guidance values is similar, regardless of the actual terminology used, and the following text generally refers to TDI for ease of reading. The TDI has traditionally been established by identifying the NOAEL for the relevant effect occurring at lowest doses, and dividing it by uncertainty (safety) factors to allow for variability between species and within the human population. By convention, a default uncertainty factor of 100 has been used. Initially, this was an arbitrary decision, but soon became defined as comprising two equal components:

- a factor of 10 for interspecies differences, i.e. to allow for possible greater sensitivity of humans compared with the animal model, due to slower elimination from the body, greater balance of activation to detoxication reactions and/or greater sensitivity to the toxic effect; and
- a factor of 10 to allow for human interindividual (intraspecies) variation, i.e. the possibility that a proportion of the population may be at greater risk because of differences in toxicokinetics or tissue sensitivity within the human population.

The overall uncertainty factor of 100 may be increased if there are important gaps in the database for a contaminant, e.g. the absence of a NOAEL or of long-term animal studies. Conversely, if the TDI is based on human data, then the uncertainty factor for interspecies differences is not required.

More recently there have been moves to refine the uncertainty factor by subdividing the ten-fold factors into factors for the toxicokinetics and toxicodynamics. Examination of various databases has indicated a differential split, with greater weight given to toxicokinetic causes of interspecies differences, whereas equal weighting may be given to toxicodynamic and toxicokinetic differences in individual variability (Fig. 8.2). If individual data on any of these components were available, they could then be incorporated into the evaluation by replacement of the appropriate default. For example, if information is available indicating that the toxicokinetics of a particular chemical are quantitatively similar in the experimental animal used to establish the NOAEL and in humans, then the default factor of 4.0 in Fig. 8.2 would be replaced by the value of 1. The factors would then be 2.5 for interspecies differences in toxicokinetics and 10 for human variability, giving an overall factor of 25.

Analysis of available data indicates that, in general, the default safety factors are appropriate; however, where data on a compound indicate that the defaults are inappropriate (too low or too high), then the subdivision of the factors allows additional data to be used to modify the defaults and introduce compound-specific data. This approach has more frequently been applied to contaminants than to residues. Examples of health-based guidance values using modified uncertainty factors and chemical-specific adjustment factors

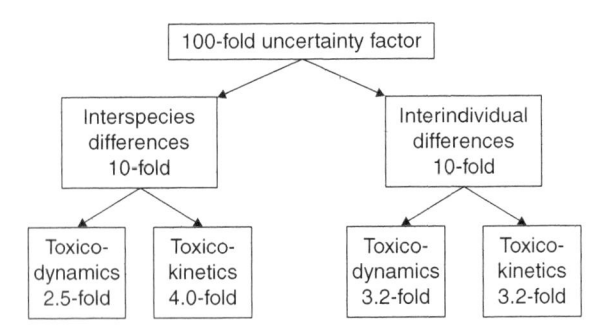

180 Persistent organic pollutants and toxic metals in foods

```
                    ┌─────────────────────────┐
                    │ 100-fold uncertainty factor │
                    └─────────────────────────┘
                         /              \
            ┌──────────────┐      ┌──────────────┐
            │ Interspecies │      │ Interindividual│
            │ differences  │      │ differences  │
            │   10-fold    │      │   10-fold    │
            └──────────────┘      └──────────────┘
              /        \            /         \
        ┌────────┐ ┌────────┐ ┌────────┐ ┌────────┐
        │Toxico- │ │Toxico- │ │Toxico- │ │Toxico- │
        │dynamics│ │kinetics│ │dynamics│ │kinetics│
        │2.5-fold│ │4.0-fold│ │3.2-fold│ │3.2-fold│
        └────────┘ └────────┘ └────────┘ └────────┘
```

Fig. 8.2 Subdivision of uncertainty factors (from IPCS 2005).

include dioxins and dioxin-like polychlorinated biphenyls (SCF 2001), methylmercury (FAO/WHO 2004) and zearalenone (EFSA 2011c).

The benchmark dose
Recently there has been increasing use of a benchmark dose (BMD) approach in preference to using the NOAEL in setting health-based guidance values, since it makes more use of the dose–response relationship and provides quantification of the uncertainty and variability in the dose–response data (IPCS 2009b; EFSA 2009e). The BMD is a dose level, derived by statistical modelling of dose–response data, associated with a specified low but measurable change in response, the benchmark response (BMR). The BMR should be in the region of the low end of the observed dose–response range, since extrapolation outside the range of observation increases the dependence on the statistical models. For quantal data, the BMR is an increase in the incidence of a lesion/response compared with the background response. The EFSA Scientific Committee recommended a default BMR value of 10% extra risk. This approach has been taken in recent evaluations of a number of carcinogens based on data from carcinogenicity studies in experimental animals, such as acrylamide and furan (FAO/WHO 2006a; 2011a), and polycyclic aromatic hydrocarbons (EFSA 2008c). However, when human data are used, different BMRs may be preferred in order to avoid extrapolation outside the observed range of the data. The JECFA used BMRs of 0.5 and 5% extra risk when analysing datasets for cancer endpoints from epidemiological studies of arsenic (FAO/WHO 2011a).

For continuous data, the EFSA Scientific Committee recommended a default BMR of 5% change in the magnitude of response, but stressed that other values may be preferred based on biological or statistical considerations (EFSA 2009e). For example, BMR values of 1% decrease in intelligence quotient (IQ) and 1% increase in systolic blood pressure, which were both considered to have significant health consequences at the population level, were used (EFSA 2010). A hybrid approach may be preferred for some types of continuous data, whereby the BMR relates to an increased incidence of a

Risk assessment of chemical contaminants and residues 181

magnitude of response considered to be abnormal, as in the EFSA opinion on cadmium (EFSA 2011a).

When using the BMD approach for setting health-based guidance values, the lower confidence bound of the BMD (the BMDL) is used in place of the NOAEL, applying the same uncertainty factors. An advantage of the BMD approach is that it can be applied to studies that have failed to identify a NOAEL. The BMDL can also be used as a reference point for calculating a margin of exposure (MOE) (see Section 8.3.2).

8.2.3 Exposure assessment

Assessment of exposure to chemicals in food requires information on the occurrence of the chemical in different types of food, and on the amounts of those foods that are consumed by different population groups.

Dietary exposure to residues can be assessed for a plant protection product or veterinary drug before it has been approved for use (pre-regulation) or after it has potentially been in the food supply for years (post-regulation). Pre-regulation, chemical concentration data are available or estimated from the manufacturer. In the case of pesticide residues, JMPR uses data generated from field trial studies performed under the proposed Good Agricultural Practice. For veterinary drugs, the data are derived by JECFA from controlled residue depletion studies carried out in compliance with Good Practice in the Use of Veterinary Drugs (GPVD). Maximum residue levels (MRL) are also generated from these data, which are then used to check that pesticides and veterinary medicines have been used in accordance with the assessed practices.

Post-regulation, and for contaminants, additional chemical concentration data can be obtained from food in the marketplace. The available data should be relevant to the purpose of the risk assessment (e.g. some market data may not be sufficient for acute exposure assessments). There may also be a need to consider particular scenarios for concentrations in food, such as the potential impact of introducing, or changing, regulatory limits (e.g. maximum levels, MRLs).

Occurrence data for contaminants and residues may be generated by monitoring programmes, targeted surveys or total diet study approaches. Monitoring programmes and targeted surveys frequently focus on foods that are expected to contain the chemical of interest, and the results therefore are not representative of levels in food in general. Total diet studies are based on analyses of foods prepared as for consumption and pooled into composite samples. Due to the dilution arising from pooling, much lower limits of detection are required than for analysis of individual foods, and they are more often used for ubiquitous contaminants than for residues.

Consumption data to be used in exposure assessments should cover the general population, as well as critical groups that are vulnerable or are expected to have exposures that are significantly different from those of the general population (e.g. infants, children, pregnant women or the elderly). In order for risk assessments to be conservative, the dietary exposure assessment should be designed in such a way that potential high dietary exposure to a specific chemical is not underestimated. The methodologies should take into consideration non-average individuals, such as those who consume large portions of specific food items or show loyalty to specific foods or brands of food containing the highest concentrations of the chemical of interest. If specific consumption data for certain foods are not available from, for instance, nutritional surveys, it may be necessary to develop scenarios based on portion sizes, particularly when considering acute exposure for comparison to an ARfD.

8.2.4 Risk characterisation

Chemicals with health-based guidance values
Risk characterisation involves comparison of the results of the exposure assessment with the health-based guidance value if one has been set. If the relevant estimates of high-level dietary exposure for different population subgroups are lower than the relevant health-based guidance value (for example, if estimates of chronic dietary exposure are lower than the TWI, or estimates of acute dietary exposure are lower than the ARfD), the result indicates no appreciable risk to health. Exceedance of a health-based guidance value by some subgroups does not inevitably indicate that adverse health effects will occur. The ADI or TWI is not a threshold for toxic effect, but aims to be health-protective for the most sensitive population. The risk assessment should aim to identify the possible impact of exceeding the health-based guidance value, taking into account the nature of the adverse effects seen at the lowest doses and the magnitude and duration of the exceedance.

Margin of exposure (MOE) approach
In circumstances where no health-based guidance value has been proposed, it may be possible to comment on the MOE between a reference point from the dose–response relationship in animals or humans and the estimated human dietary exposure. Consideration of whether the resulting MOE indicates a health effect depends on whether the chemical is likely to have a threshold mode of action. For chemicals that are not genotoxic, identification of a health-protective MOE is based on the same considerations as identifying the appropriate uncertainty factor for establishing a health-based guidance value. Hence, if the reference point was from an animal study, but with some important gaps in the database, then the minimum MOE considered to be health-protective would be typically greater than 100 and possibly up to 10 000. A lower MOE would be considered health-protective if the reference point was based on human data. The EFSA CONTAM panel has

also considered MOEs based on body burden, e.g. for non-dioxin-like polychlorinated biphenyls (EFSA 2005b) and for polybrominated diphenylethers (EFSA 2011c), which obviates the need for allowance for an uncertainty factor for toxicokinetic differences.

For chemicals that are genotoxic and carcinogenic, the traditional assumption is that there may not be a threshold dose and that some degree of risk may exist at any level of exposure. Such substances are generally not considered acceptable for use as plant protection products or veterinary drugs, but can be present as unavoidable contaminants. Estimation of a dose associated with a defined estimate of risk by extrapolation from the high doses used in animal carcinogenicity studies to low doses relevant to human dietary exposure can result in very precise estimates that are highly dependent on the statistical model used for extrapolation and subject to considerable uncertainty regarding the shape of the dose–response relationships at doses far below the observed range. Risk estimates based on epidemiological data are subject to less extrapolation and uncertainty, but in practice there are extremely few genotoxic carcinogens in food with data suitable for dose–response modelling. Aflatoxin is one exception to this (FAO/WHO 1998). Therefore, in the past, risk characterisation advice for substances that are genotoxic and carcinogenic was primarily that the exposure should be as low as reasonably achievable (ALARA). However, this approach does not take into account either human exposure or carcinogenic potency and therefore does not support prioritisation for risk management action. Therefore WHO and EFSA have applied an MOE approach (EFSA 2005a; FAO/WHO 2006a). The EFSA Scientific Committee considered that an MOE of 10 000 or more, based on a BMDL for a 10% extra risk derived from animal cancer bioassay data, 'would be of low concern from a public health point of view and might reasonably be considered as a low priority for risk management actions' (EFSA 2005a). The JECFA has taken a similar view in its evaluations (FAO/WHO 2006a; 2006b; 2011a), but there is as yet no universal consensus on the value of an MOE of low concern. Furthermore, the magnitude of the MOE gives an indication of the level of concern, but is not a precise quantification of risk: the larger the MOE, the smaller the potential risk posed by exposure to the compound under consideration, but a carcinogen with an MOE of 1000 cannot be assumed to represent ten times the cancer risk of a different carcinogen with an MOE of 10 000. Particular MOE values are not necessarily directly comparable, due to the uncertainties in the carcinogenicity data and exposure assessments, and it is important for these to be described in the narrative accompanying the MOE (Benford et al., 2010).

Threshold of toxicological concern (TTC)
The threshold of toxicological concern (TTC) has been developed as an approach to risk characterisation for chemicals with minimal available toxicological data and low human exposure. The TTC approach defines a number of generic exposure values, derived by extrapolation of toxicity data for

structurally related chemicals, below which there is a low probability of adverse effects on human health (Kroes *et al.*, 2004; Munro *et al.*, 2008). So far the approach has mainly been used for food contact materials and flavouring agents, but it also has potential for evaluation of impurities.

Combined exposure
There is increasing awareness of the need to consider any risks associated with combined exposure to mixtures of chemicals present in food, including naturally occurring substances as well as contaminants and residues. This has been the focus of considerable risk assessment activity around the world (e.g. COT 2002; EFSA 2008d; Boobis *et al.* 2011; Meek *et al.* 2011). It is beyond the scope of this chapter to consider this issue in detail. However, the currently accepted general principles are as follows (based on EC 2011):

- Chemicals with common modes of action may act jointly to produce combination effects that can be described by dose/concentration addition.
- For chemicals with different modes of action (independently acting), no robust evidence is available that exposure to a mixture of such substances is of health concern if the individual chemicals are present at or below their no-effect levels.
- Interactions (including antagonism, potentiation and synergy) generally occur at medium or high dose levels (relative to the lowest effect levels).

8.3 Role of risk assessment in risk management

Risk assessment provides the scientific advice to underpin risk management action, including:

- response to incidents of food contamination or adulteration;
- prioritisation of research needs;
- development of advice to consumers; and
- development of regulations relating to chemicals in food.

The risk assessment is not the sole consideration involved in developing regulations, which means that exceeding a regulatory limit does not inevitably entail a risk to the health of the consumer.

For residues of plant protection products, MRLs are based on Good Agricultural Practice, even if the risk assessment indicates that higher levels would not result in appreciable risk. For veterinary drugs, good practice considerations are also taken into account. However, the determining criterion is that dietary exposure estimates, based on scenarios related to a set 'food basket' of products of animal origin, should be below the ADI. In the pre-regulation phase, when proposed uses result in potential chronic or acute dietary exposures that exceed relevant health-based guidance values, the dietary exposure estimates may be refined, e.g. by restricting approved uses. For veterinary drugs, residue levels can also be decreased by extending the withdrawal period before slaughter.

For chemical contaminants, maximum levels (MLs) are established to be compatible with tolerable intake levels but also based on the lowest level of contamination that can be reasonably achieved without removing the food from the food supply. In addition, reliable measurement in the region of the ML must be feasible.

8.4 Sources of further information

EFSA opinions: http://www.efsa.europa.eu/ [accessed 13 April 2012].

JECFA procedures and evaluations: http://www.who.int/ipcs/food/jecfa/en/ [accessed 13 April 2012].

JMPR procedures and evaluations: http://www.who.int/foodsafety/chem/jmpr/en/index. html [accessed 13 April 2012].

IPCS (2009). Environmental Health Criteria 240. Principles and methods for the risk assessment of chemicals in food: http://www.who.int/foodsafety/chem/principles/en/ index1.html [accessed 13 April 2012].

8.5 References

BENFORD, D., BOLGER, P. M., CARTHEW, P., COULET, M., DINOVI, M., LEBLANC, J. C., RENWICK, A. G., SETZER, W., SCHLATTER, J., SMITH, B., SLOB, W., WILLIAMS, G. and WILDEMANN, T. (2010), 'Application of the Margin of Exposure (MOE) approach to substances in food that are genotoxic and carcinogenic', *Food Chem. Toxicol.*, **48** (Supplement 1), S2–S24.

BOOBIS, A. R., COHEN, S. M., DELLARCO, V., MCGREGOR, D., MEEK, M. E., VICKERS, C., WILLCOCKS, D. and FARLAND, W. (2006), 'IPCS framework for analyzing the relevance of a cancer mode of action for humans', *Crit. Rev. Toxicol.*, **36** (10), 781–792.

BOOBIS, A., BUDINSKY, R., COLLIE, S., CROFTON, K., EMBRY, M., FELTER, S., HERTZBERG, R., KOPP, D., MIHLAN, G., MUMTAZ, M., PRICE, P., SOLOMON, K., TEUSCHLER, L., YANG, R. and ZALESKI, R. (2011), 'Critical analysis of literature on low-dose synergy for use in screening chemical mixtures for risk assessment', *Crit. Rev. Toxicol.*, **41** (5), 369–383.

COT (2002), UK committee on toxicity of chemicals in food, consumer products and the environment, *risk assessment of mixtures of pesticides and similar substances*, Food Standards Agency, London, UK. http://cot.food.gov.uk/cotreports/cotwgreports/ cocktailreport [accessed 13 April 2012].

EC (2011), Scientific Committee on Consumer Safety (SCCS), Scientific Committee on Health and Environmental Risks (SCHER), Scientific Committee on Emerging and Newly Identified Health Risks (SCENIHR), *Toxicity and Assessment of Chemical Mixtures (Preliminary Opinion approved for Public Consultation)*. http://ec.europa.eu/health/ scientific_committees/consultations/public_consultations/scher_consultation_06_en.htm [accessed 13 April 2012].

186 Persistent organic pollutants and toxic metals in foods

EFSA (2005a), 'Opinion of the Scientific Committee on a request from EFSA related to a harmonised approach for risk assessment of substances which are both genotoxic and carcinogenic', *The EFSA Journal*, **282**, 1–137. http://www.efsa.europa.eu/en/efsajournal/ pub/282.htm [accessed 13 April 2012].

EFSA (2005b), 'Opinion of the Scientific Panel on contaminants in the food chain [CONTAM] related to the presence of non dioxin-like polychlorinated biphenyls (PCB) in feed and food', *The EFSA Journal*, **284**, 1–137.

EFSA (2008a), 'Opinion of the Scientific Panel on Contaminants in the Food chain on a request from the European Commission on marine biotoxins in shellfish – okadaic acid and analogues', *The EFSA Journal*, **589**, 1–62.

EFSA (2008b), 'Opinion of the Scientific Panel on Contaminants in the Food chain on a request from the European Commission on marine biotoxins in shellfish – azaspiracids', *The EFSA Journal*, **723**, 1–52.

EFSA (2008c), 'Scientific Opinion of the Panel on Contaminants in the Food Chain on a request from the European Commission on Polycyclic Aromatic Hydrocarbons in Food', *The EFSA Journal*, **724**, 1–114.

EFSA (2008d), 'European Food Safety Authority. Opinion of the Scientific Panel on Plant Protection products and their Residues to evaluate the suitability of existing methodologies and, if appropriate, the identification of new approaches to assess cumulative and synergistic risks from pesticides to human health with a view to set MRLs for those pesticides in the frame of Regulation (EC) 396/20052', *The EFSA Journal*, **704**, 12–84.

EFSA (2009a), 'EFSA Panel on Contaminants in the Food Chain (CONTAM); Scientific Opinion on Arsenic in Food', *The EFSA Journal*, **7** (10), 1351 (199 pp.). doi:10.2903/j. efsa.2009.1351. www.efsa.europa.eu [accessed 13 April 2012].

EFSA (2009b), 'Scientific Opinion of the Panel on Contaminants in the Food Chain on a request from the European Commission on cadmium in food', *The EFSA Journal*, **980**, 1–139. http://www.efsa.europa.eu/en/efsajournal/doc/980.pdf [accessed 13 April 2012].

EFSA (2009c), 'Scientific Opinion of the Panel on Contaminants in the Food Chain on a request from the European Commission on Marine Biotoxins in Shellfish – Saxitoxin Group', *The EFSA Journal*, **1019**, 1–76.

EFSA (2009d), 'Scientific Opinion of the Panel on Contaminants in the Food Chain on a request from the European Commission on marine biotoxins in shellfish – domoic acid', *The EFSA Journal*, **1181**, 1 – 61.

EFSA (2009e), 'Guidance of the Scientific Committee on a request from EFSA on the use of the benchmark dose approach in risk assessment', *The EFSA Journal*, **1150**, 1–72.

EFSA (2010), 'EFSA Panel on Contaminants in the Food Chain (CONTAM); Scientific Opinion on Lead in Food', *The EFSA Journal*, **8** (4), 1570 (147 pp.). doi:10.2903/j.efsa.2010.1570. www.efsa.europa.eu [accessed 13 April 2012].

EFSA (2011a), 'European Food Safety Authority; Comparison of the Approaches Taken by EFSA and JECFA to Establish a HBGV for Cadmium', *The EFSA Journal*, **9** (2), 2006 (28 pp.). doi:10.2903/j.efsa.2011.2006. www.efsa.europa.eu/efsa-journal [accessed 13 April 2012].

EFSA (2011b), 'EFSA Panel on Contaminants in the Food Chain (CONTAM); Scientific opinion on polybrominated diphenyl ethers (PBDEs) in food', *The EFSA Journal*, **9** (5), 2156 (274 pp.). doi:10.2903/j.efsa.2011.2156. www.efsa.europa.eu/efsajournal [accessed 13 April 2012].

EFSA (2011c), 'EFSA Panel on Contaminants in the Food Chain (CONTAM); Scientific opinion on the risks for public health related to the presence of zearalenone in food', *The EFSA Journal*, **9** (6), 2197 (124 pp.). doi:10.2903/j.efsa.2011.2197. www.efsa.europa.eu/efsajournal [accessed 13 April 2012].

FAO/WHO (1998), *Safety evaluation of certain contaminants in food*, Food Additive Series No. 40, World Health Organization, Geneva.

FAO/WHO (2004), *Safety evaluation of certain food additives and contaminants*, Food Additive Series No. 52, World Health Organization, Geneva. http://whqlibdoc.who.int/ publications/2004/924166052X.pdf [accessed 13 April 2012].

FAO/WHO (2006a), *Safety evaluation of certain contaminants in food*, Food Additive Series No. 56, World Health Organization, Geneva. http://whqlibdoc.who.int/publications/2006/ 9241660554_eng.pdf [accessed 13 April 2012].

FAO/WHO (2006b), *Safety evaluation of certain veterinary drugs*, Food Additive Series No. 61, World Health Organization, Geneva. http://www.who.int/ipcs/publications/jecfa/trs_954.pdf [accessed 13 April 2012].

FAO/WHO (2011a), *Safety evaluation of certain contaminants in food*, Food Additive Series No. 63, World Health Organization, Geneva. http://whqlibdoc.who.int/publications/2011/ 9789241660631_eng.pdf [accessed 13 April 2012].

FAO/WHO (2011b), *Safety evaluation of certain food additives and contaminants*, Food Additive Series No. 64, World Health Organization, Geneva. http://whqlibdoc.who.int/ publications/2011/9789241660648_eng.pdf [accessed 13 April 2012].

IPCS (2005), *Chemical-specific adjustment factors for interspecies differences and human variability: guidance document for use of data in dose/concentration–response assessment*. World Health Organization, Geneva, International Programme on Chemical Safety, Harmonization Project Document, No. 2. http://whqlibdoc.who.int/publications/ 2005/9241546786_eng.pdf [accessed 13 April 2012].

IPCS (2009a), *Environmental Health Criteria 240. Principles and methods for the risk assessment of chemicals in food*, Environmental Health Criteria 240. Geneva, World Health Organization: http://www.who.int/foodsafety/chem/principles/en/index1.html [accessed 13 April 2012].

IPCS (2009b), *Principles for modelling dose–response for the risk assessment of chemicals*, Geneva, World Health Organization, International Programme on Chemical Safety, Environmental Health Criteria, No. 239: http://whqlibdoc.who.int/publications/2009/ 9789241572392_eng.pdf [accessed 13 April 2012].

KROES, R., RENWICK, A. G., CHEESEMAN, M., KLEINER, J., MANGELSDORF, I., PIERSMA, A., SCHILTER, B., SCHLATTER, J., VAN SCHOTHORST, F., VOS, J. G. and WÜRTZEN, G. (2004), 'Structure-based thresholds of toxicological concern (TTC): guidance for application to substances present at low levels in the diet', *Food Chem. Toxicol.*, **42** (1), 65–83.

MEEK, M. E., BOOBIS, A. R., CROFTON, K. M., HEINEMEYER, G., RAAIJ, M. V. and VICKERS, C. (2011), 'Risk assessment of combined exposure to multiple chemicals: A WHO/IPCS framework', *Regul. Toxicol. Pharmacol.* [Epub ahead of print].

MUNRO, I. C., RENWICK, A. G. and DANIELEWSKA-NIKIEL, B. (2008), 'The Threshold of Toxicological Concern (TTC) in risk assessment', *Toxicol. Lett.*, **15**, 180 (2), 151–156.

SCF (2001), Scientific committee on food, *Opinion on the risk assessment of dioxins and dioxins-like PCB in food* (update based on the new scientific information available since the adoption of the SCF opinion of 22 November 2000; adopted by the SCF on 30 May 2001), http://ec.europa.eu/food/fs/sc/scf/out90_en.pdf [accessed 13 April 2012].

Part II

Particular persistent organic pollutants, toxic metals and metalloids

9
Dioxins and polychlorinated biphenyls (PCBs) in foods

D. Schrenk and M. Chopra,
University of Kaiserslautern, Germany

DOI: 10.1533/9780857098917.2.191

Abstract: This chapter discusses the properties, occurrence, sources, exposure and toxicity of two major groups of food contaminants, the polychlorinated dibenzo-*p*-dioxins and dibenzofurans (PCDD/F) and the polychlorinated biphenyls (PCBs). This chapter describes in detail their mode of action, general toxicity, carcinogenicity, developmental and reproductive toxicity, endocrine effects and adverse effects on individual organs and the immune system.

Key words: carcinogenicity, endocrine disruptors, polychlorinated dibenzo-*p*-dioxins, polychlorinated dibenzofurans, polychlorinated biphenyls.

Note: This chapter was first published as Chapter 5 'Dioxins and polychlorinated biphenyls in foods' by D. Schrenk and M. Chopra, in *Chemical contaminants and residues in food*, edited by D. Schrenk, Woodhead Publishing Ltd, 2012, ISBN: 978-0-85709-058-4.

9.1 Introduction

The toxicology of certain halogenated aromatic hydrocarbons, bearing two aromatic rings, i.e., the polychlorinated dibenzo-*p*-dioxins (PCDD), dibenzofurans (PCDF) and PCBs, has gained enormous interest because of the outstanding toxic potency of some congeners. This family of compounds can be considered as one of the most thoroughly investigated in toxicological research with respect to their sources, generation, exposure and toxicity. The toxic congeners show a pronounced chemical stability and persistence, both in the environment and when having been absorbed by organisms. They are highly lipophilic, i.e. almost insoluble in water, and mostly show only minor volatility at ambient temperatures. Their toxicity in humans has become evident during occupational exposure, industrial disasters or in cases of poisoning. Since food is the major source of PCDD/F and PCB background exposure in the general population, this family of contaminants is presented here in detail.

9.2 Properties and occurrence of polychlorinated dibenzo-*p*-dioxins and dibenzofurans (PCDD/Fs)

This section discusses the occurrence of PCDD/Fs in food, as well as their adverse effects in humans and laboratory animals. PCDDs, and structurally related PCDFs, belong to the polyhalogenated aromatic hydrocarbons and are usually termed 'dioxins'. These substances are highly lipophilic and persistent, resulting in their accumulation in the food chain. They are of great toxicological concern and cause a variety of adverse effects in laboratory animals, wildlife and humans. Most of the toxic effects are thought to be mediated by intracellular signalling via the aryl hydrocarbon receptor (AhR).

9.2.1 Physical and chemical properties

PCDDs and PCDFs (Fig. 9.1) structurally consist of a dibenzo-*p*-dioxin or a dibenzofuran scaffold, respectively. These scaffolds can be substituted with up to eight chlorine atoms at the two benzene rings. Depending on the chlorination pattern of the substances, different congeners can be denoted. For PCDDs there are 75 possible congeners and 135 for PCDFs. The number and position of the chlorine substituents determine both their degradability in the environment and their toxic properties. Those with chlorine atoms at all four lateral ring positions 2, 3, 7 and 8 are of the most concern. In total there are 17 PCDD/Fs with chlorine substituents at these positions.

PCDD/Fs are planar, highly lipophilic substances. When pure they are colourless, crystalline solids with very low vapour pressure. These substances are characterised by a high chemical and thermal stability. They are practically insoluble in water but readily soluble in most organic solvents. PCDD/Fs bioaccumulate in the food web due to both their stability and lipophilicity. These substances are listed as persistent organic pollutants (POP) by the World Health Organization (WHO). In the environment, PCDD/Fs are almost not degraded at all, the only known abiotic reactions being photolysis and photo-oxidation. Bacteria hardly metabolise these substances, while some fungi do to a very minor extent. Because of their persistence, they can be transported all around the globe bound to aerosol particles. This makes them a global problem, and PCDD/Fs can be detected in human, animal and environmental samples all over the world, even in remote polar regions.

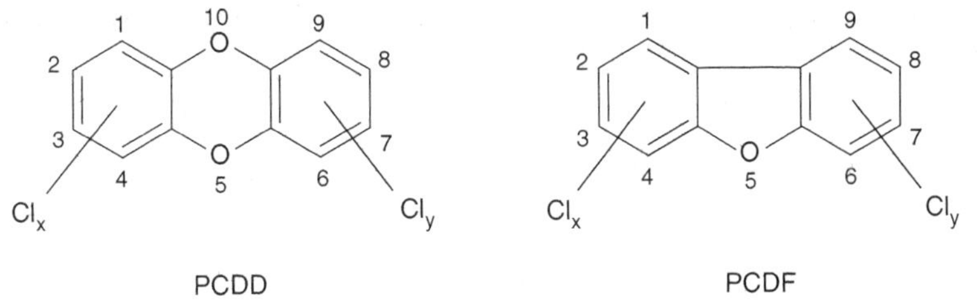

Fig. 9.1 Chemical structure of polychlorinated dibenzo-*p*-dioxins (PCDDs) and polychlorinated dibenzofurans (PCDFs).

9.2.2 Sources of PCDD/Fs

PCDD/Fs have never been intentionally produced industrially and they have no commercial application. They are synthesised solely for research and analytical purposes. Nevertheless, there are several ways in which PCDD/Fs can be formed and find their way into the environment.

These substances are formed in metal smelting facilities and during the production of chlorophenols and chlorophenoxy herbicides. With the synthesis of polychlorinated phenols, polychlorinated sodium phenolates can be converted into PCDD/Fs under the effect of heat. Additionally, PCDD/Fs are formed during paper-pulp bleaching with free chlorine, and during the incineration of organic matter in the presence of chlorine-containing compounds, e.g. in municipal waste incineration. Since this class of substances is also formed during the combustion of fossil fuel, humans have always been exposed to PCDD/Fs, but mostly since the onset of the industrial revolution (National Research Council, 2006). Starting from the 1940s, human exposure to PCDD/Fs has risen steeply because of the production and usage of chlorophenols and chlorophenoxy herbicides. Even higher exposure has resulted from accidents in these industrial sectors. From 1980 onwards, the release of PCDD/Fs into the environment has been gradually declining because of effective environmental protection measures. Values have been stalling at a rather low level since 2000. By then, total PCDD/F output into the environment had declined to approximately 10% of the initial 1987 values (National Research Council, 2006). The most toxic compound among the PCDD/Fs is 2,3,7,8-tetrachlorodibenzo-*para*-dioxin (TCDD). TCDD has attracted international attention mainly from two incidents, the Seveso disaster and the poisoning of Victor Yushchenko.

9.2.3 Occurrence in food and human exposure

PCDD/Fs are formed by a variety of industrial and other processes. Because of their persistence and chemical stability, these contaminants can be found ubiquitously in the environment, soil, sediments and air (particle-bound). Due to their lipophilicity, they accumulate in the food chain and ultimately

Table 9.1 Average levels (means) of PCDD/F (expressed in WHO-TEQ) in various food groups in Europe (EFSA, 2005)

Food group	Subgroup	Unit	Mean	95 (Percentile)
Meat and meat products	Ruminants	pg/g fat	2.61	2.92
	Poultry		0.72	1.90
	Pigs		0.47	1.17
Liver and products (terrestrial animals)		pg/g fat	3.34	13.64
Muscle meat fish	Eel	pg/g fresh weight	2.59	6.24
	Others		1.89	8.97
Milk and dairy products incl. butter		pg/g fat	0.78	2.86
Hen eggs and egg products		pg/g fat	0.94	3.04
Mixed animal fats		pg/g fat	0.89	1.17
Vegetable oils and fats		pg/g fat	0.20	0.41
Marine oils		pg/g fat	0.51	1.90
Fish liver and products		pg/g fresh weight	8.52	21.65
Fruits, vegetables and cereals		pg/g fresh weight	0.45	1.90
Infant and baby food		pg/g fat	0.20	0.73

enrich in human fatty tissue and the liver. PCDD/Fs can be readily detected in all kinds of food, especially in fatty food of animal origin. The total PCDD/F content of a sample is usually given in TCDD equivalents (TEQ), a sum parameter, which takes into account the relative toxicities of the individual congeners, using the so-called toxic equivalency factor (TEF) concept (see Section 9.3.3).

A study conducted by the European Union assessed the PCDD/F content of more than 7200 samples collected from 19 Member States in the period from 1999 to 2008. Results for selected food groups are presented in Table 9.1.

Dioxins from food sources make up 90% to 95% of total human exposure. Two-thirds of this is taken up with animal-derived food, especially dairy products (39%) and meat and meat products (28%). Fish is generally contaminated to a higher degree but consumed in smaller quantities. Fish and eggs contribute 14% and 10%, respectively, to the total PCDD/F exposure in humans.

PCDD/F load in human breast milk samples in Germany averaged 35 pg WHO-TEQ/g fat between 1985 and 1989; from 1990 onwards, levels have declined steadily to about 10 pg WHO-TEQ/g fat in 2002/2003. This trend can be observed for some food groups as well, especially fish, where a considerable reduction in the TEQ-levels from 1993 onwards could be observed. The same holds true for cow's milk e.g. in France (Durand et al., 2008), where WHO-PCDD/F-TEQ values decreased from approximately 2.3 pg/g fat in 1987 to 0.4 pg/g fat in 2007. Levels have been fairly stable since 1999.

Two incidents brought the occurrence of dioxin-like substances in food to public awareness around the world. In 1999, high levels of dioxins were found in poultry and eggs from Belgium, probably resulting from animal feed contaminated with illegally disposed PCB-based waste oil. In early December 2008, a global recall of Irish pork was initiated as a result of a subset of the national pork output being contaminated with dioxin-like PCBs (dl-PCB). This contamination was traced back to animal feed as well.

The total intake of PCDD/Fs varies considerably, depending on each subject's nutrition. A diet rich in animal products results in a higher mean intake of PCDD/Fs. For Germany, it was estimated that an adult took in an average of 1.9–2.6 pg WHO-PCDD/F-TEQ/kg body weight per day in the years from 1987 to 1990. For the years from 2000 to 2003, the respective value was estimated to be 0.27 pg WHO-PCDD/F-TEQ/kg body weight per day. For the general US population estimated mean daily intakes, including dl-PCB, were 2.4 and 2.2 pg TEQ/kg body weight for men and women, respectively. Because of the high PCDD/F content in human milk, breastfed infants are exposed to rather high levels of these substances. It was estimated that the mean daily intake of TEQ for US breastfed infants during the first year of life was 42 pg/kg body weight (Schecter et al., 2001).

Total body PCDD/F-burden in humans increases with age because of the accumulating effect of these substances and their long half-life of approximately three to more than ten years in humans. Human body burden has declined over the last decades and averages 20–30 pg PCDD/F-TEQ/g fat in adults, with TCDD accounting for 2–3 pg/g fat.

9.3 Toxicity of PCDD/Fs

The toxicity of PCDD/Fs is a consequence of their high environmental and metabolic stability, their lipophilicity and their affinity to the arylhydrocarbon receptor. The latter effect is responsible for most if not all adverse effects of PCDD/Fs including acute and chronic organ toxicity, developmental and reproductive toxicity, immunotoxicity and carcinogenicity.

9.3.1 Toxicokinetics

PCDD/Fs are readily absorbed through the skin, lung and the gastrointestinal tract. Bioavailability is generally higher for the last two than for dermal absorption. Dioxins are distributed throughout the body, mostly bound to lipoproteins, and accumulate in adipose tissue and liver.

Highly chlorinated congeners are metabolised to a far smaller degree than lower chlorinated congeners. The main routes of metabolism are reductive dechlorination, hydroxylation and subsequent conjugation by phase II-enzymes. Metabolism of PCDD/Fs seems to be detoxifying, i.e., it is not assumed that metabolised PCDD/Fs are harmful. Although metabolism of TCDD was shown for some species (rat, dog), it is unlikely that human xenobiotic metabolism is capable of detoxifying TCDD, while human CYP enzymes are capable of metabolising less chlorinated congeners (Inouye

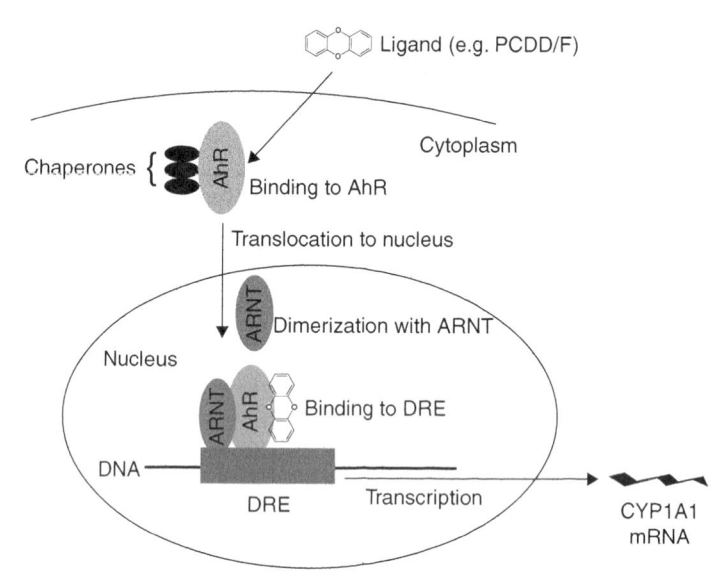

Fig. 9.2 Pathways of arylhydrocarbon receptor activation leading to CYP1A1 induction.

et al., 2002). Whether the parent compounds or their metabolites are excreted is highly dependent on the congener itself and the species to which it was administered.

Elimination half-life for TCDD in rodents ranges from 10 to 30 days, while in humans it is in the range of three to more than ten years, depending on body burden (Aylward *et al.*, 2005). The same holds true for other PCDD/F congeners. An explanation for the extremely long retention time of these substances in humans might be the virtually non-existent metabolism.

9.3.2 Mode of action

PCDD/Fs themselves are not chemically reactive, neither are they activated metabolically. Most, if not all, of the biological actions of these substances are mediated by their binding to an intracellular receptor protein, the AhR (Fig. 9.2). This transcription factor is expressed in most tissues among all vertebrates investigated so far. The AhR is a member of the basic helix-loop-helix (bHLH) protein family. It is localised in the cytosol associated with multiple chaperones.

Binding of PCDD/Fs, dioxin-like PCBs, and some natural compounds to this receptor leads to the dissociation of the chaperones. The AhR-ligand complex then translocates to the nucleus where it dimerises with a second member of the bHLH-family, the arylhydrocarbon receptor nuclear translocator (ARNT). The ligand-activated AhR/ARNT dimer binds to recognition sequences in the DNA, so-called dioxin responsive elements (DRE) (also termed AhR-(AHRE) or xenobiotic response elements (XRE)). These sequences are located in the promoter region, upstream of the transcription start of AhR target genes. The binding of the AhR-complex to these sites leads to chromatin remodelling and binding of general transcription factors to the DNA. This results in enhanced transcription of AhR target genes (Fujii-Kuriyama and Kawajiri, 2010).

The best investigated AhR target gene is the cytochrome P450-dependent monoxygenase 1A1 (CYP1A1). This gene is almost not expressed constitutively in the liver of rodents. Following activation of the AhR, e.g. by TCDD, the expression of this gene is massively induced. The induction of CYP1A1 is seen as a very sensitive marker for the activation of the AhR, and is used as an indication for the presence of PCDD/F or related substances in sample materials. A variety of different AhR target genes have been identified (Beischlag et al., 2008). There is a common response towards an activation of the AhR shared in most cell types, which includes the induction of enzymes involved in the metabolism of xenobiotics (CYPs, UGTs, GSTs). Besides these commonly shared target genes, the AhR also regulates a number of functionally unrelated genes, the pattern of which differs between organisms and cell types. This circumstance is probably due to the fact that the presence of DREs in a certain gene might not be well conserved across species, rather than to a species difference in AhR properties itself. Though the possibly severe toxic effects of PCDD/Fs seem mostly dependent on AhR function, they can probably not be reduced to the induction of specific target genes only. Today it is known that the AhR not only acts by its classic genomic pathway, i.e. its dimerisation with ARNT and subsequent effects on gene transcription. In fact, it can be shown that the receptor interferes with a variety of intra- and intercellular signalling pathways, e.g. calcium homeostasis, protein kinases and different receptors. This makes the picture of the toxic actions of PCDD/Fs more complex than was first anticipated (Matsumura, 2009).

The AhR evolved long before PCDD/Fs were released into the environment in considerable amounts. This justifies the assumption that the receptor has to have other functions beside the cellular response to environmental toxins. Studies employing AhR-deficient mice did show that the receptor plays a role in the normal development of the liver and the immune system (Gonzales and Fernandez-Salguero, 1998). The search for endogenous ligands of the AhR, as well as for alternative activation routes for the receptor, is ongoing.

9.3.3 Relative toxicities – the TEF concept

Dioxin-like contaminants usually do not occur as single substances in the environment but rather as complex mixtures composed of PCDDs, PCDFs and dioxin-like PCBs. These mixtures may vary considerably at different locations, and because of diverse exposition paths. This makes risk assessment for human exposure particularly difficult. To account for these difficulties, the WHO (van den Berg et al., 2006) and other institutions have adopted the TEF concept in order to calculate toxic equivalents (TEQ) in complex dioxin

Table 9.2 TEF values for selected PCDD, PCDF and dl-PCB (van den Berg, 2006)

PCDD	WHO-TEF (2005)	PCDF	WHO-TEF (2005)
2,3,7,8-TCDD	1.0	2,3,7,8-TCDF	0.1
1,2,3,7,8-PeCDD	1.0	2,3,4,7,8-PeCDF	0.3
1,2,3,4,7,8-HxCDD	0.1	2,3,4,7,8-PeCDF	0.03
1,2,3,4,6,7,8-HpCDD	0.1	1,2,3,4,7,8-HxCDF	0.1
OCDD	0.01	1,2,3,4,6,7,8-HpCDF	0.01
	0.0003	OCDF	0.0003
Non-ortho-PCB		Mono-ortho-PCB	
PCB 126	0.1	PCB 105	0.0003
PCB 169	0.3	PCB 156	0.0003
PCB 81	0.0003	PCB 167	0.0003
PCB 77	0.0001	PCB 189	0.0003

mixtures. Toxicities of mixtures can be calculated with this concept in relation to TCDD, the most potent PCDD/F congener, which was assigned a TEF of 1.0. All other congeners were assigned lower TEF values, as determined from their toxic potency. Table 9.2 lists TEF the same or values for selected PCDDs, PCDFs and PCBs. Relative toxicity is highest with 2,3,7,8-TCDD, increasing substitution of the benzene rings with chlorine, up to octachloro-congeners, resulting in lower toxicity and lower TEF values.

Chemicals have to meet the following requirements to be included into the TEF concept:

(a) structural similarities to TCDD;
(b) binding to the AhR;
(c) AhR-mediated biochemical and toxic effects, resembling those of TCDD;
(d) persistence and accumulation in the environment/food web.

To apply the TEF concept, TEF values for every congener detected in the dioxin mixture are multiplied with the respective amount (e.g. in ng). The products for all congeners are summarised and given as (ng) TEQ (van den Berg et al., 2006).

9.4 Toxic effects of PCDD/Fs in humans and experimental animals

The toxicity of PCDD/Fs in adult and juvenile individuals mostly affects the development and differentiation of a variety of tissues of liver, endocrine organs and immune system in the embryo. Many effects have been described for experimental animals but not for humans, where dermal effects (chloracne) and changes in certain blood parameters (clinical chemistry) are among the most well-described effects. From animal experiments, it is known that the dose level is of outstanding importance since, e.g., certain effects such as wasting (progressive loss of body weight and lethality) are observed only at high doses, while others occur at lower doses.

9.4.1 Toxicity in laboratory animals

PCDD/Fs exert a broad range of toxic effects. The effects themselves, as well as their severity, vary among different species, strains within a species, individual organs and tissues, as well as with the age and sex of the animals. The marked differences in the relative toxicities of these substances might result from genetic variations of the AhR and, thereby, from different affinities of receptor binding of ligands. For example, TCDD binds to the AhR of most rat strains with a much higher affinity than to human AhR. It has been postulated that all toxic effects of PCDD/Fs could be observed in all species if early lethality had not prevented their onset in more sensitive species.

The toxic potencies of different congeners correlate with their ability to bind to and activate the AhR. The same holds true for single congeners in different species. Although it is fairly clear that PCDD/F toxicity depends on AhR-binding, this and especially the effects on target gene transcription cannot fully explain the severe toxicities of these substances. A peculiar effect that has been observed is that PCDD/F toxicity in laboratory animals usually appears in weeks or months following exposure.

TCDD is the most toxic PCDD/F congener, i.e., most animal studies trying to elucidate the molecular mechanisms of PCDD/F toxicity have been conducted with this substance. PCDD/Fs, and especially TCDD, are extremely toxic substances. Oral LD_{50} values for TCDD vary between different species: for Guinea pigs it is 0.5, for rats 20, for mice 200 and for hamsters 1000 µg/kg body weight. Furthermore, among different strains within one species, LD_{50} values can vary by a factor of up to 100. There are no reliable LD_{50} values given for humans. However, it has been estimated that it would be in the range of 4–6 mg/kg body weight, i.e., higher than for most laboratory species.

It takes up to several weeks for the rat to decease following a lethal dosage of TCDD. A very characteristic effect of single high-dose administration of TCDD in laboratory animals is severe loss of body weight (wasting syndrome). This is accompanied by a decrease in food intake and a reduction of body fat. Death by wasting is the most frequently observed cause of death following oral exposure to TCDD in small laboratory animals. To date, the molecular mechanisms underlying death by TCDD-induced wasting syndrome are unknown. Since other PCDD/F congeners also cause wasting, it can be assumed that the AhR might be involved in this toxic effect.

PCDD/Fs, in sublethal doses, lead to hepatotoxicity, accompanied by damage to liver cells in most animal species. They also cause hepatomegaly, an abnormal liver growth resulting from hypertrophy and hyperplasia of parenchymal liver cells, as well as a rapid decrease in liver retinol levels in laboratory animals.

Furthermore, TCDD cause atrophy of lymphatic organs, such as thymus, spleen and lymph nodes, in all species tested. In rhesus monkeys and nude mice TCDD causes dermal symptoms resembling human chloracne (see below). Besides that, toxic effects on the organs of the reproductive system, including hormone producing organs (testes, prostate, uterus, thyroid), have been observed in different species.

Subchronically and chronically, PCDD/Fs exert similar effects as they do acutely. The substances are hepatotoxic and cause dermal effects. In all species assessed TCDD causes hepatomegaly. Further on, PCDD/Fs cause severe weight loss in laboratory animals. Chronic exposure to TCDDs also results in significantly decreased weight gain.

Estimated No Observed Adverse Effect Levels (NOAEL) for subchronic toxicity of TCDD are 0.6 ng/kg body weight per day in Guinea pigs, 10 in rats and 100 in mice.

There are several animal studies on the chronic effects of TCDD, the most conclusive being a 2-year study in rats with daily doses ranging from 1 to 100 ng TCDD/kg body weight. The study showed that female animals were more sensitive than males with respect to most endpoints. The most consistent effects were observed in the liver. TCDD caused porphyria and an elevation of serum amino-transferase activities. The livers of the animals showed multiple, degenerative, inflammatory and necrotic lesions. Furthermore, hyperplasia of the liver parenchymal tissue and canaliculi were observed. The NOAEL in this study was 1 ng TCDD/kg body weight per day (Kociba *et al*., 1978). In humans, toxic effects resulting from chronic exposure to high levels of PCDD/Fs are mainly chloracne and hepatotoxicity. In addition, discussion has been ongoing about a variety of other effects, including reproduction toxicity, immunotoxicity and oncogenic effects. These issues will be presented in the following sections.

9.4.2 Toxicity in humans

The main acute effect of PCDD/Fs in humans is the so-called chloracne. This dermal condition is characterised by follicular hyperkeratoses, a thickening of the *stratum corneum*. Blackheads, cysts and pustules erupt on the cheeks, behind the ears, in the armpits and groin region, but can also cover other regions of the body. The period of time from dioxin exposure to the occurrence of chloracne depends on dosage and can range from days, following very high doses, to weeks or months with lower doses.

There is no causal therapy for chloracne, but symptomatic treatment is available with antibiotics and retinoids. Nevertheless, chloracne is highly resistant to treatment and can sustain for years. One approach is lowering toxin body burdens, e.g. by giving Olestra®, a non-absorbable synthetic fat substituent to patients. This is assumed to wash out the highly lipophilic PCDD/Fs from the body.

Dioxins and PCBs in foods 201

The exact causal mechanism of chloracne is unknown to date. Discussion has been ongoing linking it to growth factors affecting the proliferation and differentiation of epidermal cells. A general inflammatory response of the skin might also be involved.

Other acute toxic effects of PCDD/Fs have been reported in addition to chloracne in some cases. They resemble the effects observed in laboratory animals, e.g. hepato- and immuno-toxic effects. TCDD at high doses also leads to long-lasting nausea, gastrointestinal symptoms and weight loss.

9.4.3 Reproductive and developmental toxicity

TCDD impairs reproduction in laboratory animals. On the one hand it directly affects reproductive organs in both males and females, while on the other hand TCDD interferes with steroid hormones. In particular, TCDD acts on estradiol homeostasis by inducing CYP1 enzymes, which can metabolise estradiol. This effect is thought to contribute to the anti-oestrogenic actions of PCDD/Fs. Furthermore, the AhR directly affects oestrogen receptors, enhancing their degradation by the proteasome, and a direct cross-talk between AhR and oestrogen receptor signalling has also been shown.

TCDD also exerts developmental toxicity and acts as a teratogen. Prenatal exposure to TCDD leads to decreased viability in virtually all animal species tested. These effects might be caused mainly by the toxicity of TCDD to the dams rather than by direct embryotoxicity. In some experiments, it has been shown that certain effects are induced in the offspring without pronounced toxicity to the dams. The offspring of exposed animals showed impaired reproduction and structural abnormalities. Prenatal exposure to TCDD led to an increase in the incidence of cleft palate, hydronephrosis and thymic atrophy in mice. In rats, TCDD caused intestinal bleeding besides other effects.

PCDD/Fs and PCBs also seem to exert developmental and reproductive toxicities in humans. Two incidents in Asia point towards this assumption: the Yusho incident in Japan and the Yu-Cheng incident in Taiwan. Mothers exposed to PCBs and PCDFs gave birth to hyperpigmented children with increased perinatal mortality. Exposure to these substances caused delayed development, both pre- and post-natally, and affected the central nervous system. Although these effects could be seen without pronounced toxicity to the mothers, nevertheless most of them suffered from chloracne. Long-term exposure to TCDD in men led to a decrease in testosterone levels.

9.4.4 Immunotoxicity

A number of animal studies have identified the immune system as a target for PCDD/F toxicity. These substances cause atrophy of lymphatic organs and cause an inhibition of both innate and adaptive immune responses. Because of the variety of immunotoxic effects observed with PCDD/Fs it seems plausible that these pollutants affect more than just a single cell type. Furthermore, toxic effects in other than lymphatic organs seem to contribute to the immunotoxicity of PCDD/Fs.

Various mechanisms are being discussed on how PCDD/Fs affect the immune system. It has to be assumed that these substances interfere with hormone signalling pathways. Another hypothesis has arisen from the observation that TCDD induces apoptosis in lymphocytes. This could help to explain both atrophy of lymphatic organs and the inhibition of immune responses.

The AhR is thought to mediate the immuno-suppressive effects of PCDD/Fs, since the relative potency of immuno-suppression of different congeners correlates to their TEF values. A rather high expression of the AhR in regulatory T-cells also suggests a role of this receptor in the immune function, i.e., activation of a T-cell subtype that represses a number of immune functions.

The human immune system appears to be affected by PCDD/F, too, although there are only very limited data on this issue. A common effect seems to be a slight decrease in CD4+-cells following exposure to these substances.

9.4.5 Carcinogenicity and genotoxicity

PCDD/Fs, and among those especially TCDD, are so-called multiple-site, multiple-species carcinogens. That means they cause cancer in different animal species in various organs. Although it is beyond debate that these substances cause cancer, the exact mechanisms of their carcinogenicity are still unknown (Knerr and Schrenk, 2006a).

Dioxins are not DNA-reactive, i.e., they do not bind covalently to nucleic acids. That is why other mechanisms have been proposed by which these substances cause tumours. A large number of studies have been conducted, most of them with the model compound TCDD. It was discussed that TCDD causes oxidative stress, and that it metabolically activates oestrogens by inducing CYP1 enzymes. Reactive oestrogen metabolites might cause DNA damage. PCDD/Fs interfere with multiple signalling pathways involved in proliferation and differentiation. By means of affecting normal cell homeostasis, these substances might shift cells towards growth and malignancy. Furthermore, TCDD inhibits apoptosis in various cellular systems, and this might also enable damaged cells to survive and proceed on their way to malignancy.

Lifelong exposure to TCDD has resulted in a pronounced increase in the incidence of hepatocellular carcinoma in female rats, whereas the carcinogenic potency of TCDD in the liver was much less pronounced in males. Furthermore, TCDD caused an increase in tumour incidence in a number of different tissues in both sexes, e.g., tongue, nasal turbinates, hard palate and lung. In addition, male rats developed thyroid tumours. Interestingly, TCDD caused a decrease in the occurrence of some oestrogen-dependent tumours, e.g. in the uterus and mammary gland (Kociba et al., 1978). It has been discussed that an anti-oestrogenic effect of TCDD, based, e.g., on the enhanced catabolism of estradiol might explain this finding.

TCDD also causes liver tumours in mice. In contrast to rats, no sex-difference could be observed. As with rats, TCDD also caused an increased number of tumours in other tissues besides the liver.

In initiation-promotion studies on the livers of female rats, TCDD by itself did not initiate tumour growth. Following initiation of the animals with genotoxic carcinogens, however, TCDD was found to be a very potent tumour promoter. In fact, this substance was evaluated the most potent tumour promoter in rodent liver.

Oestrogens might play a decisive role in tumour promotion by TCDD, since mainly female rats are responsive to the liver tumour-promoting effect of this substance in initiation-promotion studies. They lose their responsiveness when their ovaries are removed. Exposure to estradiol makes the ovariectomised female rats responsive again. TCDD is also a potent tumour promoter in mouse skin when applied topically.

There have been only very limited studies assessing the carcinogenicity of other PCDD/Fs besides TCDD, but it was proposed that all 2,3,7,8-PCDD/F congeners might be rodent carcinogens and act as liver tumour promoters. Since their relative potencies correlate with their respective TEF values, and animals bearing a low-affinity AhR lose their sensitivity towards the carcinogenicity of TCDD, the AhR seems to be involved in the carcinogenic effects of PCDD/F.

TCDD was evaluated as carcinogenic to humans (group 1 carcinogen) by the International Agency for Research on Cancer (IARC) in 1997. This classification was based on the following supporting evidence:

- TCDD is a multi-site carcinogen in experimental animals that has been shown by several lines of evidence to act through a mechanism involving the AhR;
- this receptor is highly conserved in an evolutionary sense and functions the same way in humans as in experimental animals;
- tissue concentrations are similar both in heavily exposed human populations in which an increased overall cancer risk was observed and in rats exposed to carcinogenic regimens in bioassays.
- Other PCDD/Fs are not classifiable as to their carcinogenicity to humans (group 3) due to inadequate data.

In the IARC 1997 evaluation, the strongest evidence for the carcinogenicity of TCDD from epidemiological studies of high-exposure cohorts was for all cancers combined, rather than for any specific site. The relative risk for all cancers combined in the most highly exposed and longer-latency sub-cohorts was 1.4. More recent data showed this relative risk to be in the range of 1.4–2.0. It should be borne in mind that the general population is exposed to levels far below those experienced by these cohorts (IARC, 1997).

The most recent data on the Seveso-cohort did not find an increase in the all-cancer risk. An excess of lymphatic and hematopoietic tissue neoplasms was observed. An increased risk of breast cancer has been detected in females 15 years after the accident. Interestingly, no cancer cases were observed among subjects diagnosed with chloracne shortly after the accident (Pesatori et al., 2009).

9.4.6 Case studies

Seveso accident
An accident occurred in a chemical plant where 2,4,5-trichlorophenol was produced in Seveso, Italy in 1976. Due to an exothermic reaction in a production reactor, its content was released into the environment. The mixture of chemicals was estimated to contain between one and three kilograms of TCDD. Because of the events that have taken place in Seveso, TCDD is sometimes called 'Seveso poison'. Following this incident, thousands of birds and small animals died in the vicinity of the factory over the next couple of days. Approximately 200 people had to be treated for severe chloracne. Today, almost 35 years after the Seveso accident, it appears that highly exposed subjects who lived in the direct vicinity of the plant have an excess risk for some types of cancer as well as for cardiovascular diseases.

The case of Victor Yushchenko
In September of 2004, during the presidential campaign in Ukraine, the opposition candidate Victor Yushchenko was admitted into hospital with severe symptoms. He suffered from abdominal and back pain, multiple organ inflammation, facial paralysis and chloracne. Three months later it was made public that Victor Yushchenko had been poisoned with TCDD (Sorg et al., 2009).

9.4.7 Regulations of PCDD/Fs
In 1998, the WHO established a tolerable daily intake of 1–4 pg TEQ/kg body weight for PCDD/F and dioxin-like PCB (van den Berg et al., 2000). The European Union (EU) has set the tolerable weekly intake to 14 pg TEQ/kg body weight. Furthermore, the EU has set maximum values for PCDD/Fs in foodstuff (Table 9.3) (EU, 2006). There are no tolerances or other administrative levels for dioxins in food or feed in the USA and the FDA considers all detectable levels to be of concern. In Japan the annual emission of TEQ is regulated (http://www.env.go.jp/en/laws/chemi/dioxin.pdf).

9.4.8 Summary of PCDD/Fs
PCDD/Fs exert a variety of toxic effects. Although they have never been produced industrially they can be found ubiquitously in the environment. Humans are inevitably exposed to these substances, due to bioaccumulation along the food chain.

Table 9.3 Maximum regulatory values (in WHO-TEQ) for PCDD/F in food in the EU

Food group	Subgroup	Unit	Value
Meat and meat products	Ruminants	pg/g fat	3
	Poultry		2
	Pigs		1
Liver and products (terrestrial animals)			6
Muscle meat fish and fish products excl. eel		pg/g fresh weight	4
Raw milk and dairy products incl. butter		pg/g fat	3
Hen eggs and egg products			3
Mixed animal fats			1
Oils	Vegetable oils and fats		0.75
	Marine oils		2

PCDD/Fs, and among those especially the most potent congener TCDD, cause acute toxicity, they are hepatotoxic, immunotoxic, teratogenic in rodents and carcinogenic. Most, if not all toxic effects of these substances are believed to be mediated by high-affinity binding to the cytosolic ligand-dependent transcription factor AhR. Nevertheless, effects on the transcription of specific AhR target genes have so far failed to convincingly explain the whole pattern of PCDD/F toxicities.

The human AhR appears to be less sensitive to PCDD/Fs than that of most small laboratory animals. It is beyond doubt that exposure to high levels of these substances causes acute toxic effects in humans and might also lead to a statistically significant increase in the incidence of a number of malignancies. The risk of long-term exposure to the small amounts of PCDD/Fs that humans are exposed to by food every day cannot be adequately assessed. It seems rather unlikely that average exposure to these substances in the range of below 1 pg TEQ/kg body weight per day causes adverse effects in humans.

9.5 Properties and occurrence of PCBs

The PCBs were first identified as environmental pollutants because of their oustanding tendency to persist in the environment and to accumulate in many organisms. The physico-chemical properties of this class of compounds are responsible for their chemical stability, lipophilicity, limited volatility and relative resistance towards metabolic attack. Thus, they play an important role in the adverse effects of PCBs.

9.5.1 Physical and chemical properties

Technical PCBs are colourless, odourless highly viscous liquids that were used in many technical applications, such as hydraulic and fire-protecting fluids, paintings, sealings, etc. The PCB family comprises 209 substitution isomers

Fig. 9.3 Chemical structure of PCBs. (a) 'dioxin-like' PCB congener and (b) 'non-dioxin-like' PCB congener.

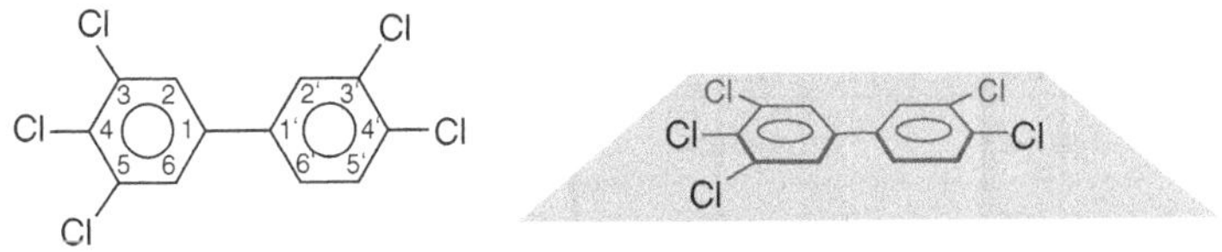

3, 3′, 4, 4′, 5-pentachlorobiphenyl (PCB 126)

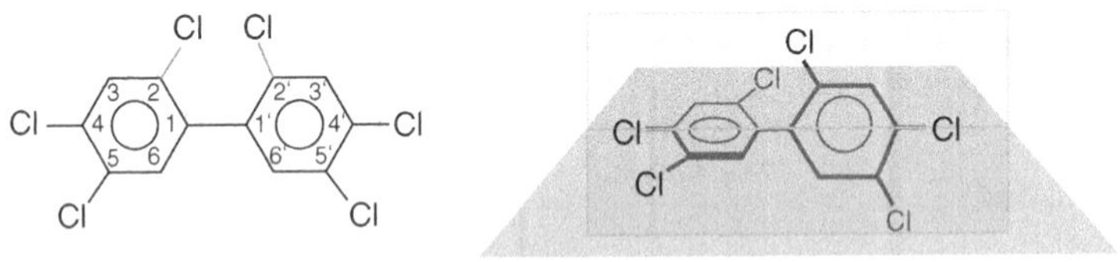

2,2′, 4,4′,5,5′-hexachlorobiphenyl (PCB 153)

(congeners), which are classified according to Ballschmiter with respect to their chlorination pattern. The general structure is given in Fig. 9.3.

According to this nomenclature, the congeners are attributed with increasing numbers in a systematic way according to an increasing number of chlorine substituents. The congener, 2,2′,3,4,4′,5,5′-heptachlorobiphenyl, e.g., is called PCB 180.

Based on biochemical considerations, the PCBs can be divided into two groups, i.e., the dioxin-like (dl) and the non-dioxin-like (ndl) PCB. The dl-PCB share several structural properties with the highly toxic PCDD/F congeners, i.e., they are likely to occur in a coplanar configuration of the two rings, and thus can bind to the Ah receptor and activate it (Fig. 9.2). The lack of chlorine substituents at the four *ortho*-positions, together with chlorination at the lateral positions, is a crucial requirement for coplanarity and relevant AhR-binding. In contrast, single or multiple chlorination at the *ortho*-position(s) lead(s) to a more or less complete loss of dioxin-like properties. The ortho-substituents prevent coplanarity in spite of the theoretical possibility of free rotation of the C–C bridge between the two aromatic rings (Fig. 9.3).

The pure PCBs are solid compounds, while the technical mixtures are odourless, highly viscous oils. Due to their pronounced lipophilicity, they accumulate in the lipid portion of organisms, foods, etc. In an aqueous environment they are bound to particles, sediments, etc. and show a marked tendency to accumulate in organisms. The volatility of the higher chlorinated congeners is very low, i.e. they occur in the atmosphere mostly in a particle-bound manner. A number of lower chlorinated congeners, however, exhibit some volatility and can be detected in the gas phase, e.g. in indoor air. Both particle-bound and gaseous PCB can be transferred through the atmosphere over long distances reaching remote areas of the world, in particular polar regions. Their thermal and chemical stability was a major reason for their widespread use but turned out to be a major disadvantage for the environment. Furthermore, most PCBs are highly resistant to metabolic degradation and, once taken up by an organism, can persist there for long periods of time. These properties are characteristic of POPs, which have been banned from production and worldwide use.

9.5.2 Sources of PCBs

About 1 million metric tons of technical mixtures (Aroclor, Kaneclor, etc.) have been manufactured worldwide since 1920. Because of their outstanding persistence in the environment, occurrence of PCBs in food has to be anticipated for many decades to come.

During the Vietnam War a chloro-organic mixture termed 'Agent Orange' was sprayed as a defoliant over wide areas. PCBs and PCDD/Fs were major contaminants in these mixtures, leading to an increased exposure of the local population. Furthermore, increased levels were also found for the US airmen and other personnel involved in the spraying procedure. Systematic epidemiological data have mainly been reported for the latter cohorts.

Due to their specific properties, PCBs show a local persistence e.g., in the vicinity of former production, storage or use of technical mixture. Furthermore, river sediments downstream of such areas are particularly contaminated with respective consequences for the levels in fish. In general, PCBs accumulate in the food chain. Relatively high levels are thus found in meat and dairy products, fish and crustaceans, while plant products show lower levels. Among marine organisms, predator fish and mammals ranking high in the food chain show the highest levels of contamination in comparison to other fish species or to terrestrial plant-eating species.

9.5.3 Occurrence in food and human exposure

As a parameter for the PCB contents in environmental samples, tissue samples, food, etc. the sum of six abundant PCB congeners is widely used. In many food samples the sum of these congeners, i.e., 28, 52, 101, 138, 153 and 180 is about half of the total amount of PCB.

Food is the major source for human background exposure to PCBs in the general population. Relatively high mean levels for the sum of the six indicator PCBs are found in food samples of animal origin. Except in fish, the sum of the six indicator PCBs in the remaining food categories of animal origin are found to range in Europe between 2.6 and 12.7 ng/g lipid. An overview of the mean levels of the six indicator congeners in various food categories is given in Table 9.4 (EFSA, 2005).

Fish oil, and fish and fishery products are examples of foods having higher contamination with mean levels for the six indicator PCBs of 70.2 ng/g lipid

Table 9.4 Average levels (means) of PCBs (sum of the six indicator congeners) in various food groups in Europe (EFSA, 2005)

Food group	Subgroups	Unit	Mean	Range[1]
Cereals and cereal products		pg/g product	21.3	0.28–89.8
Fruits and vegetables		pg/g product	49.5	11.8–101
Eggs		ng/g fat	6.60	1.03–21.9
Fats and oils	Vegetable oil		5.05	0.66–13.7
	Animal fats		2.61	0.80–6.37
	Fish oil		70.2	0.53–169
Fish and fish products		ng/g fresh weight	12.5	0.42–30.6
Meat and meat products	Poultry	ng/g fat	12.7	1.76–24.0
	Ruminants		9.53	2.80–16.8
	Pork		6.80	0.66–13.5
	Liver (terrestrial animals)		6.79	0.30–14.5
Milk and dairy products			10.7	6.50–15.0

[1] Range between maximum and minimum values.

and 12.5 ng/g whole weight, respectively. The wide range of contamination, which was found to be 0.42–30.6 ng/g wet weight for fish and fishery products, is probably due to the large variety of species analysed. These include farmed as well as wild fish, with significantly different fat contents, fish from various geographical regions, and fish belonging to different levels of the trophic web.

An analysis of PCBs in human milk pools from 18 European countries revealed a mean sum of the six indicator PCBs of 210 ng/g lipid, with a maximum value of 1009 ng/lipid (EFSA, 2006).

Recent dietary intake studies indicate that, for adults in most European countries, the average daily intake is in the range of 10–45 ng/kg body weight (EFSA, 2006). For young children up to 6 years of age, mean daily intakes of 27–50 ng/kg body weight, and for breastfed infants daily intakes of approximately 1600 ng/kg body weight were estimated.

Exposure from ambient air, dust and soil generally only contributes a few percent to the body burden of ndl-PCBs. In buildings where PCBs have been used, indoor PCB concentrations are often found to be elevated. Under special conditions, such as frequent fish consumption, in particular from regions with higher contamination such as the Baltic Sea or the Great Lakes, intakes of up to 80 ng/kg body weight per day were estimated for adults.

The dl-PCBs, expressed as mass percentage, contribute a few per cent only to the overall PCB intake. However, their toxicological properties are very different from those of the ndl-PCBs. Therefore, they have been attributed with TEFs and form part of the total intake of dl-compounds calculated as TEQ/kg body weight (see Section 9.3.3).

In background exposure calculations via food, dl-PCBs contribute in the range of 50% to the total intake of TEQ. Exposure via the respiratory tract or the skin (ambient air, dust, soil) usually plays a very minor role, except for toddlers with the possibility of relevant intake of contaminated soil from playgrounds, etc. In some cases of extraordinary levels of lower chlorinated congeners in ambient air, these could be found at elevated levels in blood of exposed individuals.

In spite of the fact that highly contaminated samples in many instances contain increased levels of both dl- and ndl-PCBs, there is no evidence for a 'fixed' ratio between the two groups of PCBs. Only in certain cases of a single point local source of contamination, a certain pattern of congeners can eventually be found in samples from this area. However, both environmental conditions (temperature, sunlight, etc.) and biological factors (absorption, metabolism) can exert profound effects on the pattern of congeners.

9.6   Toxicity of PCBs

The toxicity of PCBs depends strongly on their substitution pattern. While certain dl-PCBs bearing chlorine substituents at the lateral positions resemble PCDD/Fs in their toxicity, although being less potent than TCDD, ndl-PCBs bearing two or more chlorine substituents at the ortho-positions, do not possess dioxin-like properties. These compounds exhibit their own pattern of adverse effects which have been described mainly in experimental animals, and may even qualitatively depend on the individual ndl-PCB congener.

9.6.1   Toxicokinetics

PCBs are readily absorbed by passive diffusion from the gastrointestinal tract or the respiratory tract, lower chlorinated congeners having a better oral bioavailability than higher chlorinated ones. Distribution is directed towards the adipose tissue and the lipid portion of organs, tissues, blood and mother's milk. dl-PCBs can also accumulate in the liver after high dosage, e.g., in animal experiments. Most PCBs are metabolised very slowly, the lower chlorinated with two neighboured carbon atoms bearing no chlorine substituent being more susceptible to enzyme-catalysed oxygen attack. In general, the phenols thus finally formed can be excreted more easily than the parent PCBs, but can also bind to plasma proteins. Some metabolites, such as methylsulfonyl derivatives, show a relatively high lipophilicity and are retained longer in the body, e.g. in liver and lung, than phenolic metabolites. The low elimination rate of PCBs results in long elimination half-lives. For these, however, a broad range of values is given in the literature. Even for a single congener, half-life estimates can range between several months and many years. One reason for such discrepancies may be the fact that non-uniform elimination kinetics seem to exist, allowing a more rapid elimination at higher body burden than at lower or background body burden.

9.6.2 Mode of action

The most toxic dl-PCBs exert toxic effects in laboratory animals similar to those of potent PCDDs and PCDFs (see above). Most, if not all, adverse effects of this group of chemicals have been attributed to their binding to and activation of the AhR, whereas the ndl-PCBs are more or less inactive as AhR agonists. Individual dl-PCBs have been given TEFs in order to describe their toxicity relative to TCDD, the most toxic persistent dl-compound (van den Berg *et al.*, 2006).

In contrast, less is known about the molecular mode of action of the ndl-PCBs. In experimental animals, they lead to hepatic hypertrophy and liver tumour promotion, and can affect endocrine functions such as thyroid hormone levels and the thyroid-pituitary axis. Furthermore, some ndl-PCB congeners have been described as immunotoxic and neurotoxic with respect to the intrauterine and postnatal development (Ulbrich and Stahlmann, 2004).

Since a number of ndl-PCBs act in a way similar to the prototype liver tumour promoter phenobarbital, they have been described as 'phenobarbital'-type promoting agents (Knerr and Schrenk, 2006b). These compounds have in common a hypertrophic effect in rodent liver together with induction of drug-metabolizing enzymes and promotion of preneoplastic hepatocytes. The most prominent phenobarbital-inducible drug-metabolizing enzyme is cytochrome P450 (CYP) 2B1, which is mainly under the regulation of the constitutive active receptor (CAR), a member of the family of nuclear transcription factors (Kawamoto *et al.*, 2000). Recent studies in CAR-deficient mice have revealed a loss of the tumour-promoting effect of phenobarbital in the livers of those animals (Yamamoto *et al.*, 2004). The neurotoxic effects of ndl-PCBs have been attributed to their binding to targets in neuronal cells such as protein kinase C or the ryanodine receptor. Binding to the latter can result in a sustained increase in intracellular calcium levels, which can cause cell death. Furthermore, ndl-PCBs have been shown in animal experiments to decrease dopamine levels in vulnerable regions of the central nervous system (Seegal, 1999).

9.6.3 Toxic effects

Most toxicity studies have been carried out with technical PCB mixtures. Mink turned out to be one of the most sensitive species, showing clear reproductive toxicity after chronic exposure to 100 ng Aroclor 1254/kg body weight per day. Furthermore, single ndl-PCBs lead to adverse effects in animal experiments, e.g. in rodents, in the liver (hypertrophy, fatty degeneration) and the thyroid (decrease in serum T4, increase in TSH) as well as oestrogenic effects and reproductive toxicity. Particularly in mice, an immunotoxic (immuno-suppressive) effect is observable. dl-PCBs exhibit a pattern of toxic effects very similar to those of 2,3,7,8-substituted PCDD/Fs (see above), reflecting their relative TEF values (Table 9.2) when a dose–response analysis is performed. The risk assessment of these congeners is part of the evaluation of exposure to total TEQ.

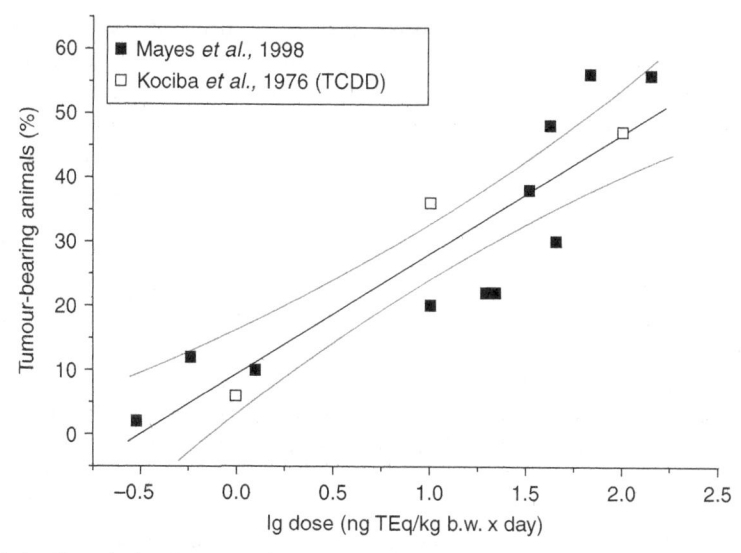

Fig. 9.4 Correlation between the number of tumour-bearing animals and the TEQ dose in Aroclors according to Mayes *et al.* (1998) or via TCDD treatment according to Kociba *et al.* (1976).

When assessing the risk of exposure to ndl-PCBs, the effect of traces of dl-contaminants (dl-PCB, PCDD/F) has to be addressed. This evaluation is made more complicated by the fact that many adverse effects of technical PCB mixtures and single 'pure' ndl-PCBs, such as alterations in the liver or thyroid, resemble those of dl-compounds.

In a variety of standard assays, PCBs are not genotoxic. However, the formation of reactive metabolites such as epoxides has been reported for some lower chlorinated congeners. In rodents, technical PCB mixtures lead to liver and thyroid tumours. A detailed analysis reveals that the dl-portion in those mixtures plays a major role in this effect (Fig. 9.4). It is not possible to decide from these data if ndl-PCBs have a carcinogenic potency of their own, while their tumour-promoting potency has been substantiated for a few congeners. It remains to be clarified what meaning a tumour-promoting potency in rodent liver has for the risk of development of cancer in humans.

For human exposure, it is even more difficult to distinguish between the effects of dl- and ndl-PCBs. This is due to the fact that human exposure at the workplace, via food, etc. even in the course of severe intoxication, is usually a mixed exposure to ndl-PCBs, dl-PCBs and other dl-compounds. Nevertheless, some differences between 'pure' PCDD/F intoxications and strong PCB exposure during the Yusho- or Yu-Cheng disasters are evident. In February 1968 a mass intoxication (more than 2000 intoxicated persons) with a technical PCB mixture, also containing PCDFs, occurred in Japan (Yusho disaster). The PCB mixture was used as a liquid in a heat exchanger during the manufacture of rice oil. A leakage in the system led to a pronounced contamination of the rice oil with the PCB mixture, which was not noted by the rice oil consumers due to the lack of colour, smell or taste of PCBs. Within a few days, chloracne, skin and mucous membrane alterations occurred in adults and children ('black babies'). Furthermore, weight loss, liver and kidney damage, immune defects and endocrine disruption (e.g. effects on the oestrous cycle) were reported. In addition, bone and joint defects with arthritis-type symptoms were observed (Yusho means 'ouch' in Japanese).

A similar disaster occurred in 1979 in Taiwan (Yu-Cheng disaster). There, chloracne, nail alterations, hyperpigmentation of the skin and mucous membranes (e.g. in new-born), hyperkeratosis, goitre, anaemia and painful bone and joint alterations (Yu-Cheng' (chin.) = 'ouch') were observed. Many of the symptoms prevailed over years as in the Yusho case. In both cases, a mixed exposure towards ndl-PCBs, dl-PCBs, and PCDFs was found. Long-term effects of these disasters include persistent affections of the oestrous cycle, miscarriages with increased foetal mortality and increased incidence of goitre.

Epidemiological studies in other cohorts, workers in the capacitor industry, allow only limited conclusions about a possible carcinogenic effects of technical PCB mixtures. A slight increase in the combined incidence of liver, bile duct and gall bladder cancers was found in some studies. An increased incidence of mammary cancer and non-Hodgkin lymphoma was claimed in some studies but not found in others.

In the children of the Yusho/Yu-Cheng victims, a decrease in intellectual, behavioural and motor-activity deficits was reported. Similar effects were also reported for children of a cohort with high fish consumption from the Great Lakes. However, other food contaminants in fish, together with socio-economic factors, were also discussed as possible explanations for the observations in the latter cohorts. A final conclusion about the significance of these effects and the role of PCB still has to be drawn.

9.6.4 Regulations of PCBs

The current Health Canada guidelines for PCB contamination are: 0.1 ppm in eggs (whole egg less shell), 0.2 ppm in meat, beef and dairy products (w/w in fat) and 0.5 ppm poultry (w/w in fat). Action levels have been set for PCBs in red meat (3 ppm/fat basis) and fish. Temporary tolerances have also been set for animal feeds and paper packaging.

9.6.5 Summary of PCBs

PCBs have been manufactured and used on a large scale as technical mixtures and are still found ubiquitously in the environment. They were used as hydraulic liquids, moisturisers in sealings, paintings, etc. and were contaminants in defoliants used in the Vietnam War. Due to their lipophilicity and persistence, they are present in the lipid portion of organisms, e.g., in the adipose tissue or in blood and mother's milk. According to their structure they can be divided into 'planar', dl-PCB and *ortho*-chlorinated, non-dioxin-like (ndl-) PCB. The toxicity of dl-PCBs resembles that of PCDD/Fs. Their relative toxicity is assessed as part of the TEF concept of dl-compounds. In general, ndl-PCBs seem to be less toxic than ndl-PCBs. Some of the most prominent congeners are immunotoxic, endocrine disruptors and liver tumour promoters in rodents. Because of the usual contamination of ndl-PCBs with dl-PCBs and other dl-compounds, a separate risk assessment of ndl-PCBs is complicated. During intoxication disasters with technical PCB mixtures, bone defects, skin alterations, liver and kidney damage and endocrine effects were observed. In these cases, a mixed exposure towards ndl- and dl-compounds occurred.

9.7 References

AYLWARD LL, BRUNET RC, CARRIER G, HAYS SM, CUSHING CA, NEEDHAM LL, PATTERSON DG JR, GERTHOUX PM, BRAMBILLA P and MOCARELLI P (2005), 'Concentration-dependent TCDD elimination kinetics in humans: toxicokinetic modeling for moderately to highly exposed adults from Seveso, Italy, and Vienna, Austria, and impact on dose estimates for the NIOSH cohort', *J Expo Anal Environ Epidemiol*, **5**, 51–65.

BEISCHLAG TV, LUIS MORALES J, HOLLINGSHEAD BD and PERDEW GH (2008), 'The aryl hydrocarbon receptor complex and the control of gene expression', *Crit Rev Eukaryot Gene Expr*, **18**, 207–250.

DURAND B, DUFOUR B, FRAISSE D, DEFOUR S, DUHEM K and LE-BARILLEC K (2008), 'Levels of PCDDs, PCDFs and dioxin-like PCBs in raw cow's milk collected in France in 2006', *Chemosphere*, **70**, 689–93.

EFSA (2005), 'Opinion of the Scientific Panel on contaminants in the food chain [CONTAM] related to the presence of non dioxin-like polychlorinated biphenyls (PCB) in feed and food', *EFSA J*. doi:10.2903/j.efsa.2005.284.

EFSA (2006) Opinion of the Scientific Panel on contaminants in the food chain [CONTAM] related to the presence of non dioxin-like polychlorinated biphenyls (PCB) in feed and food. European Food Safety Authority, Parma, Italy. /www.efsa.europa.eu/de/efsajournal/pub/284.htm

FUJII-KURIYAMA Y and KAWAJIRI K (2010), 'Molecular mechanisms of the physiological functions of the aryl hydrocarbon (dioxin) receptor, a multifunctional regulator that senses and responds to environmental stimuli', *Proc Jpn Acad Ser B Phys Biol Sci*, **86**, 40–53.

GONZALEZ FJ and FERNANDEZ-SALGUERO P (1998), 'The aryl hydrocarbon receptor: studies using the AHR-null mice', *Drug Metab Dispos*, **26**, 1194–1198.

IARC (1997), 'IARC Working Group on the Evaluation of Carcinogenic Risks to Humans: Polychlorinated Dibenzo-Para-Dioxins and Polychlorinated Dibenzofurans, Lyon, France, 4–11 February-para-1997', *IARC Monogr Eval Carcinog Risks Hum*, **69**, 1–631.

INOUYE K, SHINKYO R, TAKITA T, OHTA M AND SAKAKI T (2002), 'Metabolism of polychlorinated dibenzo-p-dioxins (PCDDs) by human cytochrome P450-dependent monooxygenase systems', *J Agric Food Chem*, **50**, 5496–5502.

KNERR S and SCHRENK D (2006a), 'Carcinogenicity of 2,3,7,8-tetrachlorodibenzo-*p*-dioxin in experimental models', *Molec Nutr Food Res*, **50**, 897–907.

KNERR S and SCHRENK D (2006b), 'Carcinogenicity of non-dioxinlike polychlorinated biphenyls', *Crit Rev Toxicol*, **36**, 663–694.

KOCIBA RJ, KEYES DG, BEYER JE, CARREON RM, WADE CE, DITTENBER DA, KALNINS RP, FRAUSON LE, PARK CN, BARNARD SD, HUMMEL RA and HUMISTON CG

(1978), 'Results of a two-year chronic toxicity and oncogenicity study of 2,3,7,8-tetrachlorodibenzo-*p*-dioxin in rats', *Toxicol Appl Pharmacol*, **46**, 279–303.

MATSUMURA F (2009), 'The significance of the nongenomic pathway in mediating inflammatory signaling of the dioxin-activated Ah receptor to cause toxic effects', *Biochem Pharmacol*, **77**, 608–626.

MAYES BA, MCCONNELL EE, NEAL BH, BRUNNER MJ, HAMILTON SB, SULLIVAN TM, PETERS AC, RYAN MJ, TOFT JD, SINGER AW, BROWN JF, MENTON RG and MOORE JA (1998), 'Comparative carcinogenicity in Sprague-Dawley rats of the polychlorinated biphenyl mixtures Aroclors 1016, 1242, 1254, and 1260', *Toxicol Sci*, **41**, 62–76.

NATIONAL RESEARCH COUNCIL (2006), *Health Risks from Dioxin and Related Compounds: Evaluation of the EPA Reassessment*. The National Academies Press, Washington, D.C., USA.

PESATORI AC, CONSONNI D, RUBAGOTTI M, GRILLO P and BERTAZZI PA (2009), 'Cancer incidence in the population exposed to dioxin after the "Seveso accident": twenty years of follow-up', *Environ Health*, **8**(39), 1–11.

SCHECTER A, CRAMER P, BOGGESS K, STANLEY J, PÄPKE O, OLSON J, SILVER A and SCHMITZ M (2001), 'Intake of dioxins and related compounds from food in the U.S. population', *J Toxicol Environ Health*, **63**, 1–18.

SEEGAL RF (1999), 'Are PCBs the major neurotoxicant in Great Lakes salmon?', *Environ Res*, **80**, S38-S45.

SORG O, ZENNEGG M, SCHMID P, FEDOSYUK R, VALIKHNOVSKYI R, GAIDE O, KNIAZEVYCH V and SAURAT JH (2009), '2,3,7,8-tetrachlorodibenzo-p-dioxin (TCDD) poisoning in Victor Yushchenko: identification and measurement of TCDD metabolites', *Lancet*, **374**, 1179–1185.

ULBRICH B and STAHLMANN R (2004), 'Developmental toxicity of polychlorinated biphenyls (PCBs): a systematic review of experimental data', *Arch Toxicol*, **78**, 252–268.

VAN DEN BERG M, BIRNBAUM L, DENISON M, DE VITO M, FARLAND W, FEELEY M, FIEDLER H, HAKANSSON H, HANBERG H, HAWS L, ROSE M, SAFE S, SCHRENK D, TOHYAMA C, TRITSCHER A, TUOMISTO J, TYSKLIND M, WALKER N and PETERSON RE (2006), 'The 2005 World Health Organization re-evaluation of human and mammalian Toxic Equivalency Factors for dioxins and dioxin-like compounds', *Tox Sci*, **93**, 223–241.

VAN DEN BERG M, VAN BIRGELEN A, BIRNBAUM L, BROUWER B, CARRIER G, CONOLLY R, DRAGAN Y, FARLAND W, FEELEY M, FÜRST P, GALLI C, DE GERLACHE J, GRIEG J, HAYASHI Y, HAYASHI Y, HERRMAN J, KOGEVINAS M, KUROKAWA, LARSEN JC, Y, LIEM D, LUIJCKX L, MATSUMURA F, MCGREGOR D, MOCARELLI P, MOORE M, MOY G, NEWHOOK R, OUANE F, PETERSON R, POELLINGER L, PORTIER C, RAPPE C, ROGAN W, SCHRENK D, SHKOLENOK G, SWEENEY M, TOHYAMA C, TUOMISTO J, UEDA H, WATERS J, VAN DE WIEL J, VAN LEEUWEN FXR, YOUNES M and ZEILMAKER M (2000), 'Revision of the tolerable daily intake of dioxin by WHO', *Food Add Contam*, **17**, 223–240.

YAMAMOTO Y, MOORE R, GOLDSWORTHY TL, NEGISHI M and MARONPOT RR (2004), 'The orphan nuclear receptor constitutive active/androstane receptor is essential for liver tumor promotion by phenobarbital in mice', *Cancer Res*, **64**, 197–200.

10

Non-dioxin-like polychlorinated biphenyls (NDL-PCBs) in foods: exposure and health hazards

L. E. Elabbas, E. Westerholm, R. Roos and K. Halldin, Karolinska Institutet, Sweden, M. Korkalainen, National Institute for Health and Welfare, Finland, M. Viluksela, National Institute for Health and Welfare, Finland and University of Eastern Finland, Finland and H. Håkansson, Karolinska Institutet, Sweden

DOI: 10.1533/9780857098917.2.215

Abstract: Non-dioxin-like polychlorinated biphenyls (NDL-PCBs) are ubiquitous in the environment. Humans are exposed to NDL-PCBs mainly via food. Exposure to NDL-PCBs is suspected to be of concern to human health. This chapter discusses the NDL-PCB sources, their environmental transport and occurrence in food items. Human groups with high exposure are addressed. Furthermore, this chapter deals with toxicokinetics and the toxicity profile of NDL-PCBs. It also provides information about the regulatory situation of NDL-PCBs, which could be useful, both for academic and government scientists. This information will also be of value for the food industry, policy makers and consumers.

Key words: non-dioxin-like polychlorinated biphenyls (NDL-PCBs), exposure pathways, NDL-PCBs in food, toxicity profile.

10.1 Introduction

Polychlorinated biphenyls (PCBs) are synthetic aromatic chemicals, not occurring naturally in the environment, which have been used commercially since 1929. The PCB molecule consists of two linked benzene rings, where some or all of the hydrogen atoms have been substituted by chlorine atoms (Fig. 10.1). Theoretically, 209 individual congeners are possible, of which about 130 have been identified in commercial products. Technical PCB

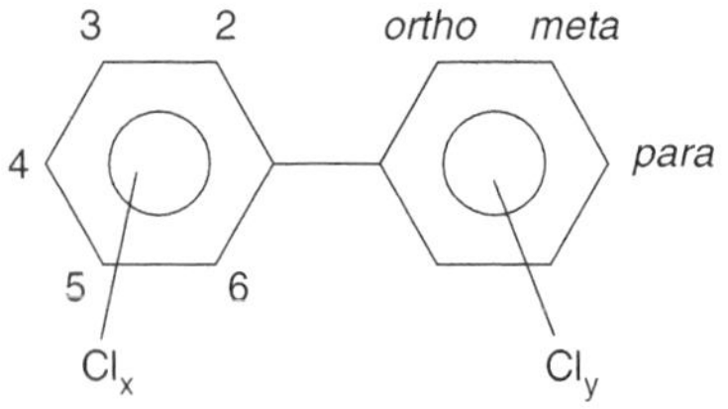

216 Persistent organic pollutants and toxic metals in foods

Fig. 10.1 The chemical structure of PCBs showing the *ortho-*, *meta-* and *para-* positions, and the numbering of the carbon atoms on the two phenolic rings.

products are always a mixture of different congeners, and their absolute composition varies from batch to batch. Impurities, such as polychlorinated dibenzofurans (PCDF) and naphthalenes (PCN), have all been identified in commercial PCB products.

PCBs resist degradation and are transported through environmental media across international boundaries, and accumulate in terrestrial and aquatic ecosystems. The presence of PCBs in human and wildlife tissues was first recognized in 1966. Since then, investigations in many parts of the world have revealed widespread distribution of PCBs in the environment, including remote areas where they have never been used or produced (Schreitmüller and Ballschmiter, 1994; Macdonald *et al.*, 2000). Today, PCBs are found in almost all compartments of the global ecosystem in at least trace amounts. The ability of PCBs to cross international borders and move long distances via air and water has made PCB contamination an international problem. Many countries and intergovernmental organizations have, over decades, banned or severely restricted the production, usage, handling, transport and disposal of PCBs.

Nevertheless, due to their persistent, accumulative and toxic nature, PCBs are still of major international regulatory concern. They are scheduled for elimination under the Stockholm Convention (SCP, 2001) and the Convention on Long Range Transboundary Air Pollution (CLRTAP, 1998). Accordingly, the development of a global, legally binding treaty to phase out and eliminate production, use and sources of PCBs is necessary to minimize PCB effects on the environment and humans.

Nowadays, close to 50 years after the first identification of PCBs in human and wildlife tissue, the scientific community is still trying to clarify the toxicological mode-of-actions and assessing the PCBs health effects arising from background exposure levels. There are several health criteria documents concerning PCB chemistry and toxicity from the early 1990s (WHO/EURO, 1987; Ahlborg *et al.*, 1992; US-EPA, 1990; WHO/IPCS, 1993) and a Concise International Chemical Assessment Document on PCBs (CICAD, 2003), which is based on the Agency for Toxic Substances and Disease Registry (ATSDR) Toxicological Profile on PCBs (ATSDR, 2000). It seems likely, based on the available toxicological information, that both dioxin-like

(DL-PCB) and non-dioxin-like (NDL-PCB) congeners can cause serious toxicological outcomes, such as carcinogenicity, immunotoxicity, neurodevelopmental disorders and developmental toxicity. However, the underlying biological mechanisms are likely to be fundamentally different. In contrast to the DL-PCB congeners, for which there is sufficient data to perform a health risk assessment (WHO, 1998; EU, 2000; US-EPA, 2012), such information is largely lacking for NDL-PCB congeners (EFSA, 2005). Both individual NDL-PCB congeners and food-relevant NDL-PCB mixtures are poorly characterized from a toxicological point of view, and to date, there is no health risk assessment for NDL-PCBs.

Since humans are always exposed to complex mixtures of individual DL-PCB and NDL-PCB congeners, whose relative contribution to toxicity is unclear for a large number of different endpoints established in toxicological studies, it has not yet been possible to ascribe the observed effects in epidemiological studies to any individual NDL-PCB congener or to any sub-group(s) of those. On these grounds, and since NDL-PCB congeners constitute a major part of the PCB congeners found both in food and human tissues, it has become an urgent matter to advance the understanding of NDL-PCB toxicology as such, and furthermore, to understand whether the NDL-PCB congeners that are present in food and tissue should be classified and assessed as one single group of chemicals with common toxicological properties, or as multiple groups of chemicals with unique toxicological properties. A major issue of discussion has focused on the possibility that the present health risk assessment for dioxins and DL-PCBs could also provide sufficient protection from any health hazards due to NDL-PCB exposure.

10.2 Sources, occurrence in foods, limit values and monitoring methods

It has been estimated that over one and a half million tonnes of PCBs have been produced in the world since production began in the late 1920s. Due to chemical and physical properties, PCBs were widely used in a number of commercial and industrial open (e.g. painting, sealing) and closed (e.g. transformers, capacitors) systems. No single PCB congener has been used as an additive, but complex technical mixtures sold under such names as Aroclor (Monsanto, USA), Clophen (Bayer, Germany), Phenoclor (Caffaro, Italy) and Kanechlor (Kanegafuchi, Japan) have been used to change chemical properties of various technical systems. Industrial PCB mixtures have been widely used in products such as capacitors, plasticizers, or transformers (Breivik *et al.*, 2002). Sources of environmental contamination with PCBs are old electric equipment, sealing in buildings, waste deposit sites, waste incineration and inadequate disposal practice. Despite the ban in almost all industrial countries during the 1980s, PCBs still enter the environment, mostly due

Table 10.1 Total NDL-PCB burden in food samples collected from 20 European countries during the period from 1998 until 2008. The values are based on the sum of the six indicator PCBs

Food group	Sum of six indicator PCBs (µg/kg)
Meat and meat products (ruminants)	4.64
Meat and meat products (poultry)	2.88
Meat and meat products (pigs)	3.69
Liver and products (terrestrial animals)	2.1
Muscle meat fish and fish products, excluding eel	23.4
Muscle meat (eel)	223
Raw milk and dairy products, including butter	9.2
Hen eggs and egg products	16.7
Fat (ruminants)	8.71
Fat (poultry)	2.85
Fat (pigs)	1.04
Vegetable oils and fats	3.21
Marine oils	117
Fish liver	163
Other products	21.1
Fruits, vegetables and cereals	0.08
Infant and baby food	5.65

Source: modified from EFSA (2010).

to improper disposal practices or to the use of older electrical equipment that still contains PCBs (La Rocca and Mantovani, 2006).

Presently, the major source of exposure could be redistribution, but primary sources could also be of significant importance. It is believed that large surfaces of water, such as, for example, the Baltic Sea and the Canadian Great Lakes, may release significant amounts of PCB residues from previous use, into the atmosphere.

The major source of human exposure is food and NDL-PCB levels in food samples collected from 20 European countries during the period 1998–2008 have revealed the highest levels in fish and fish products and the lowest levels in fruits, vegetables and cereals (Table 10.1). To protect its citizens from PCB exposure, the European Commission and individual European countries have established maximum PCB levels for different types of food, while Canada has established a tolerable daily intake (TDI) of 1 µg PCB/kg body weight (b.w.) per day (Van Oostdam et al., 1999). Foetal and neonatal PCB exposure continues to raise concerns regarding potential health effects on developing infants. Exposure to the DL-PCBs via breast milk continue to exceed the TDI level, while for NDL-PCB congeners there is not yet a health-based guidance value. Despite the high PCB contamination of human milk in some regions, breastfeeding is recognized as a healthy practice, optimizing infant growth

NDL-PCBs in foods: exposure and health hazards 219

and development (AMAP, 2002). However, there is a need to reduce contaminant levels and to provide food advice to women of childbearing age (AMAP, 2009).

To facilitate the environmental monitoring of PCBs, most countries have adopted the method of focusing on the 6 PCB congeners 28, 52, 101, 138, 153 and 180 (Fig. 10.2), referred to as 'indicator PCBs' or 'sum of the PCBs (Σ6PCB)' (La Rocca and Mantovani, 2006). These congeners, all belonging to the group of NDL-PCBs, were selected from an analytical rather than a toxicological point of view, but were considered as indicators for different PCB forms in various sample types. The measurement of the indicator PCBs covers approximately 50% of the total PCB burden found in food and feed samples (EFSA, 2005).

The introduction of capillary columns for gas chromatography separation was a breakthrough in PCB analysis (Sonchik, 1983), since it enabled congener-specific analysis in environmental, food and human samples. To avoid possible coelution problems with other organohalogen compounds, a combined capillary gas chromatography/mass spectrometry (GC/MS) in the electron impact (EI) or a negative chemical ionization (NCI) mode can be used.

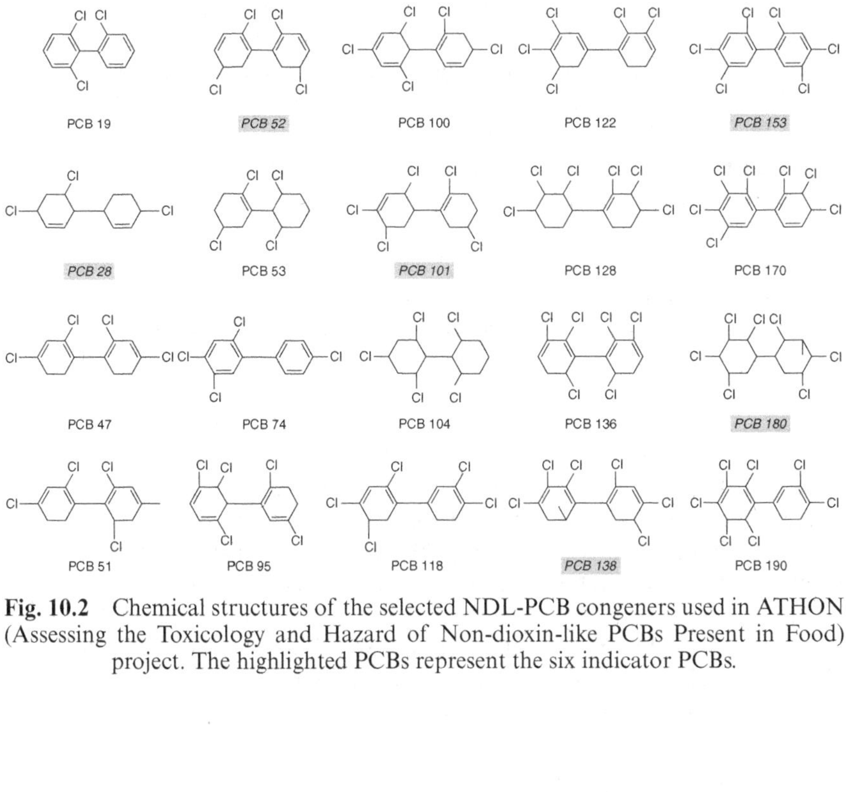

Fig. 10.2 Chemical structures of the selected NDL-PCB congeners used in ATHON (Assessing the Toxicology and Hazard of Non-dioxin-like PCBs Present in Food) project. The highlighted PCBs represent the six indicator PCBs.

220 Persistent organic pollutants and toxic metals in foods

10.3 Human exposure and tissue levels

Dietary intake is the main exposure route for most PCB congeners, while dermal and inhalation exposures are of relevance under occupational exposure situations (Smith *et al.*, 1982; Wolff, 1985; Lees *et al.*, 1987; Duarte-Davidson and Jones 1994; Wilson *et al.*, 2001).

10.3.1 Dietary intake

General population
The general population is continuously exposed to NDL-PCBs through ingestion of everyday food items such as fish, meat and dairy products, and NDL-PCB congeners constitute a major part of the PCB congeners found in human tissue and food, both number and mass-wise. Data available from industrialized countries have shown a reduction of population exposure levels over recent decades, but there are indications that this decline has leveled off. The average intake levels for adults in the Western world are estimated to be in the range of 10–45 ng/kg b.w. per day for total NDL-PCBs; meanwhile, the estimates of exposure for high-level consumers of fish were in the range of 60–80 ng/kg b.w. per day for total NDL-PCB (EFSA, 2005).

The dietary intake of NDL-PCBs is dominated by the congeners 138, 153 and 180 (Juan *et al.*, 2002), since these congeners also dominate the PCB composition of important animal-based foods such as meat, fish and dairy products (Szlinder-Richert *et al.*, 2009). Six NDL-PCBs, i.e. PCBs 28, 52, 101, 138, 153 and 180, account for more than 50% of the NDL-PCBs found in food (Table 10.1). In France, the mean dietary intake of NDL-PCBs (Σ6PCB) is 7.6, 7.7 and 12.9 ng/kg b.w. per day for women of childbearing age (19–44 years), adults (15 years and over) and children (3–14 years), respectively (Arnich *et al.*, 2009). A Dutch study calculated lifelong averaged median intakes of 5.6 ng (Σ6PCB)/kg b.w. per day (Bakker *et al.*, 2003; Baars *et al.*, 2004). This was based on the averaged intakes over 70 years, whereas children have higher exposures (12.1 ng (Σ6PCB)/kg b.w. per day for 2 years old children) than adults (Bakker *et al.*, 2003). For Germany, an average of 15.3 ng/kg b.w. per day was calculated for 3 PCB congeners (Σ PCB 138, 153, 180) (Arnold *et al.*, 1998). An Italian study calculated a daily intake of 18 ng (Σ6PCB) per kg b.w. per day (Zuccato *et al.*, 1999). Besides the geographic variations there are also age-dependent variations in the daily intakes and Fattore *et al.* (2008) observed that the highest exposure occurred in the two first years of life, where they calculated a daily intake of 111 ng (Σ6PCB) per kg b.w., while daily intakes of 24.6 ng (Σ6PCB) per kg b.w. were found for infants (0–6 years), excluding breastfed infants. Older children, 7–12 years old, had a calculated daily intake of 16.1 ng (Σ6PCB) per kg b.w. and from 13 to 94 years of age, a daily intake of 10.9 ng (Σ6PCB) per kg b.w. was determined (Fattore *et al.*, 2008). For Sweden, the daily intake was calculated for 23 PCB congeners for men and women of different age groups; these results showed that the estimated median daily intake by women was in the range 5.5–11.5 ng/kg b.w., where the group 21–30 years had the lowest intake and 61–70 years the highest. For the men, the median daily intake was 6.22–9.62 ng/kg b.w., where the age group 31–40 years had the lowest and 61–70 years the highest intake (Lind *et al.*, 2002). The daily intake of PCBs (Σ6PCB) by fishermen from the Swedish east coast from fish amounted to 41.5 ng/kg b.w., which is approximately 5–10-fold higher than the corresponding intake from fish estimated for the average male adult population in Sweden (EFSA, 2005).

Breastfed infants
Numerous studies have demonstrated that PCBs are transferred from the mother to the infant during breastfeeding, which represents a major route of exposure to infants. As a consequence, breastfed infants will exceed adult exposures by one to two orders of magnitude. The dominant NDL-PCB congeners in breast milk are PCBs 138, 153 and 180 (Guvenius *et al.*, 2003; Polder *et al.*, 2009). Based on data from human milk pools from 18 European countries, it has been estimated that the median daily intake of PCBs (Σ6PCB) for exclusively breastfed infants is 984 ng/kg b.w. (EFSA, 2005). Another study estimated that the mean intake of NDL-PCBs (28, 138, 153, 170 and 180) in 1, 3 and 6 months old infants was 418, 294 and 165 ng/kg b.w. per day, respectively, which showed a clear reduction trend of NDL-PCB intake over the breastfeeding period (Bergkvist *et al.*, 2010). Further, PCB exposure through breastfeeding decreased significantly during the past decades in the countries that adopted measures to regulate open and closed application of technical PCB mixtures. For example, in West Germany it was reported that the PCB intake by exclusively breastfed infants decreased by approximately 80% (Fürst, 2006).

10.3.2 Human tissue levels

General population
Both serum and adipose tissue levels may serve as useful indicators for the body burden of NDL-PCBs, as illustrated by the good correlations that were obtained for PCBs 74, 99, 118, 138, 146, 153, 156, 167, 170, 180, 183 and 187, in adipose tissue and serum from women in the general population (Stellman *et al.*, 1998). The predominant PCB congeners in human adipose tissue and liver are the NDL-PCB congeners 138, 153 and 180 (Schecter *et al.*, 1989; Meironyte Guvenius *et al.*, 2001; Covaci *et al.*, 2002, 2008). Lower chlorinated PCBs (e.g. PCBs 31/28, 52, 74, 95) are either not detectable or have a very low detection frequency. The mean

serum concentrations of PCB 153 and the sum of PCB congeners were 124 and 953 ng/g lipid, respectively, in a population of Native Americans exposed to PCB both via neighboring industrial pollution and frequent consumption of local fish (Goncharov *et al.*, 2009). The average PCB levels of Inuit maternal blood (measured as Aroclor 1260) are estimated to be 8.0 µg/L plasma. Concentrations of PCB congeners are found in the following order: PCB 153 > 138 > 180 > 187, with PCB 153 being the predominant congener. On the lipid weight basis, the Inuit maternal PCB levels were estimated as 454 µg/kg lipid (Van Oostdam *et al.*, 2005). Fishermen from the Swedish east coast had higher plasma concentrations of several PCBs, including PCB 138, 153 and 180 (Svensson *et al.*, 1995). The plasma concentrations of these congeners were 442, 836 and 661 ng/g lipid, respectively, for south-east coast fishermen and 186, 336 and 290 ng/g lipid, respectively, for west coast fishermen.

Placenta, cord blood and mother's milk
Levels of NDL-PCBs 138, 153 and 180 in placental tissue and umbilical cord blood have been used as indicators of PCB exposure of the developing foetus, and in breast milk as indicators of exposure in the breastfed infant (Koopman-Esseboom *et al.*, 1994; DeKoning and Karmaus, 2000). Concentrations in placenta (5027 ng/g fat) were 2.8 times higher than the highest concentration of PCB in breast milk (1770 ng/g fat) (DeKoning and Karmaus, 2000). The foetal and infant tissue levels correlated well with maternal plasma levels (Koopman-Esseboom *et al.*, 1994).

The dominant congener in cord blood samples collected from Menorca was PCB 180, while PCB 153 and PCB 138 were the prevalent congeners in cord blood samples collected from Valencia (Vizcaino *et al.*, 2011). The cohort from Valencia had median PCB concentrations of 150 ng/g lipid, analyzed as sum of 7 PCBs i.e. PCB 28, 52, 101, 118, 138, 153, and 180, while the median PCB concentrations from the cohort of Menorca was 240 ng/g lipid, analyzed as sum of 7 PCBs, which is higher than those reported in many other locations such as Sweden (Guvenius *et al.*, 2003), USA (Herbstman *et al.*, 2007) and Italy (Bergonzi *et al.*, 2009), but lower than levels reported in Slovakia (Park *et al.*, 2008) and Russia (Eik Anda *et al.*, 2007).

NDL-PCB contaminated mother's milk is a worldwide phenomenon (Ennaceur *et al.*, 2008; Polder *et al.*, 2008; Vukavić *et al.*, 2008; Devanathan *et al.*, 2009, 2012; Cerná *et al.*, 2010; Skrbić *et al.*, 2010; Thomson *et al.*, 2010; Todaka *et al.*, 2010; Tue *et al.*, 2010; Asante *et al.*, 2011; Darnerud *et al.*, 2011; Deng *et al.*, 2011; Park *et al.*, 2011; Rodríguez-Dozal *et al.*, 2011; Zhang *et al.*, 2011; Cok *et al.*, 2012). Although PCB levels in human mother's milk have decreased significantly, since the 1970s in most countries it seems that the rate of decrease has levelled off during the last couple of years (Hooper *et al.*, 2007; Abballe *et al.*, 2008; Raab *et al.*, 2008; Zietz *et al.*, 2008; Ulaszewska *et al.*, 2011). Furthermore, in some regions of the Czech Republic, the levels have not decreased at all, due to high contamination of the surrounding areas (EFSA, 2005).

Findings from an Inuit cohort study revealed that the most abundant PCB congener in all biological samples analyzed was PCB 153, which accounted for approximately 40% of the total body burden of PCB, defined as the sum of the 6 PCB congeners 99, 118, 138, 153, 180 and 187 (Ayotte *et al.*, 2003). It was also shown that Inuit women have a mean total PCB concentration in breast milk which exceeds that of southern Québec women by a factor of 7 (Dewailly *et al.*, 1993).

PCBs and their methyl-sulphonyl metabolites are excreted through human milk and may also cross the placenta. Perinatal PCB exposure is therefore a major concern with regard to human health effects, due to present background exposure levels.

10.4 Toxicokinetics and metabolism

The physicochemical properties of PCBs enable these compounds to be readily and extensively absorbed from the gastrointestinal tract by passive diffusion (Albro and Fishbein, 1972). It has been observed that the net absorption efficiencies of PCB congeners in humans are determined by several factors, such as the amount of body fat, weight loss/gain, compound properties (i.e. susceptibility to metabolism) and past exposure (Juan *et al.*, 2002). Around 80% of the NDL-PCBs 28, 52, 101 and 105 were efficiently absorbed in humans (Schlummer *et al.*, 1998).

Once absorbed, PCBs are readily distributed to all body compartments by non-specific association to blood cells and plasma proteins (Matthews and Dedrick, 1984). Due to their high lipid- and low-water-solubility, PCBs are concentrated in lipid-rich tissues and levels in blood, being particularly associated with the lipid fraction (Park *et al.*, 2005). The initial distribution of PCBs is determined by biophysical factors such as tissue volume, tissue/blood partition ratio, absorption to proteins and perfusion rate. Additional factors that contribute to the distribution pattern of different congeners include the extent of recent intake, long-term intake, fat content of the organ, protein binding and, most importantly, the potential for metabolism. Lactation or significant weight loss increases PCB release into the blood.

The initial step in the biotransformation of PCBs involves oxidation by cytochrome P-450 enzymes, and several hydroxylated as well as sulphur-containing PCB-metabolites have been identified. Tanabe and co-workers (1981) demonstrated that the rate of metabolism of PCBs is inversely associated with the degree of halogenation, suggesting that lower chlorinated PCBs are transformed more rapidly. However, PCB metabolism becomes more complex as the degree of chlorination increases (Tanabe *et al.*, 1981). Öberg *et al.* (2002) investigated the correlation between

Table 10.2 Reported half-life times of some NDL-PCB congeners plus DL-PCBs 105 and 118 in human studies

PCB congener	Half-life time	Reference
28	18/44 days[a]	Luotamo et al. (1991)
	182 days	Wolff and Schecter (1991)
	1.4 years	Brown et al. (1989)
	4.8 years	Wolff et al. (1992)
	3.0 years	Yakushiji et al. (1984)
52	5.5 years[b]	Wolff et al. (1992)
101	7/14 days[a]	Luotamo et al. (1991)
	5.7 years[c]	Wolff et al. (1992)
105	186/212 days[a]	Luotamo et al. (1991)
	3.9 years	Brown et al. (1989)
118	0.78/0.83 years[a]	Chen et al. (1982)
	0.3–0.81 years	Bühler et al. (1988)
	1.1 year	Ryan et al. (1993)
	5.8 years	Brown et al. (1989)
	9.6 years	Wolff et al. (1992)
138	0.89 years	Bühler et al. (1988)
	3.4 years	Ryan et al. (1993)
	6–7 years	Brown et al. (1989)
	16.3 years	Yakushiji et al. (1984)
	16.7 years	Wolff et al. (1992)
	20/32 years[a]	Yakushiji et al. (1984)
153	0.94 years	Bühler et al. (1988)
	3.9 years	Ryan et al. (1993)
	12.4 years	Brown et al. (1989)
	26/47 years[a]	Chen et al. (1982)
	27 years	Yakushiji et al. (1984)
170	4.5 years	Ryan et al. (1993)
	47/71 years[a]	Chen et al. (1982)
180	0.34 years	Bühler et al. (1988)
	4.8 years	Ryan et al. (1993)
	9.9 years	Wolff et al. (1992)

[a] Two values given in the study depending on sample selections.
[b] Coelution between PCB 47, 49 and 52.
[c] Coelution between PCB 99 and 101.
Source: modified from EFSA (2005).

molecular weight and bioavailability and distribution of PCB congeners, and the observations showed that high molecular weight correlated to lower uptake and slower redistribution.

Elimination of non-metabolized PCBs is low and largely depends on the excretion of polar metabolites in urine and faeces, and is therefore associated with the rates of metabolism of the individual congeners. A set of PCB half-life values in humans is presented in Table 10.2, showing that half-life time could range from one week up to several years. Highly chlorinated PCB

Table 10.3 Enzyme inducing properties of individual PCB congeners

PCB congener	MC-type induction	PB-type induction	Mixed-type induction
11	−	(+)	−
14	−	(+)	−
15	−	(+)	−
37	−	−	+
47	−	+	−
52	−	(+)	−
54	−	(+)	−
66	−	+	−
75	−	(+)	−
77	+	−	−
80	−	(+)	−
81	−	−	+
87	−	+	−
101	−	+	−
105	−	−	+
114	−	−	+
115	−	+	−
118	−	−	+
119	−	−	+
123	−	−	+
126	+	−	−
128	−	−	+
133	−	+	−
136	−	(+)	−
137	−	+	−
138	−	−	+
151	−	(+)	−
153	−	+	−
154	−	+	−
155	−	(+)	−
156	−	−	+
157	−	−	+
158	−	−	+
159	−	(+)	−
163	−	+	−
165	−	+	−
166	−	−	+
167	−	−	+
168	−	−	+
169	+	−	−
170	−	−	+
180	−	+	−
189	−	−	+
190	−	+	−
191	−	+	−
194	−	+	−
195	−	+	−
205	−	+	−

MC-type, 3-methylcholantrene-type; PB-type, phenobarbital-type; Mixed-type, mixed phenobarbital- and 3-methylcholanthrene-type microsomal enzyme inducers. −, no induction; (+), weak induction; +, strong induction.
Source: modified from McFarland and Clarke (1989).

congeners show the longest half-life, and therefore the greatest accumulation (e.g. NDL-PCBs 138, 153, 170 and 180). The main excretion routes of PCBs and their methyl-sulphonyl metabolites are via faeces (biliary excretion), urine and breast milk.

10.5 Classification of PCB congeners

Both chemical and biological/toxicological criteria are involved in the classification of PCB congeners as either DL or NDL. From a chemical point of view, NDL-PCB congeners are characterized by having one or more chlorine in *ortho*-positions of the chemical backbone of the molecule (Fig. 10.1). Substituents in *ortho*-positions cause the PCB molecule to assume a more bulky and less planar configuration, as compared to the DL-PCB congeners, which lack chlorines in *ortho*-positions and therefore are more likely to assume a planar configuration. NDL-PCB congeners bind with low or no affinity to the aryl hydrocarbon receptor (AhR), while the planar structure of the DL-PCB congeners enables these molecules to fit with high affinity into the ligand-binding domain of AhR, a key mediator of physiological as well as toxicological cell regulatory events (Lindén *et al.*, 2010).

Based on the induction of various cytochrome P-450 (CYP) enzymes, PCB congeners have been classified as CYP1A1/2 (also referred to as 3-methylcholanthrene (3-MC) type), CYP2B1/2 (also referred to as phenobarbital (PB) type), or mixed-type inducers (Table 10.3). Typically, NDL-PCB congeners with two or more *ortho*-chlorines induce CYP2B1/2 alone, and bind with low or no affinity to the AhR, while mono-*ortho*-substituted NDL-PCB congeners, which bind with somewhat higher affinity to AhR, induce both CYP1A1/2 and CYP2B1/2 (mixed-type inducers). In contrast, DL-PCB congeners, which bind to AhR with high affinity, typically induce CYP1A1/2 alone. AhR-binding and CYP-induction profiles have been used as tools for initial toxicity-ranking of individual PCB congeners, as well as congeners among the polychlorinated dibenzo-*p*-dioxins (PCDD) and PCDFs.

Comprehensive toxicity profiling studies show that DL-PCB congeners exhibit the numerous toxicological effects typical of those caused by exposure to 2,3,7,8-tetrachlorodibenzo-*p*-dioxin (TCDD), the prototypic model for DL-compounds. Health risks associated with exposure to these PCB congeners are covered by the dioxin risk assessment (WHO, 1998; EU, 2000) and their environmental monitoring is covered by regulations regarding DL-compounds (EFSA, 2005). A toxic equivalency factor (TEF) scheme established by the World Health Organization (WHO; van den Berg *et al.*, 2006; Table 10.4) is being used to allow for cumulative assessment of exposure to mixtures of DL-compounds. Since NDL-PCB congeners bind with

Table 10.4 Summary of World Health Organization TEF values

Compound	WHO 2005 TEF
Chlorinated dibenzo-*p*-dioxins	
2,3,7,8-TCDD	1
1,2,3,7,8-PeCDD	1
1,2,3,4,7,8-HxCDD	0.1
1,2,3,6,7,8-HxCDD	0.1
1,2,3,7,8,9-HxCDD	0.1
1,2,3,4,6,7,8-HpCDD	0.01
OCDD	0.0003
Chlorinated dibenzofurans	
2,3,7,8-TCDF	0.1
1,2,3,7,8-PeCDF	0.03
2,3,4,7,8-PeCDF	0.3
1,2,3,4,7,8-HxCDF	0.1
1,2,3,6,7,8-HxCDF	0.1
1,2,3,7,8,9-HxCDF	0.1
2,3,4,6,7,8-HxCDF	0.1
1,2,3,4,6,7,8-HpCDF	0.01
1,2,3,4,7,8,9-HpCDF	0.01
OCDF	0.0003
Non-*ortho*-substituted PCBs	
PCB 77	0.0001
PCB 81	0.0003
PCB 126	0.1
PCB 169	0.03
Mono-*ortho*-substituted PCBs	
105	0.00003
114	0.00003
118	0.00003
123	0.00003
156	0.00003
157	0.00003
167	0.00003
189	0.00003

Source: modified from Van den Berg *et al.* (2006).

low or no affinity to AhR, and lack the typical toxicity profile elicited by DL-compounds, there has hitherto been less emphasis on further toxicity profiling of these congeners despite their higher abundance in environmental media.

Although scientifically the toxicological profiling and classification of PCB congeners is not clear-cut, it has direct regulatory implications, since only the PCB congeners, which have been assigned a TEF-value in the WHO scheme for DL-compounds (van den Berg *et al.*, 2006) are currently covered by international health risk assessment regulations for dioxin (WHO, 1998; EU, 2000; US-EPA 2012).

10.6 NDL-PCB regulatory status

It is a serious problem for management that the majority of the available toxicological data on NDL-PCBs cannot be interpreted in a way that is useful for risk assessment purposes. The topic has been discussed under the umbrella of the WHO and the European Commission (EC), who organized several international exploratory consultations to address the issue of NDL-PCB exposure and implications for human health, in order to provide advice to regulatory organizations (WHO, 2001). The scope of these consultations was to explore scientifically justified and practically feasible concepts of addressing the overall toxicological properties of NDL-PCB congeners in the context of risk assessment and regulatory decision-making. One of the approaches was to investigate the possibility that the present health risk assessment for dioxins and DL-PCBs also covers the health hazard of NDL-PCB exposure.

In 2005, the European Food Safety Authority (EFSA) stressed that the available NDL-PCB data-base was inadequate and that more toxicological data were needed to identify any NDL-PCB hazard and to assess any risk involved with the daily consumption of NDL-PCBs (EFSA, 2005). As such, the available carcinogenicity and immunotoxicity datasets were considered unsuitable for risk assessment, as few congeners were tested; the compounds were not always tested for chemical purity (or purity reporting was insufficient), and the mixtures used were not relevant for human exposure situations. The neurobehavioral data, although considered insufficient, were regarded as relevant since behavioral effects observed in laboratory animals, following perinatal PCBs exposure, had also been observed in humans and linked to PCB exposure using epidemiological approaches. A lowest observed effect level (LOEL) was identified for subtle neurotoxic effects in infants following perinatal exposure in the range 0.014–0.9 µg PCB/kg b.w. per day. It was concluded that this exposure is in the same order of magnitude as the present PCB exposure of the general population in many countries. On these grounds, the EFSA identified the developing foetus and the neonate as a potential 'at risk' population due to increased susceptibility. Furthermore, 'as human background exposure to PCBs, including both DL and NDL-PCB congeners, may reach estimated lowest observed adverse effect levels (LOAELs) for neurodevelopmental effects in infants, the weight of evidence suggests an increased health risk due to current exposures'. Lack of congener-specific exposure and toxicity data, as well as lack of appropriate compound purity information, have limited the possibilities to conclude which congeners are the ones responsible for the observed effects.

Thus it has not yet been possible for the EFSA to establish a scientifically justified TDI of either PCB mixtures or of any individual NDL-PCB congener. Instead, the EFSA presented its concern about the current exposure based on reported findings in experimental and human studies, and voiced the immediate regulatory need to advance the health hazard characterization of NDL-PCB congeners.

10.7 ATHON R&D project dedicated to generating NDL-PCB toxicity data for regulatory use

The European Research and Development project 'Assessing the Toxicology and Hazard of Non-dioxin-like PCBs Present in Food' (ATHON; Contract FOOD-CT-2005–022923 under the 6th Framework of Research and Technological Development of the EU, Priority 5 (Food Quality and Safety) call) was tailored to address urgent data needs on qualitative and quantitative aspects of NDL-PCB toxicology, with a focus on developmental toxicology and neurotoxicity, including also multiple indicators for endocrine system disturbances, general liver toxicity, immunotoxicity and NDL-PCB metabolism, as pointed out by the EFSA (2005). The unique concept of ATHON was to establish toxicity data for highly pure (99.9999%) NDL-PCB congeners, and to design the studies to meet quantitative and qualitative requirements of toxicological data to allow for evaluation and use in health hazard assessment. Efforts were made to identify toxicological mechanisms and outcomes, which are specific for NDL-PCBs. Congeners for *in vivo* studies were selected based on abundance in food and human tissue levels, while for the *in vitro* studies physicochemical and biological properties were the basis for selection (Fig. 10.2). On these grounds repeated dose 28-day oral (OECD TG 407) and one generation reproduction (OECD TG 415) toxicity studies were performed with NDL-PCBs 52 and 180, while a mouse gene array study and a zebrafish embryotoxicity study were performed with the six indicator PCBs (Table 10.5), and all NDL-PCBs listed in Fig. 10.2 were used in the *in vitro* studies (Table 10.6). The rat studies were designed to include conventional 'apical' toxicity endpoints as well as additional 'biomarker' endpoints tailored to address also AhR as well as other endocrine and neurotoxic modes of action. The inclusion of both perinatal and adult exposure studies allowed for comparative evaluations based on life-stage as well as gender. The purity issue focused on any contamination with DL-compounds, in order to address this particular concern of the EFSA-opinion. Another major goal of the project was to define toxicological endpoints specific for NDL-PCB congeners, if present. The ATHON project as such is now over, while the evaluation and reporting work is still in progress and much of the findings, though not yet finally published, are presented in Sections 10.7 and 10.8 of this chapter.

Table 10.5 Study designs of the NDL-PCB *in vivo* experiments performed within ATHON (Assessing the Toxicology and Hazard of Non-dioxin-like PCBs Present in Food) project and NDL-PCB *in vivo* studies identified in the open literature

Type of study	Duration of study (days)	Species (strain)	Sex (no./group)	Route (vehicle)	Congener (dose)	Treatment schedule	Parameters monitored	Response level	Reference
In vivo brain microdialysis after *in utero*/ lactational exposure	Pups tested on PNDs 90 and 110	Rat (Wistar)		Oral/ corn oil (sweet jelly)	52, 180, 138 (1 mg/kg b.w./day)	Dams exposed on GD7 to PND 21	Function of the glutamate–NO–cGMP pathway; extracellular dopamine, GABA and glutamate		De Boever et al., 2010 (abstract); Fernandes et al., 2010; Lilienthal et al., 2009; Llansola et al., 2009, 2010; Wigestrand et al., 2013
In vivo motor activity and coordination after *in utero*/ lactational exposure	Pups tested on PNDs 60 and 90	Rat (Wistar)			52, 180, 138 (1 mg/kg b.w./day)	Dams exposed on GD7 to PND 21	Locomotor activity, open field and motor coordination, proteomics		Boix et al., 2010; Campagna et al., 2011; Fernandes et al., 2010
In vivo learning ability after *in utero*/ lactational exposure	Pups tested on PNDs 75 and 90	Rat (Wistar)			52, 180, 138 (1 mg/kg b.w./day)	Dams exposed on GD7 to PND 21	Ability to learn a Y-maze conditional discrimination task, proteomics		Boix et al., 2010; Campagna et al., 2011
In utero/ lactational exposure study on neurotoxicity		Rat (Long Evans)	12 dams per congener		74 (11.68 mg/ kg/day 95 (13.06 mg/ kg/day)	Developmental treatment, GD10 to PND7	Endocrine-related neurobehavior in offspring, sweet preference, haloperidol-induced catalepsy, brain stem auditory evoked potentials		Lilienthal et al., 2010 (abstract); Lilienthal et al., 2011

Study	Duration	Species	Number/sex/group	Administration	Dose levels (mg/kg unless noted)	Endpoints	Reference
In utero/lactational exposure study including follow-up of offspring	Dams terminated on PND 28. Pups terminated on PNDs 7, 35, 84	Rat (Sprague–Dawley)	Dams: F (6) Offspring: M, F 52: 5M, 5F/dam, 180: 4M, 4F/dam	Dams: oral gavage (corn oil) Offspring: *in utero* + lactational	52 (30, 100, 300, 1000, 3000 mg/kg) 180 (10, 30, 100, 300, 1000 mg/kg)	PCB52: total dose divided into ten equal subdoses, dams were exposed to the subdoses on GDs 7, 9, 11, 14, 16 and PNDs 1, 10. PCB 180: total dose divided into four equal subdoses, dams were exposed to the subdoses on GDs 7–10.	Development of the reproductive system, bone and teeth, sex steroidogenesis, limited histopathology, biochemistry; molecular biology	Roos *et al.*, PCB 52 (in preparation); Roos *et al.*, PCB 180 (in preparation)
Repeated 28-day oral gavage toxicity study	28 days	Rat (Sprague–Dawley)	M, F (5)	Oral gavage (corn oil)	52, 180 (3, 10, 30, 100, 300, 1000, 1700* mg/kg)	PCB 180: loading dose administered during the first study week in six equal subdoses + maintenance dose once a week for three weeks	Clinical signs, organ weights, clinical chemistry, hematology, histopathology, molecular biology, proteomics, bone quality	Roos *et al.*, PCB 52 Adult (in preparation); Viluksela *et al.*, PCB 180 Adult (in preparation)
Gene expression in liver	1 day 5 days 28 days	Mouse (C57BL/6)	M, F (6)	Oral gavage (corn oil)	28, 52, 101, 138, 153, 180 (0.2, 2, 10, 20 mg/kg b.w.) 126 (2, 20, 100, 200 µg/kg b.w.)	Single oral dose in male and female adults. For 28d study, animals were dosed on days 0, 3, 6, 9, 12, 15, 18, 21, 24, 27 (PCBs 28, 52, 101) and 0, 3, 7, 14, 21 (PCBs 126, 138, 153, 180)	Gene expression, microarray analysis, RT-PCR, PCB analysis, enzyme assays (CYP1A, CYP2B), proteomics	Heimeier *et al.*, 2010 (abstract); Schoenauer, 2011 (M.Sc. thesis); Thörnqvist *et al.*, 2010 (abstract)

(*Continued*)

Table 10.5 Continued

Type of study	Duration of study (days)	Species (strain)	Sex (no./group)	Route (vehicle)	Congener (dose)	Treatment schedule	Parameters monitored	Response level	Reference
Subchronic oral toxicity study	90 days	Rat (Sprague–Dawley)	M, F (10)	In diet (corn oil)	28 (0, 0.05, 0.5, 5.0, or 50 mg/kg feed)	Adult exposure daily via diet	Growth rate, food consumption, clinical toxicity, enzyme activity, tissue vitamin A content, histology	NOAEL 0.5 ppm in diet or 36 µg/kg b.w./day	Chu et al.1996a
Subchronic oral toxicity study	90 days	Rat (Sprague–Dawley)	M, F (10)	In diet (corn oil)	153 (0.05, 0.5, 5.0, or 50 mg/kg feed)	Adult exposure daily via diet	Growth rate, food consumption, clinical toxicity, enzyme activity, tissue vitamin A content, histology	NOAEL 0.5 ppm in diet or 34 µg/kg b.w./day	Chu et al.1996b
Subchronic oral toxicity study	90 days	Rat (Sprague–Dawley)	M, F (10)	In diet (corn oil)	128 (0, 0.05, 0.5, 5 or 50 ppm (mg/kg feed))	Adult exposure daily via diet	Growth rate, food consumption, organ weights clinical toxicity, enzyme activity, tissue vitamin A content, histology	NOAEL 0.5 ppm in diet or 42 µg/kg b.w. per day	Lecavalier et al.1997
Chronic oral toxicity and carcinogenesis study	2 years	Rat (Harlan Sprague–Dawley)	F (50)	Gavage (corn oil)	153 (0, 10, 100, 300, 1000 or 3000 µg/kg b.w.)	5 days per week	Hypertrophy of thyroid gland, follicular cell and liver hepatocytes. Liver enzyme activity PROD, EROD		NTP, 2006
In utero lactational exposure study	3 months	Rat (Wistar–Hannover)	M, F (3 dams with pups)	(Corn oil)	153 (0.5, 1 mg/kg b.w./day)	GD7 to PND 21	Pups; learn a conditional discrimination task in a Y-maze, function of the glutamate–nitric oxide–cGMP pathway	NOAEL 1 mg/kg b.w. per day	Piedrafita et al., 2008a
In utero lactational exposure study	8 months	Rat (Wistar–Hannover)	M, F (3 dams with pups)	(Corn oil)	153 (1 mg/kg b.w./day)	GD7 to PND 21	Pups; learn a conditional discrimination task in a Y-maze, function of the glutamate–nitric oxide–cGMP pathway	NOAEL 1 mg/kg b.w. per day	Piedrafita et al., 2008b

*Dose adjusted after signs of toxicity in dams.

Table 10.6 Assays included in the *in vitro* screening programme of ATHON (Assessing the Toxicology and Hazard of Non dioxin-like PCBs Present in Food) project

ER agonism	COMT activity
ER antagonism	Calcium homeostasis
GJIC inhibition	Interaction with the $GABA_A$ receptor
AR agonism	Dopamine uptake (synaptic vesicles)
AR antagonism	Dopamine uptake (synaptosomes)
AhR agonism	Glutamate uptake (synaptic vesicles)
AhR antagonism	Glutamate uptake (synaptosomes)
TTR binding	GABA uptake (synaptic vesicles)
E2SULT inhibition	GABA uptake (synaptosomes)
Biotransformation rates	DAT inhibition
UGT-1A1 mRNA expression	ROS
UGT-1A6 mRNA expression	Mdm2 phosphorylation
DI-1 mRNA expression	Akt phosphorylation
Aromatase activity in microsomes	ERK phosphorylation
Aromatase activity in H295R cell line	

ER, oestrogen receptor; GJIC, gap junction intracellular communication; AR, androgen receptor; AhR, aryl hydrocarbon receptor; TTR, transthyretine; E2SULT, estradiol-sulfotransferase; UGT, uridine diphosphoglucuronosyl transferase; DI-1, deiodinase-1; COMT, catechol-o-methyltransferase; GABA, gamma-aminobutyric acid; DAT, dopamine active transporter; ROS, reactive oxygen species.
Source: modified from Hamers *et al.* (2011).

10.7.1 Development and reproduction endpoints

In a perinatal toxicity study with PCB 180, maternal body-weight development was slightly decreased at the highest dose level (Miettinen *et al.*, 2009 (Abstract); Roos *et al.*, PCB 180 (in preparation)). Similarly, the body-weight development of the offspring was decreased during the first weeks of life after high dose exposure, but recovered thereafter. Neonatal mortality was slightly and dose-dependently increased at the two highest dose levels. Analysis of developmental milestones revealed slight delays in balano-preputial separation and vaginal opening, while tooth eruption, eye opening or ano-genital distance were not affected. Offspring liver weights were dose-dependently increased on postnatal days (PND) 7, 35 and 84, but not yet on PND 1. The likely explanation for this finding is hypertrophy of hepatocytes due to increased activity of several hepatic enzymes. The results emphasize the importance of lactational transfer in exposure of offspring to PCB 180. For PCB 52, an increased liver weight was observed, while the body-weight development, mortality and developmental milestones (eye opening, tooth eruption) of the offspring after perinatal exposure was not affected (Roos *et al.*, PCB 52 (in preparation); Korkalainen *et al.*, 2012 (Abstract); Roos *et al.*, 2012 (Abstract)).

A trend for decreased cortical bone mineral density was observed in female offspring on PND 35 and in male offspring on PND 84 after perinatal exposure to PCB 180. In female offspring on PND 84, the cross-sectional area and thickness of cortical bone were increased, while no effects were seen on the bone mineral density (Roos et al., PCB 180 (in preparation)). Additional data by Elabbas et al. (2011) indicated that perinatal exposure to the technical PCB Aroclor 1254 leads to shorter, thinner and weaker bones in juvenile rats at PND 35. It seems that the observed effects were mainly driven by the DL-congeners of this technical mixture; however, the contribution from NDL-congeners to the exposure outcome could not be excluded (Elabbas et al., 2011).

A zebrafish embryotoxicity test was used to rank the NDL-PCBs 28, 52, 101, 138, 153 and 180 for their potency to induce developmental toxicity. In the test, freshly fertilized eggs were dechorionated to enhance transfer of the PCBs to the embryo and monitored over 72 hours (van der Ven et al., (in preparation). In this design, the following potency ranking could be made on the basis of the fraction of embryos that showed teratogenic effects: PCB 101 = PCB 28 > PCB 52 = PCB 153 > PCB 138 > PCB 180. The DL-PCB 126 was also evaluated and had the highest potency of all tested compounds.

10.7.2 Neurotoxicity

Table 10.7 summarizes the *in vivo* neurotoxicity findings after developmental and subacute PCB exposures. Impaired function of the glutamate-nitric oxide (NO) – cyclic guanosine monophosphate (cGMP) pathway by PCBs may contribute to the cognitive alterations induced by perinatal exposure to PCBs in humans. Chronic exposure to the NDL-PCBs 138, or 180, during rat development (*in vivo*) and in rat cerebellar neurons (*in vitro*) revealed that NDL-PCBs impair the function of the glutamate-NO-cGMP pathway by different mechanisms and with different potencies (Llansola et al., 2010). PCBs 138 and 180 reduced the amount of *N*-methyl-D-aspartate (NMDA) receptors in the cerebellum in rat offspring, which reduced the function of the glutamate-NO-cGMP pathway. This in turn may contribute to the impairment in the learning ability of these rats. Exposure to PCBs 138 or 180 during pregnancy and lactation also alters the modulation of dopaminergic neurotransmission by glutamate in the nucleus accumbens *in vivo* and this could give an altered locomotor activity (Boix et al., 2010).

Developmental exposure to PCB 52 increased extracellular gamma-aminobutyric acid (GABA) in the cerebellum, which may contribute to motor coordination impairment observed in the rat offspring. *In vitro* analysis revealed that both activation and potentiation of human $GABA_A$ receptors by NDL-PCBs are newly discovered modes of action that could (partly) underlie the previously recognized NDL-PCB-induced effects on motor coordination observed *in vivo*. So far, this is the only type of neurotransmitter receptor known to be directly activated or potentiated by NDL-PCBs (Fernandes et al., 2010).

Table 10.7 Summary of *in vivo* neurotoxicity findings after developmental (perinatal and/or postnatal) and subacute (28 days) NDL-PCB exposures or perinatal exposure to DL-PCB 126

Type of study	Species (and strain)	PCB congener (exposure)																								References
		28		47		52		74		77		95		118		138		153		156		180		126		
		♀	♂	♀	♂	♀	♂	♀	♂	♀	♂	♀	♂	♀	♂	♀	♂	♀	♂	♀	♂	♀	♂	♀	♂	
A. Perinatal exposure																										
Learning ability																										
1. Y-maze	Rat^WKY					nr																				Boix et al. (2010)
	Rat^WKY							nr									→					→			→	Piedrafita et al. (2008)
Motor activity and coordination																										
1. Spontaneous	Rat^WKY					nr											→									Boix et al. (2010)
	Rat^WY/Rat^LE			↑															↓^a		nr^b		→		→	Weinand-Härer et al. (1997)
																		nr								Hany et al. (1999)
	Rat^SD															nr		nr				→ →		nr		Boix et al. (2010)
2. Vertical activity	Rat^WKY					nr		nr								←						nr				Weinand-Härer et al. (1997)
3. Jumping	Rat^WKY					nr		nr								→						nr				Hany et al. (1999)
4. Rotarod	Rat^SD					→		→										nr				nr				
Haloperidol-induced catalepsy																										
1. Dopaminergic system	Rat^WY/Rat^LE			lp↑		←				↑														→		Weinand-Härer et al. (1997)
	Rat^SD/Rat^LE					←		→		nr	nr											nr		→		Hany et al. (1999); Lilienthal et al. (2011)

(*Continued*)

Table 10.7 Continued

Type of study	Species (and strain)	PCB congener (exposure)																							References
		28		47		52		74		77		95		118		138		153		156		180		126	
		♀	♂	♀	♂	♀	♂	♀	♂	♀	♂	♀	♂	♀	♂	♀	♂	♀	♂	♀	♂	♀	♂	♀	♂	
BAEPs																										
1. Threshold	Rat[SD]/Rat[LE]					lp↑		lp↑	↑			nr	lp↑									lp↑	nr		♀ ♂	Lilienthal et al. (2011)
2. Latencies	Rat[SD]/Rat[LE]					nr	↑	↑	↑			↑	↑									nr	nr		♀ ♂	
Electroretinogram																										
1. Amplitude	Rat[LE]			nr	nr					→		nr														Kremer et al. (1999)
2. Latencies	Rat[LE]			↑						↑		nr														
B. Postnatal exposure																										
Learning ability																										
1. Operant testing:																										
PND10–20 single dose, test PND37–58																										Johansen et al. (2011)
(a) Spontaneous activity	Rat[WKY]					nr	nr											→	↑	↑		→				
(b) Attention	Rat[WKY]					nr	nr											↑	nr	↑	↑	↑	nr			
(c) Impulsivity	Rat[WKY]					nr	nr											nr	↑	↑		→				
PND8–20 dose-response, test PND25–65																										Johansen et al. (2011)
(a) Spontaneous activity	Rat[WKY]																	↑		↑						
(b) Attention	Rat[WKY]																	nr	↑	→	nr					
(c) Impulsivity	Rat[WKY]																	↑		nr						

	Species			Reference
PND8–20 dose–response, test PND25–65				
(a) Spontaneous activity	Rat[SH]			Johansen et al. (2011)
(b) Attention	Rat[SH]	↑		
(c) Impulsivity	Rat[SH]		↑	
2. Swim maze	Mouse[NMRI]	nd	nr	Eriksson and Fredriksson (1996)
		nd	nr	
		→		
3. Radial-arm maze	Mouse[NMRI]	nd	nr	Eriksson and Fredriksson (1996)
		nd	nr	
		→		
Motor activity				
1. Spontaneous activity:				
(a) Locomotion	Mouse[NMRI]	nd	nr	Fischer et al. (2008)
		nd	nr	Eriksson and Fredriksson (1996)
		→	→	
(b) Rearing	Mouse[NMRI]	nd	nr	
		nd	nr	
		→	→	
(c) Total activity	Mouse[NMRI]	nd	nr	
		nd	nr	
		→	→	
C. Subacute exposure				
Learning ability				
1. Open field: spontaneous activity	Rat[SD]	Δ	Δ	Lilienthal et al. (2011)

[a] 2–4 months old; [b] 7 months old; nr, no response; lp, less pronounced; nd, not detected; Δ, change observed; LE, Long Evans; WI, Wistar WIGA; WKY, Wistar Kyoto; SD, Sprague–Dawley; SH, spontaneously hypertensive; NMRI, Naval Medical Research Institute; BAEP, brainstem auditory evoked potential.

Haloperidol-induced cataleptic behavior was less expressed in male offspring exposed to PCB 180 during development than in females, with significant dose-dependent reduction in latencies to movement onset (Fig. 10.3). This effect is most probably due to enhanced metabolism of haloperidol in males compared to females, since males also exhibited a stronger induction of enzymes of the CYP3A family in the liver, known to be involved in the metabolism of haloperidol. On the other hand, maternal exposure to PCB 52 caused increased latencies to movement onset in female offspring, most probably due to reduced dopamine levels in the striatum induced by this NDL-PCB (Lilienthal *et al.*, 2009). PCB 74 reduced latencies to movement onset in females, whereas PCB 95 did not significantly alter this dopamine-dependent behavior in either sex (Lilienthal *et al.*, 2010 (abstract)).

Results of a sweet preference test indicated a feminization of behavior in adult male offspring by maternal PCB 74 and PCB 95 exposure, with PCB

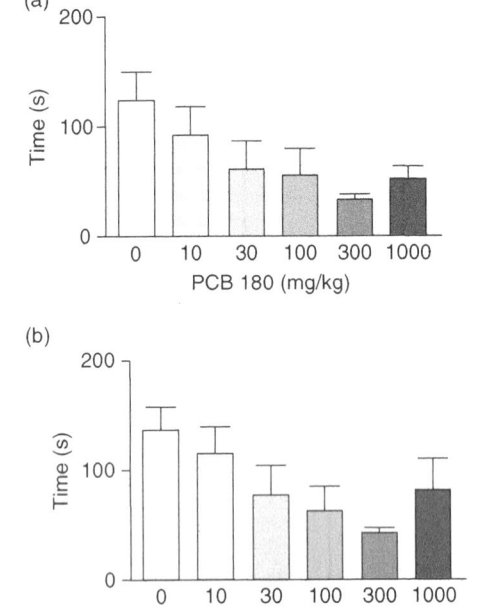

Fig. 10.3 Catalepsy neurotoxicity of PCB 180. Haloperidol-induced catalepsy was studied as a dopamine-dependent behaviour. It was induced by IP injection with haloperidol at a dose level of 0.3 mg/kg b.w. The test was performed 60 minutes after the injection with haloperidol (which is a dopamine D2 receptor blocker) by putting the rats in unusual body postures, erect position with both front paws on a horizontal bar, and by determination of time to remove the first and the second paw (descent latency to the floor). The graphs show the results for animals treated with PCB 180. Shown are the latencies to removal of paws from the bar. Significant dose-response relations were obtained for latencies of the first paw (a) and the second paw (descent latency, b) in male offspring, $p<0.05$. Source: modified from Lilienthal *et al.*, (2009).

95 being somewhat more effective than PCB 74. Developmental exposure to PCB 52 caused a slight effect at the highest dose level, whereas exposure to PCB 180 induced some signs of supernormality in exposed females (Lilienthal *et al.*, 2009).

Developmental exposure to NDL-PCBs affected auditory function in adult offspring as demonstrated by elevated thresholds of brainstem auditory evoked potentials (BAEP) and, to a lesser extent, prolonged latencies (Lilienthal *et al.*, 2011). PCB 74 was the most potent congener of the NDL-PCBs tested to induce threshold increases, followed by PCB 95 and PCB 52, which were similar in their effects. In contrast, increases by PCB 180 were smaller. Effects of PCB 74 and PCB 52 were more expressed in male compared to female offspring, whereas PCB 95 elevated thresholds only in males. The modest effects of PCB 180 on BAEP thresholds were found only in females. Since all congeners resulted in rather similar reduction in circulating thyroid hormones, other factors involved in the development of cochlear and neural structures of the auditory system are likely to contribute to the observed effects.

Gene expression changes were observed in different brain areas in rats that were perinatally exposed to PCB 52, PCB 138 or PCB 180. Some of these gene expression changes were even reflected in peripheral blood (De Boever *et al.*, 2010 (abstract)). The results indicate that the changes induced by PCB 138 exposure were more pronounced than in the case of PCB 52 or PCB 180. Detailed proteomics analysis of brain tissue after exposure to PCB 138 revealed that biological pathways associated with calcium homeostasis and androgen receptor signaling pathways were primarily disrupted (Campagna *et al.*, 2011). Generally, transcriptomics and proteomics revealed a sex-specific reaction to the PCB exposure. Expression levels of androgen receptor (AR) mRNA was down-regulated in the female but not the male brain after exposure to PCB 180, suggesting that *in utero* and lactational exposure to PCB 180 may affect the female neuroendocrine system at gene expression level.

In addition, *in vitro* studies of *ortho*-chlorinated PCBs have pointed out several factors which are important for neurotoxicity effects of NDL-PCBs *in vivo*. These include transport mechanisms for neurotransmitters, particularly dopamine (Wigestrand *et al.*, 2013). Moreover, *ortho*-chlorinated PCBs have been shown to exert dual effects on calcium homeostasis (Langeveld *et al.*, 2012). On the one hand, these PCBs increase basal intracellular calcium levels, which could relate to changes in protein phosphorylation, gene expression, cell viability and neurodevelopment. On the other hand, *ortho*-chlorinated PCBs inhibit voltage-gated calcium channels, thereby reducing calcium influx during neuronal activation and consequently inhibiting neurotransmission. The importance of effects on transport mechanism and calcium homeostasis may be observed both *in vitro* and by microdialysis *in vivo*. Additional calcium-mediated effects have been observed, including calcium–calmodulin-mediated activation of NO synthetase, which in due course produces NO and cGMP both *in vitro* and *in vivo* (Llansola *et al.*,

2009, 2010). Moreover, proteomics analysis of *in vitro* rat primary neurons has shown interference of the NDL-PCB congeners with synaptic plasticity through the modulation of different effectors related to cytoskeleton organization (Brunelli *et al.*, 2012).

For developmental exposure *in vivo*, additional factors involved in the regulation of the development of the nervous system, such as retinoids and thyroid and steroid hormones, are likely to contribute to alterations found in adult offspring.

In conclusion, evaluation of the *in vivo* and *in vitro* neurotoxicity studies added new knowledge on underlying mechanisms that may contribute to the interpretation of epidemiology reports on cognitive impairment and motor disorders in children exposed to PCBs during pregnancy and lactation.

10.8 Cell regulation and metabolism

PCBs induce a wide range of toxic effects and AhR is assumed to mediate the majority of effects induced by DL-PCB congeners, while alternative mode-of-actions are likely to predominate for the NDL-PCB congeners, based on their low or no-affinity binding to the AhR. Here we present mode-of-action relevant information for NDL-PCB congeners with a focus on some endocrine pathways i.e. thyroid and steroid hormones, and the retinoid system, the cytochrome p-450 system, kinase activation and cell-to-cell communication.

10.8.1 Endocrine system

Thyroid hormones
Studies after developmental and adult exposure *in vivo* collectively indicated that one of the characteristic effects of NDL-PCBs 52 and 180 is disruption of thyroid hormone homeostasis; both dams and the offspring were sensitive to compound exposure and developed hypothyroidism (Roos *et al.*, PCB52 (in preparation), PCB 180 (in preparation)). After adult PCB 52 and PCB 180 exposure, the decrease in circulating thyroxine (T4) was more pronounced than the decrease in triiodothyronine (T3) whereas, after perinatal PCB 52 and PCB 180 exposure, the decrease in T3 was more pronounced (Roos *et al.*, PCB 52 (in preparation), PCB 180 (in preparation), PCB 52 Adult (in preparation); Viluksela *et al.*, PCB 180 Adult (in preparation)). Hypothyroidism is known to interfere with the development of the central nervous system (CNS) and is associated with thyroid follicle depletion and activation of hepatic clearance of thyroid hormones as indicated by induction of uridine diphosphoglucuronosyl transferases (UGTs) in liver (Fig. 10.4). PCB 180 can induce thyroid hormone metabolism via up-regulation of the constitutive active (androstane) receptor (CAR) dependent enzymes UGT1A1 and UGT1A6, thus enhancing the elimination of thyroid hormones from

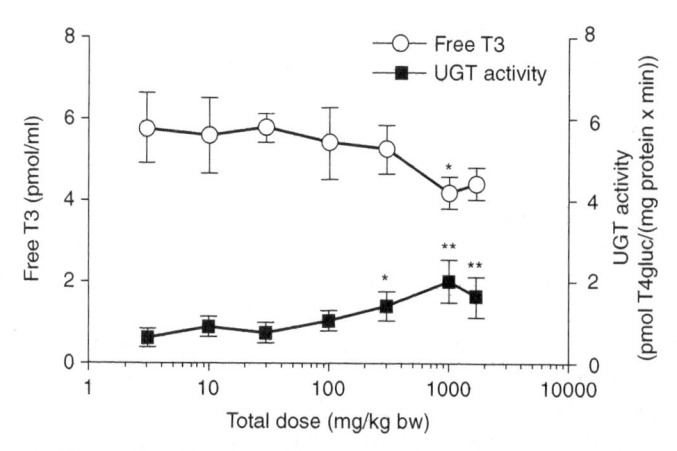

Fig. 10.4 An illustration showing uridine diphosphoglucuronosyl transferase (UGT) activity in liver with free triiodothyronine (T3) concentration in adult Sprague–Dawley male rats treated with PCB 180 for 28 days. Significance: *$p < 0.05$; **$p < 0.01$. Source: modified from Roos et al., (2011a).

the circulation. The potency of different congeners for UGT induction in liver therefore most probably reflects their ability to cause hypothyroidism in rats (Roos et al., 2011). For the human situation, however, it is not clear whether PCB exposure leads to similar effects, because in humans the main T4-carrying protein, thyroxine-binding globulin (TBG), protects the circulating thyroid hormones from degradation.

An increased thyroid gland weight was observed in adult male rats after a 28-day subacute exposure to high purity PCB 180, and histopathology examination revealed increased thyroid follicle cell vacuolization, suggestive of thyroid activation (Viluksela et al., PCB 180 Adult (in preparation)). The histological findings are, as well as the observed decreases in circulating thyroid hormone levels, most likely due to increased elimination of thyroid hormones. Figure 10.5(a) shows human adipose tissue concentrations of PCB 180, as well as changes in serum free T3 observed in rats following perinatal exposure to PCB 180 (Roos et al., PCB 180 (in preparation)). A decreased ratio of large thyroid follicles was also observed in histopathology samples, from males in particular, after subacute exposure to PCB 52 (Roos et al., PCB 52 Adult (in preparation)). The decreased thyroid follicle ratio was not accompanied with follicle cell activation. This suggests that the increased peripheral demand of thyroid hormone was still within the homeostatic capacity of the animals.

NDL-PCBs and hydroxyl (OH)-PCBs are able to affect thyroid hormone status through different mechanisms. *In vitro* experiments indicated that especially OH-metabolites of NDL-PCBs inhibit the binding of T4 to transthyretine (TTR) (Cheek et al., 1999), which is an important transport protein for both T4 and T3 mainly in rats. Formation of hydroxyl-metabolites is therefore suggested to be a key factor in determining the significance of TTR displacement *in vivo*. Furthermore, T4-displacement from TTR by OH-PCBs is considered to have human relevance especially for the developing foetal brain, since TTR is involved in T4 transport across the placenta and the blood-brain barrier (Viluksela *et al.*, 2010 (abstract)).

Steroid hormones
Decreases in testosterone and increases in luteinising hormone and follicle stimulating hormone levels were observed in PCB 180-treated male offspring, together with decreased prostate weight and decreased epididymal sperm counts at the high exposure level, after developmental exposure in rats (Roos *et al.*, PCB 180 (in preparation)). These changes are suggestive of testicular damage and consistent with the observed anti-androgenic activity observed for all tested PCBs *in vitro* (Hamers *et al.*, 2011). Figure 10.5(b) shows effects in serum testosterone levels noticed in rats perinatally exposed to PCB 180 together with human adipose tissue concentrations of PCB 180 (Roos *et al.*, PCB 180 (in preparation)). There were also minor changes in aromatase activity of several organs at high exposure levels and, at high test concentrations, OH-PCBs had androgenic effects via aromatase inhibition *in vitro* (Fernandes *et al.*, 2011). In addition, *in vitro* tests showed that lower chlorinated NDL-PCBs can activate the oestrogen receptor (ER), while higher chlorinated NDL-PCBs were anti-oestrogenic (Hamers *et al.*, 2011). OH-PCBs can exert oestrogenic activities via ER activation and via inhibition of estradiol sulfonation. Similar to TTR binding, inhibition of estradiol-sulfotransferase occurs at (sub)nanomolar OH-PCB concentrations, making these two modes of action the most potent *in vitro* effects described to date regarding endocrine disrupting activity of PCBs or their metabolites. An analysis of protein expression in livers from adult female rats suggested an oestrogenic effect of NDL-PCB 180 also *in vivo*. In contrast, serum steroid hormone and gonadotropin levels were not affected in adult rats following subacute exposure to PCB 52 or PCB 180 (Roos *et al.*, PCB 52 Adult (in preparation); Viluksela *et al.*, PCB 180 Adult (in preparation)). Nor was there an effect on these hormone levels in offspring following *in utero*/lactational exposure to PCB52 (Roos *et al.*, PCB 52 (in preparation)).

Retinoid system
Retinoid levels were altered in livers from male and female rats after both perinatal and adult exposure to PCB 180 and to PCB 52 (Table 10.8; Borg *et al.*, 2010 (abstract); Roos *et al.*, 2011a,b; Roos *et al.*, PCB 52 (in preparation), PCB 180 (in preparation); Roos *et al.*, 2012 (abstract), Borg *et al.*, 2010 (abstract)). Additionally, Fig. 10.5(c) illustrates the effects on liver retinol levels in rats after perinatal exposure to PCB 180 as well as human adipose tissue concentrations of PCB 180 (Roos *et al.*, 2012 (abstract); (in preparation)). *All-trans* retinoic acid was the most sensitive retinoid form in both perinatal studies, with $BMDL_{05}$ of 2 and 3 mg/kg b.w. respectively, for PCB 52 and PCB 180 (Roos *et al.*, PCB 52 (in preparation), PCB 180 (in preparation)).

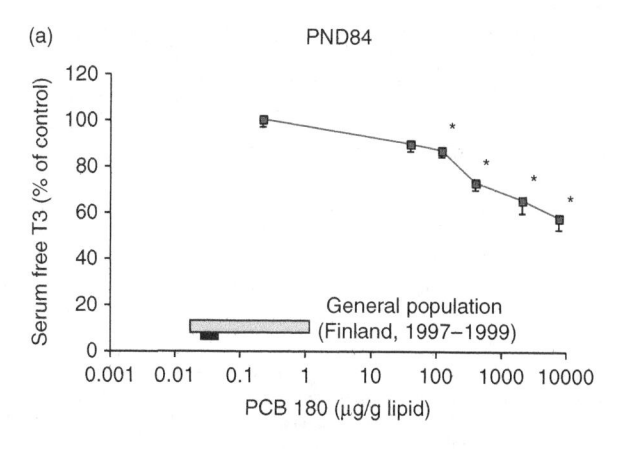

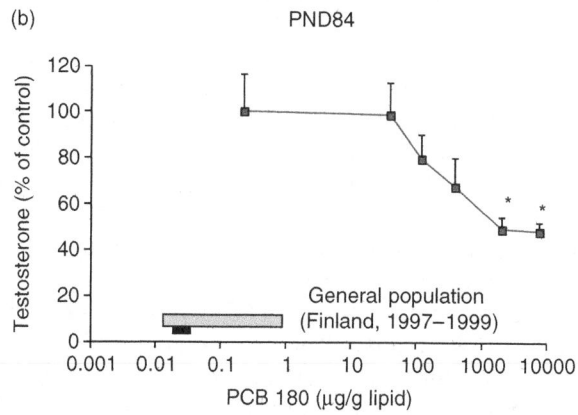

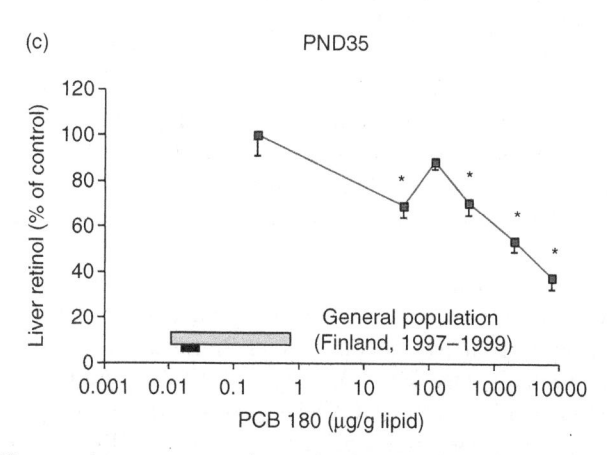

Fig. 10.5 Human tissue concentrations of PCB 180 and effects seen in serum free triiodothyronine (T3) levels (a), serum testosterone levels (b) and liver retinol levels (c) after perinatal exposure of rat to PCB 180 (male offspring). Source: Roos *et al.*, PCB 180 (in preparation).

Table 10.8 Schematic summary of alterations in retinoid concentrations in tissue samples from 28 days and perinatal studies with PCB 180 and PCB 52

Study	Tissue	Retinol				*All-trans* retinoic acid				Retinyl esters			
		28 days		Perinatal[a]		28 days		Perinatal[a]		28 days		Perinatal[a]	
		♂	♀	♂	♀	♂	♀	♂	♀	♂	♀	♂	♀
PCB 180	Liver	↓	↓	↓	↓	↑	/	↑	↑	↓	↓	/	/
	Kidney	↑	↑	↑	/	/	↑	/	/	↑	/	↑	↑
	Serum	/	↑	nd	nd	/	/	nd	nd	/	/	nd	nd
PCB 52	Liver	↓	/	/	/	/	/	↓	↓	↓	↓	↓	/
	Kidney	nd	nd	/	/	nd	nd	↓	/	nd	nd	/	↓

[a] Retinoid concentrations on PND 35. '↓,↑' decrease or increase in concentration; '/', no effect; 'nd', not done in these studies. Source: derived from Roos *et al.*, 2011a,b; Roos *et al.*, PCB 52 (in preparation), PCB 180 (in preparation); Roos *et al.*, 2012 (abstract).

Furthermore, reduced hepatic vitamin A levels were among the most sensitive effects observed following dietary exposure to the NDL-PCBs 28, 128 and 153, as well as the DL-PCBs 77, 105, 118 and 156, used in a subchronic toxicity study protocol (OECD TG 408) in male and female rats (Kalantari *et al.*, 2012).

Additionally, significant expression changes of genes involved directly or indirectly with retinoic acid synthesis and metabolism were observed in a mouse-liver whole genome array study (Thörnqvist *et al.*, 2010 (abstract); Schoenauer, 2011 (M.Sc. thesis)). The retinoic acid receptors (RAR) alpha and beta, as well as lecithin-retinol acyltransfesrase (LRAT) expression, were up-regulated by most tested PCBs and mainly in male mice. In female mice, the retinol binding protein 1 (RBP1) was induced by most PCBs (Thörnqvist *et al.*, 2010 (abstract); Schoenauer, 2011 (M.Sc. thesis)). Together with alterations in liver retinoid levels, these gene expression data suggest a significant change in hepatic retinoic acid homeostasis five days after NDL-PCB exposure. Thus, taken together, alterations in tissue retinoid levels and retinoid system-related gene expression data are suggestive of a significant change in retinoic acid homeostasis after NDL-PCB as well as DL-PCB exposure.

Cytochrome p-450 system
In rats exposed to PCB 180, a variety of cytochrome P-450 (CYP)-related changes were observed in liver within 28 days. Significant increases in liver pentoxyresorufin O-dealkylase (PROD) activity, CYP2B1 and CYP3A1 mRNA levels, as well as CYP2B1/2 and CYP3A1 protein levels were found in both male and female rats, with males being more sensitive (Roos *et al.*, 2011a,b). The increases were histopathologically translated as centrilobular hepatocellular hypertrophy, which was also observed with higher sensitivity in males. A significant induction of hepatic 7-ethoxyresorufin O-deethylase (EROD) activity was observed both in male and female rats, while CYP1A1 mRNA

and protein levels were unchanged. In contrast, CYP1B1 and CYP1A2 showed slight inductions in females only. These findings suggest that PCB 180 acts as a CAR and pregnane X receptor (PXR) agonist but not as a typical inducer of AhR-dependent changes (Roos et al., 2011a, b). In human hepatoma cell line (HepG2), 1μM PCB 180 marginally affected mRNA levels of CYP1A1, whereas CYP1B1 levels were markedly increased. In contrast, OH-PCB 180 significantly elevated CYP1A1 mRNA but did not affect CYP1B1 mRNA levels (Al-Anati et al., in preparation).

CYP2B1/2 mRNA levels were significantly induced at high dose of PCB 52 in both male and female rats, and the corresponding protein levels were also highly induced especially in males (Roos et al., PCB 52 Adult (in preparation)). In contrast, levels of CYP1A1, CYP1A2 and CYP1B1 mRNA and protein remained unchanged after 28 days of PCB 52 exposure. CYP3A1 and UGT1A1 mRNA and protein levels were not affected by the treatment, while UGT1A6 mRNA and protein levels were induced. PROD activity was significantly induced in both males and females, while EROD activity was significantly induced in males only. The associated hepatocellular hypertrophy was particularly pronounced in males, but only at high doses (Roos et al., PCB 52 Adult (in preparation)).

Cytochrome P450-related gene expression was also studied in a mouse-liver global microarray study 5 days after the exposure to NDL-PCBs 28, 52, 101, 138, 153 or 180, or DL-PCB 126. A broad spectrum of changes in gene expression, with sex playing a crucial role in the response, was observed (Heimeier et al., 2010 (abstract)). Multivariate analyses revealed different clustering patterns for the seven PCBs, as seen in the Venn diagram in Fig. 10.6. The analyses suggest that most congeners tested are altering the expression of CYP3A44 and CYP2B10, which are known to be mediated by CAR and PXR, respectively. PCB 52 was found to induce only CYP2B10, while PCB138 induces CYP2B10 and CYP3A44, and to some minor extent also the expression of CYP1A1, a hallmark for AhR induction. PCB 28 did not induce any CYP genes. For the CYP enzyme induction, it was observed that the highly chlorinated NDL-PCBs 138, 153 and 180 induced hepatic PROD activity significantly in both sexes, but at a lower dose in females. The lower chlorinated NDL-PCBs induced hepatic PROD activity in a sex-specific way. According to their CYP-induction pattern, the studied PCBs can be classified into CAR (CYP2B10) inducers, such as PCB 52, CAR/PXR (CYP2B10/CYP3A44), mixed-type inducers such as PCBs 101, 153 and 180 and CAR/PXR plus weak AhR (CYP2B10/CYP3A44/CYP1A1) inducers such as PCB 138.

10.8.2 Kinase activation and cell-to-cell communication

To investigate the tumor promotive activity of PCB 180 the tumor suppressor p53 was analyzed together with DNA damage markers such as

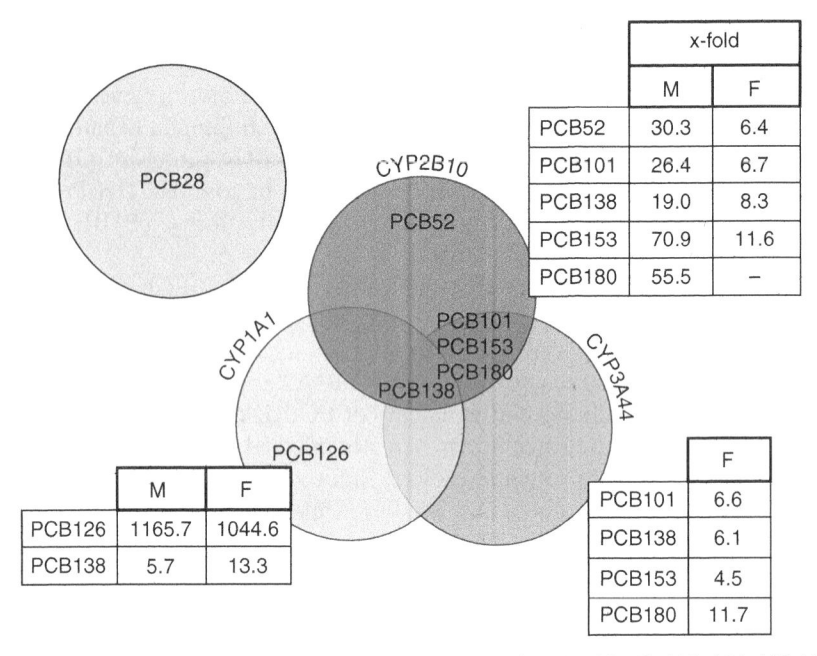

	x-fold	
	M	F
PCB52	30.3	6.4
PCB101	26.4	6.7
PCB138	19.0	8.3
PCB153	70.9	11.6
PCB180	55.5	–

	M	F
PCB126	1165.7	1044.6
PCB138	5.7	13.3

	F
PCB101	6.6
PCB138	6.1
PCB153	4.5
PCB180	11.7

Fig. 10.6 Venn diagram of CYP-induction pattern of PCBs 28, 52, 101, 126, 138 153 or 180-treated male and female mice over a period of five days. Shown are x-fold gene expression data for CYP1A1, CYP2B10 and CYP3A44 resulting from microarray experiments. PCBs in intersection areas indicate a mixed-type inducer of the corresponding CYPs. PCB 28 treatment had no effect on selected CYP gene expressions.

MDM2 and PARP (Al-Anati et al., 2010). Furthermore, in female but not in male rats treated with PCB 180 for 28 days, a dose-dependent increase was observed for hepatic p53 levels. The elevation of p53 was correlated with DNA damage markers (pChk1 Ser317, pChk2 Thr68, p53 Ser15 and γH2AX Ser139 (Viluksela et al. (in preparation); Al-Anati et al. (in preparation)). Similar observations were made in vitro when HepG2 cells were exposed to 20 NDL-PCBs. From these 20 NDL-PCBs, six of the congeners lowered the basal levels of the tumor suppressor p53, as well as attenuated the p53 response after treatment with inducers (Al-Anati et al., 2009). Similar effects were induced by the DL-PCB 126 and it was concluded that both NDL-PCBs and DL-PCBs in low concentrations can induce alterations in p53 signaling, an effect that can be correlated to rat liver carcinogenesis. The inhibition of gap junction intracellular communication (GJIC) was studied in vitro for the 20 highly purified NDL-PCBs. The quantitative dose–response relationship reported for this endpoint might represent an AhR-independent tumor promoting mode of action of NDL-PCBs, since DL-PCBs were found to have no acute GJIC inhibitory effects. (Simecková et al., 2009a, b).

10.9 Classification of NDL-PCB congeners

A classification strategy for NDL-PCB congeners based on *in vitro* effect biomarker information was developed using Quantitative Structure Activity Relationship (QSAR) models and following the OECD validation procedure (Stenberg *et al.*, 2011). The models were established based on a careful selection of a training set of 20 NDL-PCBs (Fig. 10.2), rendering a wide representation of the chemical structures, a defined applicability domain and the biological responses derived from *in vitro* assays (Stenberg and Andersson, 2008). The models for antagonistic effects on the oestrogen receptor (ER), inhibition of the dopamine active transporter (DAT) in striatum, inhibition of GJIC and induction of formation of reactive oxygen species (ROS), reached the set statistical quality limit of a cross-validated explained variance above 0.5. PCBs 44 and 49 are examples of non-tested NDL-PCBs found in food or human tissues with high predicted activity in the DAT, ROS and GJIC models. PCBs 157 and 189 were found to have a relatively high predicted activity in the ER antagonistic assay (Stenberg *et al.*, 2011).

A multivariate classification analysis showed that the NDL-PCBs can be divided into at least two groups based on their different activities in the *in vitro* assays (Fig. 10.7; Stenberg *et al.*, 2011). NDL-PCB congeners with four to five chlorine substituents, with a high proportion of *ortho*-chlorines, and few *para*-substituents, have a higher level of activity in bioassays describing e.g. inhibited uptake of various neurotransmitters and GJIC. NDL-PCB congeners with six to seven chlorine atoms and a substitution pattern evenly distributed over the phenyl group do not show activity in these bioassays. Environmentally abundant congeners, such as NDL-PCBs 18, 25, 26, 44, 47, 49 and 52, were predicted into the biologically most active group of NDL-PCBs (Stenberg *et al.*, 2011).

Hamers *et al.* (2011) classified 24 congeners according to their agonistic/antagonistic properties with regard to ER and AR, and protein binding assays, the thyroid related transthyretin protein, and were able to cluster the NDL-PCBs in three separate groups. Out of these 24 PCBs, PCB 168 is recommended as an additional indicator congener, given its relatively high abundance and anti-androgenic and transthyretin binding characteristics. (Hamers *et al.*, 2011).

Experiments performed *in vivo* enabling classification include the zebrafish embryo toxicity study and the mouse differential gene expression study, which were performed on the six indicator PCBs as well as the DL-PCB 126. Partial least squares (PLS) modeling based on these data sets remains to be done.

10.10 Conclusions and future trends

The qualitative toxicological effect data of ATHON clearly show that NDL-PCBs have endocrine system modulating properties. Effects on several

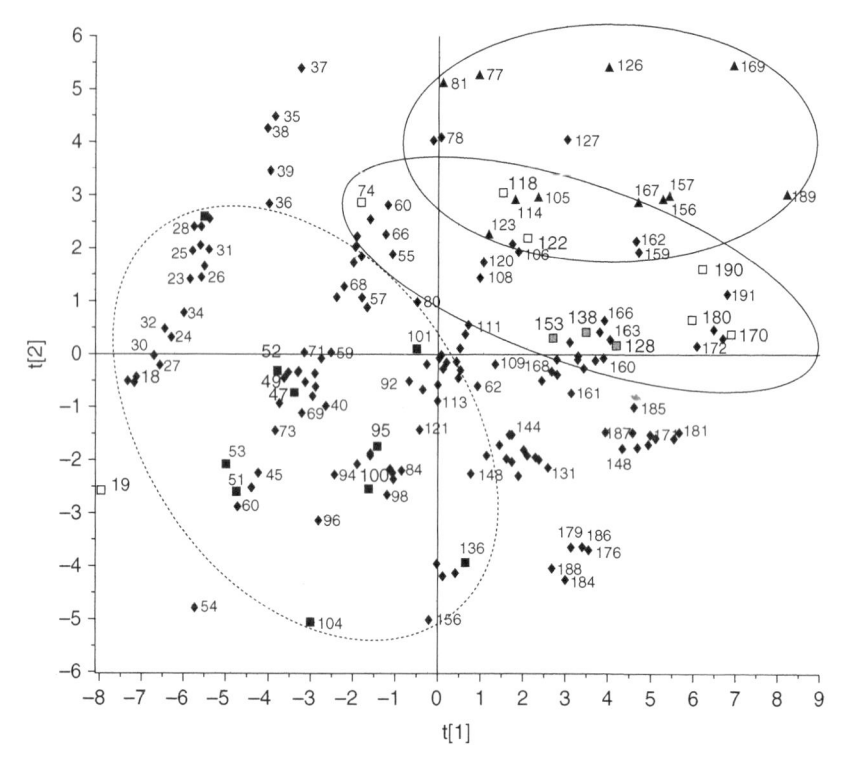

Fig. 10.7 The principal component analysis (PCA) score plot illustrates the relationship between all tri- to hepta-congeners. The homology groups can be seen in order from the left: tri-CBs (including training set PCB 19 and 28), tetra-CBs (including PCB 47, 51, 52, 53 and 74), penta-CBs (including PCB 95, 100, 101, 104, 118 and 122), hexa-CBs (including PCB 128, 136, 138 and 153) and hepta-CBs (including PCB170, 180 and 190, on the right-hand side). The different markings in the plot are as follows; triangles = DL-PCBs; open squares = the group with higher chlorination degree and fewer *ortho* substituents; black squares = the lower chlorinated, *ortho*-substituted group; diamonds = congeners predicted as active in a QSAR model, i.e. PCB in 45 and PCB 18, 25 and 49 in ROS. Source: Stenberg *et al.*, (2011).

hormonal systems including the thyroid, steroid and retinoid systems were observed. In addition, several neurotoxicity modes of action influencing cognitive impairment and motor disorders have been confirmed both *in vivo* and *in vitro*. Exposure to NDL-PCBs during development-induced long-lasting behavioral alterations. The findings *in vivo* that NDL-PCBs perturb neurotransmitter transport and signaling pathways essential for neuronal differentiation, growth and function are also supported by *in vitro* studies. Sex differences in effects after exposure to NDL-PCBs have been observed for multiple endpoints including endocrine disruption, gene expression profiles, induction of hepatic enzyme activities and changes in bone geometry. The results also indicate that specific aspects of NDL-PCB toxicity can be assigned to the *in vivo* formation of PCB-metabolites. QSAR-based classification approaches have been (or will be) applied including *in vitro* effect biomarkers as well as in *vivo* data on toxicokinetics, differential gene expression and zebrafish embryo toxicity. Further detailed quantitative effect and margin of exposure estimations for multiple effect parameters and multiple NDL-PCB congeners are also in progress.

In conclusion, observations from the ATHON studies show that NDL-PCBs of high purity cause a partially different pattern of effects as compared to the effects observed for DL-PCBs; yet the observed outcomes share many similarities. Study results indicate that PCB congeners elicit a continuum of effects not always easily separated into categories, such as DL versus NDL, or PB versus 3-MC types (Table 10.3). Furthermore, the results generated so far clearly show that NDL-PCBs are not a homogeneous group of compounds. Taken together, these findings, lend further support to the notion of developing endpoint specific relative potency (REP) values and additional novel strategies for coping with complex mixtures of compounds.

10.11 Acknowledgements

Most of the experimental work presented here has been supported by funds from the Commission of the European Communities project 'Assessing the Toxicology and Hazard of Non-dioxin-like PCBs Present in Food' (ATHON, grant number FOOD-CT-2005–022923). Additional funding sources are the Swedish Research Council for Environment Agricultural Sciences and Spatial Planning (FORMAS), the National Fund Research, Luxembourg (FNR) and the Marie Curie Actions of the EC (FP7-COFUND). Last but not least, the authors would like to thank all partners/participants and subcontractors of the ATHON project for their dedicated contribution to the ATHON project and for generating much of the data referred to in this chapter, which is an extension of the project summary report to the EC:

- Karolinska Institutet (www.ki.se): Helen Håkansson coordinator, Krister Halldin project manager, Emma Westerholm scientific administrator, Oliver Adfeldt-Still Daniel Borg, Javier Esteban, Rachel Heimeier, Maria Herlin, Per-Ove Thörnqvist/Sandra Ceccatelli, Roshan Tofighi/Johan Högberg, Ulla Stenius, Lauy Al-Anati
- Flemish Institute of Technological Research (www.vito.be): Greet Schoeters, Patrick De Boever
- Veterinary Research Institute (www.vri.cz): Miroslav Machala, Jan Vondracek
- Foundation Centro Investigacion Principe Felipe (www.cipf.es): Vicente Felipo, Jordi Boix, Marta Llansola, Omar Cauli, Ana Agusti

- National Institute for Health and Welfare (www.thl.fi): Matti Viluksela, Merja Korkalainen, Päivi Heikkinen, Hanna Miettinen
- Institute for Prevention and Occupational Medicine, Ruhr University: Hellmuth Lilienthal
- University of Kaiserslautern (www.uni-kl.de): Dieter Schrenk, Robert Roos
- Institute of Pharmacological Research "Mario Negri" (www.marionegri.it): Roberto Fanelli, Roberta Pastorelli
- Utrecht University (www.iras.uu.nl): Martin van den Berg, Majorie van Duursen, Remco Westerink, Elsa Antunes-Fernandes
- VU University Amsterdam – Institute for Environmental Studies (IVM) (www.ivm.vu.nl): Jacob de Boer, Heather Leslie, Timo Hamers
- University of Oslo (www.uio.no): Frode Fonnum
- Umeå University (www.umu.se): Mats Tysklind, Patrik Andersson, Mia Stenberg
- University Hospitals Bristol NHS Foundation Trust (www.uhbristol.nhs.uk): Margaret Saunders, Laura Cartwright, Sara Correia Carreira
- Health Canada (www.hc-sc.gc.ca): Wayne Bowers
- National Institute of Public Health and the Environment of the Netherlands (www.rivm.nl): Leo van der Ven
- Justus Liebig University Giessen (http://www.uni-giessen.de): Gerd Hamscher.

10.12 References

ABBALLE, A., BALLARD, T. J., DELLATTE, E., DI DOMENICO, A., FERRI, F., FULGENZI, A. R., GRISANTI, G., IACOVELLA, N., INGELIDO, A. M., MALISCH, R., MINIERO, R., PORPORA, M. G., RISICA, S., ZIEMACKI, G. and DE FELIP, E. (2008). Persistent environmental contaminants in human milk: concentrations and time trends in Italy. *Chemosphere*, **73**, S220–7.

AHLBORG, U. G., HANBERG, A. and KENNE, K. (1992). Nordic risk assessment of PCBs. *Nord* 1992:26.

AL-ANATI, L., HÖGBERG, J. and STENIUS, U. (2009). Non-dioxin-like-PCBs phosphorylate Mdm2 at Ser166 and attenuate the p53 response in HepG2 cells. *Chem Biol Interact*, **182** (2–3), 191–8.

AL-ANATI, L., HOGBERG, J. and STENIUS, U. (2010). Non-dioxin-like PCBs interact with benzo[a]pyrene-induced p53-responses and inhibit apoptosis. *Toxicol Appl Pharmacol*, **249**, 166–77.

AL-ANATI, L., HÖGBERG, J., VILUKSELA, M., STRID, A. and STENIUS, U. Hydroxylation of PCB 180 mediated DNA damage and enhanced benzo[a]pyrene-induced genotoxicity. (in preparation).

ALBRO, P. W. and FISHBEIN, L. (1972). Quantitative and qualitative analysis of polychlorinated biphenyls by gas-liquid chromatography and flame ionization detection. I. One to three chlorine atoms. *J Chromatogr*, **69**, 273–83.

AMAP (2002). Arctic Monitoring and Assessment Programme (www.amap.no). *Oslo, Norway*, **XIV**, 137.

AMAP (2009). Arctic Monitoring and Assessment Programme, (www.amap.no) (AMAP), P.O. Box 8100 Dep, N-0032. *Oslo, Norway, Human Health in the Arctic.* Chapter 10, 209–10.

ARNICH, N., TARD, A., LEBLANC, J. C., LE BIZEC, B., NARBONNE, J. F. and MAXIMILIEN, R. (2009). Dietary intake of non-dioxin-like PCBs (NDL-PCBs) in France, impact of maximum levels in some foodstuffs. *Regul Toxicol Pharmacol*, **54**, 287–93.

ARNOLD, R., KIBLER, R. and BRUNNER, B. (1998). Alimentary intake of selected pollutants and nitrate – results of a duplicate study in Bavarian homes for youth and seniors. *Z Ernahrungswiss*, **37**, 328–35.

ASANTE, K. A., ADU-KUMI, S., NAKAHIRO, K., TAKAHASHI, S., ISOBE, T., SUDARYANTO, A., DEVANATHAN, G., CLARKE, E., ANSA-ASARE, O. D. and DAPAAH-SIAKWAN, S. (2011). Human exposure to PCBs, PBDEs and HBCDs in Ghana: Temporal variation, sources of exposure and estimation of daily intakes by infants. *Environ. Int*, **37** (5), 921–28.

ATSDR (2000). *U.S. Department of Health and Human Services Toxicological Profile for Polychlorinated Biphenyls (PCBs)*. Public Health Service, Agency for Toxic Substances and Disease Registry, USA.

AYOTTE, P., MUCKLE, G., JACOBSON, J. L., JACOBSON, S. W. and DEWAILLY, E. (2003). Assessment of pre- and postnatal exposure to polychlorinated biphenyls: lessons from the Inuit Cohort Study. *Environ Health Persp*, **111**, 1253–8.

BAARS, A. J., BAKKER, M. I., BAUMANN, R. A., BOON, P. E., FREIJER, J. I., HOOGENBOOM, L. A., HOOGERBRUGGE, R., VAN KLAVEREN, J. D., LIEM, A. K., TRAAG, W. A. and DE VRIES, J. (2004). Dioxins, dioxin-like PCBs and non-dioxin-like PCBs in foodstuffs: occurrence and dietary intake in The Netherlands. *Toxicol Lett*, **151**, 51–61.

BAKKER, M., BAARS, A., BAUMANN, B., BOON, P. and HOOGERBRUGGE, R. (2003). Indicator PCBs in foodstuffs: occurrence and dietary intake in The Netherlands at the end of the 20th century.

BERGKVIST, C., LIGNELL, S., SAND, S., AUNE, M., PERSSON, M., HAKANSSON, H. and BERGLUND, M. (2010). A probabilistic approach for estimating infant exposure to environmental pollutants in human breast milk. *J Environ Monit*, **12**, 1029–36.

BERGONZI, R., SPECCHIA, C., DINOLFO, M., TOMASI, C., DE PALMA, G., FRUSCA, T. and APOSTOLI, P. (2009). Distribution of persistent organochlorine pollutants in maternal and foetal tissues: data from an Italian polluted urban area. *Chemosphere*, **76**, 747–54.

BOIX, J., CAULI, O. and FELIPO, V. (2010). Developmental exposure to polychlorinated biphenyls 52, 138 or 180 affects differentially learning or motor coordination in adult rats. Mechanisms involved. *Neuroscience*, **167**, 994–1003.

BORG, D., RENDEL, F., HEIKKINEN, P., SIMANAINEN, U., VILUKSELA, M., GIESE, N., HAMSCHER, G., HALLDIN, K. and HÅKANSSON, H. (2010). *Effects on retinoids and body/organ weights in rats after in utero and lactational exposure to purified PCB180 or TCDD*. 6th International PCB Workshop, Visby, Sweden, May 30–June 2, 2010.

BREIVIK, K., SWEETMAN, A., PACYNA, J. M. and JONES, K. C. (2002). Towards a global historical emission inventory for selected PCB congeners – a mass balance approach. 1. Global production and consumption. *Sci Total Environ*, **290**, 181–98.

BROWN JR, J., LAWTON, R., ROSS, M., FEINGOLD, J., WAGNER, R. and HAMILTON, S. (1989). Persistence of PCB congeners in capacitor workers and Yusho patients. *Chemosphere*, **19**, 829–34.

BÜHLER, F., SCHMID, P. and SCHLATTER, C. (1988). Kinetics of PCB elimination in man. *Chemosphere*, **17**, 1717–26.

BRUNELLI, L., LLANSOLA, M., FELIPO, V., CAMPAGNA, R., AIROLDI, L., DE PAOLA, M., FANELLI, R., MARIANI, A., MAZZOLETTI, M. and PASTORELLI, R. (2012) Food-relevant non-dioxin like polychlorinated biphenyls alter the proteome of cerebellar neurons in culture by different key functional networks. *J Proteomics*, **75**, 2417–30.

CAMPAGNA, R., BRUNELLI, L., AIROLDI, L., FANELLI, R., HAKANSSON, H., HEIMEIER, R. A., DE BOEVER, P., BOIX, J., LLANSOLA, M. and FELIPO, V. (2011). Cerebellum Proteomics Addressing the Cognitive Deficit of Rats Perinatally Exposed to the Food-Relevant Polychlorinated Biphenyl 138. *Toxicol Sci*, **123**, 170–9.

CHEEK, A. O., KOW, K., CHEN, J. and MCLACHLAN, J. A. (1999). Potential mechanisms of thyroid disruption in humans: interaction of organochlorine compounds with thyroid receptor, transthyretin, and thyroid-binding globulin. *Environ Health Persp*, **107**, 273.

CHEN, P., LUO, M., WONG, C. and CHEN, C. (1982). Comparative rates of elimination of some individual polychlorinated biphenyls from the blood of PCB-poisoned patients in Taiwan. *Food Chem Toxicol*, **20**, 417–25.

CHU, I., VILLENEUVE, D. C., YAGMINAS, A., LECAVALIER, P., POON, R., FEELEY, M., KENNEDY, S. W., SEEGAL, R. F., HAKANSSON, H., AHLBORG, U. G., VALLI, V. E. and BERGMAN, A. (1996a). Toxicity of 2,2',4,4',5,5'-hexachlorobiphenyl in rats: effects following 90-day oral exposure. *J Appl Toxicol*, **16**, 121–8.

CHU, I., VILLENEUVE, D. C., YAGMINAS, A., LECAVALIER, P., POON, R., HAKANSSON, H., AHLBORG, U. G., VALLI, V. E., KENNEDY, S. W., BERGMAN, A., SEEGAL, R. F. and FEELEY, M. (1996b). Toxicity of 2,4,4'-trichlorobiphenyl in rats following 90-day dietary exposure. *J Toxicol Environ Health*, **49**, 301–18.

CICAD (2003). *Concise International Chemical Assessment Document 55: Polychlorinated biphenyls: Human health aspects*. World Health Organization, Geneva.

CLRTAP (1998). Protocol to the 1979 convention on long-range transboundary air pollution on persistent organic pollutants. http://www.unece.org/fileadmin/DAM/env/lrtap/full%20text/1998.POPs.e.pdf

COK, I., MAZMANCI, B., MAZMANCI, M. A., TURGUT, C., HENKELMANN, B. and SCHRAMM, K. W. (2012). Analysis of human milk to assess exposure to PAHs, PCBs and organochlorine pesticides in the vicinity Mediterranean city Mersin, Turkey. *Environ Int*, **40**, 63–9.

COVACI, A., DE BOER, J., RYAN, J. J., VOORSPOELS, S. and SCHEPENS, P. (2002). Distribution of organobrominated and organochlorinated contaminants in Belgian human adipose tissue. *Environ Res*, **88**, 210–8.

COVACI, A., VOORSPOELS, S., ROOSENS, L., JACOBS, W., BLUST, R. and NEELS, H. (2008). Polybrominated diphenyl ethers (PBDEs) and polychlorinated biphenyls (PCBs) in human liver and adipose tissue samples from Belgium. *Chemosphere*, **73**, 170–5.

DARNERUD, P. O., AUNE, M., LARSSON, L., LIGNELL, S., MUTSHATSHI, T., OKONKWO, J., BOTHA, B. and AGYEI, N. (2011). Levels of brominated flame retardants and other pesistent organic pollutants in breast milk samples from Limpopo province, South Africa. *Sci Total Environ*, **409** (19), 4048–53.

DE BOEVER, P., BOIX, J., HOLLANDERS, K. and FELIPO, V. (2010). Effects of developmental exposure to non-dioxin like PCBs on gene expression in brain and blood. 6th International PCB Workshop, Visby, Sweden, May 30–June 2, 2010.

DEKONING, E. P. and KARMAUS, W. (2000). PCB exposure in utero and via breast milk. a review. *J Expo Anal Environ Epidemiol*, **10**, 285–93.

DENG, B., ZHANG, J. Q., ZHANG, L., JIANG, Y. S., ZHOU, J., FANG, D., ZHANG, H. and HUANG, H. Y. (2011). Levels and profiles of PCDD/Fs, PCBs in mothers' milk in Shenzhen of China: Estimation of breast-fed infants' intakes. *Environ. Int.*, **42**, 47–52.

DEWAILLY, E., AYOTTE, P., BRUNEAU, S., LALIBERTE, C., MUIR, D. C. and NORSTROM, R. J. (1993). Inuit exposure to organochlorines through the aquatic food chain in arctic quebec. *Environ Health Persp*, **101**, 618–20.

DEVANATHAN, G., SUBRAMANIAN, A., SOMEYA, M., SUDARYANTO, A., ISOBE, T., TAKAHASHI, S., CHAKRABORTY, P. and TANABE, S. (2009). Persistent organochlorines in human breast milk from major metropolitan cities in India. *Environ Pollut*, **157**, 148–54.
DEVANATHAN, G., SUBRAMANIAN, A., SUDARYANTO, A., TAKAHASHI, S., ISOBE, T. and TANABE, S. (2012). Brominated flame retardants and polychlorinated biphenyls in human breast milk from several locations in India: potential contaminant sources in a municipal dumping site. *Environ Int*, **39**, 87–95.
DUARTE-DAVIDSON, R. and JONES, K. C. (1994). Polychlorinated biphenyls (PCBs) in the UK population: estimated intake, exposure and body burden. *Sci Total Environ*, **151**, 131–52.
EFSA (2005). *Opinion of the scientific panel on contaminants in the food chain on a request from the commission related to the presence of non dioxin-like polychlorinated biphenyls (PCB) in feed and food*. European Food Safety Authority, Parma, Italy, http://www.efsa.europa.eu/en/efsajournal/doc/284.pdf.
EFSA (2010). *Results of the monitoring of non dioxin-like PCBs in food and feed*. European Food Safety Authority, Parma, Italy. http://www.efsa.europa.eu/en/efsajournal/doc/1701.pdf
EIK ANDA, E., NIEBOER, E., DUDAREV, A. A., SANDANGER, T. M. and ODLAND, J. O. (2007). Intra- and intercompartmental associations between levels of organochlorines in maternal plasma, cord plasma and breast milk, and lead and cadmium in whole blood, for indigenous peoples of Chukotka, Russia. *J Environ Monit*, **9**, 884–93.
ELABBAS, L. E., HERLIN, M., FINNILA, M. A., RENDEL, F., STERN, N., TROSSVIK, C., BOWERS, W. J., NAKAI, J., TUUKKANEN, J., VILUKSELA, M., HEIMEIER, R. A., AKESSON, A. and HAKANSSON, H. (2011). In utero and lactational exposure to Aroclor 1254 affects bone geometry, mineral density and biomechanical properties of rat offspring. *Toxicol Lett*, **207**, 82–8.
ENNACEUR, S., GANDOURA, N. and DRISS, M. (2008). Distribution of polychlorinated biphenyls and organochlorine pesticides in human breast milk from various locations in Tunisia: Levels of contamination, influencing factors, and infant risk assessment. *Environ. Res*, **108**, 86–93.
ERIKSSON, P. and FREDRIKSSON, A. (1996). Developmental neurotoxicity of four ortho-substituted polychlorinated biphenyls in the neonatal mouse. *Environ Toxicol Pharmacol*, **1**, 155–65.
EU (2000). Opinion of the SCF on the Risk Assessment of Dioxins and Dioxin-like PCBs in Food SCF/CS/CNTM/DIOXIN/8 Final, 23 November 2000.
FATTORE, E., FANELLI, R., DELLATTE, E., TURRINI, A. and DI DOMENICO, A. (2008). Assessment of the dietary exposure to non-dioxin-like PCBs of the Italian general population. *Chemosphere*, **73**, S278–83.
FERNANDES, E. C. A., HENDRIKS, H. S., VAN KLEEF, R. G. D. M., RENIERS, A., ANDERSSON, P. L., VAN DEN BERG, M. and WESTERINK, R. H. S. (2010). Activation and Potentiation of Human GABAA Receptors by Non-Dioxin-Like PCBs Depends on Chlorination Pattern. *Toxicol. Sci*, **118**, 183–190.
FISCHER, C., FREDRIKSSON, A. and ERIKSSON, P. (2008). Neonatal co-exposure to low doses of an ortho-PCB (PCB 153) and methyl mercury exacerbate defective developmental neurobehavior in mice. *Toxicology*, **244**, 157–65.
FÜRST, P. (2006) Dioxins, polychlorinated biphenyls and other organohalogen compounds in human milk. Levels, correlations, trends and exposure through breast-feeding. *Mol Nutr Food Res*, **50** (10), 922–33.
GONCHAROV, A., REJ, R., NEGOITA, S., SCHYMURA, M., SANTIAGO-RIVERA, A., MORSE, G. and CARPENTER, D. O. (2009). Lower serum testosterone associated with elevated polychlorinated biphenyl concentrations in Native American men. *Environ Health Perspect*, **117**, 1454–60.

GUVENIUS, D. M., ARONSSON, A., EKMAN-ORDEBERG, G., BERGMAN, A. and NOREN, K. (2003). Human prenatal and postnatal exposure to polybrominated diphenyl ethers, polychlorinated biphenyls, polychlorobiphenylols, and pentachlorophenol. *Environ Health Perspect*, **111**, 1235–41.

HAMERS, T., KAMSTRA, J. H., CENIJN, P. H., PENCIKOVA, K., PALKOVA, L., SIMECKOVA, P., VONDRACEK, J., ANDERSSON, P. L., STENBERG, M. and MACHALA, M. (2011). In vitro toxicity profiling of ultrapure non-dioxin-like polychlorinated biphenyl congeners and their relative toxic contribution to PCB mixtures in humans. *Toxicol Sci*, **121**, 88–100.

HANY, J., LILIENTHAL, H., SARASIN, A., ROTH-HARER, A., FASTABEND, A., DUNEMANN, L., LICHTENSTEIGER, W. and WINNEKE, G. (1999). Developmental exposure of rats to a reconstituted PCB mixture or aroclor 1254: effects on organ weights, aromatase activity, sex hormone levels, and sweet preference behavior. *Toxicol Appl Pharmacol*, **158**, 231–43.

HEIMEIER, R., THÖRNQVIST, P., ROOS, R., SCHRENK, D., DE BOEVER, P. and SCHOETERS, G. (2010). *Genome-wide analysis of murine liver toxicity profiles reveal gender specific changes following exposure to individual PCB congeners*. 6th International PCB Workshop Visby, Sweden, May 30–June 2, 2010.

HERBSTMAN, J. B., SJODIN, A., APELBERG, B. J., WITTER, F. R., PATTERSON, D. G., HALDEN, R. U., JONES, R. S., PARK, A., ZHANG, Y., HEIDLER, J., NEEDHAM, L. L. and GOLDMAN, L. R. (2007). Determinants of prenatal exposure to polychlorinated biphenyls (PCBs) and polybrominated diphenyl ethers (PBDEs) in an urban population. *Environ Health Persp*, **115**, 1794–800.

HOOPER, K., SHE, J., SHARP, M., CHOW, J., JEWELL, N., GEPHART, R. and HOLDEN, A. (2007). Depuration of polybrominated diphenyl ethers (PBDEs) and polychlorinated biphenyls (PCBs) in breast milk from California first-time mothers (primiparae). *Environ Health Persp*, **115**, 1271–5.

JOHANSEN, E. B., KNOFF, M., FONNUM, F., LAUSUND, P. L., WALAAS, S. I., WOIEN, G. and SAGVOLDEN, T. (2011). Postnatal exposure to PCB 153 and PCB 180, but not to PCB 52, produces changes in activity level and stimulus control in outbred male Wistar Kyoto rats. *Behav Brain Funct*, **7**, 18.

JUAN, C. Y., THOMAS, G. O., SWEETMAN, A. J. and JONES, K. C. (2002). An input-output balance study for PCBs in humans. *Environ Int*, **28**, 203–14.

KALANTARI, F., WESTERHOLM, E., FATTORE, E., ÖBERG, M., SAND, S. and HÅKANSSON, H. (2012). Estimation of relative potency values for polychlorinated biphenyl (PCB) congeners based on hepatic endpoints of toxicity. Manuscript in Kalantari F thesis book (2012), "Quantitative Approaches for Health Risk Assessment of Environmental Pollutants. Estimation of Differences in Sensitivity, Relative Potencies, and Margins of Exposure", ISBN: 978-91-7457-868-3. Karolinska Institutet, Sweden.

KOOPMAN-ESSEBOOM, C., HUISMAN, M., WEISGLAS-KUPERUS, N., VAN DER PAAUW, C. G., TH. TUINSTRA, L. G. M., BOERSMA, E. R. and SAUER, P. J. J. (1994). PCB and dioxin levels in plasma and human milk of 418 dutch women and their infants. Predictive value of PCB congener levels in maternal plasma for fetal and infant's exposure to PCBs and dioxins. *Chemosphere*, **28**, 1721–32.

KORKALAINEN, K., HÅKANSSON, H. and VILUKSELA, M. (2012). Risk assessment of non-dioxin like PCB 180 and PCB 52 present in food. Eurotox 2012, Stockholm, Sweden. *Toxicol. Lett*, **211** (Supplement), p. S124.

KREMER, H., LILIENTHAL, H., HANY, J., ROTH-HÄRER, A. and WINNEKE, G. (1999). Sex-Dependent Effects of Maternal PCB Exposure on the Electroretinogram in Adult Rats1. *Neurotoxicol. teratol*, **21**, 13–19.

LANGEVELD, W.T., MEIJER, M. and WESTERINK, R.H. (2012). Differential effects of 20 non-dioxin-like PCBs on basal and depolarization-evoked intracellular calcium levels in PC12 cells. *Toxicol Sci*, **126** (2), 487–96.

LA ROCCA, C. and MANTOVANI, A. (2006). From environment to food: the case of PCB. *Ann Ist Super Sanita*, **42**, 410–6.

LECAVALIER, P., CHU, I., YAGMINAS, A., VILLENEUVE, D. C., POON, R., FEELEY, M., HAKANSSON, H., AHLBORG, U. G., VALLI, V. E., BERGMAN, A., SEEGAL, R. F. and KENNEDY, S. W. (1997). Subchronic toxicity of 2,2',3,3',4,4'-hexachlorobiphenyl in rats. *J Toxicol Environ Health*, **51**, 265–77.

LEES, P. S., CORN, M. and BREYSSE, P. N. (1987). Evidence for dermal absorption as the major route of body entry during exposure of transformer maintenance and repairmen to PCBs. *Am Ind Hyg Assoc J*, **48**, 257–64.

LILIENTHAL, H., HEIKKINEN, P., ANDERSSON, P. L., VAN DER VEN, L. T. M. and VILUKSELA, M. (2011). Auditory effects of developmental exposure to purity-controlled polychlorinated biphenyls (PCB52 and PCB180) in rats. *Toxicol. Sci*, **122**, 100–11.

LILIENTHAL, H., HEIKKINEN, P., DANIELSSON, C., ANDERSSON, P. and VILUKSELA, M. (2009). Effects of purity-controlled PCB52 and PCB180 on dopamine-dependent behavior in rat offspring after maternal exposure. *Toxicol. Lett*, **189**, 229–30.

LILIENTHAL, H., KORKALAINEN, M., DANIELSSON, C., ANDERSSON, A. and VILUKSELA, M. (2010). *Auditory effects of developmental exposure to purity controlled PCB74 or PCB95 in rats*. 6th International PCB Workshop Visby, Sweden, May 30–June 2, 2010.

LIND, Y., DARNERUD, P. O., AUNE, M. and BECKER, W. (2002). *Exposure to organic environmental pollutants in the food. Intake estimations of ΣPCB, PCB-153, ΣDDT, PCDD/DF, dioxin-like PCB, PBDE and HBCD based on consumption data from Riksmaten 1997–98 in Swedish*. Report 26–2002, the National Food Administration, Sweden, pp.103

LINDÉN, J., LENSU, S., TUOMISTO, J. and POHJANVIRTA, R. (2010). Dioxins, the aryl hydrocarbon receptor and the central regulation of energy balance. *Front. Neuroendocrin.*, **31**, 452–78.

LLANSOLA, M., MONTOLIU, C., BOIX, J. and FELIPO, V. (2010). Polychlorinated biphenyls PCB 52, PCB 180, and PCB 138 impair the glutamate − nitric oxide − cGMP pathway in cerebellar neurons in culture by different mechanisms. *Chem Res Toxicol*, **23**, 813–20.

LLANSOLA, M., PIEDRAFITA, B., RODRIGO, R., MONTOLIU, C. and FELIPO, V. (2009). Polychlorinated biphenyls PCB 153 and PCB 126 impair the glutamate–nitric oxide–cGMP pathway in cerebellar neurons in culture by different mechanisms. *Neurotox. Res*, **16**, 97–105.

LUOTAMO, M., JARVISALO, J. and AITIO, A. (1991). Assessment of exposure to polychlorinated biphenyls: analysis of selected isomers in blood and adipose tissue. *Environ Res*, **54**, 121–34.

MACDONALD, R., BARRIE, L., BIDLEMAN, T., DIAMOND, M., GREGOR, D., SEMKIN, R., STRACHAN, W., LI, Y., WANIA, F. and ALAEE, M. (2000). Contaminants in the Canadian Arctic: 5 years of progress in understanding sources, occurrence and pathways. *Sci. Total. Environ*, **254**, 93–234.

MATTHEWS, H. B. and DEDRICK, R. L. (1984). Pharmacokinetics of PCBs. *Annu Rev Pharmacol Toxicol*, **24**, 85–103.

MCFARLAND, V. A. and CLARKE, J. U. (1989). Environmental occurrence, abundance, and potential toxicity of polychlorinated biphenyl congeners: considerations for a congener-specific analysis. *Environ Health Perspect*, **81**, 225–39.

MEIRONYTE GUVENIUS, D., BERGMAN, A. and NOREN, K. (2001). Polybrominated diphenyl ethers in Swedish human liver and adipose tissue. *Arch Environ Contam Toxicol*, **40**, 564–70.

MIETTINEN, H., HEIKKINEN, P., KORKALAINEN, K., ANTUNES-FERNANDES, E., VAN DUURSEN, M., LESLIE, H., HAMSCHER, G., ANDERSSON, P., HALLDIN, K., HÅKANSSON, H. and VILUKSELA, M. (2009). Effects of perinatal exposure to high purity PCB180 on rat offspring. Eurotox 2009, Dresden, Germany. *Toxicol Lett*, **189** (supplement), S104.

NTP (2006). NTP technical report on the toxicology and carcinogenesis studies of 2,2',4,4',5,5'-hexachlorobiphenyl (PCB 153) (CAS No. 35065-27-1) in female Harlan Sprague-Dawley rats (Gavage studies). National Institutes of Health (NIH) Publication No. 06-4465, NIH, Public Health Service, U.S. Department of Health and Human Services, 4–168.

ÖBERG, M., SJÖDIN, A., CASABONA, H., NORDGREN, I., KLASSON-WEHLER, E. and HÅKANSSON, H. (2002). Tissue distribution and half-lives of individual polychlorinated biphenyls and serum levels of 4-hydroxy-2, 3, 3, 4, 5-pentachlorobiphenyl in the rat. *Toxicol. Sci*, **70**, 171–82.

PARK, J. S., BERGMAN, A., LINDERHOLM, L., ATHANASIADOU, M., KOCAN, A., PETRIK, J., DROBNA, B., TRNOVEC, T., CHARLES, M. J. and HERTZ-PICCIOTTO, I. (2008). Placental transfer of polychlorinated biphenyls, their hydroxylated metabolites and pentachlorophenol in pregnant women from eastern Slovakia. *Chemosphere*, **70**, 1676–84.

PARK, J. S., SHE, J., HOLDEN, A., SHARP, M., GEPHART, R., SOUDERS-MASON, G., ZHANG, V., CHOW, J., LESLIE, B. and HOOPER, K. (2011). High postnatal exposures to polybrominated diphenyl ethers (PBDEs) and polychlorinated biphenyls (PCBs) via breast milk in California: does BDE-209 transfer to breast milk? *Environ Sci Tech*, **45**, 4579–85.

PARK, M. J., LEE, S. K., YANG, J. Y., KIM, K. W., LEE, S. Y., LEE, W. T., CHUNG, K. H., YUN, Y. P. and YOO, Y. C. (2005). Distribution of organochlorines and PCB congeners in Korean human tissues. *Arch Pharm Res*, **28**, 829–38.

PIEDRAFITA, B., ERCEG, S., CAULI, O., MONFORT, P. and FELIPO, V. (2008a). Developmental exposure to polychlorinated biphenyls PCB153 or PCB126 impairs learning ability in young but not in adult rats. *Eur J Neurosci*, **27**, 177–82.

PIEDRAFITA, B., ERCEG, S., CAULI, O. and FELIPO, V. (2008b). Developmental exposure to polychlorinated biphenyls or methylmercury, but not to its combination, impairs the glutamate–nitric oxide–cyclic GMP pathway and learning in 3-month-old rats. *Neuroscience*, **154** (4), 1408–16.

POLDER, A., SKAARE, J. U., SKJERVE, E., LOKEN, K. B. and EGGESBO, M. (2009). Levels of chlorinated pesticides and polychlorinated biphenyls in Norwegian breast milk (2002–2006), and factors that may predict the level of contamination. *Sci Total Environ*, **407**, 4584–90.

POLDER, A., THOMSEN, C., LINDSTROM, G., LOKEN, K. B. and SKAARE, J. U. (2008). Levels and temporal trends of chlorinated pesticides, polychlorinated biphenyls and brominated flame retardants in individual human breast milk samples from Northern and Southern Norway. *Chemosphere*, **73**, 14–23.

RAAB, U., PREISS, U., ALBRECHT, M., SHAHIN, N., PARLAR, H. and FROMME, H. (2008). Concentrations of polybrominated diphenyl ethers, organochlorine compounds and nitro musks in mother's milk from Germany (Bavaria). *Chemosphere*, **72**, 87–94.

RODRÍGUEZ-DOZAL, S., RODRÍGUEZ, H. R., HERNÁNDEZ-ÁVILA, M., VAN OOSTDAM, J., WEBER, J. P., NEEDHAM, L. L. and TRIP, L. (2011). Persistent organic pollutant concentrations in first birth mothers across Mexico. *J. Expo. Sci. Env. Epid*, **22**, 60–9.

ROOS, R., ANDERSSON, P. L., HALLDIN, K., HAKANSSON, H., WESTERHOLM, E., HAMERS, T., HAMSCHER, G., HEIKKINEN, P., KORKALAINEN, M., LESLIE, H. A., NIITTYNEN, M., SANKARI, S., SCHMITZ, H. J., VAN DER VEN, L. T., VILUKSELA, M. and SCHRENK, D. (2011a). Hepatic effects of a highly purified 2,2',3,4,4',5,5'-heptachlorbiphenyl (PCB 180) in male and female rats. *Toxicology*, **284**, 42–53.

ROOS, R., BORG, D., RENDEL, F., HEIKKINEN, P., SIMANAINEN, U., VILUKSELA, M., SCHRENK, D., GIESE, N., HAMSCHER, G., HALLDIN, K. and HÅKANSSON, H. (2011b). *A comparative toxicity study in rats after in utero and lactational exposure to purified PCB180 or TCDD*. Dioxin, Brussels, Belgium.

ROOS, R., KORKALAINEN, K., RENDEL, F., HERLIN, M., HALDIN, K., HÅKANSSON, H., SCHRENK, D. and VILUKSELA, M. (2012). Toxicological profiles of two abundant PCBs in rat offspring. Eurotox 2012, Stockholm, Sweden. *Toxicol. Lett*, **211** (supplement), S85–S86.

ROOS, R., HALLDIN, K., WESTERHOLM, E., HAMERS, T., HAMSCHER, G., HEIKKINEN, P., KORKALAINEN, M., LESLIE, H. A., NIITTYNEN, M., SANKARI, S., VAN DER VEN, L. T., SCHRENK, D., VILUKSELA, M. and HÅKANSSON, H. (PCB52Adult). Toxicological Profile of High Purity 2,2′,5,5′-Heptachlorbiphenyl (PCB 52) in a subacute oral dose toxicity study in adult rats. (in preparation).

ROOS, R., KORKALAINEN, K., RENDEL, F., HERLIN, M., HÅKANSSON, H., SCHRENK, D. and VILUKSELA, M. (PCB52). Toxicological Profile of 2,2′,5,5′-Heptachlorbiphenyl (PCB 52) in a one generation oral dose study in rats. (in preparation).

ROOS, R., KORKALAINEN, K., RENDEL, F., HERLIN, M., HÅKANSSON, H., SCHRENK, D. and VILUKSELA, M. (PCB180). Toxicological Profile of 2,2′,3,4,4′,5,5′-Heptachlorbiphen yl (PCB 180) in a one generation oral dose study in rats. (in preparation).

RYAN, J. J., LEVESQUE, D., PANOPIO, L. G., SUN, W. F., MASUDA, Y. and KUROKI, H. (1993). Elimination of polychlorinated dibenzofurans (PCDFs) and polychlorinated biphenyls (PCBs) from human blood in the Yusho and Yu-Cheng rice oil poisonings. *Arch. Environ. Contam. toxicol*, **24**, 504–12.

SCHECTER, A., CONSTABLE, J., BANGERT, J. V., WIBERG, K., HANSSON, M., NYGREN, M. and RAPPE, C. (1989). Isomer specific measurement of polychlorinated dibenzodioxin and dibenzofuran isomers in human blood from American Vietnam veterans two decades after exposure to Agent Orange. *Chemosphere*, **18**, 531–38.

SCHLUMMER, M., MOSER, G. A. and MCLACHLAN, M. S. (1998). Digestive tract absorption of PCDD/Fs, PCBs, and HCB in humans: mass balances and mechanistic considerations. *Toxicol Appl Pharmacol*, **152**, 128–37.

SCHOENAUER SEBAG, A. (2011). *Non-dioxin like PCBs and the retinoid system in Mus musculus: combining chemical analysis and bioinformatics to build toxicogenomic profiles*. M.Sc., Institute of Environmental Medicien, Karolinska Institutet.

SCHREITMÜLLER, J. and BALLSCHMITER, K. (1994). Levels of polychlorinated biphenyls in the lower troposphere of the North-and South-Atlantic Ocean. *Fresen. J. anal. Chem.*, **348**, 226–39.

SCP (2001). *Stockholm Convention on Persistent Organic Pollutants (POPs)*. Stockholm. http://chm.pops.int/default.aspx http://www.pops.int/documents/convtext/convtext_en.pdf

SIMECKOVÁ, P., VONDRÁCEK, J., PROCHÁZKOVÁ, J., KOZUBÍK, A., KRCMÁR, P. and MACHALA, M. (2009a). 2,2',4,4',5,5'-hexachlorobiphenyl (PCB 153) induces degradation of adherens junction proteins and inhibits beta-catenin-dependent transcription in liver epithelial cells. *Toxicology*, **260** (1–3), 104–11.

SIMECKOVÁ, P., VONDRÁCEK, J., ANDRYSÍK, Z., ZATLOUKALOVÁ, J., KRCMÁR, P., KOZUBÍK, A. and MACHALA, M. (2009b). The 2,2',4,4',5,5'-hexachlorobiphenyl-enhanced degradation of connexin 43 involves both proteasomal and lysosomal activities. *Toxicol Sci*, **107** (1), 9–18.

SMITH, A. B., SCHLOEMER, J., LOWRY, L. K., SMALLWOOD, A. W., LIGO, R. N., TANAKA, S., STRINGER, W., JONES, M., HERVIN, R. and GLUECK, C. J. (1982). Metabolic and health consequences of occupational exposure to polychlorinated biphenyls. *Br J Ind Med*, **39**, 361–9.

SONCHIK, S. M. (1983). Environmental applications of capillary GC columns. *J Chromatogr Sci*, **21**, 106–110.

STELLMAN, S. D., DJORDJEVIC, M. V., MUSCAT, J. E., GONG, L., BERNSTEIN, D., CITRON, M. L., WHITE, A., KEMENY, M., BUSCH, E. and NAFZIGER, A. N. (1998). Relative abundance of organochlorine pesticides and polychlorinated biphenyls in adipose tissue and serum of women in Long Island, New York. *Cancer Epidem Biomar*, **7**, 489–96.

STENBERG, M., and ANDERSSON, P. L. (2008). Selection of non-dioxin-like PCBs for in vitro testing on the basis of environmental abundance and molecular structure. *Chemosphere*, **71** (10), 1909–15.

STENBERG, M., HAMERS, T., MACHALA, M., FONNUM, F., STENIUS, U., LAUY, A. A., VAN DUURSEN, M., WESTERINK, R. H. S., FERNANDES, E. C. A. and ANDERSSON, P. L. (2011). Multivariate toxicity profiles and QSAR modeling of non-dioxin-like PCBs-An investigation of in vitro screening data from ultra-pure congeners. *Chemosphere*, **85**, 1423–9.

SVENSSON, B. G., NILSSON, A., JONSSON, E., SCHUTZ, A., AKESSON, B. and HAGMAR, L. (1995). Fish consumption and exposure to persistent organochlorine compounds, mercury, selenium and methylamines among Swedish fishermen. *Scand J Work Environ Health*, **21**, 96–105.

SZLINDER-RICHERT, J., BARSKA, I., MAZERSKI, J. and USYDUS, Z. (2009). PCBs in fish from the southern Baltic Sea: levels, bioaccumulation features, and temporal trends during the period from 1997 to 2006. *Mar Pollut Bull*, **58**, 85–92.

TANABE, S., NAKAGAWA, Y. and TATSUKAWA, R. (1981). Absorption efficiency and biological half-life of individual chlorobiphenyls in rats treated with kanechlor products. *Agric. Biol. Chem.*, **45**, 717–26.

THÖRNQVIST, P., HEIMEIER, R., ROOS, R., SCHRENK, D., DE BOEVER, P., SHOETERS, G., Håkansson, H. and HALDIN, K. (2010). *Retinoid system related gene expression in murine liver after exposure to PCB 28, 52, 101, 126, 138, 153 or 180*. 6th International PCB Workshop Visby, Sweden, May 30–June 2, 2010.

TODAKA, T., HIRAKAWA, H., KAJIWARA, J., HORI, T., TOBIISHI, K., YASUTAKE, D., ONOZUKA, D., SASAKI, S., MIYASHITA, C. and YOSHIOKA, E. (2010). Relationship between the concentrations of polychlorinated dibenzo-p-dioxins, polychlorinated dibenzofurans, and polychlorinated biphenyls in maternal blood and those in breast milk. *Chemosphere*, **78**, 185–92.

TUE, N. M., SUDARYANTO, A., MINH, T. B., ISOBE, T., TAKAHASHI, S., VIET, P. H. and TANABE, S. (2010). Accumulation of polychlorinated biphenyls and brominated flame retardants in breast milk from women living in Vietnamese e-waste recycling sites. *Sci Total Environ*, **408**, 2155–62.

ULASZEWSKA, M. M., ZUCCATO, E., CAPRI, E., IOVINE, R., COLOMBO, A., ROTELLA, G., GENEROSO, C., GRASSI, P., MELIS, M. and FANELLI, R. (2011). The effect of waste combustion on the occurrence of polychlorinated dibenzo-p-dioxins (PCDDs), polychlorinated dibenzofurans (PCDFs) and polychlorinated biphenyls (PCBs) in breast milk in Italy. *Chemosphere*, **82**, 1–8.

US-EPA (1990). *Drinking water criteria document for polychlorinated biphenyls (PCBs)*. US Environmental Protection Agency, Cincinatti, OH.

US-EPA (2012). 2,3,7,8-Tetrachlorodibenzo-p-dioxin (TCDD) IRIS Assessment Summary. U.S. Environmental Protection Agency Washington, DC. Available at: http://www.epa.gov/iris/subst/1024.htm

VAN DEN BERG, M., BIRNBAUM, L. S., DENISON, M., DE VITO, M., FARLAND, W., FEELEY, M., FIEDLER, H., HAKANSSON, H., HANBERG, A., HAWS, L., ROSE, M., SAFE, S., SCHRENK, D.,

TOHYAMA, C., TRITSCHER, A., TUOMISTO, J., TYSKLIND, M., WALKER, N. and PETERSON, R. E. (2006). The 2005 World Health Organization reevaluation of human and Mammalian toxic equivalency factors for dioxins and dioxin-like compounds. *Toxicol Sci*, **93**, 223–41.

VAN DER VEN, L. Effect of NDL-PCBs in Zebrafish. (In preparation).

VAN OOSTDAM, J., DONALDSON, S. G., FEELEY, M., ARNOLD, D., AYOTTE, P., BONDY, G., CHAN, L., DEWAILY, E., FURGAL, C. M., KUHNLEIN, H., LORING, E., MUCKLE, G., MYLES, E., RECEVEUR, O., TRACY, B., GILL, U. and KALHOK, S. (2005). Human health implications of environmental contaminants in Arctic Canada: a review. *Sci Total Environ*, 351–352, 165–246.

VAN OOSTDAM, J., GILMAN, A., DEWAILLY, E., USHER, P., WHEATLEY, B., KUHNLEIN, H., NEVE, S., WALKER, J., TRACY, B., FEELEY, M., JEROME, V. and KWAVNICK, B. (1999). Human health implications of environmental contaminants in Arctic Canada: a review. *Sci Total Environ.*, **230**, 1–82.

VILUKSELA M., HEIKKINEN P., VAN DER VEN L., RENDEL F., ROOS R., KORKALAINEN M., LENSU S., MIETTINEN H. M., SAVOLAINEN K., SANKARI S., LILIENTHAL H., ADAMSSON A., TOPPARI J., HERLIN M., FINNILÄ M., TUUKKANEN J., LESLIE H., HAMERS T., HAMSCHER G., AL-ANATI L., STENIUS U., DERVOLA K. S., BOGEN I. L., ANDERSSON P. L., SCHRENK D., HALLDIN K., HÅKANSSON H. Toxicological profile of ultrapure 2,2´,3,4,4´,5,5´-heptachlorbiphenyl (PCB 180) in young adult rats. (in preparation).

VIZCAINO, E., GRIMALT, J. O., CARRIZO, D., LOPEZ-ESPINOSA, M. J., LLOP, S., REBAGLIATO, M., BALLESTER, F., TORRENT, M. and SUNYER, J. (2011). Assessment of prenatal exposure to persistent organohalogen compounds from cord blood serum analysis in two Mediterranean populations (Valencia and Menorca). *J Environ Monitor.*, **13**, 422–32.

WEINAND-HARER, A., LILIENTHAL, H., BUCHOLSKI, K. A. and WINNEKE, G. (1997). Behavioral effects of maternal exposure to an ortho-chlorinated or a coplanar PCB congener in rats. *Environ Toxicol Pharmacol.*, **3**, 97–103.

WHO (1998). *Assessment of the health risk of dioxins: re-evaluation of the tolerable daily intake (TDI)*. World Health Organization, Geneva.

WHO/EURO (1987). *PCBs, PCDDs, and PCDFs: Prevention and control of accidental and environmental exposures*. Environmental Health Series No 23, Geneva.

WHO/IPCS (1993). *Polychlorinated biphenyls and terphenyls (second edition)*. World Health Organization, Environmental Health Criteria 140, Geneva.

WHO (2001). *WHO consultation on risk assessment of non-dioxin-like PCBs*. Federal Institute for Health Protection of Consumers and Veterinary Medicine (BGVV), Berlin, Germany, 3–4 September 2001, http://www.who.int/pcs/pubs/pub_othr.htm

WIGESTRAND, M.B., STENBERG, M., WALAAS, S.I., FRODE FONNUM, F. and ANDERSSON, P.L. (2013). Inhibition of [3H]WIN-35,428 binding to dopamine active transporter by non-dioxin-like PCBs – a structure-activity relationship study. *Archives of Pharmacology* (submitted).

WILSON, N. K., CHUANG, J. C. and LYU, C. (2001). Levels of persistent organic pollutants in several child day care centers. *J Expo Anal Env. Epid.*, **11**, 449–58.

WOLFF, M. S. (1985). Occupational exposure to polychlorinated biphenyls (PCBs). *Environ Health Persp.*, **60**, 133–8.

WOLFF, M. S., FISCHBEIN, A. and SELIKOFF, I. J. (1992). Changes in PCB serum concentrations among capacitor manufacturing workers. *Environ. res.*, **59**, 202–216.

WOLFF, M. S. and SCHECTER, A. (1991). Accidental exposure of children to polychlorinated biphenyls. *Arch Environ Contam Toxicol*, **20**, 449–53.

YAKUSHIJI, T., WATANABE, I., KUWABARA, K., TANAKA, R., KASHIMOTO, T., KUNITA, N. and HARA, I. (1984). Rate of decrease and half-life of polychlorinated biphenyls (PCBs) in the blood of mothers and their children occupationally exposed to PCBs. *Arch Environ Contam Toxicol.*, **13**, 341–5.

ZHANG, L., LI, J., ZHAO, Y., LI, X., YANG, X., WEN, S., CAI, Z. and WU, Y. (2011). A national survey of polybrominated diphenyl ethers (PBDEs) and indicator polychlorinated biphenyls (PCBs) in Chinese mothers' milk. *Chemosphere*, **84** (5), 625–33.

ZIETZ, B. P., HOOPMANN, M., FUNCKE, M., HUPPMANN, R., SUCHENWIRTH, R. and GIERDEN, E. (2008). Long-term biomonitoring of polychlorinated biphenyls and organochlorine pesticides in human milk from mothers living in northern Germany. *Int J Hyg Environ. Health.*, **211**, 624–38.

ZUCCATO, E., CALVARESE, S., MARIANI, G., MANGIAPAN, S., GRASSO, P., GUZZI, A. and FANELLI, R. (1999). Level, sources and toxicity of polychlorinated biphenyls in the Italian diet. *Chemosphere*, **38**, 2753–65.

11
Brominated flame retardants in foods

R. J. Law, The Centre for Environment, Fisheries and Aquaculture Science, UK

DOI: 10.1533/9780857098917.2.261

Abstract: This chapter reviews recently published information concerning brominated flame retardants (BFR) in food. Most of the published information relates to the polybrominated diphenyl ethers (PBDEs) and hexabromocyclododecane (HBCD), with scant attention given so far to novel/emerging compounds which are likely to find more common use as alternative flame-retardant compounds in the future. The incident involving polybrominated biphenyl (PBB) contamination of cattle feed and consequently the human food chain in Michigan, USA, in 1973 is described. Analytical methodology and toxicology are briefly described and likely future trends and knowledge gaps outlined.

Key words: brominated flame retardants, food, dietary intake, polybrominated diphenyl ethers, hexabromocyclododecane.

11.1 Introduction

Flame-retardant compounds are added to a wide range of products, including plastics, textiles, surface coatings, foams, building insulation materials and man-made fibres, in order to make them more fire resistant (Rose and Fernandes, 2012). Some of these (e.g. polybrominated diphenyl ethers (PBDEs) and hexabromocyclododecane) are additive; that is, they are simply mixed into products at the time of manufacture; others are reactive (e.g. tetrabromobisphenol A) and are covalently bonded into the material being flame-retarded. The former are more likely to leach into the local environment during use, although all may be released during e-waste recycling of electronic products, and this process is causing increasing concern as a result, particularly in China (Law *et al.*, 2008). Transport of e-waste from developed countries continues, despite regulation (BBC, 2010). A broad range of compounds have been developed as flame retardants, often organic compounds which are brominated, chlorinated or contain phosphorus (Bergman *et al.*, 2012). Their use has been reported to have led to a reduction in human injury and death resulting from fires (Spiegelstein, 2001; Emsley *et al.*, 2002), but the real benefits of their use have recently been questioned (DiGangi *et al.*, 2010; Hull *et al.*, submitted). As flame-retarded products are extremely widely distributed, multiple diffuse inputs have resulted in widespread environmental contamination. They have been detected in all matrices studied (including food) and are distributed across the globe from the Arctic to the Antarctic as a result of long-range atmospheric transport processes (Vorkamp *et al.*, 2011; Yogui *et al.*, 2011). As the physicochemical properties of BFRs generally make them persistent and bioaccumulative, they are accumulated through food chains and are found in top predators (Law *et al.*, 2006a, 2006b) and food items consumed by people (Rose and Fernandes, 2012), which is the subject of this chapter.

11.2 Sources, occurrence in foods and human exposure

Exposure of humans to PBDEs occurs mainly via a combination of diet, ingestion of indoor dust and inhalation of indoor air (Domingo, 2012) and this probably also applies to most, if not all, of the other BFRs currently in use. In a recent paper, Domingo (2012) reviewed levels of PBDEs (the most commonly studied family of BFR compounds) in food, and the information available on human dietary exposure. He reported a lack of published data from complete market-based studies in a considerable number of developed countries and a total absence of data from important emerging countries, such as Brazil, India, Mexico, Argentina and South Africa (among others). Information on human total dietary intake was limited to Europe (though not France, Italy, Portugal or Greece), the USA, China and Japan. Identified knowledge gaps included the need for epidemiological studies to determine whether current contamination levels result in adverse health effects, particularly in relation to neurobehavioural development and reproductive effects. In this chapter the literature assessed by Domingo (2012) will be taken as a starting point, followed by a focus on data published subsequently, and on data from studies of particular interest and importance.

11.2.1 Polybrominated diphenyl ethers (PBDEs) in foods

The available data have been extensively reviewed by Domingo (2012). In Europe, the human dietary intake of PBDEs (sum of concentrations of the congeners determined) ranged from 23 to 98 ng day^{-1}. In general, fish, dairy products and meat were the major contributors to dietary intake. In Japan, Ashizuka *et al.* (2008) estimated a dietary intake from fish and shellfish consumption of 29 ng day^{-1}, whilst in the USA, Huwe and Larsen (2005) reported an intake of 15–45 ng day^{-1}. Further, Lorber (2008) estimated a dietary intake of PBDEs for US citizens as 91 ng day^{-1} vs a total of 539 ng day^{-1} from all sources. This suggests that diet represents only about one-fifth of total intake.

There is no standard suite of brominated diphenyl ether (BDE) congeners to be determined, in food or other matrices, although one has been suggested (Law et al., 2006a). This comprised eight congeners which are frequently determined and which could identify contributions from all three PBDE commercial products: BDE28, BDE47, BDE99, BDE100, BDE153, BDE154, BDE183 and BDE209. In this chapter, total BDE congener concentrations determined in specific studies are cited as for example, ΣBDE_{10}, which represents the sum of the ten individual BDE congeners determined in that study.

Rose and Fernandes (2012) also summarised available information on BFRs in food. For BDEs, higher concentrations were generally seen in fish (from the USA, Canada, Spain, Japan and the UK) than in meat and dairy products, as also noted by Domingo (2012).

Roosens et al. (2010a) determined BDEs in eels from Belgium during 2000–2006. ΣBDE_{11} concentrations (the sum of the concentrations of the 11 congeners determined in this study) ranged from 10 to 5810 μg kg^{-1} lipid weight, with a median concentration of 81 μg kg^{-1} lipid weight. BDE47 dominated in all but six of the fifty samples, in the others BDE209 was the dominant congener. For average consumers (2.9 g eel day^{-1}) ΣBDE intakes ranged between 3 and 2300 ng day^{-1}. For recreational fishermen consuming 12 or 86 g eel day^{-1}, estimated intakes ranged from 13 to 9500 ng day^{-1} and 94 to 68 000 ng day^{-1}, respectively, above reference doses described in the literature which may induce adverse effects (National Research Council, 2000). Roosens et al. (2010b) concluded that intake from all sources was dominated by the diet, which contributed over 90% of the total intake. HBCD concentrations in eels were reported to be 16–4400 μg kg^{-1} lipid weight (see below), similar to those of the BDEs in the same study (Roosens et al., 2010a). Tapie et al. (2011) determined ΣBDE_4 concentrations in glass and silver eels from the Gironde estuary, France, during 2004–2005. Mean concentrations ranged from 24 to 237 μg kg^{-1} lipid weight. In eel muscle tissue from five catchments in Ireland, McHugh et al. (2010) reported ΣBDE_{11} and ΣPBB_5 concentrations ranging from 1.0 to 7.0 μg kg^{-1} wet weight and 0.005 to 0.01 μg kg^{-1} wet weight, respectively. MacGregor et al. (2010) reported ΣBDE_{13} concentrations in eel muscle tissue from 30 sites across Scotland. These ranged from 1.3 to 132 μg kg^{-1} wet weight.

Kim and Stapleton (2010) reported ΣBDE_{10} concentrations in squid from Korean offshore waters to be 21–292 μg kg^{-1} lipid weight. Takahashi et al. (2010) determined ΣBDE_{14} concentrations to be 1.3–53 μg kg^{-1} lipid weight in 12 species of deep-sea fish off the coast of Japan. Tri- to hexa-BDE congeners predominated, consistent with contamination resulting from the penta-PBDE product. HBCD concentrations were < 0.05–110 μg kg^{-1} lipid weight in the same samples. In another study in Japanese coastal waters, Ueno et al. (2010) analysed mussels and oysters, reporting ΣBDE_{14} concentrations of 3.1–86 μg kg^{-1} lipid weight. Concentrations of HBCDs ranged from 12 to 5200 μg kg^{-1} lipid weight. The dietary exposure of PBDEs and HBCDs to human consumers was estimated to be 0.05–6.8 ng kg body weight^{-1} day^{-1} and 0.45–34 ng kg body weight^{-1} day^{-1}, respectively.

Montory et al. (2010) reported ΣBDE_{14} concentrations of 0.3–1.05 μg kg^{-1} wet weight in muscle tissue of wild Chinook salmon from Chile. Ondarza et al. (2011) reported a mean ΣBDE_9 concentration of 1.1 μg kg^{-1} wet weight in brown trout from Andean Patagonia, Argentina, very similar to the levels reported from Chile in the earlier study. In juvenile Chinook salmon from the Pacific Northwest of the USA, Sloan et al. (2010) reported gutted body concentrations up to 13 000 μg kg^{-1} lipid weight.

In smallmouth and striped bass from the Hudson River, NY, USA, taken in 2003, mean ΣBDE_{40} concentrations ranged from 2.1 to 138 μg kg^{-1} wet weight and 35 to 169 μg kg^{-1} wet weight, respectively (Skinner, 2011). In muscle tissue of mullet and bass from Bizerte Lagoon, Tunisia, ΣBDE_{10} mean concentrations were 45 and 96 μg kg^{-1} lipid weight (Ben Ameur et al., 2011).

In four areas of China, Liu et al. (2011) determined ΣBDE_8 concentrations in muscle tissue of marine fish. Mean concentrations ranged from 0.8 to 388 μg kg^{-1} wet weight. BDE209 comprised 23–70% (mean 54%) of the total, possibly a reflection of the fact that the deca-mix PBDE product is the most commonly used BFR in China (Zhou, 2006). ΣBDE_9 concentrations in two fish species (yellow croaker and silver pomfret) from nine Chinese coastal cities taken in 2008 averaged 3.0 μg kg^{-1} lipid weight (Xia et al., 2011). Guo et al. (2010) analysed 602 seafood samples from Guandong Province, South China, collected on a market basis during 2004–2005. ΣBDE_{10} concentrations ranged from undetectable in all sample types to maximum values of 1.1 (shrimp), 1.6 (crab), 2.2 (molluscs) and 5.9 (fish) μg kg^{-1} wet weight. Human dietary intakes were dominated by fish and molluscs, each representing 45% of the total intake. Jin et al. (2010) determined ΣBDE_{11} concentrations in 11 species of fish, shellfish and shrimps taken from markets close to Laizhou Bay, Shandong Province, China, in a PBDE production area. Concentrations ranged from 2.7 to 42 μg kg^{-1} wet weight. The mean dietary intake of PBDEs in this area was estimated to be 218 ng day^{-1}, whilst the daily intake via inhalation was 612 ng for men and 455 ng for women. BDE209 represented, on average, 80% of this intake. Human exposure (by both routes) was the highest reported to date.

Edible marine products collected in fish markets in Adelaide, Australia, were analysed for ΣBDE_{11} (Shanmuganathan et al., 2011). Concentrations ranged from 1.0 to 45 μg kg^{-1} wet weight and BDE209 predominated. The highest levels were seen in imported silver fish and prawns from Thailand and Vietnam. In muscle tissue of farmed turbot in Spain, Blanco et al. (2011) reported ΣBDE_{27} concentrations of 0.5–2.1 μg kg^{-1} lipid weight, reflecting concentrations in their feed (2.4–4.8 μg kg^{-1} lipid weight). The need to source fish food with lower levels of PBDE contamination in order to reduce human dietary intake was emphasised.

Concentrations of BDEs were determined in 299 vegetable- and animal-based food samples of 31 species collected in Shanghai, China, by Yu *et al.* (2011). In addition, they determined the bioaccessibility of the BDEs. ΣBDE_{13} concentrations ranged from not detected to 1.25 µg kg^{-1} wet weight, with animal-based food containing higher concentrations than found in vegetables. The bioaccessibility of the BDEs, measured using a technique which simulated the human gastrointestinal digestion process, was 2.6–40% in vegetables and 5.2–105% in animal-based food. For animal-based food, this correlated well with its fat content. The total daily intake by an average resident was estimated to be 13.2–13.7 ng day^{-1}, reducing to 2.7–4.3 ng day^{-1} after bioaccessibility was taken into account. Vegetables were the major contributors to this intake figure (49%) followed by fish (34%); however, this was reversed when bioaccessibility was accounted for (38% and 52%, respectively).

Zheng *et al.* (2012) determined BDEs in free range chicken eggs collected from three locations close to electronic waste recycling facilities in South China. Mean levels of ΣBDE_{10} ranged from 2640 to 14 100 µg kg^{-1} lipid weight, one to two orders of magnitude higher than those seen at a reference site. Mean total daily intakes for humans consuming eggs from the vicinity of the e-waste recycling sites were 4200–20 000 ng day^{-1}.

Schecter *et al.* (2011) highlighted a potential packaging contamination problem. In a survey utilising ten butter samples, it was found that one had much higher levels of higher brominated BDEs than the others, and that they had probably been transferred into the butter from contaminated wrapping paper. With such a small sample size it is impossible to assess the likely scale of this problem, but it should not be overlooked in future studies.

Law *et al.* (2008) noted that BDEs accumulate in fish, and that the *per capita* consumption of fish in Asian countries is high relative to that in other parts of the world. In Hong Kong (as an example), fish or shellfish are consumed more than four times a week by most of the population, and the annual average consumption of marine products is approximately 60 kg per person (Dickman and Leung, 1998). These populations may therefore be particularly vulnerable to high levels of BDEs (and potentially other BFRs) in seafood.

11.2.2 Hexabromocyclododecanes (HBCDs) in foods

Rose and Fernandes (2012) also summarised available information on BFRs in food. For HBCDs, the data are too sparse to establish norms between food groups, but these authors cited studies in which a correlation was observed between fat-rich marine fish and human serum HBCD levels.

Roosens *et al.* (2010a) determined HBCDs in eels from Belgium during 2000–2006. ΣHBCD concentrations (the sum of the concentrations of the α-, β- and γ-isomers) ranged from 16 to 4400 µg kg^{-1} lipid weight, with a median concentration of 73 µg kg^{-1} lipid weight. α-HBCD dominated in all of the 50 samples. For average consumers (2.9 g eel day^{-1}) ΣHBCD intakes ranged between 3 and 1110 ng day^{-1}. For recreational fishermen consuming 12 or 86 g eel day^{-1}, estimated intakes ranged from 13 to 4600 ng day^{-1} and 95 to 32 800 ng day^{-1}, respectively, above reference doses described in literature, which may induce adverse effects (National Research Council, 2000).

Goscinny et al. (2011) conducted an assessment of exposure of the adult Belgian population to HBCDs, analysing 45 composite samples from five major food groups (milk and dairy products, meat and meat products, eggs, fish and fishery products and an 'other foods' group that comprised cakes and vegetable products among others) sampled in 2008. The medium bound estimate of their average daily intake of HBCDs was 0.99 ng kg body weight^{-1} day^{-1} (about 70–80 ng day^{-1}, in line with the above estimate for average eel consumers). γ-HBCD predominated, as in the two food groups that contributed most to the daily intake estimate (meat and other foods).

Nakagawa et al. (2010) determined HBCDs in 54 wild and 11 farmed seafood samples collected from four regions of Japan during 2003–2008. For fish classified as Anguilliformes, Perciformes, Clupeiformes and farmed Salmoniformes, ΣHBCD concentrations ranged from 0.05 to 37 (median 2.1), not detected to 26 (median 0.8), 0.09–77 (median 0.1) and 1.1–1.3 (median 1.3) μg kg^{-1} wet weight, respectively. For Anguilliformes, farmed Salmoniformes and Perciformes, the daily intake by an average Japanese adult was estimated to be 3.7, 2.3 and 1.3 ng kg body weight^{-1} day^{-1}, respectively (or approximately 259, 161 and 91 ng day^{-1} assuming a 70 kg body weight, somewhat lower than those reported for average consumers of eels from Belgium, as described above). α-HBCD mostly dominated in both wild and farmed fish but, in wild fish with concentrations greater than 20 μg kg^{-1} wet weight, γ-HBCD was predominant, suggesting an influence due to recent discharges from an industrial plant nearby.

Zheng et al. (2012) determined HBCDs in free range chicken eggs collected from three locations close to electronic waste recycling facilities in South China. Mean levels of ΣHBCD ranged from 44 to 350 μg kg^{-1} lipid weight, slightly higher than those seen at a reference site. Mean total daily intake estimates for humans consuming eggs from the vicinity of the e-waste recycling sites were 80–490 ng day^{-1}.

Meng et al. (2012) determined HBCDs in 12 consumer fish species from South China. ΣHBCD concentrations ranged from not detected to 0.19 μg kg^{-1} wet weight, with HBCD being detected in 70% of the samples. Concentrations were higher in carnivorous species than in herbivorous and detritivorous species, indicating a potential for biomagnifications of HBCDs via the food chain. Also, ΣHBCD concentrations were higher in farmed fish from both freshwater and seawater farms than in wild marine fish, indicating that human activities were probably an important source of HBCD in aquaculture. In the worst-case exposure, the mean estimated daily intake of ΣHBCD via fish consumption for residents of South China ranged from 0.13 to 0.16 ng kg body weight^{-1} day^{-1} for various age groups.

11.2.3 Other brominated flame retardants (BFRs) in foods

Zheng et al. (2012) determined polybrominated biphenyls (PBB), decabromodiphenyl ethane and 1,2-*bis*(2,4,6-tribromophenoxy)ethane in free range chicken eggs collected from three locations close to electronic waste recycling facilities in South China. Mean levels of ΣPBB ranged from 700 to 1620 µg kg^{-1} lipid weight, compared to a mean level of 28 µg kg^{-1} lipid weight at a reference site. Mean levels of decabromodiphenyl ethane and 1,2-*bis*(2,4,6-tribromophenoxy)ethane ranged from 6 to 38 µg kg^{-1} lipid weight and 37 to 264 µg kg^{-1} lipid weight, respectively, compared to 14 and < 6 µg kg^{-1} lipid weight at a reference site.

Fernandes et al. (2010) determined three emerging BFRs in 115 UK and 100 Irish food samples. Hexabromobenzene and decabromodiphenylethane were not detected in any of the food samples analysed. Concentrations of 1,2-*bis*(2,4,6-tribromophenoxy)ethane ranged from 0.01 µg kg^{-1} wet weight in potatoes to 0.18 µg kg^{-1} wet weight in strawberry jam. This BFR compound was also detected in samples of meat, fish, eggs and vegetables.

11.3 Methods of analysis and monitoring of brominated flame retardants in foods

A fully validated method for the determination of PBDEs (including BDE209) has been in place for more than 10 years at the time of writing, which can readily be applied to food analysis (de Boer et al., 2001). For biota, the method comprises *Soxhlet* extraction with *n*-hexane/acetone (1:1), clean-up using alumina column chromatography or gel permeation chromatography followed by further clean-up using silica gel column chromatography and sulphuric acid treatment. The final determination utilises gas chromatography with electron capture negative ion mass spectrometry (GC-ECNIMS) at *m/z* 79 and 81 to detect the two bromine isotopes. For HBCD and tetrabromobisphenol A (TBBP-A), the preferred methods utilise LC-MS with a variety of ionisation modes (Morris et al., 2006; Zhou et al., 2010). Subsequent developments have been reviewed by Covaci et al. (2007). Most recently, Pena-Abaurrea et al. (2011) have published a multi-residue method for the determination of a number of classes of BFRs (PBDEs, PBBs and eight novel BFRs) using comprehensive two-dimensional gas chromatography coupled with time-of-flight mass spectrometry, validated in bluefin tuna samples. For the novel and emerging BFRs, two recent reviews have considered the applicability of current methodologies to their detection and quantification (Covaci et al, 2011; Papachlimitzou et al., 2012). Papachlimitzou et al. (2012) noted that method performance was limited by the fact that, in most cases, methods employed had usually been optimised for the determination of the major BFR groups and so were suboptimal for many novel/emerging BFRs, depending on their physicochemical properties relative to those of BDEs and HBCDs (an exception is the study reported by Zhou *et al.* (2010)). In addition, availability of standard materials and appropriate certified reference materials was a further limitation. Few countries, if any, have routine monitoring in place for BFRs in food samples – generally occasional one-off surveys, whether undertaken by government agencies or research groups, are used in order to assess the risk to human health. This is also limited by the paucity of toxicological information.

11.4 Toxicity of brominated flame retardants

Hull *et al.* (submitted) reviewed the available knowledge on the harmful effects of BFRs. PBBs are considered in detail in the next section of this chapter, which outlines the Michigan feed contamination incident in the USA, and will not be considered here. For most BFRs, human toxicological data are inadequate for a robust assessment of their potential for harm to be undertaken. PBDEs have been linked to neurodevelopmental effects, such as learning difficulties, to thyroid and liver cell effects, and to reproductive impacts including effects on sperm, and deca-PBDE has been shown to induce cancer in male rats (reviewed in Hull *et al.*, submitted). Additional concerns include the transplacental transfer of BDEs to the developing foetus (particularly as some effects may have discrete exposure windows at specific developmental stages) and the formation of strongly thyroidogenic BDE metabolites that are retained in human blood (Hull *et al.*, submitted; EFSA, 2011a). In his review of PBDEs in food and of human dietary exposure, Domingo (2012) concluded that human health risks derived from dietary exposure to PBDEs are probably limited.

HBCD has a low acute toxicity, but exhibits a number of chronic effects (Hull *et al.*, submitted; EFSA, 2011b). Some studies have indicated reproductive, developmental and behavioural effects in mammals (Ema *et al.*, 2008). An EU risk assessment concluded that HBCD may cause reproductive toxicity and long-term toxicity (ECB, 2008). Exposure to the technical product in developing rodents affects the nervous system and results in behavioural changes, with impacts on thyroid hormone regulation and immune function (Hull *et al.*, submitted). There are major gaps in the knowledge of HBCD toxicology, not least because much of the work has been conducted using the technical mixture, in which γ-HBCD predominates whilst, in humans, α-HBCD is the most prevalent isomer.

TBBP-A is the most abundant BFR and is readily taken up from the diet, but has a short half-life in humans (days) and is readily metabolised and excreted (Hull *et al.*, submitted). Exposure is also limited as TBBP-A is a reactive flame retardant that is chemically bound within the flame-retarded products and so less likely to leach and become environmentally available.

It does not seem to cause the neurodevelopmental effects of the PBDEs and HBCD, but exhibits thyroid hormone disruption effects due to the similarity of its structure and that of thyroxin (Hull *et al.*, submitted). Toxicological information for TBBP-A derivatives, which are also used as flame retardants, and for the majority of the novel/emerging BFRs, is almost entirely absent.

11.5 Major incidences of brominated flame retardant contamination of foods

In 1973, 1000–2000 pounds of a flame-retardant product containing PBBs (product FireMaster BP-6) were accidentally added to a cattle feed in place of magnesium oxide, an animal feed supplement, in the US state of Michigan (Chanda *et al.*, 1982; Miceli *et al.*, 1985; Yard *et al.*, 2011). Although over 35 000 cattle were culled once the contamination was identified (Di Carlo *et al.*, 1978), it has been estimated that 85% of the state's nine million population received some exposure to PBBs as a result, particularly as dairy product marketing involved mixing milk from many farms (Fries, 1985). As a result, numerous health studies have been conducted, both on cattle (Cook *et al.*, 1978; Wastell *et al.*, 1978) and humans in the affected area. The latter were confounded to some degree by the fact that, as Yard *et al.* (2011) have noted, US residents have detectable levels of most persistent organic pollutants (POP) in their tissues, including other BFRs (Chen *et al.*, 2011; Foster *et al.*, 2011; Marvin *et al.*, 2011; Park *et al.*, 2011; Petreas *et al.*, 2011; Zota *et al.*, 2011; Johnson *et al.*, 2012). Long-term studies of human health impacts were conducted (Terrell *et al.*, 2008; Yard *et al.*, 2011). Health effects noted included a significantly higher prevalence of skin, neurological and musculoskeletal symptoms (Anderson *et al.*, 1978; Valciukas *et al.*, 1978, 1979), neurobehavioural complaints (Selikoff, 1979), multiple symptoms consistent with immunosuppression (Meester, 1979), immunological abnormalities (Bekesi *et al.*, 1979; Roboz *et al.*, 1985), lymphocyte function (Bekesi *et al.*, 1978) and an increased risk of breast and digestive cancer (Henderson *et al.*, 1995; Hoque *et al.*, 1998). As the PBB fat elimination half-life in humans was estimated to be at least 7.8 years, PBBs will have persisted in contaminated individuals throughout their lives (Miceli *et al.*, 1985). These findings have been the subject of much debate, but a recent assessment undertaken by the European Food Safety Authority has linked PBBs to cancer and neurobehavioural, thyroidogenic, embryotoxic and teratogenic effects (EFSA, 2010). Largely as a result of this incident in Michigan in 1973, production of PBBs has been banned or severely limited in most countries since the 1970s (Silberhorn *et al.*, 1990). Further information on food levels of PBBs, their toxicity and derived risk levels are given in Rose and Fernandes (2012).

270 Persistent organic pollutants and toxic metals in foods

11.6 Implications for the food industry and policy makers for prevention and control of contamination

BFRs are released into the environment via releases during manufacture and use, leaching from products into indoor and outdoor air, via discharges from wastewater treatment plants, from landfill sites, and following remobilisation from environmental sinks such as soils and sediments. These move through the environment using a variety of transport pathways and the levels observed in food reflect these inputs and the persistence and bioaccumulation potential of the individual compounds, and their ability to be metabolised. For the most part, there is little that can be done to influence this process apart from cutting off the inputs at source, as has been done for the penta- and octa-mix PBDE products by banning their production and use within the EU. The consideration of the designation of HBCD as a POP within the Stockholm Convention may also lead to restrictions being applied to that compound in time (Hull *et al.*, submitted). An exception is the production of farmed fish, whether freshwater or marine. As Hites *et al.* (2004) noted, there is an association between the concentrations of BDEs in farmed salmon and those in the feed provided. Hence, sourcing less contaminated feed is a route to producing farmed fish with lower levels of BFR contamination, if it can be done sustainably and cost-effectively.

11.7 Future trends

Following the bans on the use of the penta- and octa-mix PBDE formulations within the EU and restrictions on the use of the deca-PBDE formulation in the EU, and ongoing or pending restrictions on production and use of these products elsewhere, concentrations of these compounds should begin to decline, as has already been seen for penta-PBDE congeners in blubber of UK harbour porpoises (Law *et al.*, 2010). As the requirements for addition of flame-retardant compounds in order to meet safety standards still remain, however, other compounds (already in use and classified as novel/emerging BFRs) will be substituted. Information regarding environmental levels of these compounds and their significance is sparse (Covaci *et al.*, 2011) (though not as sparse as the information about the identity of the compounds being marketed, the applications in which they are being used and their production volumes) and are, for food, almost non-existent. Accruing environmental occurrence data and toxicological information will help in targeting those compounds of most concern with regard to human health, and will drive the development of food safety programmes in relation to BFRs over the next decade. Another major concern relates to the vast reservoir of BDE209 which is building up in sediments worldwide. Whilst it remains as the parent compound, it is of relatively little concern as its acute toxicity is low, its mobility is low as it is firmly attached to sediment particles due to its high log K_{ow}, and its bioaccumulation potential is also low. However, debromination of BDE209 to yield more mobile, more toxic and more bioaccumulative BDE congeners has been demonstrated in the laboratory and observed in various fish species and inferred in marine sediments (Salvadó *et al.*, 2012). Future concerns centre on how important this process will prove to be in the environment. There is the potential for this to represent a chronic input to aquatic systems of compounds, which regulators have sought to eliminate, with likely impacts on marine food chains and seafood. Also, contamination of the environment and food sources around e-waste recycling operations and enhanced uptake of BFRs by local residents remains a concern and merits further study (see, for example, Qin *et al.*, 2011; Zheng *et al.*, 2012).

Concerns have also been raised in relation to polybrominated dibenzo-*p*-dioxins and furans (PBDD/F). These compounds occur as contaminants in the PBDE formulations, and can also be formed during thermal processing of materials (such as extrusion, moulding and recycling) and by degradation within the temperature range 350–400°C (Rose and Fernandes, 2010). Their toxicity is poorly understood at present, but they are significantly more toxic than the BDEs and have also been detected in food samples from the UK and Ireland (Rose and Fernandes, 2010).

A number of future needs have been identified in this and other studies. For novel and emerging contaminants, including BFR compounds, Rose and Fernandes (2012) have identified a kind of 'Catch-22' situation (Heller, 2011) in relation to the funding of studies, as funding bodies with limited resources tend to want evidence of exposure before funding work on toxicology, and evidence on toxicology before funding studies to assess exposure. This loop can be difficult to break, as the authors noted, but both types of studies are essential if the risks posed by these novel and/or emerging compounds are to be assessed and their significance determined. Also, as mentioned in the introduction, Domingo (2012) has highlighted the need for epidemiological studies to determine whether current contamination levels of PBDEs (and other BFRs, particularly HBCD) result in adverse health effects, particularly in relation to neurobehavioural development and reproductive effects. In addition, he noted that, from a toxicological viewpoint, mechanistic studies would provide important information for a better assessment of the likelihood of adverse health effects as a result of PBDE exposure. Such studies should also define the toxicity of individual BDE congeners, and investigate interactions between different BDE congeners and between BDEs and other environmental POPs. In relation to novel and emerging BFRs, Covaci *et al.* (2011) has noted that concentrations in food are poorly known at present and that few toxicity studies have addressed the human impacts of these compounds to date.

11.8 Sources of further information and advice

The European Food Safety Authority has recently published a series of scientific opinion documents on BFRs in food (PBBs, PBDEs, HBCD and

TBBP-A and novel and emerging BFRs) (EFSA, 2010, 2011a, 2011b, 2011c, 2012). These contain information relating to analysis, sources, use and environmental fate, occurrence in food, toxicity and assessments of human dietary exposure. In addition, the EU has conducted risk assessments of the continued production and use of PBDEs and HBCD (ECB, 2001, 2003, 2004, 2008) and these reports provide a ready source of further useful information. A further report for the Health and Consumers Directorate General of the European Commission (Arcadis, 2011) provides information on physicochemical properties, human health effects, consumer exposure and a human health risk assessment for a range of BFRs (decabromodiphenyl ether, hexabromocyclododecane, *tris*(tribromoneopentyl)phosphate, tetrabromobisphenol A *bis*(2,3-dibromopropyl ether), *tris*(2,4,6-tribromophenoxy)triazine, *bis*(2-ethylhexyl)tetrabromophthalate, decabromodiphenyl ethane, ethylene *bis*(tetrabromophthalimide) and 1,2-*bis*(2,4,6-tribromophenoxy)ethane), as well as a range of other chlorinated and organophosphorus flame retardants.

11.9 References

ANDERSON, H.A., LILIS, R., SELIKOFF, I.J., 1978. Unanticipated prevalence of symptoms among dairy farmers in Michigan and Wisconsin. *Environmental Health Perspectives* **23**, 217–226.

ARCADIS, 2011. Evaluation of data on flame retardants in consumer products – final report. Arcadis, Belgium. European Commission Health and Consumers DG contract no. 17.020200/09/549040. http://ec.europa.eu/consumers/safety/news/flame_retardant_substances_study-en.pdf accessed 11 June 2012. 401 pp.

ASHIZUKA, Y., NAKAGAWA, R., HORI, T., YASUTAKE, D., TOBIISHI, K., SASAKI, K., 2008. Determination of brominated flame retardants and brominated dioxins in fish collected from three regions of Japan. *Molecular Nutrition and Food Research* **52**, 273–283.

BBC, 2010. Europe still exporting electronic waste. www.bbc.co.uk/news/world-europe-10846395 accessed 4 August 2010.

BEKESI, J.G., HOLLAND, J.F., ANDERSON, H.A., 1978. Lymphocyte function of Michigan dairy farmers exposed to polybrominated biphenyls. *Science* **199**, 1207–1209.

BEKESI, J.G., ROBOZ, J., ANDERSON, H.A., 1979. Impaired immune function and identification of polybrominated biphenyls (PBBs) in blood compartments of exposed Michigan dairy farmers and chemical workers. *Drug and Chemical Toxicology* **2**, 179–191.

BEN AMEUR, W., HASSINE, S.B., ELJARRAT, E., EL MEGDICHE, Y., TRABELSI, S., HAMMAMI, B., BARCELÓ, D., DRISS, M.R., 2011. Polybrominated diphenyl ethers and their methoxylated analogs in mullet (*Mugil cephalus*) and sea bass (*Dicentrarchus labrax*) from Bizerte Lagoon, Tunisia. *Marine Environmental Research* **72**, 258–264.

BERGMAN, Å., RYDÉN, A., LAW, R.J., DE BOER, J., COVACI, A., ALAEE, M., BIRNBAUM, L., PETREAS, M., ROSE, M., SAKAI, S., VAN DEN EEDE, N., VAN DER VEEN, I., 2012. A novel abbreviation standard for organobromine, organochlorine and organophosphorus flame retardants and some characteristics of the chemicals. *Environment International* **49**, 57–82. doi: 10.1016/j.envint.2012.08.003

BLANCO, S.L., MARTINEZ, A., PORRO, C., VIEITES, J.M., 2011. Dietary uptake of polybrominated diphenyl ethers (PBDEs), occurrence and profiles. *Chemosphere* **85**, 441–447.

CHANDA, J.J., ANDERSON, H.A., GLAMB, R.W., LOMATCH, D.L., WOLFF, M.S., VOORHEES, J.J., SELIKOFF, I.J., 1982. Cutaneous effects of exposure to polybrominated biphenyls (PBBs): The Michigan PBB incident. *Environmental Research* **29**, 97–108.

CHEN, A., CHUNG, E., DEFRANCO, E.A., PINNEY, S.M., DIETRICH, K.N., 2011. Serum PBDEs and age at menarche in adolescent girls: analysis of the National Health and Nutrition Examination Survey 2003–2004. *Environmental Research* **111**, 831–837.

COOK, H., HELLAND, D.R., VANDERWEELE, B.H., DEJONG, R.J., 1978. Histotoxic effects of polybrominated biphenyls in Michigan dairy cattle. *Environmental Research* **15**, 82–89.

COVACI, A., HARRAD, S., ABDALLAH, M.A. -E., ALI, N., LAW, R.J., HERZKE, D., DE WIT, C.A., 2011. Novel brominated flame retardants: A review of their analysis, environmental fate and behaviour. *Environment International* **37**, 532–556.

COVACI, A., VOORSPOELS, S., RAMOS, L., NEELS, H., BLUST, R., 2007. Recent developments in the analysis of brominated flame retardants and brominated natural compounds. *Journal of Chromatography A* **1153**, 145–171.

DE BOER, J., ALLCHIN, C., LAW, R., ZEGERS, B., BOON, J.P., 2001. Method for the analysis of polybrominated diphenyl ethers in sediments and biota. *Trends in Analytical Chemistry* **20**, 591–599.

DI CARLO, F.J., SEIFTER, J., DECARLO, V.J., 1978. Assessment of the hazards of polybrominated biphenyls. *Environmental Health Perspectives* **23**, 351–365.

DICKMAN, M.D., LEUNG, K.M.C., 1998. Mercury and organochlorine exposure from fish consumption in Hong Kong. *Chemosphere* **37**, 991–1015.

DIGANGI, J., BLUM, A., BERGMAN, Å., DE WIT, C.A., LUCAS, D., MORTIMER, D., SCHECTER, A., SCHERINGER, M., SHAW, S.D., WEBSTER, T.F., 2010. San Antonio statement on brominated and chlorinated flame retardants: Do the fire safety benefits justify the risks? *Environmental Health Perspectives* **118**, A 516–536.

DOMINGO, J.L., 2012. Polybrominated diphenyl ethers in food and human dietary exposure: A review of the recent scientific literature. *Food and Chemical Toxicology* **50**, 238–249.

ECB (EUROPEAN CHEMICALS BUREAU), 2001. *European Union Risk Assessment Report. Diphenyl Ether, Pentabromo Derivative.* CAS No. 32534–81–9. EINECS No: 251–084–2. Office for Official Publications of the European Communities, Brussels, Belgium.

ECB (EUROPEAN CHEMICALS BUREAU), 2003. *European Union Risk Assessment Report. Diphenyl Ether, Octabromo Derivative.* CAS No. 32536–52–0. EINECS No: 251–087–9. Office for Official Publications of the European Communities, Brussels, Belgium.

ECB (EUROPEAN CHEMICALS BUREAU), 2004. *European Union Risk Assessment Report. Update of the Risk Assessment of bis(pentabromophenyl) Ether (decabromodiphenyl ether).* CAS No. 1163–19–5. EINECS No: 214–604–9. Office for Official Publications of the European Communities, Brussels, Belgium.

ECB (EUROPEAN CHEMICALS BUREAU), 2008. *European Union Risk Assessment Report. Hexabromocyclododecane.* CAS No. 3194–55–6. EINECS No: 247–148–4. Office for Official Publications of the European Communities, Brussels, Belgium.

EFSA, 2010. Scientific opinion on polybrominated biphenyls (PBBs) in food. EFSA Panel on Contaminants in the Food Chain. *EFSA Journal* **8**, 1789 (151 pp.).

EFSA, 2011a. Scientific opinion on polybrominated diphenyl ethers (PBDEs) in food. EFSA Panel on Contaminants in the Food Chain. *EFSA Journal* **9**, 2156 (274 pp.).

EFSA, 2011b. Scientific opinion on hexabromocyclododecanes (HBCDDs) in food. EFSA Panel on Contaminants in the Food Chain. *EFSA Journal* **9**, 2296 (118 pp.).

EFSA, 2011c. Scientific opinion on tetrabromobisphenol A (TBBPA) and its derivatives in food. EFSA Panel on Contaminants in the Food Chain. *EFSA Journal* **9**, 2477 (61 pp.).

EFSA, 2012. Scientific opinion on emerging and novel brominated flame retardants (BFRs) in food. EFSA Panel on Contaminants in the Food Chain. *EFSA Journal* **10**, 2908 (125pp.).

EMA, M., FUJII, S., HIRATA-KOIZUMI, M., MATSUMOTO, M., 2008. Two-generation reproductive toxicity study of the flame retardant hexabromocyclododecane in rats. *Reproductive Toxicology* **25**, 335–351.

EMSLEY, A., LIM, L., STEVENS, G., 2002. *International Fire Statistics and the Potential Benefits of Fire Counter-measures*. Paper presented at Flame Retardants 2002. Interscience Communications Ltd., London.

FERNANDES, A., SMITH, F., PETCH, R., PANTON, S., CARR, M., MORTIMER, D., TLUSTOS, C., ROSE, M., 2010. *The Emerging BFRs Hexabromobenzene, bis(246-tribromophenoxy) ethane (BTBPE), and Decabromodiphenylethane (DBDPE) in UK and Irish Foods*. Proceedings of BFR2010, Kyoto, Japan, 7–9 April 2010. #90028. 4 pp.

FOSTER, W.G., GREGOROVICH, S., MORRISON, K.M., ATKINSON, S.A., KUBWABO, C., STEWART, B., TEO, K., 2011. Human maternal and umbilical cord blood concentrations of polybrominated diphenyl ethers. *Chemosphere* **84**, 1301–1309.

FRIES, G.F., 1985. The PBB episode in Michigan: an overall appraisal. *Critical Reviews in Toxicology* **16**, 105–156.

GOSCINNY, S., VANDEVIJVERE, S., MALEKI, M., VAN OVERMEIRE, I., WINDAL, I., HANOT, V., BLAUDE, M.-N., VLEMINCKX, C., VAN LOCO, J., 2011. Dietary intake of hexabromocyclododecane diastereoisomers (α-, β- and γ-HBCD) in the Belgian adult population. *Chemosphere* **84**, 279–288.

GUO, J., WU, F., SHEN, R., ZENG, E.Y., 2010. Dietary intake and potential health risk of DDTs and PBDEs via seafood consumption in South China. *Ecotoxicology and Environmental Safety* **73**, 1812–1819.

HELLER, J., 2011. Catch-22: 50th anniversary edition. Vintage Classics, 544 pp. ISBN 0099529122.

HENDERSON, A.K., ROSEN, D., MILLER, G.L., FIGGS, L.W., ZAHM, S.H., SIEBER, S.M., HUMPHREY, H.E.B., SINKS, T., 1995. Breast cancer among women exposed to polybrominated biphenyls. *Epidemiology* **6**, 544–546.

HITES, R.A., FORAN, J.A., SCHWAGER, S.J., KNUTH, B.A., HAMILTON, M.C., CARPENTER, D.O., 2004. Global assessment of polybrominated diphenyl ethers in farmed and wild salmon. *Environmental Science and Technology* **38**, 4945–4949.

HOQUE, A., SIGURDSON, A.J., BURAU, K.D., HUMPHREY, H.E.B., HESS, K.R., SWEENEY, A.M., 1998. Cancer among a Michigan cohort exposed to polybrominated biphenyls in 1973. *Epidemiology* **9**, 373–378.

HULL, T.R., LAW, R.J., BERGMAN, Á, submitted. Environmental drivers for replacement of halogenated flame retardants. In *Polymer Green Flame Retardants: A Comprehensive Guide to Additives and Their Applications*. Editors C.D. PAPASPYRIDES and P. KILARIS. Elsevier B.V.

HUWE, J.K., LARSEN, G.L., 2005. Polychlorinated dioxins, furans, and biphenyls, and polybrominated diphenyl ethers in a U.S. meat market basket and estimates of dietary intake. *Environmental Science and Technology* **39**, 5606–5611.

JIN, J., WANG, Y., YANG, C., HU, J., LIU, W., CUI, J., 2010. Human exposure to polybrominated diphenyl ethers at production area, China. *Environmental Toxicology and Chemistry* **29**, 1031–1035.

JOHNSON, P.I., ALTSHUL, L., CRAMER, D.W., MISSMER, S.A., HAUSER, R., MEEKER, J.D., 2012. Serum and follicular fluid concentrations of polybrominated diphenyl ethers and *in-vitro* fertilization outcome. *Environment International* **45**, 9–14.

KIM, G.B., STAPLETON, H.M., 2010. PBDEs, methoxylated PBDEs and HBCDs in Japanese common squid (*Todarodes pacificus*) from Korean offshore waters. *Marine Pollution Bulletin* **60**, 935–940.

LAW, R.J., ALLCHIN, C.R., DE BOER, J., COVACI, A., HERZKE, D., LEPOM, P., MORRIS, S., TRONCZYNSKI, J., DE WIT, C.A., 2006 a. Levels and trends of brominated flame retardants in the European environment. *Chemosphere* **64**, 187–208.

LAW, R.J., BARRY, J., BERSUDER, P., BARBER, J.L., DEAVILLE, R., REID, R.J., JEPSON, P.D., 2010. Levels and trends of brominated diphenyl ethers in blubber of harbor porpoises (*Phocoena phocoena*) from the UK, 1992–2008. *Environmental Science and Technology* **44**, 4447–4451.

LAW, R.J., HERZKE, D., HARRAD, S., MORRIS, S., BERSUDER, P., ALLCHIN, C.R., 2008. Levels and trends of HBCD and BDEs in the European and Asian environments, with some information for other BFRs. *Chemosphere* **73**, 223–241.

LAW, R.J., KOHLER, M., HEEB, N.V., GERECKE, A.C., SCHMID, P., VOORSPOELS, S., COVACI, A., BECHER, G., JANÁK, K., THOMSEN, C., 2006b. Response to 'HBCD: facts and insinuations'. *Environmental Science and Technology* **40**, 2.

LIU, Y.-P., LI, J.-G., ZHAO, Y.-F., WEN, S., HUANG, F.-F., WU, Y.-N., 2011. Polybrominated diphenyl ethers (PBDEs) and indicator polychlorinated biphenyls (PCBs) in marine fish from four areas of China. *Chemosphere* **83**, 168–174.

LORBER, M., 2008. Exposure of Americans to polybrominated diphenyl ethers. *Journal of Exposure Science and Environmental Epidemiology* **18**, 2–19.

MACGREGOR, K., OLIVER, I.W., HARRIS, L., RIDGWAY, I.M., 2010. Persistent organic pollutants (PCB, DDT, HCH, HCB &BDE) in eels (*Anguilla anguilla*) in Scotland: current levels and temporal trends. *Environmental Pollution* **158**, 2402–2411.

MARVIN, C.H., TOMY, G.T., ARMITAGE, J.M., ARNOT, J.A., MCCARTY, L., COVACI, A., PALACE, V., 2011. Hexabromocyclododecane: Current understanding of chemistry, environmental fate and toxicology and implications for global management. *Environmental Science and Technology* **45**, 8613–8623.

MCHUGH, B., POOLE, R., CORCORAN, J., ANNINOU, P., BOYLE, B., JOYCE, E., FOLEY, M.B., MCGOVERN, E., 2010. The occurrence of persistent chlorinated and brominated organic contaminants in the European eel (*Anguilla anguilla*) in Irish waters. *Chemosphere* **79**, 305–313.

MEESTER, W.D., 1979. The effect of polybrominated biphenyls on man: the Michigan PBB disaster. *Veterinary and Human Toxicology* **21**, 131–135.

MENG, X.-Z., XIANG, N., DUAN, Y.-P., CHEN, L., ZENG, E.Y., 2012. Hexabromocyclododecane in consumer fish from South China: Implications for human exposure via dietary intake. *Environmental Toxicology and Chemistry* **31**, 1424–1430.

MICELI, J.N., NOLAN, D.C., MARKS, B., HANRAHAN, M., 1985. Persistence of polybrominated biphenyls (PBB) in human post-mortem tissue. *Environmental Health Perspectives* **60**, 399–403.

MONTORY, M., HABIT, E., FERNANDEZ, P., GRIMALT, J.O., BARRA, R., 2010. PCBs and PBDEs in wild Chinook salmon (*Oncorhyncus tshawytscha*) in the Northern Patagonia, Chile. *Chemosphere* **78**, 1193–1199.

MORRIS, S., BERSUDER, P., ALLCHIN, C.R., ZEGERS, B., BOON, J.P., LEONARDS, P.E.G., DE BOER, J., 2006. Determination of the brominated flame retardant, hexabromocyclododecane, in sediments and biota by liquid chromatography-electrospray ionisation mass spectrometry. *Trends in Analytical Chemistry* **25**, 343–349.

NAKAGAWA, R., MURATA, S., ASHIZUKA, Y., SHINTANI, Y., HORI, T., TSUTSUMI, T., 2010. Hexabromocyclododecane determination in seafood samples collected from Japanese coastal areas. *Chemosphere* **81**, 445–452.

NATIONAL RESEARCH COUNCIL, 2000. *Toxicological Risks of Selected Flame-retardant Chemicals*. National Academy Press, Washington, USA.

ONDARZA, P.M., GONZALEZ, M., FILLMANN, G., MIGLIORANZA, K.S.B., 2011. Polybrominated diphenyl ethers and organochlorine compound levels in brown trout (*Salmo trutta*) from Andean Patagonia, Argentina. *Chemosphere* **83**, 1597–1602.

PAPACHLIMITZOU, A., BARBER, J.L., LOSADA, S., BERSUDER, P., LAW, R.J., 2012. A review of the analysis of novel brominated flame retardants. *Journal of Chromatography A* **1219**, 15–28.

PARK, J.-S., SHE, J., HOLDEN, A., SHARP, M., GEPHART, R., SOUDERS-MASON, G., ZHANG, V., CHOW, J., LESLIE, B., HOOPER, K., 2011. High postnatal exposures to polybrominated diphenyl ethers (PBDEs) and polychlorinated biphenyls (PCBs) via breast milk in California: Does BDE-209 transfer to breast milk? *Environmental Science and Technology* **45**, 4579–4585.

PENA-ABAURREA, M., COVACI, A., RAMOS, L., 2011. Comprehensive two-dimensional gas chromatography-time-of-flight mass spectrometry for the identification of organobrominated compounds in bluefin tuna. *Journal of Chromatography A* **1218**, 6995–7002.

PETREAS, M., NELSON, D., BROWN, F.R., GOLDBERG, D., HURLEY, S., REYNOLDS, P., 2011. High concentrations of polybrominated diphenyl ethers (PBDEs) in breast adipose tissue of California women. *Environment International* **37**, 190–197.

QIN, X., QIN, Z., LI, Y., ZHAO, Y., XIA, X., YAN, S., TIAN, M., ZHAO, X., XU, X., YANG, Y., 2011. Polybrominated diphenyl ethers in chicken tissues and eggs from an electronic waste recycling area in southeast China. *Journal of Environmental Sciences* **23**, 133–138.

ROBOZ, J., GREAVES, J., BEKESI, J.G., 1985. Polybrominated biphenyls in model and environmentally contaminated human blood: Protein binding and immunotoxicological studies. *Environmental Health Perspectives* **60**, 107–113.

ROOSENS, L., CORNELIS, C., D'HOLLANDER, W., BERVOETS, L., REYNDERS, H., VAN CAMPENHOUT, K., VAN DEN HEUVEL, R., NEELS, H., COVACI, A., 2010b. Exposure of the Flemish population to brominated flame retardants: Model and risk assessment. *Environment International* **36**, 368–376.

ROOSENS, L., GEERAERTS, C., BELPAIRE, C., VAN PELT, I., NEELS, H., COVACI, A., 2010a. Spatial variations in the levels and isomeric patterns of PBDEs and HBCDs in the European eel in Flanders. *Environment International* **36**, 415–423.

ROSE, M., FERNANDES, A., 2012. Other environmental contaminants in food. In *Chemical Contaminants and Residues in Food*. Editor D. SCHRENK. Woodhead Publishing Ltd., Cambridge. Chapter 6, pp. 124–147.

ROSE, M., FERNANDES, A., 2010. Are BFRs Responsible for Brominated Dioxins and Furans (PBDD/Fs) in Food? *Proceedings of BFR2010*, Kyoto, Japan, 7–9 April 2010. #90029. 6 pp.

SALVADÓ, J.A., GRIMALT, J.O., LÓPEZ, J.F., DURRIEU DE MADRON, X., HEUSSNER, S., CANALS, M., 2012. Transformation of PBDE mixtures during sediment transport and resuspension in marine environments (Gulf of Lion, NW Mediterranean Sea). *Environmental Pollution* **168**, 87–95.

SCHECTER, A., SMITH, S., COLACINO, J., MALIK, N., OPEL, M., PAEPKE, O., BIRNBAUM, L., 2011. Contamination of U.S. butter with polybrominated diphenyl ethers from wrapping paper. *Environmental Health Perspectives* **119**, 151–154.

SELIKOFF, I.J., 1979. Polybrominated biphenyls in Michigan. *Proceedings of the Royal Society of London* – Biological Sciences **205**, 153–156.

SHANMUGANATHAN, D., MEGHARAJ, M., CHEN, Z., NAIDU, R., 2011. Polybrominated diphenyl ethers (PBDEs) in marine foodstuffs from Australia: Residue levels and contamination status of PBDEs. *Marine Pollution Bulletin* **63**, 154–159.

SILBERHORN, E.M., GLAUERT, H.P., ROBERTSON, L.W., 1990. Carcinogenicity of polyhalogenated biphenyls: PCBs and PBBs. *Critical Reviews in Toxicology* **20**, 439–496.

SKINNER, L.C., 2011. Distributions of polyhalogenated compounds in Hudson River (New York, USA) fish in relation to human uses along the river. *Environmental Pollution* **159**, 2565–2574.

SLOAN, C.A., ANULACION, B.F., BOLTON, J.L., BOYD, D., OLSON, O.P., SOL, S.Y., YLITALO, G.M., JOHNSON, L.L., 2010. Polybrominated diphenyl ethers in outmigrant juvenile Chinook salmon from the lower Columbia River and estuary and Puget Sound, Washington. *Archives of Environmental Contamination and Toxicology* **58**, 403–414.

SPIEGELSTEIN, M., 2001. *Proceedings of the BFR2001 Symposium, Stockholm, Sweden.* p. 41. AB Firmatryck, Stockholm, Sweden.

TAKAHASHI, S., OSHIHOI, T., RAMU, K., ISOBE, T., OHMORI, K., KUBODERA, T., TANABE, S., 2010. Organohalogen compounds in deep-sea fishes from the western North Pacific, off-Tohoku, Japan: Contamination status and bioaccumulation profiles. *Marine Pollution Bulletin* **60**, 187–196.

TAPIE, N., LE MENACH, K., PASQUAUD, S., ELIE, P., DEVIER, M.H., BUDZINSKI, H., 2011. PBDE and PCB contamination of eels from the Gironde estuary: From glass eels to silver eels. *Chemosphere* **83**, 175–185.

TERRELL, M.L., MANATUNGA, A.K., SMALL, C.M., CAMERON, L.L., WIRTH, J., BLANCK, H.M., LYLES, R.H., MARCUS, M., 2008. A decay model for assessing polybrominated biphenyl exposure among women in the Michigan long-term PBB Study Decay model: Polybrominated biphenyl exposure. *Journal of Exposure Science and Environmental Epidemiology* **18**, 410–420.

UENO, D., ISOBE, T., RAMU, K., TANABE, S., ALAEE, M., MARVIN, C., INOUE, K., SOMEYA, T., MIYAJIMA, T., KODAMA, H., NAKATA, H., 2010. Spatial distribution of hexabromocyclododecanes (HBCDs), polybrominated diphenyl ethers (PBDEs) and organochlorines in bivalves from Japanese coastal waters. *Chemosphere* **78**, 1213–1219.

VALCIUKAS, J.A., LILIS, R., ANDERSON, H.A., WOLFF, M.S., PETROCCI, M., 1979. The neurotoxicity of polybrominated biphenyls: Results of a medical field survey. *Annals of the New York Academy of Sciences* **320**, 337–367.

VALCIUKAS, J.A., LILIS, R., WOLFF, M.S., ANDERSON, H.A., 1978. Comparative neurobehavioral study of a polybrominated biphenyl exposed population in Michigan and a nonexposed group in Wisconsin. *Environmental Health Perspectives* **23**, 199–210.

VORKAMP, K., RIGÉT, F.F., BOSSI, R., DIETZ, R., 2011. Temporal trends of hexabromocyclododecane, polybrominated diphenyl ethers and polychlorinated biphenyls in ringed seals from East Greenland. *Environmental Science and Technology* **45**, 1243–1249.

WASTELL, M.E., MOODY, D.L., PLOG JR., J.F., 1978. Effects of polybrominated biphenyl on milk production, reproduction, and health problems in Holstein cows. *Environmental Health Perspectives* **23**, 99–103.

XIA, C., LAM, J.C.W., WU, X., SUN, L., XIE, Z., LAM, P.K.S., 2011. Levels and distribution of polybrominated diphenyl ethers (PBDEs) in marine fishes from Chinese coastal waters. *Chemosphere* **82**, 18–24.

YARD, E.E., TERRELL, M.L., HUNT, D.R., CAMERON, L.L., SMALL, C.M., MCGEEHIN, M.A., MARCUS, M., 2011. Incidence of thyroid disease following exposure to polybrominated biphenyls and polychlorinated biphenyls, Michigan, 1974–2006. *Chemosphere* **84**, 863–868.

YOGUI, G.T., SERICANO, J.L., MONTONE, R.C., 2011. Accumulation of semivolatile organic compounds in Antarctic vegetation: A case study of polybrominated diphenyl ethers. *Science of the Total Environment* **409**, 3902–3908.

YU, Y.-X., HUANG, N.-B., ZHANG, X.-Y., LI, J.-L., YU, Z.-Q., HAN, S.-Y., LU, M., VAN DE WIELE, T., WU, M.-H., SHENG, G.-Y., FU, J.-M., 2011. Polybrominated diphenyl ethers in food and associated human daily intake assessment considering bioaccessibility measured by simulated gastrointestinal digestion. *Chemosphere* **83**, 152–160.

ZHENG, X.-B., WU, J.-P., LUO, X.-J., ZENG, Y.-H., SHE, Y.-Z., MAI, B.-X., 2012. Halogenated flame retardants in home-produced eggs from an electronic waste recycling region in

South China: Levels, composition profiles, and human dietary exposure assessment. *Environment International* **45**, 122–128.

ZHOU, S.N., REINER, E.J., MARVIN, C., KOLIC, T., RIDDELL, N., HELM, P., DORMAN, F., MISSELWITZ, M., BRINDLE, I.D., 2010. Liquid chromatography-atmospheric pressure photoionization tandem mass spectrometry for analysis of 36 halogenated flame retardants in fish. *Journal of Chromatography A* **1217**, 633–641.

ZHOU, Z.M., 2006. Implement of administrative measures on the control of pollution caused by electronic information products and the exemption of deca-BDE mixture. *Flame Retardant Materials Technology* **4**, 15–16 (in Chinese).

ZOTA, A.R., PARK, J.-S., WANG, Y., PETREAS, M., ZOELLER, R.T., WOODRUFF, T.J., 2011. Polybrominated diphenyl ethers, hydroxylated polybrominated diphenyl ethers, and measures of thyroid function in second trimester pregnant women in California. *Environmental Science and Technology* **45**, 7896–7905.

12
Human dietary exposure to per- and poly-fluoroalkyl substances (PFASs)

R. Vestergren and I. T. Cousins, Stockholm University, Sweden

DOI: 10.1533/9780857098917.2.279

Abstract: Per- and polyfluoroalkyl substances (PFASs) are a class of emerging contaminants with numerous industrial and commercial applications. Within this class the best studied substances are the perfluoroalkane sulfonates (PFSAs) and perfluoroalkyl carboxylates (PFCAs). PFSAs and PFCAs have been detected in human serum samples from all around the world and are ubiquitous in the global environment and wildlife. As well as being completely resistant to environmental degradation, some PFSAs and PFCAs are bioaccumulative and potentially toxic, which raises a concern about population-wide exposure to this group of substances. Dietary intake has been suggested as a major pathway of human exposure to the two most widely studied substances among PFSAs and PFCAs, namely perfluorooctanoic acid (PFOA) and perfluorooctane sulfonic acid (PFOS). However, the difficulties associated with the analysis of PFSAs, PFCAs and related PFASs at ultra-trace levels in food samples have hampered the understanding of human exposure. Recent advances in analytical chemistry have dramatically improved the ability to measure these substances and other PFASs in food matrices, and method detection limits down to low picogram per gram food can now be reached. Worldwide interlaboratory studies also indicate that the accuracy and precision of analytical methods have significantly improved over the last decade. These modern methods have been applied to quantify human dietary exposure to PFCAs and PFSAs in several European countries. Overall, the exposure to PFOS and PFOA from diet is typically a factor of 6 to 10 higher than the exposure from other known exposure pathways for the general adult population of Western countries. Furthermore, application of toxicokinetic models indicates that present day serum concentrations of PFOS and PFOA can largely be explained by the estimated dietary exposures. Despite the recent advances in analytical techniques, the sources of food contamination are not very well characterized. It has been demonstrated that bioaccumulation and biomagnification in aquatic food webs is a primary transfer mechanism for PFOS and several long-chain perfluoroalkyl carboxylic acids to the human diet. However, more research is needed to understand the accumulation of PFASs in terrestrial food webs and the transfer of a range of PFASs from food-contact materials.

Key words: food, diet, human exposure, perfluoroalkyl substances, PFOA, PFOS.

12.1 Introduction

PFASs are a large group of synthetic chemicals that contain a carbon chain where all hydrogen atoms attached to at least one carbon atom have been substituted by fluorine atoms so that they contain the perfluoroalkyl moiety (C_nF_{2n+1}–). The perfluoroalkyl moiety provides hydrophobic and oleophobic properties to the molecule and makes it resistant to degradation, which has made PFASs useful in numerous industrial processes and consumer product applications (Kissa, 2001; Buck et al., 2011). Low molecular weight monoalkylated PFASs are primarily used as industrial surfactants whereas di- and tri-alkylated PFASs and polymeric PFASs have many direct consumer applications in products such as greaseproof food-contact paper and textile stain and soil repellents respectively (Kissa, 2001; Buck et al., 2011).

During the last decade PFASs have been identified by scientists and regulatory bodies as a class of problematic contaminants. Particular attention has been directed to perfluoroalkane sulfonic acids (PFSAs) and perfluoroalkyl carboxylic acids (PFCAs), which have been found to be globally present in wildlife (Giesy and Kannan, 2001; Martin et al., 2004a) and human serum samples (Hansen et al., 2001; Houde et al., 2006a). PFSAs and PFCAs are extremely persistent in the environment (Lemal et al., 2004) and long-chain PFSAs ($C_nF_{2n+1}SO_3H$, $n \geq 6$) and PFCAs ($C_nF_{2n+1}COOH$, $n \geq 7$) display a potential to bioaccumulate/biomagnify in food webs (Martin et al., 2003a, 2003b; Conder et al., 2008; Mueller et al., 2011). Furthermore, toxicological studies with rodents have shown that perfluorooctanoic sulfonic acid (PFOS) and perfluorooctanoic acid (PFOA) (the most widely studied PFSAs and PFCAs) may induce a range of different tumors, reduce immune responses, reduce birth weight and cause neonatal death (Lau et al., 2004, 2007). Recent epidemiological studies have further suggested that exposure to PFOA and PFOS is associated with increased levels of blood lipids (Steenland et al., 2010), reduced humoral immune response to routine childhood immunization (Grandjean, 2012) and delayed sexual maturity (Lopez-Espinoza et al., 2011) in the human population.

The concerns regarding adverse environmental and human health effects have led to a series of actions by the regulatory community and manufacturers. In the year 2000, the major manufacturer of PFOS in Europe and North America decided to phase out its production of all long-chain PFASs including PFOS and PFOA (USEPA, 2000). In 2009 PFOS was also added to Annex B of the Stockholm Convention on Persistent Organic Pollutants which restricts its use globally (UNEP, 2009). Other manufacturers of PFASs have also agreed to stewardship programs, which include a 95% reduction of emissions and product content by 2010 and complete elimination by 2015 of PFOA and related substances (USEPA, 2006; Environment Canada, 2010). However, due to their persistence (Lemal, 2004), and potential secondary sources of PFSAs and PFCAs to the environment (D'eon et al., 2009; Washington et al., 2009), exposure to these contaminants can be expected for some time to come. Furthermore, as production of PFOS and PFOA is gradually being phased-out from North America and Europe, manufacture has increased substantially in other parts of the world, notably Asia (Chen *et al.*, 2009). More recently, an increasing level of scrutiny has also been directed to the replacement PFASs, which are generally short-chain versions of their predecessors (Buck *et al.*, 2011). Thus, PFASs continue to be environmental contaminants of scientific and regulatory interest.

Human exposure to PFASs can occur via a large number of different pathways resulting from the use of consumer products, contact with environmental media, and intake of drinking water and food (Washburn *et al.*, 2005; Trudel *et al.*, 2008). Exposure modeling studies have identified intake of contaminated food to be a major exposure pathway of PFOA and PFOS for the general population of industrialized countries (Trudel *et al.*, 2008; Vestergren *et al.*, 2008; Fromme *et al.*, 2009). However, difficulties associated with the analysis of many food matrices have hampered a reliable quantification of dietary exposure to a wide range of PFASs. Furthermore, there is a lack of understanding of how food is contaminated with PFASs, and of which transfer mechanisms are of the greatest importance for human exposure.

The overarching objective of this chapter is to review recent advances in understanding food contamination of PFASs and resulting dietary exposure. Remaining uncertainties and gaps of knowledge will be identified, in order to provide recommendations for further research. Due to the prevalence of data for PFSAs and PFCAs (particularly for PFOS and PFOA) an emphasis will be placed on these compounds. Although drinking water is often included in dietary intake assessments, the widespread occurrence of PFASs in drinking water and groundwater has been described by others (Saito *et al.*, 2004; Quinones and Snyder, 2009; Rayne and Forrest, 2009) and will not be discussed in detail here.

12.2 Analytical methods for PFASs in foods

PFASs display a wide range of physicochemical properties and numerous analytical methods have been developed for the different compound classes and sample matrices (van Leeuwen and de Boer, 2007; Jahnke and Berger, 2009). Extraction techniques for PFCAs and PFSAs most commonly include ion-pair extraction (IPE) or solid–liquid extraction (SLE). The IPE method was first presented by Ylinen *et al.* (1985) using tetrabutylammonium counter ion at alkaline pH and ethyl acetate as the extraction solvent. The IPE method was later adapted by Hansen *et al.* (2001) using methyl *tert*-butyl ether as the extraction solvent and has more recently been applied to food samples (Guruge *et al.*, 2008; Vestergren *et al.*, 2012). A drawback of the IPE method is the co-extraction of lipophilic matrix constituents, which can adversely affect the instrumental analysis (Powley *et al.*, 2005; van Leeuwen *et al.*, 2009). SLE extraction techniques, employing medium polar solvents including methanol and acetonitrile (Powley *et al.*, 2005; Berger *et al.*, 2009), has therefore often been the preferred technique for analysis of PFSAs and PFCAs in fat-rich food matrices. Recently, a novel SLE method using a 75:25 mixture of tetrahydrofuran and water was found to provide the optimal hydrogen bonding, dispersion and dipole–dipole interactions to efficiently extract PFSAs and PFCAs with 4 to 14 carbon perfluoroalkyl chains (Ballestéros-Gomez *et al.*, 2011). In order to increase the amount of sample for extraction and to prevent co-extraction of water from food samples, SLE methods often require freeze drying of the samples prior to extraction (Kärrman *et al.*, 2009; Ballestéros-Gomez *et al.*, 2011). In addition to IPE and SLE, alkaline or acidic digestion, precipitation and back-extraction techniques have occasionally been applied to analyze food samples (Taniyasu *et al.*, 2005; Kärrman *et al.*, 2009; Sundström *et al.*, 2011). Although the majority of extraction methods employed in food analysis have been developed and optimized for PFSAs and PFCAs (Pico *et al.*, 2011; Tittlemier and Braekevelt, 2011), a few techniques have been applied for analysis of fluorotelomer alcohols (FTOH), perfluorooctane sulfonamides (PFOSA), perfluorooctane sulfonamidoethanols (PFOSE) and polyfluorinated phosphate surfactants (PAPS) in food. Neutral N-alkylated PFOSAs were first extracted using Soxhlet with a 2:1 hexane:acetone mixture (Tittlemier *et al.*, 2006). However, more recently, SLE has been employed to simultaneously extract neutral PFOSAs, PFOSEs and anionic PFSAs, PFCAs and FTUCAs (Ostertag *et al.*, 2009; Lacina *et al.*, 2011).

A commonly observed problem in mass spectrometry (MS) with electrospray ionization (ESI) (see below) is the phenomenon of ion suppression or ion enhancement due to co-eluting matrix constituents (Mallet *et al.*, 2004). In order to avoid analytical interferences from matrix constituents, a range of different purification methods have been developed. These typically involve dispersive clean-up with graphitized carbon (Powley *et al.*, 2005; Lacina *et al.*, 2011) or solid-phase extraction using ion-exchange, graphitized carbon or silica type cartridges (Taniyasu *et al.*, 2005; Kärrman *et al.*, 2009; Ballesteros-Gomez *et al.*, 2010; Vestergren *et al.*, 2012). The majority of these clean-up steps report quantitative recoveries around 100% for a wide range of PFSAs and PFCAs and efficient reduction of matrix effects (Taniyasu *et al.*, 2005; Powley *et al.*, 2007; Kärrman *et al.*, 2009; Ballesteros-Gomez *et al.*, 2010; Vestergren *et al.*, 2012a).

Quantification of PFCAs and PFSAs was for a long time hampered by the lack of suitable analytical technology. The low volatility of PFCAs and PFSAs made direct analysis by gas chromatography (GC) impossible, whereas a lack of a suitable chromophore made liquid chromatography (LC) analysis with ultraviolet detection also impractical (Martin *et al.*, 2004b). To circumvent these problems, derivatization of PFCAs and PFSAs into methyl esters and subsequent analysis by GC coupled electron capture detector (Ylinen *et al.*, 1981) or MS was the method of choice in early research on PFASs. The advances of the ESI techniques for LC-MS applications made the analysis of PFCAs much more sensitive and facilitated the first detection of PFCAs and PFSAs in human blood serum and wildlife samples (Giesy and Kannan, 2001; Hansen *et al.*, 2001). High-performance liquid chromatography (HPLC) coupled to quadrupole tandem mass spectrometry (MS/MS) applying ESI is today the standard instrumental method for separation and detection of PFSAs and PFCAs (Jahnke and Berger, 2009). Although, ion-trap MS (Louqe *et al.*, 2011) and high resolution (HR) MS (Berger and Haukas, 2005) have been successfully used, the widespread availability of MS/MS instruments has made this the most frequently used detection technique. Due to its better separation power compared to LC, derivatization and GC separation coupled to MS have also been exploited for analysis of individual isomers (Langlois *et al.*, 2007). However, more recently the application of a perfluorooctyl stationary phase and an acidified mobile phase has shown that a full separation of many PFCA and PFSA isomers can be achieved using LC-MS/MS (Benskin *et al.*, 2007). For neutral and volatile PFASs, including fluorotelomer alcohols (FTOHs), fluorotelomeracrylates (FTO) and sulfonamidoethanols (FOSE), analysis is usually performed by GC-MS in combination with electron impact or chemical ionization (Jahnke and Berger, 2009). However, analysis of neutral PFASs has also been successfully conducted using LC-MS (Taniyasu *et al.*, 2005; Szostek *et al.*, 2006; Lacina *et al.*, 2011). Recently, a screening type LC-MS method for a suite of nonionic and anionic fluorinated surfactants used in food-contact materials has also been developed (Trier *et al.*, 2011).

The rapid development of LC-MS instrumentation and increased availability of high quality standards have significantly improved the performance of PFAS analysis. Over the past decade, method detection limits for PFSAs and PFCAs have improved by almost three orders of magnitude and the number of per- and poly-fluoroalkyl analytes being analyzed simultaneously have increased several fold (Hansen *et al.*, 2001; Ballesteros-Gomez *et al.*, 2010; Haug *et al.*, 2010a; Lacina *et al.*, 2011; Sundström *et al.*, 2011; Vestergren *et al.*, 2012b). The research efforts to find efficient extraction and clean-up methods have also allowed complex and heterogeneous samples to be analyzed without matrix effects (Ballesteros-Gomez *et al.*, 2010; Lacina *et al.*, 2011; Sundström *et al.*, 2011; Vestergren *et al.*, 2012a). As the majority of food items of non-animal origin contain concentrations in the low to sub pg g^{-1} range (see Section 12.3), these breakthroughs represent a prerequisite for a reliable quantification of PFCAs and PFSAs in food items. The advances in analytical methodologies have also been reflected in the gradual improvements made in interlaboratory comparability studies (van Leeuwen *et al.*, 2006; Lindström *et al.*, 2008; Longnecker *et al.*, 2008; van Leeuwen, 2009; van der Veen *et al.*, 2012; Vestergren *et al.*, 2012). In the first interlaboratory studies, comparability between reported concentrations of PFASs in environmental samples was generally low and many analytical problems were identified (van Leeuwen *et al.*, 2006; Lindström *et al.*, 2008; Longnecker *et al.*, 2008). During the fifth worldwide interlaboratory study, 31–100% (depending on the analyte) of the participating laboratories reported concentrations in satisfactory agreement with the assigned value in a spiked fish muscle sample (van der Veen *et al.*, 2012). However, a lower number of laboratories achieved satisfactory results for fish, vegetable and pig liver samples containing low concentrations of the analytes (van der Veen *et al.*, 2012). As reduced precision and accuracy are often observed at concentrations close to the method limits of quantification (van Leeuwen *et al.*, 2009), the relatively high comparability observed in highly contaminated fish samples may therefore not apply to the majority of food samples intended for human consumption. Thus, ultra-trace analysis of PFASs in many food samples remains a delicate task requiring specific attention to quality assurance issues such as procedural blank contamination, total method recovery and matrix effects (Vestergren *et al.*, 2012a).

Among the analytical challenges related to PFASs in food, special care should be taken to avoid the interference of endogenous compounds present in biological matrices that co-elute with the analytes and share the same MS/MS transitions (Benskin *et al.*, 2007). If not properly dealt with, these interferences could give rise to false positive detection of the analytes. The problem has been well described for PFOS and PFHxS (Benskin *et al.*, 2007; Ballesteros-Gómez *et al.*, 2010), but may also be present for some short-chain PFSAs and PFCAs (Lacina *et al.*, 2011). In general, the use of LC columns with perfluorooctyl stationary phases, including confirmatory daughter ions or use of high resolution-MS, have been shown to resolve the problem with false positives. As PFASs are highly surface-active, special care should also be taken to avoid adsorption to surfaces during different steps throughout the entire analytical procedure (from sampling to instrumental analysis) (Martin *et al.*, 2004b; Berger *et al.*, 2011). Biases due to surface activity may be of particular importance for di- and tri-alkylated PFASs that are typically present in food-contact materials (Begley *et al.*, 2005, 2007; Trier *et al.*, 2011).

12.3 Levels of PFASs in foods

Fish and shellfish are probably the best characterized food categories with respect to PFSAs and PFCAs as the concentrations are generally higher than other food items, and numerous fish samples of various species from different regions have been analyzed (Hoff *et al.*, 2003; Kannan *et al.*, 2005; Gulkowska *et al.*, 2006; Furdui *et al.*, 2007; Tittlemier *et al.*, 2007; Bossi *et al.*, 2008; Ericson *et al.*, 2008; Berger *et al.*, 2009; Del Gobbo *et al.*, 2009; Nania *et al.*, 2009; Ostertag *et al.*, 2009; Hölzer *et al.*, 2010). PFOS has been shown to be the predominant PFAS in the majority of fish samples at concentrations ranging from < 500 to 23 000 pg g^{-1} in fish samples from nonpoint source areas (Gulkowska *et al.*, 2006; Furdui *et al.*, 2007; Tittlemier *et al.*, 2007; Bossi *et al.*, 2008; Ericson *et al.*, 2008; Berger *et al.*, 2009; Del Gobbo *et al.*, 2009; Nania *et al.*, 2009; Ostertag *et al.*, 2009; van Leeuwen *et al.*, 2009;

Hölzer et al., 2010; Shi et al., 2010; Hradkova et al., 2011; Murakami et al., 2011). In general, the highest concentrations of PFOS have been reported in lean predatory fish (Bossi et al., 2008; Berger et al., 2009; van Leeuwen et al., 2009) and trophic position in the food chain may certainly influence the observed concentrations (see also Section 12.4). Furthermore, it appears that farmed fish and shell fish samples display lower concentrations of PFOS compared to wild fish (van Leeuwen et al., 2009). However, for the large amount of data reporting PFOS in fish, proximity to emission sources seems to be the primary reason for elevated levels. In general, a high anthropogenic input has been associated with elevated concentrations of PFOS (Berger et al., 2009) and a few studies from locally polluted areas have reported PFOS concentrations as high as > 1 00 000 pg g^{-1} in fish muscle samples (Hoff et al., 2003; Kannan et al., 2005; Hölzer et al., 2010; Hradkova et al., 2011). Concentrations of PFOA in marine and freshwater fish fillets have consistently been found to be lower than those of PFOS with a high frequency of non-detects. Reported concentrations typically range between < 100 and 250 pg g^{-1} and trends in concentrations due to geographical origin, species and other factors are sparse (Furdui et al., 2007; Berger et al., 2009). The most frequent detection of PFOA has been observed in mussels and locally elevated concentrations have been observed in fish and shellfish near firefighting training sites (Hoff et al., 2003). When analyzed, long-chain PFCAs have typically been measured at higher concentrations than PFOA (< 80–1900 pg g^{-1}). The pattern of long-chain PFCAs displays some variability between different species and regions (Moody et al., 2003; Gulkowska et al., 2006). However, PFNA, PFUnDA and PFTriDA are typically found at the highest concentrations (Berger et al., 2009). The higher concentration of long-chain PFCAs compared to PFOA has been attributed to an increasing bioaccumulation potential with increasing chain-length (Martin et al., 2003a, 2003b), which is discussed in this section.

Although several studies have been performed to measure PFASs in consumer food items (3M 2001; Food Standards Agency 2006; Tittlemier et al., 2007; Ericson et al., 2008; Ericson-Jogsten et al., 2009; Clarke et al., 2010; Schecter et al., 2011), reported concentration data for PFASs in many food items have remained fairly limited. Due to the insufficient analytical sensitivity of previously used methods (described in Section 12.2), a majority of the analyzed samples have been reported below the method detection limit. Studies measuring PFSAs and PFCAs in store-bought food items have been performed in the US (3M 2001; Schecter et al., 2011), Canada (Tittlemier et al., 2007), the UK (Food Standards Agency 2006; Clarke et al., 2010) and Spain (Ericson et al., 2008; Ericson-Jogsten et al., 2009). In the USA, most food samples have levels reported below the limit of detection (20–500 pg g^{-1} depending on the homologue and sample). The highest levels have been observed for PFOA in butter (1070 pg g^{-1}), olive oil (1800 pg g^{-1}), apples (2350 pg g^{-1}) and PFOS in milk (850 pg g^{-1}) (Schecter et al., 2010). Several samples of meat and fish products contained PFOA between < 20 and 300 pg g^{-1} (Schecter et al., 2010). In Canada, PFOA was detected at the highest concentrations in microwave popcorn (3600 pg g^{-1}) and roast beef (2600 pg g^{-1}) and PFOS was detected at the highest concentrations in beef steak (2700 pg g^{-1}) and saltwater fish (2600 pg g^{-1}) (Tittlemier et al., 2007). A total diet study in the UK found PFOS at 10 000 pg g^{-1} and PFOA at 1000 pg g^{-1} in a potato composite sample (Food Standards Agency, 2006). In Spain, PFOS has been detected in many food items of primarily animal origin ranging from < 3–654 pg g^{-1} (Ericson et al., 2008; Ericson-Jogsten et al., 2009), whereas PFOA was only detected in whole milk (56 pg g^{-1}) and lettuce (164–179 pg g^{-1}). PFOS and PFOA have also been measured in duplicate diet samples consisting of homogenates from all food consumed by one individual during a day. In a German duplicate diet study, PFOS, PFOA, PFHxS and PFHxA were detected in 70, 97, 6 and 19 out of 214 samples respectively, with mean concentrations of PFOS and PFOA at 60 and 690 pg g^{-1}, respectively (Fromme et al., 2007). A more recent duplicate diet study from Japan detected PFOS (8–87 pg g^{-1}) and PFOA (8–40 pg g^{-1}) in all 20 samples that were analyzed. However, no other PFSAs and PFCAs were detected in duplicate diet samples (Kärrman et al., 2009).

With the improvements in analytical methods (as described in Section 12.2), a few recent studies have reported a significantly higher frequency of detection in consumer food samples compared to previous studies (Haug et al., 2010a; Lacina et al., 2011; Noorlander et al., 2011; Vestergren, 2011). The food samples included in these studies differed somewhat in the way they were collected. Food samples from Sweden were part of a food basket survey where frequently consumed food items were collected and pooled in different categories (Vestergren et al., 2012b), whereas the samples from Norway (Haug et al., 2010), Czech (Lacina et al., 2010) and the Netherlands (Noorlander et al., 2011) were randomly collected food items from grocery stores. We decided to present the results from a few selected recent studies in which, in our opinion, the methods used could be considered 'state of the art', and provide reliable and current concentration data for PFCAs and PFSAs, in a variety of food items at the low pg g^{-1} level. The reported concentrations from these studies are presented in Table 12.1.

Despite different sampling strategies and countries of origin, the reported concentrations presented in Table 12.1 display a rather consistent picture of food contamination of PFSAs and PFCAs. In agreement with previous studies, fish samples display higher concentrations of PFOS (13–5400 pg g^{-1}) and PFNA (< 11–803 pg g^{-1}), PFUnDA (5–803 pg g^{-1}) and PFTriDA (41–530) compared to the majority of other food items. Meat, meat products and chicken eggs display the second highest concentrations of PFOS (13–1281 pg g^{-1}) with one pooled egg sample from Sweden displaying particularly high concentrations. Long-chain PFCAs have also been detected in meat and egg samples although at significantly lower concentrations and with a different homologue pattern (predominance of PFOA, PFNA, PFDA and PFUnDA) compared to that observed in fish. In contrast to PFOS, PFOA displays

Table 12.1 Concentrations of PFSAs and PFCAs (pg g^{-1}) in food samples from Norway (Haug et al., 2010), Holland (Noorlander et al., 2010) and Sweden (Vestergren, 2011)

Food sample	Concentrations of PFSAs and PFCAs in food (pg g^{-1} fresh weight)														Year	Country	Reference
	PFHxA	PFHpA	PFOA	PFNA	PFDA	PFUnDA	PFDoDA	PFTrDA	PFTeDA	PFBS	PFHxS	PFOS					
Fish and fish products																	
Fish sticks	<18	21	49	<11	17	18	<13			5	1.6	13		2008	Norway	Haug et al., 2010a
Canned mackerel	<18	<24	24	<11	<31	19	<12			<5.5	<3	43		2008	Norway	Haug et al., 2010a
Salmon	11	16	46	10	26	4.5	<12			2.2	5.5	55		2008	Norway	Haug et al., 2010a
Cod	<11	<15	30	5.9	13	21	<7.5			<3.4	2.8	100		2008	Norway	Haug et al., 2010a
Cod liver	<48	<66	51	14	39	230	<33			<15	<8.2	310		2008	Norway	Haug et al., 2010a
Fish products	<3.3	3	50	72	92	316	72	123	12		9	1290		2010	Sweden	Vestergren, 2012b
Fish products	<3.3	16	32	90	79	214	53	113	9		9	780		2005	Sweden	Vestergren, 2012b
Fish products	<3.3	28	110	90	43	130	36	68	10		22	1095		1999	Sweden	Vestergren, 2012b
Fatty fish	<5	3	8	5	4	36	10	41	3	<1	9	61		2009	Holland	Noorlander et al., 2011
Lean fish	<3	2	34	77	48	177	56	229	24	<1	23	308		2009	Holland	Noorlander et al., 2011
Crustaceans	<4	5	46	58	90	157	45	268	45	<1	44	582		2009	Holland	Noorlander et al., 2011
Canned mackerel	11	14	15	35	52	206	68	245	50			215		2010	Czech Republic	Lacina et al., 2011
Canned sardine		12	57	194	98	187	62	126	27		83	3164		2010	Czech Republic	Lacina et al., 2011
Canned cod liver		6	15	460	394	803	261	530	77		44	5400		2010	Czech Republic	Lacina et al., 2011
Meat and meat products																	
Pork meat	<4.3	2.8	15	5.5	16	<8.2	<8			0.81	1.2	17		2008	Norway	Haug et al., 2010a
Beef	<3.3	7.6	12	15	23	<6.4	<6.2			<0.63	<0.28	60		2008	Norway	Haug et al., 2010a
Chicken meat	<13	20	52	6.8	<23	13	<9.2			3.2	<2.3	21		2008	Norway	Haug et al., 2010a
Meat products	<3.3	<1.9	12	6	6	2	1	<3	<2		5	25		2010	Sweden	Vestergren, 2012b
Meat products	<3.3	<1.9	15	9	6	8	2	4	<2		5	86		2005	Sweden	Vestergren, 2012b
Meat products	<3.3	<1.9	24	7	5	5	2	<3	<1		8	192		1999	Sweden	Vestergren, 2012b
Pork	<11	6	15	2	2	<4	<3	<23	<1	<3	<5	14		2009	Holland	Noorlander et al., 2011
Beef	<5	<0.2	<5	4	6	2	<2	<14	<0.7	<2	<4	82		2009	Holland	Noorlander et al., 2011

(*Continued*)

Table 12.1 Continued

Food sample	Concentrations of PFSAs and PFCAs in food (pg g⁻¹ fresh weight)												Year	Country	Reference
	PFHxA	PFHpA	PFOA	PFNA	PFDA	PFUnDA	PFDoDA	PFTrDA	PFTeDA	PFBS	PFHxS	PFOS			
Chicken/poultry	<7	1	<5	1	<1	<3	<2	<17	<0.8	<2	3	<5	2009	Holland	Noorlander et al., 2011
Eggs															
Eggs	13	<16	30	<7.4	12	9.9	<8.1				3.5	39	2008	Norway	Haug et al., 2010a
Eggs	3.6	<1.9	39	<4.5	3	<2.2	<2.5	<3.0	<2.0	2	3	39	2010	Sweden	Vestergren, 2012b
Eggs	5.1	<1.9	7	6	5	3	<2.5	<3.0	<2.0		<1.9	13	2005	Sweden	Vestergren, 2012b
Eggs	5.1	<1.9	31	22	14	38	10	14	<2.0		39	1281	1999	Sweden	Vestergren, 2012b
Eggs	<54	<2	<32	6	11	<19	<13	<107	<5	<3	<6	29	2009	Holland	Noorlander et al., 2011
Fats															
Margarine	2.5	<5.6	12	<13	<8.6	<16	<16			<1.6	1.3	2	2008	Norway	Haug et al., 2010a
Fats	4.3	<2.3	<2.3	<5.4	<2.5	5.8	<3.0	<3.6	<2.4		<2.3	13	2010	Sweden	Vestergren, 2012b
Fats	<3.9	<2.3	15	<5.4	<2.5	<2.6	<3.0	<3.6	<2.4		0.9	<2.3	2005	Sweden	Vestergren, 2012b
Fats	<3.9	<2.3	10	<5.4	3.8	<2.6	<3.0	<3.6	<2.4		<2.3	10	1999	Sweden	Vestergren, 2012b
Butter	20	5	16	2	6	<3	2	<19	<1	<3	16	33	2009	Holland	Noorlander et al., 2011
Vegetable oil	<3	1	<3	<0.1	<0.6	<2	<1	<11	<0.6	<0.9	<2	<3	2009	Holland	Noorlander et al., 2011
Industrial oil	<5	3	6	<0.3	2	<3	<2	<16	<0.8	<3	7	12	2009	Holland	Noorlander et al., 2011
Dairy products															
Cheese	<7.7	7.4	13	16	7	4	<15			<1.5	<0.65	12	2008	Norway	Haug et al., 2010a
Milk	1.5	<0.87	5	<2.1	4	<2.5	<2.4			<0.24	<0.11	7	2008	Norway	Haug et al., 2010a
Dairy products	<4.5	<2.7	28	<6.2	<2.9	<3.0	<3.4	<4.1	<2.7		1.0	5.6	2010	Sweden	Vestergren, 2012b
Dairy products	<4.5	<2.7	16	<6.2	6.6	<3.0	<3.4	<4.1	<2.7		<2.7	4.0	2005	Sweden	Vestergren, 2012b
Dairy products	<4.5	<2.7	<2.7	<6.2	<2.9	<3.0	<3.4	<4.1	<2.7		<2.7	<2.7	1999	Sweden	Vestergren, 2012b
Cheese	<9	7	<19	7	8	<16	<11	<92	<5	<12	<25	<85	2009	Holland	Noorlander et al., 2011
Milk	<6	<3	1	<1	1	<0.5	<0.5	<0.5	<2	<4	<2	10	2009	Holland	Noorlander et al., 2011
Vegetables, root crops and fruit															
Lettuce	0.98	0.43	2	<1	0.78	<1.3	1.3			<0.12	<0.06	0.2	2008	Norway	Haug et al., 2010a
Carrot	<1.3	<0.89	2	<2.1	<1.4	<2.5	<2.4			<0.25	<0.11	0.7	2008	Norway	Haug et al., 2010a
Potato	3.1	1.1	5	<4.1	3	2.2	<4.8			<0.48	<0.22	1.0	2008	Norway	Haug et al., 2010a
Vegetables	3.2	1.8	22	<2.3	2.5	<1.1	<1.3	<1.5	<1.0		1.2	4.1	2010	Sweden	Vestergren, 2012b

Food													Year	Country	Reference
Vegetables	3.0	1.7	52	<2.3	<1.1	<1.3	<1.1	<1.5	<1.0			24	2005	Sweden	Vestergren, 2012b
Vegetables	5.2	2.2	36	<2.3	<1.1	1.6	<1.1	<1.5	<1.0		1.0	4.6	1999	Sweden	Vestergren, 2012b
Potatoes	2.6	1.5	57	<2.3	2.6	<1.3	<1.1	<1.5	<1.0		1.2	6.9	2010	Sweden	Vestergren, 2012b
Potatoes	2.0	2.0	12	<2.3	<1.1	<1.3	<1.1	<1.5	<1.0		<1.0	5.8	2005	Sweden	Vestergren, 2012b
Potatoes	<1.7	1	5	<2.3	1.7	<1.3	<1.1	<1.5	<1.0		<1.0	1.9	1999	Sweden	Vestergren, 2012b
Fruit	2.8	<1.0	15	<2.3	2.4	<1.3	<1.1	<1.6	<1.0		<1.0	2.2	2010	Sweden	Vestergren, 2012b
Fruit	<1.7	<1.0	16	<2.3	<1.1	<1.3	<1.1	<1.6	<1.0		<1.0	4.4	2005	Sweden	Vestergren, 2012b
Fruit	<1.7	<1.0	8	<2.3	1.8	<1.3	<1.1	<1.6	<1.0		<1.0	1.9	1999	Sweden	Vestergren, 2012b
Vegetables/fruit	<4	<0.2	5	1	2	<2	<2	<14	<0.7	<6	<12	<47	2009	Holland	Noorlander et al., 2011
Flour, cereal products, bread and pastries															
Bread	14	11	51	10	17	<15	<15	<3.1	<2.0	<1.5	1.7	17	2008	Norway	Haug et al., 2010a
Cereal products	4	<2.0	62	<4.6	<2.1	<2.6	<2.2	<3.1	<2.0		<2.0	2	2010	Sweden	Vestergren, 2012b
Cereal products	11	3	11	<4.6	<2.1	<2.6	<2.2	<3.1	<2.0		<2.0	4	2005	Sweden	Vestergren, 2012b
Cereal products	8	3.0	12	<4.6	<2.1	<2.6	<2.2	<3.1	<2.0		<2.0	4	1999	Sweden	Vestergren, 2012b
Pastries	4	<1	18	<2.3	2.5	<1.3	<1.0	<1.5	<1.0		<1.0	21	2010	Sweden	Vestergren, 2012b
Pastries	4	<1	36	<2.3	1.9	<1.3	1.5	<1.5	<1.0		1.3	17	2005	Sweden	Vestergren, 2012b
Pastries	6	2.1	47	<2.3	2.0	1.6	<1.0	<1.5	<1.0		1.2	11	1999	Sweden	Vestergren, 2012b
Flour	11	14	17	15	9	4	4	<9	<0.4	<1	18	<9	2009	Holland	Noorlander et al., 2011
Bakery products	<9	<0.2	5	1	1	<0.7	<1	<6	<0.3	<1	6	4	2009	Holland	Noorlander et al., 2011
Sweets															
Strawberry jam	<7	<4.7	14	4	8.7	<13	<13	<1.5	<1.0	<1.3	<0.59	3	2008	Norway	Haug et al., 2010a
Sugar and sweets	3.2	<1.0	13	<2.3	2.0	<1.3	<1.1	<1.5	<1.0		1.5	4	2010	Sweden	Vestergren, 2012b
Sugar and sweets	3.4	<1.0	47	<2.3	2.0	1.1	1.1	<1.5	<1.0		1.3	4	2005	Sweden	Vestergren, 2012b
Sugar and sweets	4.0	<1.0	23	<2.3	1.7	<1.3	<1.1	<1.5	<1.0		<1.0	4	1999	Sweden	Vestergren, 2012b
Soft drinks															
Soft drinks	1.4	<0.5	3	<1.2	1.0	<0.6	<0.6	<0.8	<0.5		<0.5	<0.5	2010	Sweden	Vestergren, 2012b
Soft drinks	1.7	<0.5	3	<1.2	<0.5	<0.6	<0.6	<0.8	<0.5		<0.5	<0.5	2005	Sweden	Vestergren, 2012b
Soft drinks	1.6	<0.5	6	<1.2	<0.5	<0.6	<0.6	<0.8	<0.5		0.7	1.2	1999	Sweden	Vestergren, 2012b

similar concentrations in the wide range of food samples of both vegetable and animal origin (< 3–102 pg g^{-1}). The differences in homologue patterns in food samples of animal origin may reflect different sources of contamination or different uptake and elimination pathways in terrestrial and aquatic food webs (see Section 12.4).

In the light of these more recent measurements (Table 12.1), it appears as if many of the early measurements of PFSAs and PFCAs above 1000 pg g^{-1} concentrations represent anomalies in food concentration data sets (with the exception of PFOS in fish). Although the reasons for this discrepancy cannot be entirely elucidated, it has been hypothesized that earlier methods to some extent may have overestimated the concentrations of PFOA and PFOS in dietary samples due to less rigorous quality control measures compared to state-of-the-art methods today (Vestergren *et al.*, 2012a). However, it is also possible that the discrepancies between different studies may be explained by the different sampling schemes applied. In contrast to the food samples presented in Table 12.1, which were generally unprepared staple food samples, some studies have included food stored or prepared in greaseproof packaging materials. For instance, high concentrations of PFOAs (3600 pg g^{-1}) in microwave popcorn and high concentrations of FOSAs (1160–22 600 pg g^{-1}) in chicken nuggets, French fries and pizza were reported in composite samples from the Canadian market (Tittlemier *et al.*, 2006, 2007). Thus, the high concentrations of PFASs, which have been observed in some samples, may also be attributable to additional contamination pathways during processing and packaging (see Section 12.4).

12.4 Pathways of food contamination

Food is thought to be contaminated with PFASs primarily via two different processes, namely bioaccumulation in aquatic and terrestrial food chains and transfer of PFASs from food processing and packaging. Whereas pathways related to packaging and processing reflect current production and use of PFASs, bioaccumulation pathways will to a large extent reflect legacy emissions of PFASs. Thus the mechanisms of food contamination have important implications for predicting future exposure and deciding on actions to mitigate dietary exposure.

The physical chemical properties of PFSAs and PFCAs suggest that water is the primary environmental compartment in which these compounds reside. A relatively high water solubility (> 0.5 g/L for PFOA) (Kissa, 2001) and low volatility of the PFSAs and PFCAs in their deprotonated form, indicate that the aqueous environment acts as both a transport medium and final repository for these compounds (Armitage *et al.*, 2006). Subsequently, the exposure of aquatic organisms, either directly via water or indirectly through trophic transfer, is to be expected. Current understanding of bioaccumulation processes for organic chemicals is generally based on octanol–water equilibrium coefficients which describe the partitioning of hydrophobic compounds into the lipid fraction of organisms. In contrast to lipophilic bioaccumulative chemicals, such as polychlorinated biphenyls (PCBs) and polybrominated diphenylethers (PBDEs), it has been shown that PFSAs and PFCAs partition to serum proteins (Han et al., 2003; Jones et al., 2003; Conder et al., 2008; Bischel et al., 2010). Furthermore, the protein affinity of PFSAs and PFCAs has been found to increase with the length of the perfluoroalkyl chain (Jones et al., 2003; Bischel et al., 2010). Bioaccumulation/biomagnification factors (BAFs/BMFs) determined in the laboratory and the field are generally consistent with the proteinophilicity of PFSAs and PFCAs (Martin et al., 2003a, b; Kelly et al., 2009). However, bioaccumulation of PFSAs and PFCAs also require consideration of species and sex specific depuration kinetics (Andersen et al., 2006). Aquatic food web studies have found that biomagnification increases with chain-length and the highest BMFs have been reported for PFNA, PFUnDA and PFOS (Martin et al., 2004c; Houde et al., 2006b; Powley et al., 2008; Kelly et al., 2009). BMFs approximately ten times higher have been observed for air-breathing top predators in arctic marine mammals compared to piscivorous food webs (Houde et al., 2006b; Powley et al., 2008; Kelly et al., 2009). As PFCAs and PFSAs are relatively water soluble (increasing water solubility with decreasing chain-length) gill-breathing animals can eliminate the chemicals via respiration. Air-breathing animals, on the other hand, cannot eliminate PFCAs and PFSAs via respiration, which may explain the higher BMFs in air-breathing predators. The importance of biomagnification pathways for human dietary exposure has also been demonstrated by correlating measured concentrations of PFSAs and PFCAs in human serum with consumption of fish and offal (Falandysz et al., 2006; Weihe et al., 2008; Haug et al., 2010b; Hölzer et al., 2010; Rylander et al., 2010). The strongest associations have been found for PFOS serum concentrations and consumption of fish from locally contaminated lakes (Hölzer et al., 2010). However, significant associations between serum concentrations of PFOS, PFNA, PFDA, PFUnDA and PFDoDA and consumption of fish and marine mammals have also been observed in the general population (Falandysz et al., 2006; Weihe et al., 2008; Rylander et al., 2010; Haug et al., 2011).

For terrestrial and agricultural food chains, the understanding of bioaccumulation processes is more limited. A study by Stahl et al. (2008) demonstrated that PFOA and PFOS are taken up from soil via the roots and accumulate in the vegetative portion of plants. In line with the accumulation experiments in fish (Martin et al., 2003a), Felizeter et al. (2012) showed that PFSAs and PFCAs display an increasing bioaccumulation potential with chain-length in hydroponically grown plants. However, as sorption to organic matter also increases with chain-length (Higgins and Luthy, 2003) the uptake of long-chain PFSAs and PFCAs may become limited by the fraction of freely dissolved chemical in the pore water. Competitive sorption processes to soil may also explain the higher uptake of PFOA compared to PFOS observed by Stahl et al. (2008). The strong bioaccumulation of PFOA, PFOS and PFHxS in humans (Olsen *et al.*, 2007) and other mammals (Houde *et al.*, 2006a, 2011) indicate that bioaccumulation of PFSAs and PFCAs in animals for meat production may also be expected. Blood samples from chicken, beef cattle, goats and pigs in Japan demonstrate that PFOS, PFOA, PFNA and PFDA accumulate in food producing animals (Guruge *et al.*, 2008). Biomagnification has also been demonstrated for PFOS, PFNA, PFDA, PFUnDA and PFDoDA in an Arctic terrestrial food chain from lichen to caribou to wolf (Mueller *et al.*, 2011). Although biomagnification factors were highly tissue and compound specific, it was concluded that trophic magnification factors in the terrestrial food web were approximately two times lower than in the marine environment (Mueller *et al.*, 2011).

In addition to contamination from the environment, food can be contaminated via contact with cooking or packaging materials containing PFASs. As PFCAs are used as polymerization aids to manufacture polytetrafluoroethylene (PTFE) (Kissa, 2001), nonstick cooking materials have been investigated as a potential source of food contamination (Washburn *et al.*, 2005; Sinclair *et al.*, 2007). Some studies have also shown that residual levels of PFOAs may be present in the nonstick coating of cookware (Sinclair *et al.*, 2007). However, under normal cooking conditions and repeated use of the material this transfer pathway is likely to be insignificant (Washburn *et al.*, 2005; Sinclair *et al.*, 2007). A potentially more important pathway of food contamination is the transfer from food paper and board treated with PFASs. In contrast to PTFE treated cookware, the initial concentration of loosely bound PFASs in treated food papers and boards is typically 0.1–4%, which indicate that migration to food may be significant (Begley *et al.*, 2005, 2008; Trier *et al.*, 2011). In a study by Trier *et al.* (2011), more than 115 different PFASs were identified in commercial coating mixtures used for food packaging. Migration tests have also confirmed that PAPS can reach concentrations of 0.1–1.4 mg kg^{-1} in food simulants under normal preparation procedures (Begley *et al.*, 2008; Trier *et al.*, 2011). PFOA and PFOS have also been found in food packaging materials at 6–3490 and 13–283 ppb, respectively (Begley *et al.*, 2005; Tittlemier *et al.*, 2007). Although PFOA and PFOS are not intentionally added to food packaging papers, they may be present as impurities from the synthesis of dialkylated or polymeric PFASs (Kissa, 2001; Buck *et al.*, 2011). Despite the relatively low concentrations present in food-contact materials, migration tests have shown that PFOA can be transferred from microwave popcorn bags to food simulants at mg kg^{-1} levels (Begley *et al.*, 2005). As the phosphate ester linkages in dialkylated PAPS molecules are susceptible to hydrolytic degradation, it is also possible that the observed concentrations of PFOA correspond to hydrolysis products of PAPS (Rayne and Forest, 2010; D'eon and Mabury, 2011a).

The migration tests of food papers (Begley *et al.*, 2008; Trier *et al.*, 2011) clearly illustrates that packaging can lead to significantly increased concentrations of PFASs in food. For instance, the high concentrations of PFOA in popcorn and pizza samples from Canada (Tittlemier *et al.*, 2007) could potentially be explained by migration from packaging materials. However, due to the difficulties associated with analyzing commercial mixtures of PAPS and performing representative migration tests, it is difficult to determine the resulting dietary exposure from food packaging materials. The recent changes in production of PFASs also suggest that dietary exposure via food packaging is changing rapidly. This supposition is corroborated by the study of Tittlemier *et al.* (2006) who observed a rapid decrease of N-EtFOSAs (which are intermediate chemicals in the production of food packaging materials) in composite samples of French fries between 1992 and 2000. It may be expected that the replacement PFASs that are introduced to the market are polymeric materials based on short perfluoroalkyl chains (Buck *et al.*, 2011; Trier *et al.*, 2011).

12.5 Estimated exposure from food and other exposure media

Human exposure to PFSAs and PFCAs can occur from a large number of different exposure media including house dust, ambient air, drinking water and diet (Washburn *et al.*, 2006; Trudel *et al.*, 2008; Vestergren et al., 2008). In addition, a wide range of polyfluoroalkyl substances (PFASs) present in commercial products can be metabolized to PFCAs or PFSAs after being absorbed into the human body (Martin *et al.*, 2010; D'Eon and Mabury 2011a) which complicate the apportionment of different exposure sources. Numerous studies using different approaches have been performed in order to estimate the total human exposure and relative importance of different exposure sources.

12.5.1 Dietary exposure estimates

The dietary intake of PFASs has been estimated by different approaches, each with different advantages and shortcomings. Most commonly, dietary exposure has been estimated from analysis of composite foods (sometimes referred to as food basket samples) combined with national food intake data (Food Standards Agency, 2006; Tittlemier *et al.*, 2007; Ostertag *et al.*, 2008; Schecter *et al.*, 2011; Vestergren *et al.*, 2012b). However, some studies have used a limited number of individual food items combined with consumption data to estimate exposure (Ericson *et al.*, 2008; Ericson-Jogsten *et al.*, 2009; Haug *et al.*, 2010; Noorlander *et al.*, 2011). Composite samples have clear advantages when estimating general background dietary exposure as they are representative averages of the most frequently consumed food items (Tittlemier *et al.*, 2007; Vestergren *et al.*, 2012b). Single food items, on the other hand, may be useful to identify highly contaminated items that lead to an elevated exposure. A third approach to estimate dietary intake is the

Table 12.2 Estimated average dietary intakes of PFOS and PFOA (pg kg^{-1} day^{-1}) for adults from different countries

Country	Year	Average dietary intake (pg kg^{-1} day^{-1})		Reference
		PFOA	PFOS	
Canada	1998	100–200	200–400	Tittlemier et al., 2007
Canada	2004	800–2000	100–400	Ostertag et al., 2009
USA	2009	857		Schecter et al., 2010
Spain	2006		890–1070	Ericson et al., 2006
UK	2007–2008	< 50–10 000	1000–10 000	Clarke et al., 2009
Norway	2007–2008	270–360	640–770	Haug et al., 2010a
Netherlands	2009	200	300	Noorlander et al., 2011
Germany	2005	1100–11 600	600–4400	Fromme et al., 2007
Japan	2004	720–1280	1080–1470	Kärrman et al., 2009
Sweden	1999	320	1430	Vestergren et al., 2012b
Sweden	2005	370	800	Vestergren et al., 2012b
Sweden	2010	510	970	Vestergren et al., 2012b

duplicate diet method where all food consumed during one day is pooled and analyzed (Fromme et al., 2007; Kärrman et al., 2009). An advantage of the duplicate diet approach is that measured concentrations will reflect all different contamination pathways (Section 12.4). However, accurate analysis of duplicate diet samples may be more challenging, due to the dilution of the concentrations and heterogeneous matrices (Vestergren et al., 2012a). Furthermore, estimations of total dietary intake for the general population using the duplicate diet method may be biased by a narrow selection of volunteering individuals. The estimated dietary intakes from these studies are presented in Table 12.2.

Estimated dietary intakes of PFOA and PFOS for the populations in Europe, North America and Japan range over almost two orders of magnitude (Table 12.2). Although some of the variability can be attributed to regionally different contamination and different dietary preferences, the recent improvements in analytical sensitivity may strongly influence estimated intakes (as discussed in Sections 12.2 and 12.3). Comparable estimated intakes for PFOA and PFOS were observed in the studies applying more sensitive analytical methods (Kärrman et al., 2009; Haug et al., 2010; Noorlander et al., 2011; Vestergren 2011), which possibly indicates that the low detection frequency and assumptions about non-detects may have led to overestimations of exposure in some studies. Another possible explanation of the observed variability in dietary intake estimates is that the concentrations of PFOA and PFOS in certain food have decreased as a consequence of the phaseout actions taken by industry (US EPA, 2000, 2006). Since the studies applying more sensitive methods are typically the more recent ones, it is possible that there is a decreasing temporal trend in the concentrations of PFASs in certain food items. Some support for this explanation is provided by the decreasing concentrations of PFOSAs in fast food composite food samples in Canada after 1998 (Tittlemier *et al.*, 2006). However, a similar trend has not been observed for PFOA and PFOS in food basket studies from Canada and Sweden (Ostertag *et al.*, 2009; Vestergren *et al.*, 2012). Additional studies on temporal trends in food items are therefore needed to elucidate how food contamination pathways may have changed as a consequence of the industrial phaseout actions.

12.5.2 The relative importance of dietary and non-dietary exposure pathways

In addition to intake of contaminated food, inhalation of air, incidental ingestion of dust, intake of drinking water, or direct contact with consumer products containing PFASs contribute to the total exposure (Washburn *et al.*, 2005; Strynar and Lindstrom, 2008; Trudel *et al.*, 2008; Egeghy and Lorber, 2011; Lorber and Egeghy, 2011). It should also be noted that many PFASs including PFOSAs, PFOSEs, FTOHs and PAPS have been shown to produce PFSAs or PFCAs as persistent metabolites (Martin *et al.*, 2011; D'eon and Mabury, 2011b). As many of these 'precursor compounds' have been found in the gas phase of indoor and outdoor air, indoor dust and food-contact materials exposure and subsequent metabolism, constitute an indirect exposure to PFSAs and PFCAs. Several studies have attempted to comprehensively estimate the aggregated exposure and determine the relative importance of multiple exposure pathways for the total exposure (Fromme *et al.*, 2008; Trudel *et al.*, 2008; Vestergren *et al.*, 2008; Egeghy and Lorber 2011; Lorber and Egeghy 2011). Despite the different modeling approaches and different data sets for the concentrations in various exposure media, the results reached by these studies are relatively consistent with each other (Fromme *et al.*, 2008; Trudel *et al.*, 2008; Vestergren *et al.*, 2008; Egeghy and Lorber, 2011; Lorber and Egeghy, 2011). For the adult population, average or median aggregated exposures have been in the range of 2.3 to 15.3 ng kg^{-1} day^{-1} for PFOS and 1–3 ng kg^{-1} day^{-1} for PFOA. Dietary intake has been found to be the primary exposure pathway (65–95% of the total exposure) for both PFOA and PFOS (Fromme *et al.*, 2008; Trudel *et al.*, 2008; Vestergren *et al.*, 2008; Egeghy and Lorber, 2011; Lorber and Egeghy, 2011). The contribution from the ingestion of drinking water follows that of dietary intake with a contribution up to 22% and 24 % for PFOS and PFOA, respectively (Fromme *et al.*, 2008; Trudel *et al.*, 2008; Vestergren *et al.*, 2008; Egeghy and Lorber, 2011; Lorber and Egeghy, 2011). Ingestion of house dust appears to be far less important than dietary intake, contributing up to 6% and 9% of the total for PFOS and PFOA respectively (Fromme *et al.*, 2008; Trudel *et al.*, 2008; Vestergren *et al.*, 2008; Egeghy and Lorber, 2011; Lorber and Egeghy, 2011). Other exposure pathways, including inhalation, contact with consumer products and indirect exposure to PFOS and PFOA, via metabolism of precursor compounds, have been found to be of minor importance for the general adult population (Fromme et al., 2008; Trudel et al., 2008; Vestergren et al., 2008; Egeghy and Lorber, 2011; Lorber and Egeghy, 2011).

However, if other exposure scenarios are considered, the total exposure and relative importance of exposure pathways may be substantially different. For instance, locally elevated concentrations of PFOS and PFOA in drinking water (100–1000 times the background concentrations) have been reported from some parts of the world, including Little Hocking, Ohio (Emmet et al., 2006) and the Ruhr area in Germany (Hölzer et al., 2008). Under these circumstances the total exposure has been estimated to increase by a factor of 10–100 and the intake of drinking water would comprise > 90% of the exposure (Vestergren and Cousins, 2009; Egeghy and Lorber, 2011; Lorber and Egeghy, 2011). Given the great variability observed in concentrations of PFOS, PFOA and their respective precursor compounds in indoor air and dust (Björklund et al., 2009; Shoeib et al., 2011), subgroups of the general population may experience higher non-dietary exposure. At the 95th percentile of estimated exposures, dust ingestion is approximately double that of dietary intake for both PFOS and PFOA (Egeghy and Lorber, 2011; Lorber and Egeghy, 2011) and indirect exposure from precursor compounds via inhalation and dust ingestion makes a significant contribution to the total exposure (Vestergren et al., 2008; Egeghy and Lorber, 2011). Recently, evidence that extensive use of PFASs containing carpet protector products may lead to elevated serum concentrations of PFOS has also been shown by Beesoon et al. (2011).

The total exposure and relative importance of exposure pathways may also change drastically between different age groups of the population. For infants, the body weight normalized exposure to PFOS and PFOA from the ingestion of breast milk has been estimated to be < 2.1–33.7 ng kg^{-1} day^{-1} and < 2.7–31.5 ng kg^{-1} day^{-1}, respectively (Fromme et al., 2010; Llorca et al., 2010). Despite a transfer efficiency from serum to breast milk of only a few percent (Kärrman et al., 2010), the dietary exposure of infants is significantly higher compared to that of adults (compare with Table 12.2). Median exposures to PFOS and PFOA for 2-year old children also appear to be slightly higher than those for adults (3.8 and 1.9 ng kg^{-1} day^{-1}). The higher exposure for young children may be explained by the higher ingestion of house dust comprising 36% and 50% of the total median exposure (Egeghy and Lorber, 2011; Lorber and Egeghy, 2011). Although exposure at early life stages may be important from the perspective of adverse toxicological effects, relatively little work has been performed to characterize the dietary exposure to young children.

12.5.3 Toxicokinetic modeling and exposure reconstruction

The concentrations of PFASs, particularly PFOS and PFOA, in human serum samples from many parts of the world, are quite well characterized and have been summarized in several review articles (Houde et al., 2006a; Fromme et al., 2008; Vestergren and Cousins 2009). By applying toxicokinetic (TK) models (Harada et al., 2005; Washburn et al., 2005; Loccisano et al., 2011) that relate the concentration in blood serum to an internal exposure, the relative importance of estimated dietary exposures has been evaluated. A good agreement between reconstructed exposures using TK models and external exposures has been observed for subpopulations exposed to PFOA from highly contaminated drinking water (Vestergren et al., 2009) or to PFOS from locally contaminated fish (Hölzer et al., 2010). Several studies have also reported an agreement between estimated dietary exposure and reconstructed exposures to PFOA and PFOS for the background population (Fromme et al., 2007; Ericson et al., 2008; Kärrman et al., 2009; Vestergren and Cousins 2009; Egeghy and Lorber, 2011; Lorber and Egeghy, 2011). A recent study by Loccisano et al. (2011) estimated total exposures to PFOS and PFOA for the American population in the range 850–1200 pg kg^{-1} day^{-1} and 200–520 pg kg^{-1} day^{-1}. The general agreement between reconstructed exposures from serum concentrations and estimated dietary exposures (Table 12.2) corroborates the conclusion that dietary intake is the major ongoing exposure pathway for PFOA and PFOS in the average population.

12.5.4 Future perspectives on human exposure

Biomonitoring studies from several countries have shown that the concentrations of PFOS and PFOA have decreased since the year 2000 (Calafat et al., 2007; Olsen et al., 2008; Haug et al., 2009; Glynn et al., 2011). The decreasing concentrations of PFOS and PFOA have generally been attributed to the phaseout of POSF and related chemicals from 2000 to 2002 (Olsen et al., 2008; Vestergren and Cousins, 2009; D'eon and Mabury, 2011b). However, little is known about how the external exposure has changed over this time period. A few studies have observed slightly decreasing concentrations of PFOS after 1999 in egg and fish (Murakami et al., 2010; Vestergren, 2012b). However, the rate of decline in dietary exposure is not sufficient to explain the rapid decrease in serum concentrations of PFOS (Vestergren, 2012). One hypothesis, which would be consistent with the observed time trends in human serum, is that there has been a shift in the major exposure pathways from consumer product pathways to background contaminated food (Vestergren and Cousins, 2009). A rapid decrease in PFOSAs after 1998 in fast food composite samples from Canada provides an indication that food packaging materials may have made a significant contribution to the PFOS exposure prior to the year 2000.

Predictions of future dietary exposure will depend on the pathways of food contamination. If transfer from food packaging materials makes a major contribution of PFASs in the diet, a rapid change in exposure could be expected as the major producers phaseout long-chain PFASs (US EPA,

298 Persistent organic pollutants and toxic metals in foods

2006; Environment Canada, 2010). However, food contamination resulting from environmental emissions may have a much longer response time than changes in the production and use of PFASs. It seems likely that levels will drop in food produced in temperate/industrial regions as sources of PFSAs and PFCAs are reduced (Cousins *et al.*, 2011). However, due to the high persistence and lack of efficient sinks for PFSAs and PFCAs, it is expected that they will be present in the environment for decades to come (Armitage *et al.*, 2006). Thus, we can expect a low level exposure from food items derived from the marine environment for the foreseeable future.

12.6 Conclusions and future trends

Dietary exposure has been hypothesized to be the main explanation for the ubiquitous presence of PFCAs and PFSAs in the average European and North American population. The recent advances in analytical methods have resulted in more reliable and consistent data sets on PFSAs and PFCAs in representative food samples. Based on state-of-the-art exposure studies and TK modeling, there is substantial support for the conclusion that dietary intake is a major exposure pathway for PFOS and PFOA. However, given the uncertainties in dietary exposure estimations and TK modeling, it cannot be firmly concluded that all exposure pathways have been accounted for. It should also be noted that certain sub groups of the population may receive an elevated exposure from locally contaminated drinking water or ingestion of house dust. As high quality concentration data for long-chain PFCA homologues in a range of exposure media have so far been scarce, external exposure modeling for PFNA and higher homologues has not been attempted to date. However, the observed correlations with consumption of fish and marine mammals provide support for the hypothesis that dietary exposure is an important exposure pathway also for these compounds. A remaining caveat in understanding human exposure to PFCAs and PFSAs is the possible exposure via metabolic production of PFCAs from PAPS (D'eon *et al.*, 2007, 2009; D'eon and Mabury, 2011b), which have so far not been included in the assessment of aggregated exposure.

As more dietary exposure data become available, a comprehensive re-analysis of human exposure from multiple pathways should be undertaken in the near future. At the same time, recent advances in TK modeling of PFOA and PFOS indicate that the uncertainty in back-calculated exposures can be reduced. Thus, a combination of an updated external exposure modeling approach and a reverse dosimetry approach using physiologically-based TK modeling could be used to further explore the importance of different exposure pathways. The improvements in analysis of PFCAs and PFSAs in food, open up opportunities for an expanded analysis of spatial and temporal differences in food concentrations, for refinement of the external exposure assessment to several PFCAs and PFSAs.

The improved analytical methodologies also open up opportunities to track the origin of food contamination from bioaccumulation in plants and animals and food packaging.

12.7 Acknowledgements

The funding by the European Commission Seventh Framework Project PERFOOD (PERFluorinated Organics in Our Diet) (KBBE-227525) is gratefully acknowledged by the authors.

12.8 References

3M. 2001. Analysis of PFOS, FOSA, and PFOA from Various Food Matrices using HPLC Electrospray/Mass Spectrometry. 3M study conducted by Centre Analytical Laboratories, Inc. St. Paul, MN:3M Environmental Laboratory.

ANDERSEN, M. E.; CLEWELL, H. J.; TAN, Y.-M.; BUTENHOFF, J. L.; OLSEN, G. W. 2006. Pharmacokinetic modelling of saturable, renal resorption of perfluoroalkylacids in monkeys–probing the determinants of long plasma half-lives. *Toxicol.* **227**, 156–164.

ARMITAGE, J.; COUSINS, I. T.; BUCK, R. C.; PREVEDOUROS, K.; RUSSELL, M. H.; MACLEOD, M.; KORZENIOWSKI, S. H. 2006. Modeling global-scale fate and transport of perfluorooctanoate emitted from direct sources. *Environ. Sci. Technol.* **40**, 6969–6975.

BALLESTÉROS-GÓMEZ, A.; RUBIO, S.; VAN LEEUWEN, S. 2010. Tetrahydrofuran-water extraction, in-line clean-up and selective liquid chromatography/tandem mass spectrometry for the quantitation of perfluorinated compounds in food at the low picogram per gram level. *J. Chromatogr. A.* **1217**, 5913–5921.

BEESOON, S.; WEBSTER, G. M.; SHOEIB, M.; HARNER, T.; BENSKIN, J. P.; MARTIN, J. P. 2011. Isomer profiles of perfluorochemicals in matched maternal, cord, and house dust samples: Manufacturing sources and transplacental transfer. *Environ. Health Perspect.* **119**, 1659–1664.

BENSKIN, J.; BATAINEH, M.; MARTIN, J. W. 2007. Simultaneous characterization of perfluoroalkyl carboxylate, sulfonate, and sulfonamide isomers by liquid chromatography–tandem mass spectrometry. *Anal. Chem.* **79**, 6455–6464.

BISCHEL, H. N.; MACMANUS-SPENCER, L. A.; LUTHY, R. G. 2010. Noncovalent interactions of long-chain perfluoroalkyl acids with serum albumin. *Environ. Sci. Technol.* **44**, 5263–5269.

BEGLEY, T. H.; WHITE, K.; HONIGFORT, P.; TWAROSKI, M. L.; NECHES, R.; WALKER, R. A. 2005. Perfluorochemicals: Potential sources of and migration from food packaging. *Food. Addit. Contam.* **22**, 1023–1031.

BEGLEY, T. H.; HSU, W.; NOONAN, G.; DIACHENKO, G. 2008. Migration of fluorochemical paper additives from food-contact paper into foods and food stimulants. *Food. Addit. Contam.* **25**, 384–390.

BERGER, U.; HAUKÅS, M. 2005. Validation of a screening method based on liquid chromatography coupled to high-resolution mass spectrometry for analysis of perfluoroalkylated substances in biota. *J. Chromatogr. A.* **1081**, 210–217.

BERGER, U; GLYNN, A.; HOLMSTRÖM, K. E.; BERGLUND, M.; HALLDIN ANKARBERG, E.; TÖRNKVIST, A. 2009. Fish consumption as a source of human exposure to perfluorinated alkyl substances in Sweden – Analysis of edible fish from Lake Vättern and the Baltic Sea. *Chemosphere*, **76**, 799–804.

BOSSI, R.; STRAND, J.; SORTKJAER, O.; LARSEN, M. M. 2008. Perfluoroalkyl substances in Danish waste water treatment plants and aquatic environments. *Environ. Int.* **34**, 443–450.

BJÖRKLUND, J. A.; THURESSON, K.; DE WIT, C. A. 2009. Perfluoroalkyl compounds (PFCs) in indoor dust: Concentrations, human exposure estimates, and sources. *Environ. Sci. Technol.* **43**, 2276–2281.

BUCK, R. C.; FRANKLIN, J.; BERGER, U.; CONDER, J. M.; COUSINS, I. T.; DE VOOGT, P.; JENSEN, A. A.; KANNAN, K.; MABURY, S. A.; VAN LEEUWEN, S. P. J. 2011. Perfluoroalkyl and polyfluoroalkyl substances in the environment: Terminology, classification and origins. *Integr. Environ. Assess. Manag.* **7**, 513–541.

CALAFAT, A. M.; WONG, L.-Y.; KUKLENYIK, Z.; REIDY, J. A.; NEEDHAM, L. 2007. Polyfluoroalkyl chemicals in the U.S. population: Data from the National Health and Nutrition survey (NHANES) 2003–2004 and comparisons to NHANES 1999–2000. *Environ. Health. Perspect.* **115**, 1596–1602.

CHEN, C.; LU, Y. L.; ZHANG, X.; GENG, J.; WANG, T. Y.; SHI, Y. J.; HU, W. Y.; LI, J. 2009. A review of spatial and temporal assessment of PFOS and PFOA contamination in China. *Chem. Ecol.* **25**, 163–177.

CLARKE, D. B.; BAILEY, V. A.; ROUTLEDGE, A.; LOYD, A. S.; HIRD, S.; MORTIMER, D. N.; GEM, M. 2010. Dietary intake estimate for perfluorooctanesulphonic acid (PFOS) in UK retail foods following determination using standard addition LC-MS/MS. *Food. Addit. Contam.* **27**, 530–545.

CONDER, J.M.; HOKE, R. A.; DE WOLF, W.; RUSSELL. M. H.; BUCK. R. C. 2008. Are PFCAs bioaccumulative? A critical review and comparison with regulatory criteria and persistent lipophilic compounds. *Environ. Sci. Technol.* **42**, 995–1002.

CORLEY, J. 2003. Best practices in establishing detection and quantification limits for pesticide residues in foods. In *Handbook of Residue Analytical Methods for Agrochemicals*, Aizawa, H and Lee, P. W. (Ed.), John Wiley & Sons Inc., pp. 1–18.

COUSINS, I. T.; KONG, D.; VESTERGREN, R. 2011. Reconciling measurement and modelling studies of the sources and fate of perfluorinated carboxylates. *Environ. Chem.* **8**, 339–354.

DEL GOBBO, L.; TITTLEMIER, S. A.; DIAMOND, M.; PEPPER, K.; TAGUE, B.; YEUDALL, F.; VANDERLINDEN, L. 2008. Cooking decreases observed perfluorinated compound concentrations in fish. *J. Agric. Food Chem.* **56**, 7551–7559.

D'EON, J. C.; CROZIER, P. W.; FURDUI, V. I.; REINER, E. J.; LIBELO, E. L.; MABURY, S. A. 2009. Observation of a commercial fluorinated material, the polyfluoroalkyl phosphoric acid diesters, in human sera, wastewater treatment plant sludge, and paper fibers. *Environ. Sci. Technol.* **43**, 4589–4594.

D'EON, J. C.; MABURY, S. A. 2011a. Exploring indirect sources of human exposure to perfluoroalkyl carboxylates (PFCAs): Evaluating uptake, elimination and biotransformation of polyfluoroalkyl phosphate esters (PAPs) in the rat. *Environ. Health Perspect.* **119**, 344–350.

D'EON, J. C.; MABURY, S. A. 2011b. Is indirect exposure a significant contributor to the burden of perfluorinated acids observed in humans. *Environ. Sci. Technol.* **75**, 7974–7984.

EGEGHY, P.; LORBER, M. 2011. An assessment of the exposure of Americans to perfluorooctane sulfonate: A comparison of estimated intake with values inferred from the NHANES data. *J. Expo. Sci. Env. Epid.* **21**, 150–168.

EMMETT, E. A.; SHOFER, F. S.; ZHANG, H.; FREEMAN, D.; DESAI, C.; SHAW, L. S. 2006. Community exposure to perfluorooctanoate: Relationships between serum concentrations and exposure sources. *J. Occup. Environ. Med.* **48**, 759–770.

ENVIRONMENT CANADA. 2010. Environmental performance agreement respecting perfluorinated carboxylic acids (PFCAs) and their precursors in perfluorochemical products sold in Canada. [cited 2011 February 8]. http://ec.gc.ca/epe-epa/default. asp?lang%BCEn&n%BC

ERICSON, I.; MARTI-CID, R.; NADAL, M.; VAN BAVEL, B.; LINDSTROM, G.; DOMINGO, J. L. 2008. Human exposure to perfluorinated chemicals through the diet: Intake of perfluorinated compounds in foods from the Catalan (Spain) Market. *J. Agri. Food Chem.* **56**, 1787–1794.

ERICSON-JOGSTEN, I.; PERELLO, G.; LLEBARIA, X.; BIGAS, E.; MARTI-CID. R.; KARRMAN, A.; DOMINGO, J. L. 2009. Exposure to perfluorinated compounds in Catalonia, Spain, through consumption of various raw and cooked foodstuffs, including packaged food. *Food Chem. Toxicol.* **47**, 1577–1583.

FALANDYSZ, J.; TANIYASU, S.; GULKOWSKA, A.; YAMASHITA, N.; SCULTE-OEHLMANN, U. 2006. Is fish major source of fluorinated surfactants and repellants in humans living on the Baltic coast? *Environ. Sci. Technol.* **40**, 748–751.

FELIZETER, S. T.; DE VOOGT, P. 2011. Uptake of perfluorinated alkyl acids by hydroponically grown lettuce (*Latuca Sativa*) and tomato (*Solanum lycopersicum*). *Organohal. Comp.* **73**, 947–948.

FOOD STANDARDS AGENCY. 2006. *Fluorinated Chemicals: UK Dietary Intakes. Food Survey Information Sheet 11/06*. London:Food Standards Agency.

FROMME, H.; SCHLUMMER, M.; MÖLLER, A.; GRUBER, L.; WOLZ, G.; UNGEWISS, J.; BOHMER, S.; DEKANT, W.; MAYER, R.; LIEBL, B.; TWARDELLA, D. 2007. Exposure of an adult population to perfluorinated substances using duplicate diet portions and biomonitoring data. *Environ. Sci. Technol.* **41**, 7928–7933.

FROMME, H.; TITTLEMIER, S. A.; VOLKL, W.; WILHELM, M.; TWARDELLA, D. 2009. Perfluorinated compounds – Exposure assessment for the general population in western countries. *Int. J. Hyg. Environ. Health.* **212**, 239–270.

FROMME, H.; MOSCH, C.; MOROVITZ, M.; ALEJANDRE, A. I.; BOEHMER, S.; KIRANOGLU, M.; FABER, F.; HANNIBAL, I. GENZEL-BOROVICZÉNY, O.; KOLETZKO, B.; VÖLKEL, W. 2010. Pre- and postnatal exposure to perfluorinated compounds (PFCs). *Environ. Sci. Technol.* **44**, 7123–7129.

FURDUI, V. I.; STOCK, N. L.; ELLIS, D. A.; BUTT, C. M.; WHITTLE, D. M.; CROZIER, P. W.; REINER, E. J.; MUIR, D. C. G.; MABURY, S. A. 2007. Spatial distribution of perfluoroalkyl contaminants in lake trout from the great lakes. *Environ. Sci. Technol.* **41**, 1554–1559.

GIESY, J. P.; KANNAN, K. 2001. Global distribution of perfluorooctane sulfonate in wildlife. *Environ. Sci. Technol.* **35**, 1339–1342.

GLYNN, A.; BERGER, U.; BIGNERT, A.; ULLAH, S.; LIGNELL, S.; AUNE, M.; DARNERUD, P.O. Temporal trends in the serum of Swedish first time mothers. 2011. *Swedish Environmental Protection Agency* (NV report 215 0906, 2011).

GRANDJEAN, P.; WREFORD ANDERSEN, E.; BUDTZ-JØRGENSEN, E.; NIELSEN, F.; MØLBAK, K.; WEIHE, P.; HEILMANN, C. 2012. Serum vaccine antibody concentrations in children exposed to perfluorinated compounds. *J. Am. Med. Assoc.* **307**, 391–397.

GULKOWSKA, A.; JIANG, Q.; SO, M. K.; TANIYASU, S.; LAM, P. K.; YAMASHITA, N. 2006. Persistent perfluorinated acids in seafood collected from two cities of China. *Environ. Sci. Technol.* **40**, 3736–3741.

GURUGE, K. S.; MANAGE, P. M.; YAMANAKA, N.; MIYAZAKI, S.; TANIYASU, S.; YAMASHITA, N. 2008. Species-specific concentrations of perfluoroalkyl contaminants in farm and pet animals in Japan. *Chemosphere*, **73**, S210–S215.

HAN, X.; SNOW.; T. A.; KEMPER, R. A.; JEPSON, G. W. 2003. Binding of perfluorooctanoic acid to rat and human plasma proteins. *Chem. Res. Toxicol.* **16**, 775–781.

HANSEN, K. J.; CLEMEN, L. A.; ELLEFSON, M. E.; JOHNSON, H. O. 2001. Compound-specific, quantitative characterization of organic fluorochemicals in biological matrices. *Environ. Sci. Technol.* **35**, 766–770.

HARADA, K.; INOUE, K.; MORIKAWA, A.; YOSHINAGA, T.; SAITO, N.; KOIZUMI, A. 2005. Renal clearance of perfluorooctane sulfonate and perfluorooctanoate in humans and their species-specific excretion. *Environ. Res.* **99**, 253–261.

HAUG, L. S.; SALIHOVIC, S.; JOGSTEN, I. E.; THOMSEN, C.; VAN BAVEL, B.; LINDSTROM, G.; BECHER, G. 2010a. Levels in food and beverages and daily intake of perfluorinated compounds in Norway. *Chemosphere*, **80**, 1137–1143.

HAUG, L. S.; THOMSEN, C.; BRANTSAETER, A. L.; KVALEM, H. E. HAUGEN, M.; BECHER, G.; MELTZER, H. M.; KNUTSEN, H. K. 2010b. Diet and particularly seafood are major sources of perfluorinated compounds in humans. *Environ. Int.* **36**, 772–778.

HAUG, L. S.; THOMSEN, C.; BECHER, G. 2009. Time trends and the influence of age and gender on serum concentrations of perfluorinated compounds in archived human samples. *Environ. Sci. Technol.* **43**, 2131–2136.

HIGGINS, C. P.; LUTHY, R. G. 2006. Sorption of perfluorinated surfactants on sediment. *Environ. Sci. Technol.* **40**, 7251–7256.

HOFF, P. T.; VAN DE VIJVER, K.; VAN DONGEN, W.; ESMANS, E. L.; BLUST, R.; DE COEN, W. M. 2003. Perfluorootane sulfonic acid in bib (*Trisopterus luscus*) and plaice (*Pleuronectes platessa*) from the WesternScheldt and the Belgian North Sea: Distribution and biochemical effects. *Environ. Toxicol. Chem.* **22**, 608–614.

HÖLZER, J.; MIDASCH, O.; RAUCHFUSS, K.; KRAFT, M.; REUPERT, R.; ANGERER, J.; KLEESCHULTE, P.; MARSCHALL, N.; WILHELM, M. Biomonitoring of perfluorinated compounds in children and adults exposed to perfluorooctanoate-contaminated drinking water. *Environ Health Perspect.* 2008 May; **116**, 651–657.

HÖLZER, J.; GOEN, T.; JUST, P.; REUPERT, R.; RAUCHFUSS, K.; KRAFT, M.; MULLER, J.; WILHELM, M. 2011. Perfluorinated compounds in fish and blood of anglers at Lake Mohne, Sauerland area, Germany. *Environ. Sci. Technol.* **45**, 8046–8052.

HOUDE, M.; MARTIN, J. W.; LETCHER, R. J.; SOLOMON, K. R.; MUIR, D. C. G. 2006a. Biological monitoring of perfluoroalkyl substances: A review. *Environ. Sci. Technol.* **40**, 3463–3473.

HOUDE, M.; BUJAS, T. A. D.; SMALL, J.; WELLS, R. S.; FAIR, P. A.; BOSSART, G. D.; SOLOMON, K. R.; MUIR, D. C. G. 2006b. Biomagnification of perfluoroalkyl compounds in the bottlenose dolphin (Tursiops truncatus) food web. *Environ. Sci. Technol.* **40**, 4138–4144.

HOUDE, M.; DE SILVA, A. O.; MUIR, D.C.G.; LETCHER, R. J. 2011. Monitoring of perfluorinated compounds in aquatic biota: An updated review PFCs in aquatic biota. *Environ. Sci. Technol.* **45**, 7962–7973.

HRADKOVA, P.; PULKRABOVA, J.; KALACHOVA, K.; HLOUSKOVA, V.; TOMANIOVA, M.; POUSTKA, J.; HAJSLOVA, J. 2012. Occurrence of halogenated contaminants in fish from selected river localities and ponds in the Czech Republic. *Arch. Environ. Contam. Toxicol.* **62**, 85–96.

JAHNKE, A.; BERGER, U. 2009. Trace analysis of per- and polyfluorinated alkyl substances in various matrices-How do current methods perform? *J. Chromatogr. A.* **1216**, 410–421.

JONES, P. D.; HU, W.; DE COEN, W.; NEWSTED, J. L.; GIESY, J. P. 2003. Binding of perfluorinated fatty acids to serum proteins. *Environ. Toxicol. Chem.* **22**, 2639–2649.

KANNAN, K.; TAO, L.; SINCLAIR, E.; PASTVA, S. D.; JUDE, D. J.; GIESY, J. P. 2005. Perfluorinated compounds in aquatic organisms in the Great Lakes food chain. *Arch. Environ. Contam. Toxicol.* **48**, 559–566.

KÄRRMAN, A.; HARADA, K. H.; INOUE, K.; TAKASUGA, T.; OHI, E.; KOIZUMI, A. A. 2009. Relationship between dietary exposure and serum perfluorochemical (PFC) levels – A case study. *Environ. Int.* **35**, 712–717.

KÄRRMAN, A.; DOMINGO, J. L.; LLEBARIA, X.; NADAL, M.; BIGAS, E.; VAN BAVEL, B.; LINDSTRÖM, G. 2010. Biomonitoring perfluorinated compounds in Catalonia, Spain:

concentrations and trends in human liver and milk samples. *Environ. Sci. Poll. Res.* **17**, 750–758.

KELLY, B. C.; IKONOMOU, M. G.; BLAIR, J. D.; SURRIDGE, B.; HOOVER, D.; GRACE, R.; GOBAS, F. A. P. C. 2009. Perfluoroalkyl contaminants in an arctic marine food web: Trophic magnification and wildlife exposure. *Environ. Sci. Technol.* **43**, 4037–4043.

KISSA E. 2001. *Fluorinated Surfactants and Repellents* (2nd edition revised and expanded) (Surfactant science series 97). New York (NY): Marcel Dekker. 640 p.

LACINA, O.; HRADKOVA, P.; PULKRABOVA, J.; HAJSLOVA, J. 2011. Simple, high throughput ultra-high performance liquid chromatography/tandem mass spectrometry trace analysis of perfluorinated alkylated substances in food of animal origin: Milk and fish. *J. Chromatogr. A.* **1218**, 4312–4321.

LANGLOIS, I.; BERGER, U.; ZENCAK, Z.; OEHME, M. 2007. Mass spectral studies of perfluorooctane sulfonate derivatives separated by high-resolution gas chromatography. *Rapid. Commun. Mass. Spec.* **21**, 3547–3553.

LAU, C.; BUTENHOFF, J. L.; ROGERS, J. M. 2004. The developmental toxicology of perfluoroalkyl acids and their derivatives. *Toxicol. Appl. Pharmacol.* **198**, 231–241.

LAU, C.; ANITOLE, K.; HODES, C.; LAI, D.; PFAHLES-HUTCHENS, A.; SEED, J. 2007. Perfluoroalkyl acids: A review of monitoring and toxicological findings. *Toxicol. Sci.* **99**, 366–394.

LEE, H.; MABURY, S. A. 2011. A pilot survey of legacy and current commercial fluorinated chemicals in human sera from United States donors in 2009. *Environ. Sci. Technol.* **45**, 8067–8074.

LEMAL, D.M. 2004. Perspective on fluorocarbon chemistry. *J. Org. Chem.* **69**, 1–11.

LLORCA, M.; FARRÉ, M.; YOLANDA PICÓ, Y. TEIJÓN, M. L.; ÁLVAREZ, J. G.; BARCELÓ, D. 2010. Infant exposure of perfluorinated compounds: Levels in breast milk and commercial baby food. *Environ. Int.* **36**, 584–592.

LOCCISANO, A. E.; CAMPBELL, J. E.; ANDERSEN, M. E.; CLEWELL, H. J. 2011. Evaluation and prediction of pharmacokinetics of PFOA and PFOS in the monkey and human using a PBPK model. *Regul. Toxicol. Pharm.* **59**, 157–175.

LONGNECKER, M. P.; SMITH, C, S.; KISSLING, G. E.; HOPPIN, J. E.; BUTENHOFF, J. L.; DECKER, E.; EHRESMAN, D. J.; ELLEFSON, M. E.; FLAHERTY, J.; GARDNER, M. J.; LANGLOIS, E.; LEBLANC, A.; LINDSTROM, A. B.; REAGEN, W. K.; STRYNAR, M. J.; STUDABAKER, W. B. 2008. An interlaboratory study of perfluorinated alkyl compound levels in human plasma. *Environ. Res.* **107**, 152–159.

LOPEZ-ESPINOSA, M. J.; FLETCHER, T.; ARMSTRONG, B.; GENSER, B.; DHATARIYA, K.; MONDAL, D.; DUCATMAN, A.; LEONARDI, G. 2011. Association of perfluorooctanoic acid (PFOA) and perfluorooctane sulfonate (PFOS) with age of puberty among children living near a chemical plant. *Environ. Sci. Technol.* **45**, 8160–8166.

LORBER, M.; EGEGHY, P. 2011. Simple intake and pharmacokinetic modeling to characterize exposure of Americans to perfluorooctanoic acid, PFOA. *Environ. Sci. Technol.* **45**, 8006–8014.

LINDSTRÖM, G.; KÄRRMAN, A.; VAN BAVEL, B. 2008. Accuracy and precision in the determination of perfluorinated chemicals in human blood verified by interlaboratory comparisons. *J. Chromatogr. A.* **1216**, 394–400.

LUPTON, S. J.; HUWE, J. K.; SMITH, D. J.; DEARFIELD, K. L.; JOHNSTON, J. J. 2012. Absorption and excretion of 14C-perfluorooctanoic acid (PFOA) in Angus cattle (Bos taurus). *Environ. Sci. Technol.* **46**, 1128–1134.

LUQUE, N.; BALLESTEROS-GÓMEZ, A.; VAN LEEUWEN, S.; RUBIO, S. 2010. Analysis of perfluorinated compounds in biota by microextraction with tetrahydrofuran and liquid chromatography/ion isolation-based ion-trap mass spectrometry. *J. Chromatogr. A* **1217**, 3774–3782.

MALLET, C. R.; LU, Z.; MAZZEO, J. R. 2004. A study of ion suppression effects in electrospray ionization from mobile phase additives and solid-phase extracts. *Rapid Commun. Mass Spectrum.* **18**, 49–58.

MARTIN, J. W.; MABURY, S. A.; SOLOMON, K. R.; MUIR, D. C. G. 2003a. Bioconcentration and tissue distribution of perfluorinated acids in rainbow trout (*Oncorhynchus mykiss*). *Environ. Toxicol. Chem.* **22**, 196–204.

MARTIN, J. W.; MABURY, S. A.; SOLOMON, K. R.; MUIR, D. C. G. 2003b. Dietary accumulation of perfluorinated acids in juvenile rainbow trout (*Oncorhynchus mykiss*). *Environ. Toxicol. Chem.* **22**, 189–195.

MARTIN, J. W.; SMITHWICK, M. M; BRAUNE, B. M.; HOEKSTRA, P. F.; MUIR, D. C. G.; MABURY, D. C. G. 2004a. Identification of long-chain perfluorinated acids in biota from the Canadian arctic. *Environ. Sci. Technol.* **38**, 373–380.

MARTIN, J. W.; KANNAN, K.; BERGER, U.; DE VOOGT, P.; FIELD, J. A.; FRANKLIN, J.; GIESY, J. P.; HARNER, T.; MUIR, D. C. G.; SCOTT, B.; KAISER, M.; , J. A JARNBERG, U.; JONES, K. C.; MABURY, S. A.; SCHROEDER, H.; SIMCIK, M.; SOTTANI, C.; VAN BAVEL, B.; KARRMAN, A.; LINDSTROM, G.; VAN LEEUWEN, S. 2004b. Analytical challenges hamper perfluoroalkyl research. *Environ. Sci. Technol.* **38**, 248A–255A.

MARTIN, J. W.; WHITTLE, D. M.; MUIR, D. C. G.; MABURY, S. A. 2004c. Perfluoroalkyl contaminants in a food web from Lake Ontario. *Environ. Sci. Technol.* **38**, 5379–5385.

MARTIN, J. W.; ASHER, B. J.; BEESOON, S.; BENSKIN, J. P.; ROSS, M. S. 2010. PFOS or PreFOS? Are perfluorooctane sulfonate precursors (PreFOS) important determinants of human and environmental perfluorooctane sulfonate (PFOS) exposure? *J. Environ. Monit.* **12**, 1979–2004.

MOODY, C. A.; MARTIN, J. W.; CHIKWAN, W.; MUIR, D. C. G.; MABURY, S. A. 2003. Monitoring perfluorinated surfactants in biota and surface water samples following an accidental release of fire-fighting foam into etobicoke creek. *Environ. Sci. Technol.* **36**, 545–551.

MUELLER, C. E.; DE SILVA, A. O.; SMALL, J.; WILLIAMSON, M.; WANG, X. W. MORRIS, A.; KATZ, S.; GAMBERG, M.; MUIR, D. C. G. 2011. Biomagnification of perfluorinated compounds in a remote terrestrial food chain: Lichen-Caribou-Wolf. *Environ. Sci. Technol.* **45**, 8665–8673.

MURAKAMI, M.; ADACHI, N.; SAHA, M.; MORITA, C.; TAKADA, H. 2011. Levels, temporal trends, and tissue distribution of perfluorinated surfactants in freshwater fish from Asian countries. *Arch. Environ. Contam. Toxicol.* **61**, 631–641.

NANIA, N.; PELLEGRINI, G. N.; FABRIZI, L.; SESTA, G.; DE SANCTIS, P.; LUCCHETTI, D.; Di PASQUALE, M.; CONI, E. 2009. Monitoring of perfluorinated compounds in edible fish from the Mediterranean Sea. *Food Chem.* **115**, 951–957.

NOORLANDER, C. W.; VAN LEEUWEN, S. P. J.; DE BIESEBEK, J. D.; MENGLERS, M. J. B.; ZEILMAKER, M. J. 2011. Levels of perfluorinated compounds in food and dietary intake of PFOS and PFOA in the Netherlands. *J. Agric. Food Chem.* **59**, 7496–7505.

OLSEN, G. W.; BURRIS, J. M.; EHRESMAN, D. J.; FROEHLICH, J. W.; SEACAT, A. M.; BUTENHOFF, J. L.; ZOBEL, L. R. 2007. Half-life of serum elimination of perfluorooctanesulfonate, perfluorohexanesulfonate, and perfluorooctanoate in retired fluorochemical production workers. *Environ. Health Perspect.* **115**, 1298–1305.

OLSEN, G. W.; MAIR, D. C.; CHURCH, T. R.; ELLEFSON, M. E.; REAGEN, W. K.; BOYD, T. M.; HERRON, Z. M.; NOBLIETTI, J. B.; RIOS, J. A.; BUTENHOFF, J. L.; ZOBEL, L. R. 2008. Decline in perfluroooctanesulfonate and other perfluoroalkyl chemicals in American Red Cross adult blood donors, 2000–2006. *Environ. Sci. Technol.* **42**, 4989–4995.

OSTERTAG, S. K.; CHAN, M. H.; MOISEY, J.; DABEKA, R.; TITTLEMIER, S. A. 2009. Historic dietary exposure to perfluorooctane sulfonate, perfluorinated carboxylates, and fluorotelomer unsaturated carboxylates from the consumption of store-bought and restaurant foods for the Canadian population. *J. Agric. Food Chem.* (57) 8534–8544.

PICO, Y.; FARRÉ, M.; LLORCA, M.; BARCELÓ, D. 2011. Perfluorinated compounds in food: A global perspective. *Crit. Rev. Food Sci.* **51**, 605–625.

POWLEY, C.; GEORGE, S. W.; RYAN, T. W.; BUCK, R. C. 2005. Matrix effect-free analytical methods for determination of perfluorinated carboxylic acids in environmental matrices. *Anal. Chem.* **77**, 6353–6358.

POWLEY, C. R.; GEORGE, S. W.; RUSSELL, M. H.; HOKE, R. A.; BUCK, R. C. 2008. Polyfluorinated chemicals in a spatially and temporally integrated food web in the Western Arctic. *Chemosphere* **70**, 664–672.

QUINONES, O.; SNYDER, S. A. 2009. Occurrence of perfluoroalkyl carboxylates and sulfonates in drinking water utilities and related waters from the United States. *Environ. Sci. Technol.* **43**, 9089–9095.

RAYNE, S.; FOREST, K. 2009. Perfluoroalkyl sulfonic and carboxylic acids: A critical review of physicochemical properties, levels and patterns in waters and wastewaters, and treatment methods. *J. Environ. Sci. Health. A* **44**, 1145–1149.

RYLANDER, C.; SANDANGER, T. M.; FRØYLAND, L.; LUND, E. 2010. Dietary patterns and plasma concentrations of perfluorinated compounds in 315 Norwegian women: the NOWAC postgenome study. *Environ. Sci. Technol.* **44**, 5225–5232.

SAITO, N.; HARADA, K.; INOUE, K.; SASAKI, K.; YOSHINAGA, T.; KOIZUMI, A. 2004. Perfluorooctanoate and perfluorooctane sulfonate concentrations in surface water in Japan. *J. Occup. Health.* **46**, 49–59.

SCHECTER, A.; COLACINO, J.; HAFFNER, D.; PATEL, K.; OPEL, M.; PÄPKE, O.; BIRNBAUM, L. 2010. Perfluorinated compounds, polychlorinated biphenyls, and organochlorine pesticide contamination in composite food samples from Dallas, Texas, USA. *Environ. Health Perspect.* **118**, 796–802.

SHI, Y. L.; PAN, Y. Y.; YANG, R. Q.; WANG, Y. W.; CAI, Y. Q. 2010. Occurrence of perfluorinated compounds in fish from Qinghai-Tibetan Plateau. *Environ. Int.* **36**, 45–50.

SHOEIB, M.; HARNER, T.; WEBSTER, G. M.; LEE, S. C. 2011. Indoor sources of poly- and perfluorinated compounds (PFCS) in Vancouver, Canada: Implications for human exposure. *Environ. Sci. Technol.* **45**, 7999–8005.

SINCLAIR, E.; Kyu KIM, S.; AKINLEYE, H. B.; KANNAN, K. 2006. Quantitation of gas-phase perfluoroalkyl surfactants and fluorotelomer alcohols released from nonstick cookware and microwave popcorn bags. *Environ. Sci. Technol.* **41**, 1180–1185.

STAHL, T.; HEYN, J.; THIELE, H.; HUTHER, J.; FAILING, K.; GEORGII, S.; BRUNN, H. 2009. Carryover of perfluorooctanoic acid (PFOA) and perfluorooctane sulfonate (PFOS) from soil to plants. *Arch. Environ. Contam. Toxicol.* **57**, 289–298.

STEENLAND, K.; FLETCHER, T.; SAVITZ, D. A. 2010. Epidemiologic evidence on the health effects of perfluorooctanoic acid (PFOA). *Environ. Health Perspect.* **118**, 1100–1108.

STRYNAR, M. J.; LINDSTROM, A. B. 2008. Perfluorinated compounds in house dust from Ohio and North Carolina, USA. *Environ. Sci. Technol.* **42**, 3751–3756.

SUNDSTRÖM, M.; EHRESMAN, D. J.; BIGNERT, A.; BUTENHOFF, J. L.; OLSEN, G. W.; CHANG, S. C.; BERGMAN, Å. 2011. A temporal trend study (1972–2008) of perfluorooctanesulfonate, perfluorohexanesulfonate, and perfluorooctanoate in pooled human milk samples from Stockholm, Sweden. *Environ. Int.* **37**, 178–183.

SZOSTEK, B.; PRICKETT, K.; BUCK, R. C. 2006. Determination of fluorotelomer alcohols by liquid chromatography/tandem mass spectrometry in water. *Rapid Coummun. Mass Spec.* **20**, 2837–2844.

TANIYASU, S.; KANNAN, K.; SO, M. K.; GULKOWSKA, A.; SINCLAIR, E.; OKAZAWA, T.; YAMASHITA, N. 2005. Analysis of fluorotelomer alcohols, fluorotelomer acids, and short- and long-chain perfluorinated acids in water and biota. *J. Chromatogr. A* **1093**, 89–97.

TITTLEMIER, S. A.; PEPPER, K.; EDWARDS, L. 2006. Concentrations of perfluorooctanesulfonamides in canadian total diet study composite food samples collected between 1992 and 2004. *J. Agric. Food Chem.* **54**, 8385–8389.

TITTLEMIER, S. A.; PEPPER, K.; SEYMOUR, C.; MOISEY, J.; BRONSON, R.; CAO, X.; DABEKA, R. W. 2007. Dietary exposure of Canadians to perfluorinated carboxylates and perfluoroctane sulfonate via consumption of meat, fish and fast food items prepared in their packaging. *J. Agric. Food Chem.* **55**, 3203–3210.

TITTLEMIER, S. A.; BRAEKVELT, E. 2011. Analysis of polyfluorinated compounds in foods. *Anal. Bioanal. Chem.* **399**, 221–227.

TRIER, X.; GRANBY, K.; CHRISTIENSEN, J. H. 2011. Tools to discover anionic and non-ionic polyfluorinated alkyl surfactants by liquid chromatography electrospray ionisation mass spectrometry. *J. Chromatogr. A* **1218**, 7094–7104.

TRUDEL, D.; HOROWITZ, L.; WORMUTH, M.; SCHERINGER, M.; COUSINS, I. T.; HUNGERBUHLER, K. 2008. Estimating consumer exposure to PFOS and PFOA. *Risk Anal.* **28**, 251–269.

UNEP UN ENVIRONMENT PROGRAMME. 2009. Governments unite to step-up reduction on global DDT reliance and add nine new chemicals under international treaty. Available from: http://chm.pops.int/Convention/Pressrelease/COP4Geneva8May2009/tabid/542/language/en-US/Default.aspx

ULLAH, S.; ALSBERG, T.; BERGER, U. 2011. Simultaneous determination of perfluoroalkyl phosphonates, carboxylates, and sulfonates in drinking water. *J. Chromatogr. A* **1218**, 6388–6395.

USEPA US ENVIRONMENTAL PROTECTION AGENCY. 2000. EPA and 3M announce phase out of PFOS. Available from: http://yosemite.epa.gov/opa/admpress.nsf/0/33aa946e6cb11f35852568e1005246b4

USEPA US ENVIRONMENTAL PROTECTION AGENCY. 2006. 2010/2015 PFOA Stewardship Program. Available from: http://www.epa.gov/opptintr/pfoa/pubs/stewardship/

VAN DER VEEN, I.; WEISS, J.; VAN LEEUWEN, S. P. J.; COFINO, W.; CRUM, S. 2012. 5th Interlaboratory study on perfluoroalkyl substances (PFAS) in food and environmental samples. Report W-12/09 Institute of Environmental Studies (IVM), University of Amsterdam.

VAN LEEUWEN, S. P. J.; KÄRRMAN, A.; VAN BAVEL, B.; DE BOER, J.; LINDSTRÖM, G. 2006. Struggle for quality in determination of perfluorinated contaminants in environmental and human samples. *Environ. Sci. Technol.* **40**, 7854–7860.

VAN LEEUWEN, S. P. J.; DE BOER, J. 2007. Extraction and clean-up strategies for the analysis of poly- and perfluoroalkyl substances in environmental and human matrices. *J. Chromatogr. A* **1153**, 172–185.

VAN LEEUWEN, S. P. J.; VAN VELZEN, M. J. M.; SWART, C. P.; VAN DER VEEN, I.; TRAAG W. A. DE BOER, J. 2009. Halogenated contaminants in farmed Salmon, Trout, Tilapia, Pangasius, and Shrimp. *Environ. Sci. Technol.* **43**, 4009–4015.

VESTERGREN, R.; COUSINS, I. T.; TRUDEL, D.; WORMUTH, M.; SCHERINGER, M. 2008. Estimating the contribution of precursor compounds in consumer exposure to PFOS and PFOA. *Chemosphere* **73**, 1617–1624.

VESTERGREN, R.; COUSINS, I. T. 2009. Tracking the pathways of human exposure to perfluorinated carboxylates. *Environ. Sci. Technol.* **43**, 5565–5575.

VESTERGREN, R.; ULLAH, S.; COUSINS, I. T.; BERGER, U. 2012a. A matrix effect-free method for reliable quantification of perfluoroalkyl carboxylic acids and perfluoroalkane sulfonic acids at low parts per trillion levels in dietary samples. *J. Chromatogr. A* **1237**, 64–71.

VESTERGREN, R.; BERGER, U.; GLYNN, A.; COUSINS, I. T. 2012b. Dietary exposure to perfluoroalkyl acids for the Swedish population in 1999, 2005 and 2010. *Environ. Int.* **49**, 120–127.

WASHBURN, S. T.; BINGMAN, T. S.; BRAITHWAITE, S. K.; BUCK, R. C.; BUXTON, L. W.; CLEWELL, H. J.; HAROUN, L. A.; KESTER, J. E.; RICKARD, R. W.; SHIPP, A. M. 2005. Exposure assessment and risk characterization for perfluorooctanoate in selected consumer articles. *Environ. Sci. Technol.* **39**, 3904–3910.

WASHINGTON, J. W.; ELLINGTON, J. J.; JENKINS, T. M.; EVANS, J. J.; YOO, H.; HAFNER, S. C. 2009. Degradability of an acrylate-linked, fluorotelomer polymer in soil. *Environ. Sci. Technol.* **43**, 6617–6623.

WEIHE, P.; KATO, K.; CALAFAT, A. M.; NIELSEN, F.; WANITUNGA, A. A.; NEEDHAM, L. L.; GRANDJEAN, P. 2008. Serum concentrations of polyfluoroalkyl compounds in Faraose whale meat consumers. *Environ. Sci. Technol.* **42**, 6291–6295.

YLINEN, M.; HAHNHIJÄRVI, H.; PEURA, P.; RÄMÖ, O. 1985. Quantitative gas chromatographic determination of perfluorooctanoic acid as the benzyl ester in plasma and urine. *Arch. Environ. Contam. Toxicol.* **14**, 713–717.

13

Polycyclic aromatic hydrocarbons (PAHs) in foods

L. Duedahl-Olesen, National Food Institute – Technical University of Denmark, Denmark

DOI: 10.1533/9780857098917.2.308

Abstract: Polycyclic aromatic hydrocarbons (PAHs) are found in our environment. The major route of human exposure for nonsmokers is however via food. PAHs in food originate from environmental deposits or arise from food processing. Food processing techniques contributing to increased PAH concentrations include smoking, drying, roasting and barbecuing. Sixteen PAHs are believed to result in human health risk. European Union (EU) legislation on maximum limits for benzo[*a*]pyrene and PAH4 on selected food types applies. Various analytical techniques exist and challenges in detection of PAHs in food still apply, even after more than 60 years of research in the field.

Key words: benzo[*a*]pyrene, polycyclic aromatic hydrocarbons (PAHs), analytical techniques, human dietary exposure, EU legislation, food processing.

13.1 Introduction

Polycyclic aromatic hydrocarbons (PAHs) are composed of hydrogen and carbon atoms placed in two or more benzene rings. More than a hundred different PAH compounds exist and almost always in mixtures. The term polycyclic aromatic compounds (PACs) covers PAHs with additional components such as nitrogen, sulfur or oxygen found in for example oil, and are not considered here. PAHs are found in environmental depositions, as a result of environmental pollution (e.g. oil spills), and can arise as a result of human activity (e.g. traffic, incineration, etc.) as well as during processing of food (e.g. smoking, drying, barbecuing, etc.). The major route of human exposure to PAHs for nonsmokers is food. Traditionally, analytical methods using high-performance liquid chromatography (HPLC) with fluorescence detection (FLD), HPLC-FLD, gas chromatography flame ionization detection (GC-FID) and gas chromatography mass spectrometry (GC-MS) have been applied, with recent technical applications including dopant-assisted (DA) LC-MS developing. PAHs have been found in elevated concentrations in smoked fish and meat, dried fruit and vegetable oil.[1] Some of these foods, except dried fruits, have been included in European legislation since 2005 with maximum benzo[*a*]pyrene limits.[2] Since September 2012, maximum limits have also included a sum of four PAHs, namely benz[*a*]anthracene, benzo[*a*]pyrene, benzo[*b*]fluoranthene and chrysene.[3] Dietary exposure estimates have approximated the daily benzo[*a*]pyrene intake at 240 ng per day. Food contamination incidences in Europe include uncontrolled processing[4] contamination by material for storage or transport[5] and fraud for example the presence of mineral oil in vegetable oil.[6]

13.2 Sources and formation of PAHs in foods

The general characteristics of PAHs are high melting- and boiling-points, low vapor pressure, and very low water solubility, which tend to decrease with increasing molecular mass. PAHs are soluble in many organic solvents and are highly lipophilic. They are chemically rather inert. Reactions that are of interest with respect to their environmental fate and possible sources of loss during atmospheric sampling are photodecomposition and reactions with nitrogen oxides, nitric acid, sulfur oxides, sulfuric acid, ozone and hydroxyl radicals. Structures of four selected compounds can be seen in Fig. 13.1. The United States Environmental Protection Agency (US-EPA) list of organic priority pollutants from 1984 includes 16 PAHs, termed 16 US-EPA PAHs[90] (compounds listed in Table 13.1). In 2002, the Scientific Committee of Food (SCF) of the EU (now the European Food Safety Authority (EFSA)) concluded that 15 out of 33 evaluated compounds were found to be genotoxic or mutagenic in animal experiments.[7] These 15 compounds, with an additional compound (benzo[c]fluorene) evaluated by the Joint FAO/WHO Committee on Food Additives (JECFA), are today termed the EU 15 + 1 priority PAHs considered to have a health risk on humans. For comparison, compounds included in US-EPA 16 and EU 15 + 1 are listed in Table 13.1. Notice also the compounds included in the sum of PAH2, PAH4 and PAH8 in Table 13.1.

PAH can be found in the environment due to incomplete combustion of natural sources, such as volcanic eruptions and forest fires.[8] The main source of PAHs in the environment is, however, due to human activity. PAHs are formed by pyrolysis or incomplete combustion of organic matter, and a linear increase of the PAH concentration for heating temperatures between 400°C and 1000°C has been reported.[9] Environmental sources of PAHs from human activities include contributions from combustion of fossil fuels (e.g. energy production or traffic) or wood (e.g. stoves), municipal waste incineration and environmental tobacco smoke. Monitoring programs for PAHs exist

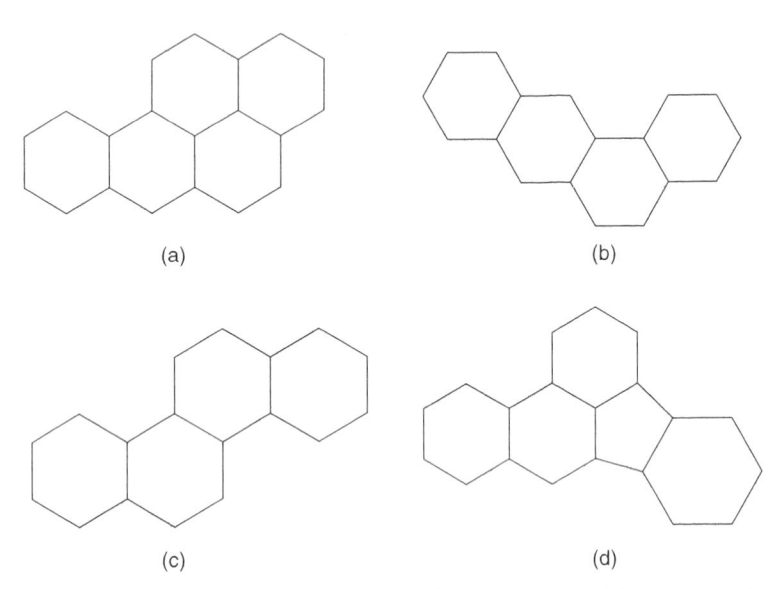

Fig. 13.1 Structure of four PAHs included in PAH4. (a) Benzo[*a*]pyrene. (b) Benz[*a*] anthracene. (c) Chrysene. (d) Benzo[*b*]fluoranthene.

in institutions that specialize in environmental studies. These include sample matrices, such as air, soil, water, sediments and fish species. Background levels of PAHs in fish from many sea areas are therefore available (e.g. International Council for the Exploration of the Sea (ICES)). Use of such data for evaluation of PAHs in food requires knowledge on collection sites, since environmental monitoring might be done at polluted areas that are of no interest for monitoring of PAHs in food.

Other more exotic sources of PAHs have been researched, for example, there are studies aiming at an indicator for environmental pollution from traffic and incineration determining the content of PAH in bees and honey.[10]

The contribution to PAH exposure from the inhalation route may increase markedly for smokers and persons exposed to passive smoking, smoke from stoves, etc. These sources have not been included here.

Food can be contaminated by PAHs present in the environment, that is, PAHs can be deposited on particles in air and thereby deposited onto crops (e.g. grain and vegetable seeds), and PAHs can accumulate on the waxy surface of vegetables and fruits. For vegetable contamination by PAHs a possible route is traffic exhaust, whereas PAH uptake from the soil is more contradictory and unclear even though environmental studies show that PAHs bind to particles in soil.[103] For cocoa seeds, drying on asphalt, on bitumen in the sun, or by the use of direct drying processes can contaminate the final product.[104]

Other products, such as seafood and fish can be exposed to PAHs, present in water and sediment. The PAH content in seafood depends on the ability of the aquatic organisms to metabolize them for example bivalve molluscs

Polycyclic aromatic hydrocarbons (PAHs) in foods 311

Table 13.1 PAH compounds with molecular formula, molecular weight (M_w) and compounds included in US-EPA 16[90] and EU 15 + 1[86]

Compound	Molecular formula	M_w	US-EPA 16	EU 15 + 1
Naphthalene	$C_{10}H_8$	128.2	X	
Acenaphthylene	$C_{12}H_8$	152.2	X	
Acenaphthene	$C_{12}H_{10}$	154.2	X	
Fluorene	$C_{13}H_{10}$	166.2	X	
Phenanthrene	$C_{14}H_{10}$	178.2	X	
Anthracene	$C_{14}H_{10}$	178.2	X	
Fluoranthene	$C_{16}H_{10}$	202.3	X	
Pyrene	$C_{16}H_{10}$	202.3	X	
Benzo[c]fluorene	$C_{17}H_{12}$	216.3		X
Cyclopenta[cd]pyrene	$C_{18}H_{10}$	226.3		X
Benz[a]anthracene[b,c]	$C_{18}H_{12}$	228.3	X	X
Chrysene[a,b,c]	$C_{18}H_{12}$	228.3	X	X
5-Methyl chrysene	$C_{19}H_{14}$	242.3		X
Benzo[b]fluoranthene[b,c]	$C_{20}H_{12}$	252.3	X	X
Benzo[j]fluoranthene	$C_{20}H_{12}$	252.3		X
Benzo[k]fluoranthene[c]	$C_{20}H_{12}$	252.3	X	X
Benzo[a]pyrene[a,b,c]	$C_{20}H_{12}$	252.3	X	X
Indeno[1,2,3-cd]pyrene[c]	$C_{22}H_{12}$	276.3	X	X
Benzo[ghi]perylene[c]	$C_{22}H_{12}$	276.3	X	X
Dibenz[a,h]anthracene[c]	$C_{22}H_{14}$	278.4	X	X
Dibenzo[a,e]pyrene	$C_{24}H_{14}$	302.4		X
Dibenzo[a,h]pyrene	$C_{24}H_{14}$	302.4		X
Dibenzo[a,i]pyrene	$C_{24}H_{14}$	302.4		X
Dibenzo[a,l]pyrene	$C_{24}H_{14}$	302.4		X

Compounds included in [a]Sum PAH2, [b]Sum PAH4 and [c]Sum PAH8.[86]

accumulate PAHs whereas vertebrate fish metabolize PAHs. The PAH content is particularly high for waters with petroleum oil contamination, illustrated by the findings for oil spill incidences such as Exxon Valdez in Alaska in 1989.[11]

For nonoccupationally exposed humans, the main source of PAHs is food, where food processing such as drying, roasting, smoking and barbecuing contributes to body burdens. These processes are important sources of PAH contamination due to a direct contact between food and combustion products. A different form of contamination may derive from the practice of storing and transporting oil seeds[12] or for example, cocoa beans in jute or sisal bags treated with mineral oils (before spinning the jute fibers),[5] with a possible migration of PAHs into the seeds[5,12] followed by an increased PAH content.

Roasting can accelerate the formation of PAHs, and the amount formed depends upon time and temperature, often with increasing concentrations, due to an increase in the parameter settings. Roasting is a crucial step for the production of coffee, due to the development of color, aroma and flavor, which are essential for the quality of coffee. Coffee roasting includes an intense thermal process, which can be applied by direct (flame-toasting, coal-grilling or gas oven-toasting) or indirect (electric oven-toasting) heating at temperatures of 220–500°C.[13] As a result of intense thermal processes, partial carbonizations can take place in coffee resulting in the formation of PAHs. Also, partial transformations from low molecular PAH (e.g. three-ring structure) to high molecular PAH (e.g. six-ring structure) as the roasting degree is increased has been suggested.[13] In addition, PAHs found in roasted duck have been reported to be due to transformation (e.g. degradation) of fatty acids, triglycerides and cholesterol within the food.[14] The PAH transfer from coffee beans to infusions is generally low (< 35%, average 5%), with a slightly lower extractability for dark-roasted coffee as compared to light-roasted coffee.[13]

Smoking of food is one of the oldest human food processing technologies. The ancient practice of hanging meat over a fire, to protect the food against animals stealing human catches, has been developed to today's smoking processes. The technology is not only used to impart the characteristic organoleptic profiles of smoked products but also for the deactivating effect of smoke (and heat) on enzymes and microorganisms.[15] Sensory active compounds, such as phenol derivates, carbonyls, organic acids and their esters, lactones, pyrazines, pyrols and furan derivates, contribute to the specific organoleptic profile of smoked products.[15]

Smoke is generated by thermal combustion of wood components at temperatures of 180–300°C (cellulose), 260–350°C (hemicellulose) and 300–500°C (lignin) with limited access to oxygen. The smoke produced at 650–700°C is richest in desirable organoleptic components.[15] The amount of PAH in smoke, formed during pyrolysis, increases linearly with the smoking temperature within the interval 400–1000°C. The temperature of smoke therefore plays an important role for PAH formation, and can be decreased by increasing the humidity of the wood.[9]

In general, smoke is a polydispersed mixture of liquid and solid components in the gaseous phase of air, carbon oxide, carbon dioxide, water vapor, methane and other gases. Smoke has a variable composition that depends on various conditions, such as procedure and temperature of smoke generation, origin and composition of wood, water content in wood, etc. The PAH composition profile of smoke depends not only on the temperature of the smoke generation, but also on the type of wood used (soft or hard), the excess of oxygen and other external factors (e.g. smoke cleaning). Direct exposure to smoke brings about higher concentrations of PAHs, as compared to indirect methods whereby the PAH content is reduced by temperature reduction that results in a partial condensation of smoke components before reaching the food product. Also, hot smoking, with product temperatures above 50°C and used for certain fish species (e.g. mackerel and herring), brings about higher concentrations of PAH than cold smoking.[16] Cold smoking techniques are used for fish species such as salmon and halibut, products where a product temperature below 35°C is desired. Uncontrolled technological conditions

Polycyclic aromatic hydrocarbons (PAHs) in foods 313

resulting in deep-brown coloring especially for household products, and uncontrolled smoking in developing countries,[4] lead to very high PAH content in smoked foods.

The highest concentration of PAHs in smoked products occurs immediately after the smoking process terminates; then it decreases, due to light decomposition and interactions with other compounds in the product, and after some time the concentration stabilizes at a certain level. The highest concentration is found in the outer layers (often the skin of e.g. fish) and has also been found to be reduced by contact with packing material.[18]

Possible sources of PAH contamination of vegetable oils are (a) contamination of plant material, primarily through the air, (b) drying of the plant material with smoke before extraction, and (c) contamination through the extraction solvent. Vegetable oils are a very heterogeneous group of foodstuffs, consisting of a variety of raw materials processed in different ways. Some of the raw materials are dried by direct exposure to burning natural gas, oil or wood, which contaminates the material with PAH. Drying of grapeseed oil using rotary, direct-fired drum driers considerably increases the PAH content, due to incomplete combustion of grape skins and stalks, contaminating the oil with gases at 650–750°C at the entrance of the rotating dryer.[19] The source of PAH in oils and fats has been shown to be due to combustion gases during drying operations.

Drying and roasting of other types of seeds have also been found to increase the PAH content of the final oil for example pumpkin seed[12] and rapeseed oil.[20] Also, combustion of pine wood during the drying stage of tea leaves has resulted in high PAH content in black tea.[21]

Lijinsky and his coworkers[22–24] first reported the formation of PAHs in charcoal-grilled foods. As mentioned with regard to the previous heating techniques, the PAH content in the barbecued products also depends upon the type of heat source (e.g. charcoal or gas) and whether the food is prepared by direct or indirect heating. In general, the higher the grilling temperature, the greater the formation of PAHs, except for products heated with electrical heaters where only small amounts of PAHs are formed. It has been suggested that PAHs are formed in smoke from the pyrolysis of melted fat dripping on the hot charcoal, and then deposited on the meat surface as the smoke rises.[24] The correlation with the concentration of fat in the grilled product, however, is not clear. Also, the type of product barbecued or grilled and the presence or absence of skin, have been found to be relevant for the final PAH content, with higher content for fatty products and products with skin.[14] For barbecuing, an important physical feature for contamination of the product is the distance between the heat source and the food product. For shorter distances, higher PAH content is commonly seen. Also, the lower PAH concentrations found in vertical, compared to horizontal, barbecues[17] confirm the importance of prevention of the dripping of fat onto the heat source.

In order to reduce the PAH content in different types of food, both the authorities and the industry have set guidelines for the use of drying, heating and smoking processes. These include guidelines set by the vegetable oil industry (FEDIOL) on testing of new heating materials for contaminants, before applying these for food production, and the Codex Alimentarius Code of practice for the reduction of contamination of food with PAHs, from smoking and direct drying processes.[25] The main guidelines includes use of (a) indirect heating in comparison to direct heating, (b) as low a temperature as possible, and (c) proper heating material (e.g. no tyres, painted wood or other unsuitable fuel).

Mitigation of PAH in oils can be done by subsequent treatment with active carbon or by steam distillation processes for example deodorization. Deodorization of vegetable oil has been found to reduce the PAH content to one-tenth of the initial concentration, whereas active carbon has been found to remove benzo[a]pyrene during the refining of oils.[26-28] Whether refining processes effectively remove all PAHs of concern is unclear. The recommendation, however, is that production and processing methods that prevent the initial contamination of oils with PAH should be used.

In some foods, such as dried fruit and food supplements, benzo[a]pyrene has been found,[1] but available data are inconclusive on how low a level can be achieved through conventional production processes. Further investigation is needed to clarify the levels in these foods.

13.3 Methods of analysis of PAHs in foods

Analysis of PAH in food has traditionally relied on tedious, time-consuming three stage procedures. These methods generally consist of an extraction step (e.g. liquid–liquid partition, caffeine complexing, or saponification) followed by one or more purification steps (e.g. column chromatography, thin-layer chromatography, solid-phase extraction) before the analytical determination of PAHs by HPLC and FLD, or by capillary gas chromatography (GC) coupled with FID or mass spectrometry (MS).

Information in the literature varies as to the question of PAH stability. Ampouled PAH solutions stored at room temperature and in the freezer (−18°C) for 2, 6 or 12 months have been reported to show unchanged concentrations[29] whereas others report that PAHs are light sensitive and that they can decompose by photo-irradiation and oxidation.[30] A general precaution is therefore to protect the solutions against light. At the same time, concentration to dryness should be avoided in order to diminish possible losses due to evaporation of the lower molecular weight compounds.

Quantification of mainly benzo[a]pyrene in food has been reported since the 1960s (e.g. References 22–24). Reviews on analytical methods have been published on PAHs in foods,[31-33] vegetable oils and fats,[34] smoked meat products,[22] and food and beverages.[35] Historically, benzo[a]pyrene has been the most commonly studied PAH compound and for many years has been used as an indicator of total PAH contamination. Fifty years of analysis have

Polycyclic aromatic hydrocarbons (PAHs) in foods   315

introduced modifications of different extraction methods and solvents, resulting in a reduction of turnover time, solvent volumes and types.

Challenges in analysis of PAH in food matrices refer to the challenges with co-extraction of fat from the matrix and the determination of nonpolar compounds with low charge. For extraction of PAH from foods in for example vegetable oils or meat, three procedures in particular have traditionally been used. These are liquid–liquid extraction (LLE), often with dimethylformamide (DMF) or dimethyl sulfoxide (DMSO), caffeine complexation by addition of a caffeine-formic acid solution, and saponification with alcoholic potassium hydroxide (KOH). All extraction principles involve removal of interfering compounds and are based on principles such as partition coefficients, complexation or degradation of for example fat.

The continued investigation into extraction methods that are faster than the techniques mentioned above–more efficient, and generate less unwanted solvent waste–has led to the introduction of new methods, some with extraction efficiency comparable to the above techniques. Use of extraction techniques, cleanup and detection of PAHs in food with examples of references can be found in Table 13.2. Supercritical fluid extraction (SFE), extensively used for environmental samples, has been successfully used for foods, too. Application of pressurized liquid extraction (PLE) enhancing extraction efficiency by operating automatically at elevated pressure and temperature, using small volumes of traditional organic solvent, has also been reported to be successful for food matrices. The use of microwave-assisted extraction (MAE) has also been proven to offer a fast and safe alternative to traditional saponification. Also, a multi-residue method, previously applied for pesticide analysis (quick, easy, cheap, effective, rugged and safe (QuEChERS)) has recently been developed for the analysis of PAHs in fish (Table 13.2).

Extracts obtained with one of the extraction methods above commonly contain substantial amounts of material other than PAHs, which may interfere with the following analytical determination. Depending on the degree of purification needed, different cleanup procedures, such as column chromatography on different adsorbent materials (e.g. silica, Florisil™) or size separation (e.g. Sephadex™, Biobeds S-X3™), are traditionally applied. Preparation and use of these columns adds time to sample preparation, and the possibility of using SPE cartridges, instead of the packed columns in the purification step, has proven successful in combination with different extraction techniques with various matrices.[37,61,68]

Detection methods for PAHs were historically based on paper chromatography or thin-layer chromatography (TLC), followed by ultraviolet (UV) or fluorescence spectroscopy (FL). These techniques have been partly replaced by high resolution GC on capillary columns and HPLC.

For GC techniques, the use of FID and its non-selectivity has been more or less replaced by the use of MS in selected ion monitoring (SIM) mode. GC-MS analysis of PAH is challenged by the low volatility of the high molecular weight PAHs (i.e. dibenzopyrenes). Challenges are also faced in terms of separation

Table 13.2 Examples of extraction techniques for analysis of PAHs with food type, type of cleanup, detection method and number of PAHs (No. PAH) detected, including at least one reference in the last column (See text for abbreviations)

Food	Extraction	Cleanup	Detection	No. PAH	References
Oils	LLE	Adsorption chromatography/TLC	GC-FID	28	36
Oils, fats		Adsorption chromatography/SPE	GC-MS	16 US-EPA to 42	26, 30, 32, 37, 38
Food supplements, edible oil		Adsorption chromatography DACC	HPLC-UV and FLD	15 + 1 EU	39, 40
Vegetables, fruit juice, vegetable oils smoked flavours, coffee, cereals, meat, fish tea, food supplements		Adsorption chromatography/SPE	HPLC-FLD	8–12	41–47
Food	Caffeine complexation	–	GC-FID	11–12	34, 48, 49
Oils, fats		Adsorption chromatography	HPLC-FLD	5–16	41, 50, 51
Meat, poultry, fish, oil and fat	Saponification	LLE/GPC	GC-FID	12–19	52–54
Meat, sausages		LLE/SPE	GC-MS	16 US-EPA	30
Fats, oils			HPLC-UV	11	34
Fish		GPC	HPLC-FLD	16 US-EPA	55
Fish		LLE/SPE	HRGC-MS	15+1 EU	105
Bivalve mollusc/meat	SFE	Adsorption chromatography	GC-MS	8	56–58
Bread		–	HPLC-FLD	11	59
Fish, meat, vegetable oils	PLE	GPC/SPE	GC-MS	15+1EU to 23	12, 60–64
Vegetable oil	MAE	LLE/adsorption chromatography	GC-MS	19	12
Cookies		LLE/SPE	HPLC-FLD	15 EU	65
Fish	QuEChERS	QuEChERS	HPLC-FLD	16 US-EPA	66
Fish		QuEChERS	GC-MS	33	67

– Not relevant or not stated.

of groups of PAHs on capillary columns, such as 'triphenylene–chrysene', 'benzo[*b*]fluoranthene–benzo[*k*]fluoranthene–benzo[*j*]fluoranthene' and 'cyclopenta[*cd*]pyrene–benz[*a*]anthracene–chrysene'. Different columns have been applied to overcome these challenges.[69] Injection techniques used are often splitless, and over the last decade programmed temperature vaporization (PTV) injection has also been applied, with improved detection limits for dibenzopyrenes. Development of appropriate stationary phases for PAH separation and use of more accurate mass detection, for example, of time of flight (TOF) MS detection, are still options for improvements of GC-MS detection of PAHs.

HPLC is somewhat faster than GC (25–40 min compared to 1–2 h on GC) and, even though it offers lower resolution efficiency in separating low molecular weight PAHs, it can be used in combination with a diode array detector (DAD). As a lot of PAHs exhibit strong fluorescence, HPLC coupled to FLD represents a powerful technique providing both sensitivity and selectivity. This technique does not suffer from separating problems caused by the low volatility of PAHs, but from the low fluorescence activity of a few target compounds. The HPLC technique also offers the advantage of allowing detection of high molecular mass PAHs, which cannot be detected by GC methods due to thermal decomposition. Only nonfluorescent compounds, such as cyclopenta[*cd*]pyrene, which also lacks selectivity by ultraviolet (UV) readers, compromises the detection of PAHs. Results obtained with HPLC-UV/FLD, however, need a recovery correction for accurate quantification as isotopic dilution (used for GC analysis) cannot be used.

New techniques include purification by preparative HPLC based on donor–acceptor complex chromatography (DACC). This cleanup is based on the specific interactions between the electron of the PAH and the electron of the stationary phase, which results in a very selective step. The methods using DACC have been developed for analysis of PAH in oil and fat by the US-EPA, and two approaches can be found in the literature. The first one couples the cleanup DACC column directly to the analytical HPLC column equipped with a FLD detector.[45,46] Automatic loading of the sample on the DACC column, elution of matrix components with the following back-flush elution of the retained PAH from the DACC column, directly onto the analytical HPLC column was applied. The second approach used the DACC cleanup step off-line and performed the analysis of the concentrated extract by HPLC-FLD in a second step.[47,70]

HPLC with mass spectrometric detection is a promising option, except that PAHs are very stable molecules and poorly ionizable with conventional ionization techniques such as electrospray ionization (ESI) or atmospheric pressure chemical ionization (APCI). Recent developments on atmospheric pressure photo ionization (APPI), as a novel ionization technique for LC-MS,[71] permits the analysis of some nonpolar and low molecular mass compounds that cannot be ionized by ESI. Compared to ESI, APPI also allows the application of higher mobile phase flow rates of up to 2 mL/min and therefore a shortening of analysis time. Different authors have studied the use of APPI for different compounds[71–73] and have looked at vacuum UV (VUV) lamp efficiencies and solvent options. Compounds having lower ionization energies (IE) than the energy of the photons emitted from the VUV lamp are ionized. Conventional mobile phase constituents, such as acetonitrile or methanol, remain un-ionized due to their much higher ionization energy. To further enhance the ionization of the analytes, a dopant can be used in combination with the APPI (dopant-assisted APPI, DA-APPI). The dopant is usually infused post-column into the eluent of the HPLC column. Toluene and anisole have previously been found to result in the highest peak areas for some lower molecular weight PAHs and to be the most suitable dopants for the analysis of PAHs. In 2012, Ahmed and coworkers[73] suggested that the increased sensitivity by the use of the dopant toluene could be explained by toluene acting as a source of protons, and that breakage of C–H bonds in the toluene molecule is important for the overall proton reaction of PAH. The mechanism can best be explained as a combination of two reactions, an electron transfer followed by a hydrogen transfer.[74]

Another interesting ionization technique is atmospheric pressure laser ionization (APLI). APLI uses the step-wise two-photon ionization, in contrast to the one-step VUV ionization process in APPI. The two-photon ionization strongly enhances the selectivity of the ionization process, and the high photon flux during an ionization event drastically increases the sensitivity for nonpolar compounds over that of APPI.[75]

Matrix-assisted laser desorption ionization (MALDI), combined with the time of flight (TOF) MS, holds potential for a novel and possibly quantitative analysis for low-polarity compounds using graphene as a matrix instead of conventional matrix.[76]

Potential in the analysis of food matrices for PAH might be found in hyphenated techniques such as on-line LC-GC-MS techniques, as reported by Vreuls and coworkers in 1991.[77] Also GC-GC techniques used in the PAH profiling of cigarette smoke, mineral oils and human blood samples might hold potential for food analysis.

For food analysis, the option to use LC-MS techniques is advantageous compared to GC-MS, due to the lower requirements of purification steps before analysis. Purification can be done on HPLC columns simultaneously to MS analysis by LC-MS, in contrast to GC-MS, using capillary columns. For control purposes of PAH, EU regulations include requirements for analytical methods.[78,79] It seems to be possible to meet these criteria with both techniques, at least for the PAH4 selected for food control. For low molecular PAHs, for example for identification of the possible source of contamination, GC-MS still seems to be the better technique, whereas higher molecular PAHs are more easily separated and detected by LC-MS.

In general, future development in instrumental techniques should be followed, especially for developing techniques with more efficient extraction and purification steps that will minimize the analysis time. For a thorough understanding of the different analysis techniques, please refer to books on relevant analytical topics.

13.4 Human dietary exposure to PAHs from foods

Estimates on human dietary exposure always depend on a range of factors. For all dietary exposure estimates, large uncertainties arise from the collection of human dietary data, for example, basket study, duplicate diets or questionnaires. For PAH dietary exposure, background information on the samples (e.g. are they collected as suspicion samples or representative samples for the market) and preparation methods in households (e.g. electric oven, fried or grilled before consumption) are crucial factors in the calculation of a realistic human dietary exposure.

Human dietary exposure studies in the 1980s, for example, on total diets in the United Kingdom (UK), indicated that both cereals and edible oils including fats each contributed one-third of the total dietary exposure to PAH. Fruits, sugars and vegetables provided much of the remainder, with smoked fish and meat making negligible contributions due to their low consumption.[80] A follow-up study showed that margarine was the major dietary source of PAHs in the oils and fats group. Also, in cereal-derived processed products, PAH contamination could frequently be determined to stem from the oil ingredients.[80] The total dietary intake of PAHs was estimated to be 3.70 µg/person per day with benzo[a]pyrene contributing some 250 ng/person per day.[80] Many European studies have confirmed that cereals are one of the largest contributors to dietary PAH intake.

In a Dutch total diet study[81] the mean daily intake of total PAHs for 18-year old males was between 5 and 17 µg/person/day. The dietary intake of the carcinogenic PAH fraction was roughly half these amounts, with the largest contributions coming from sugar and sweets, cereals, oils, fats and nuts.

In an American total diet study from 2001,[82] based on a food frequency questionnaire (FFQ) for 228 subjects and benzo[a]pyrene analysis of 200 food items, the highest mean contribution of benzo[a]pyrene was bread and cereals with 29%, followed by barbecued meat (21%) and vegetables (11–13%). The smallest contribution to benzo[a]pyrene dietary intake was the food groups of fats, sweets and dairy products. The distribution of benzo[a]pyrene dietary intake indicated that the mean intake in this study falls within the group of 31% of the subjects with a daily dietary intake in the range of 40–60 ng/person/day.

In a Catalonian study based on data from 2000,[83] the mean dietary intake from cereals was reduced to 23% with a mean benzo[a]pyrene dietary intake of 128 ng/person/day for male adults and a total of the sum of 16 US-EPA PAH of 8418 ng/person/day. The sum of carcinogenic PAH (benz[a]anthracene, benzo[a]pyrene, benzo[b]fluoranthene, benzo[k]fluoranthene, chrysene, dibenz[a,h]anthracene and indeno[1,2,3-cd]pyrene) was determined, and a follow-on study in 2008[84] illustrated that even though the total dietary intake of PAHs increased for all age groups, the sum of carcinogenic PAHs and benzo[a]pyrene decreased. In 2008, the benzo[a]pyrene dietary intake was reduced to 89 ng/person/day for male adults, whereas the sum of 16 US-EPA PAH increased to 12 045 ng/person/day.[84]

The JECFA dietary intake estimate, for 13 carcinogenic and genotoxic PAHs considered,[85] included foods that were 'ready to eat' and therefore included concentrations of PAHs arising from the cooking of food. The mean dietary intakes of benzo[a]pyrene ranged from 1.4 to 420 ng/day, and JECFA selected the value of 240 ng/day (4 ng/kg b.w./day) as being representative of a mean intake for use in the risk characterization,[85] with an upper range of consumption being 2–2.5 times higher, that is, 8–10 ng/kg b.w./day. As had many others, JECFA also identified the major contributors to dietary intakes of PAH to be cereals/cereal products and vegetable fats/oils.[85]

Based on a collection of approximately 10 000 PAH analyses from 17 European countries with data for the 15 SCF PAH compounds, and dietary intake data from EFSA, the intake of benzo[a]pyrene and PAH2 (benzo[a]pyrene and chrysene), PAH4 (PAH2, benz[a]anthracene and benzo[b]fluoranthene) and PAH8 (PAH4, benzo[k]fluoranthene, benzo[ghi]perylene,

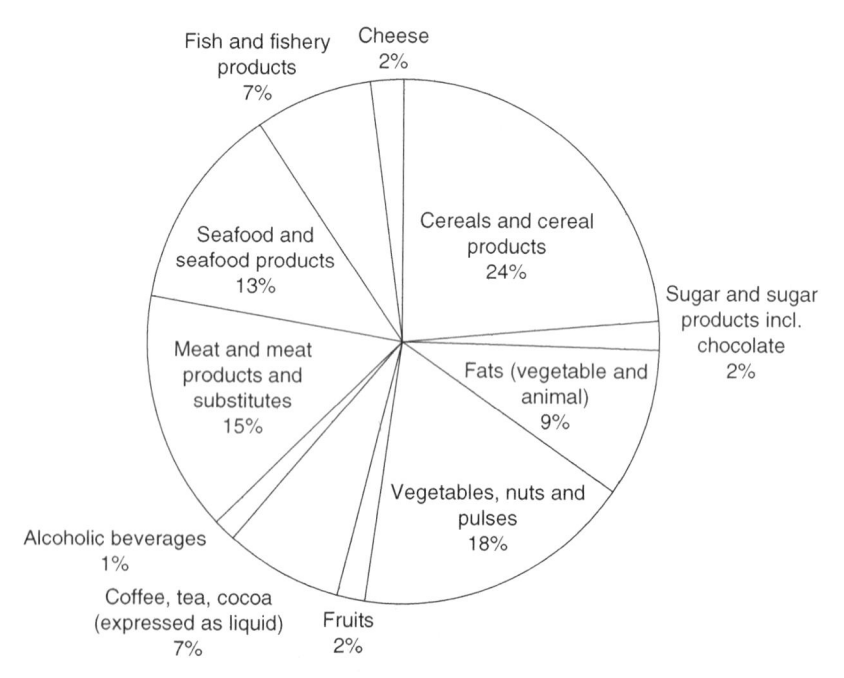

Fig. 13.2 European consumer dietary exposure to benzo[a]pyrene based on data from Table 19 in EFSA, 2008.[86]

dibenz[*a,h*]anthracene and indeno[1,2,3-*cd*]pyrene) was calculated[86] (see also Table 13.1). The mean dietary intake value for benzo[*a*]pyrene was calculated as 235 ng/person/day,[86] approximately the same as that calculated by JECFA.[85] However, JECFA's assumption for high consumption of benzo[*a*]pyrene (480–600 ng/person/day) is up to twice the value calculated by EFSA (389 ng/person/day). For PAH4 the mean dietary exposure was calculated as 1168 ng/person/day.[86] Data on benzo[*a*]pyrene and PAH4 dietary exposure for consumers only are given for food groups as pie illustrations in Figs 13.2 and 13.3, respectively. It is important to note that, because the contributions are estimated for consumers only, the calculated percentage gives an overall dietary exposure, due to the fact that the same person will not necessarily be a consumer of all the food groups.[86] For benzo[*a*]pyrene, the highest contributor to the dietary exposure was cereals and cereal products (24%) and for PAH4, seafood and seafood products (19%).

Dietary exposure to PAH4 (Fig. 13.3) was mainly found to be similar to dietary exposure to benzo[*a*]pyrene (Fig. 13.2). However, differences, such as those mentioned above, between cereal and seafood contributions to exposure, were detected. The reason for the higher values for PAH4 in seafood

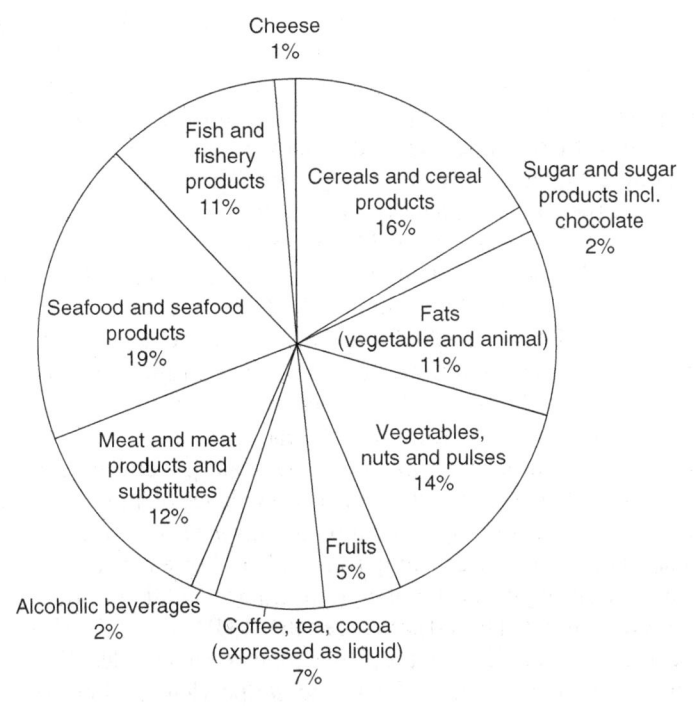

Fig. 13.3 European consumer dietary exposure to PAH4 based on data from Table 19 in EFSA, 2008.[86]

was the increased concentrations of, especially, chrysene and benzo[*b*]fluoranthene in fresh molluscs and crustaceans.[86]

Some food samples, for example, food supplements, spices and dried fruits, were found to contain high levels of PAHs.[86] However, due to the low consumption of these foods, high levels did not influence the final dietary exposure. In comparison, high concentrations of PAHs found in canned sprats in oil contributed significantly to the dietary exposure from fish and fish products. Other types of food, for example, vegetables and cereals, had low concentrations of PAHs (often benzo[*a*]pyrene concentrations below 0.3 μg/kg).[86] The high consumption of these types of products, however, makes them an important source of the overall dietary exposure.

Smoke flavoring products are produced from smoke condensates and are therefore another significant source of PAHs in food. Smoke flavoring products are being applied widely in innumerable variations of taste and odour in solid and liquid states. This type of products belongs to additives and special legislation applies to these products (see Section 13.7).

13.5 Risk assessment of PAHs

The first correlation between soot and carcinogenic effects was made in England by Percivall Pott in 1775,[87] correlating soot exposure during childhood to scrotal cancer in chimney sweeps. Other skin cancers and cancer in internal organs of chimney sweeps were correlated to soot exposure by Butlin in 1892.[88] Since then, numerous studies have been carried out on the toxicological effects of, especially, benzo[*a*]pyrene. These studies resulted in the conclusion by the International Agency for Research on Cancer (IARC) in 1987 that benzo[*a*]pyrene is a probable carcinogen.[89]

In 1998, PAHs were evaluated by the International Programme on Chemical Safety (IPCS) Environmental Health Criteria in their document on selected PAH.[90] In 2002 the SCF in the EU evaluated 33 PAHs, and 15, namely, benz[*a*]anthracene, benzo[*b*]fluoranthene, benzo[*j*]fluoranthene and benzo[*k*]fluoranthene, benzo[*ghi*]perylene, benzo[*a*]pyrene, chrysene, cyclopenta[*c,d*]pyrene, dibenz[*a,h*]anthracene, dibenzo[*a,e*]pyrene, dibenzo[*a,h*]pyrene, dibenzo[*a,i*]pyrene, dibenzo[*a,l*]pyrene, indeno[1,2,3-*cd*]pyrene and 5-methylchrysene, showed clear evidence of mutagenicity/genotoxicity in somatic cells in experimental animals *in vivo*.[7] With the exception of benzo[*ghi*]perylene, these compounds also showed clear carcinogenic effects in various types of bioassays in experimental animals. In 2005, JECFA re-evaluated PAHs and concluded that 13 PAHs are clearly genotoxic and carcinogenic.[85] JECFA at the same time recommended the inclusion of benzo[*c*]fluorene and, except for benzo[*ghi*]perylene and cyclopenta[*c,d*]pyrene, the compounds were the same as those stated by SCF. Both SCF and the JECFA concluded that a toxic equivalency factor (TEF) approach to the assessment of PAHs was not appropriate, due to limitations in the available data and because of different modes of action among different PAHs. Both evaluations also concluded that benzo[*a*]pyrene could be used as a marker for PAHs in evaluating the risks to human health of PAHs in food, and both identified a study of the carcinogenicity of coal tar mixtures and benzo[*a*]pyrene in mice[91] as critical for the risk evaluation.[7,85]

These 16 PAH compounds are therefore regarded as potentially genotoxic and carcinogenic to humans and, after collection of almost 10 000 samples for PAH analysis in 33 food categories in EU Member States, the EFSA made an evaluation of the use of benzo[*a*]pyrene as a marker of occurrence of the carcinogenic PAHs in food.[86] In about 30% of the samples other compounds were detected, even though the samples were found not to contain benzo[*a*]pyrene. In order to study the data set of different food categories for sums that would better reflect the occurrence of carcinogenic and genotoxic PAHs, individual compounds were grouped into PAH8, PAH4 and PAH2.[86] The selection of the individual PAHs was based on the frequency of their results above the limit of detection (LOD). The sum of the eight PAHs studied by Culp and coworkers,[91] were termed PAH8.[86] Also PAH4 and PAH2 were evaluated (see also Table 13.1 for compounds included in the sums). The EFSA thereby introduced the TEF value of 1 for all compounds included in the sums. They found similar correlations for all sums and concluded that PAH4 was the more appropriate sum to consider, and this is now included in the EU regulations.[3]

Both JECFA and EFSA used a Margin of Exposure (MOE) approach. Due to lack of data, JECFA based their study on dietary exposure to the benzo[*a*]pyrene and the benzo[*a*]pyrene content of coal tar mixtures to reach conclusions on a level of concern for human health.[85]

The estimation of MOEs based on the benchmark dose lower confidence limit (BMDL) for a 10% increase in the number of tumor-bearing animals, compared to control animals ($BMDL_{10}$), was used by EFSA, using the lowest $BMDL_{10}$ value among the accepted values. EFSA used an MOE approach based on dietary exposure for average and high level consumers of benzo[*a*]pyrene, PAH2, PAH4 and PAH8, respectively and their corresponding $BMDL_{10}$ values derived from the two coal tar mixtures used in the carcinogenicity studies of Culp and coworkers.[91] The $BMDL_{10}$ of 0.07 mg/kg b.w./day was chosen for benzo[*a*]pyrene and the $BMDL_{10}$ of 0.34 mg/kg b.w./day for PAH4.[86] MOEs were calculated by dividing the $BMDL_{10}$ values with the mean and high level estimates of dietary exposure to benzo[*a*]pyrene and PAH4. For the mean dietary exposures MOEs of 17 900 for benzo[*a*]pyrene and 17 500 for PAH4 were calculated.[86] For the 97.5th percentile estimates of dietary exposure, MOE values obtained were 10 800 for benzo[*a*]pyrene and 9900 for PAH4.[86] These MOEs indicate a low concern for consumer health at the mean estimated dietary exposure. However, for high level consumers, the MOEs are close to or less than 10 000, which as proposed by EFSA Scientific Committee[92] indicates a potential concern for consumer health.

No home food preparation was included and since for example barbecuing increases the PAH concentration, it is still relevant to keep the intake of PAH from food as low as reasonably achievable (ALARA), as recommended by SCF.[7]

13.6 Food scandals

Food scandals within the field of PAH have been few, and no acute effects have so far been seen. Virgin olive oil has from time to time been found to contain high concentrations of benzo[*a*]pyrene.[93] Since virgin oil is produced without heat treatment, this raised concern among producers as well as consumers. Often the problem has been found to be direct drying of olive oil with wet olive grape material or other direct firing, causing increased formation of PAHs in the processed oil. The industry now has codes of practice, including tests for PAH content in new types of heating material, before it can be used for drying.

Previously, oil-treated sisal and jute bags were reported to contaminate, for example, cocoa beans during transport.[5] The European rapid alert system has reported high benzo[*a*]pyrene concentrations in smoked sprats in oil, as well as findings of mineral oil in sunflower oil products. In general, food contamination with PAHs are either due to uncontrolled processes[4] or fraud for example mineral oil addition to sunflower oils[6] or dilution of example virgin oils not supposed to contain PAHs from heating or drying processes.[93]

13.7 Legislation of PAHs in foods within the EU

In general, EU legislation on contaminants in food are set to ensure that food placed on the market is safe to eat and does not contain contaminants at levels that could damage human health.[94] Maximum levels are set for contaminants of greatest concern to EU consumers, either due to their toxicity or their potential prevalence in the food chain, including for PAHs.

The SCF concluded, in 2002, that 15 out of 33 risk-assessed PAHs may be regarded as potentially genotoxic and carcinogenic (except for one of the 15) to humans.[7] At the same time, SCF recommended that levels of PAHs in foods should be reduced to an ALARA level. SCF also stated that until further data, not available at that time, estimating the relative proportions of the 15 PAH in foods had been evaluated, benzo[*a*]pyrene could be used as a marker for the occurrence and effect of carcinogenic PAHs in food. Combined with the collection of data in EU Member States,[1] maximum limits were set for benzo[*a*]pyrene for certain foods in 2005.² In 2006 these limits were included in Commission Regulation No. 1881/2006, setting maximum levels for certain contaminants in foodstuffs.⁹⁵ These maximum limits were also a harmonization of accepted levels for benzo[*a*]pyrene within the EU, where certain Member States had their own maximum limits for PAH in certain types of food.⁹⁶ In order to protect public health, maximum limits of benzo[*a*]pyrene were given for certain smoked, dried and environmentally polluted foods, with focus on products for infants.⁹⁵

Following a Commission Recommendation of 4 February 2005 on the further investigation of PAH levels in certain foods,⁹⁷ a total of 9714 samples, divided into 33 food categories, were evaluated by the EFSA Panel on Contaminants in the Food Chain (CONTAM Panel). The CONTAM Panel concluded that benzo[*a*]pyrene was not a suitable indicator for the occurrence of PAHs in food, and concluded that the sum of four PAHs (PAH4), namely benzo[*a*]pyrene, chrysene, benz[*a*]anthracene and benzo[*b*]fluoranthene, were more suitable indicators of PAHs in food.⁸⁶ Besides maximum levels for benzo[*a*]pyrene, maximum levels for the sum of PAH4 were therefore included in the amendments to Commission Regulation 1881/2006 given in Commission Regulation 835/2011 shown in Table 13.3.³ At the same time additional food categories, that is, cocoa beans, coconut oil, heat treated meat and meat products, were included.³

The European Commission set criteria for sampling methods and methods of analysis for the official control of the levels of benzo[*a*]pyrene in food stuffs in 2005.⁹⁸ Provisions laid down in Regulation No. 333 in 2007 included i.e. minimum values for repeatability, recovery, LOD and quantification for benzo[*a*]pyrene,⁹⁹ whereas amendments in 2011 added provisions for PAH4 analysis.⁷⁹

As an alternative to traditional smoking, liquid smoke flavorings can be added to a range of different foods to give a 'smoked' flavor. These includes foods which are not traditionally smoked, such as soups, sauces or confectionery. Smoke flavorings are regulated separately from other flavorings, and Commission Regulation (EC) No. 2065/2003 includes maximum limits of 10 µg/kg for benzo[*a*]pyrene and 20 µg/kg for benz[*a*]anthracene in the smoke condensate.¹⁰⁰ EFSA guidelines for the evaluation of smoke flavorings primary products were given in 2004, and in 2009 EFSA completed the review of the safety of 11 smoke flavorings used in the EU, with only two products passing the evaluation, with the conclusion of not being a food safety concern.¹⁰¹ Additional new information on two further smoke flavorings were finally evaluated in 2011, adding one more to the list of no-safety-concern. Based on EFSA's work, the European Commission will establish a list of permitted primary products for smoke flavorings.

Quality criteria for analytical methods for PAH analysis of primary smoke products include minimum values of repeatability, recovery, LOD, limit of quantification and analytical ranges for 15 PAH compounds.¹⁰²

Table 13.3 Maximum levels of benzo[a]pyrene and the maximum levels of the sum of PAH4 (benzo[a]pyrene, benz[a]anthracene, chrysene, benzo[b]fluoranthene) in food, specified in Commission Regulation No. 1881/2006[95] with latest amendments to Annex 6 from Commission Regulation No. 835/2011[3]

Product	Benzo[a]pyrene (µg/kg)	Sum of PAH4 (µg/kg)
Oils and fats (excluding cocoa butter and coconut oil) intended for direct human consumption or use as an ingredient in food	2.0	10.0
Cocoa beans and derived products	5.0 µg/kg fat as from 1.4.2013	35.0 µg/kg fat as from 1.4.2013 until 31.3.2015 30.0 µg/kg fat as from 1.4.2015
Coconut oil intended for direct human consumption or use as an ingredient in food	2.0	20.0
Smoked meat and smoked meat products	5.0 until 31.8.2014 2.0 as from 1.9.2014	30.0 until 31.8.2014 12.0 as from 1.9.2014
Muscle meat of smoked fish and smoked fishery products	5.0 until 31.8.2014 2.0 as from 1.9.2014	30.0 until 31.8.2014 12.0 as from 1.9.2014
Smoked sprats and canned smoked sprats; bivalve molluscs (fresh, chilled or frozen); heat treated meat and heat treated meat products sold to the final consumer	5.0	30.0
Bivalve molluscs (smoked)	6.0	35.0
Processed cereal-based foods and baby foods for infants and young children	1.0	1.0
Infant formulae and follow-on formulae, including infant milk and follow-on milk	1.0	1.0
Dietary foods for special medical purposes intended specifically for infants	1.0	1.0

13.8 References

1. EUROPEAN COMMISSION (2004). Report on experts participating in Task 3.2.12. Collection of occurrence data on polycyclic aromatic hydrocarbons in food. Available: http://ec.europa.eu/food/food/chemicalsafety/contaminants/scoop_3-2-12_final_report_pah_en.pdf. Accessed 18 August 2011.

2. European Commission (2005). Commission Regulation No. 208/2005/EC of 4 February 2005 amending Regulation (EC) No. 4667/2001 as regards polycyclic aromatic hydrocarbons. *OJEU*. **L34**: 3–5.
3. European Commission (2011). Commission Regulation (EU) No.835/2011 of 19 August 2011 amending Regulation (EC) No. 1881/2006 as regards maximum levels for polycyclic aromatic hydrocarbons in foodstuffs. *OJEU*. **L215**: 4–8.
4. AFOLABI, O.A., ADESULU, E.A. and OKE, O.L. (1983). Polynuclear aromatic hydrocarbons in some Nigerian preserved freshwater fish species. *J. Agric. Food Chem.* **31**(5): 1083–1090.
5. GROB, K., ARTHO, A., BIEDERMANN, M. and MIKLE, H. (1993). Contamination of hazelnuts and chocolate by mineral-oil from jute and sisal bags. *Z. Lebensm. Unters Forsch A*. **197**: 370–374.
6. BIEDERMANN, M. and GROB, K. (2009). How 'white' was the mineral oil in the contaminated Ukrainian sunflower oils? *Eur. J. Lipid Sci. Technol*. **111**: 313–319.
7. SCF (2002). Opinion of the Scientific Committee on Food on the risks to human health of Polycyclic Aromatic Hydrocarbons in food expressed on 4 December 2002. SCF/CS/CNTM/PAH/29 Final. Available at: http://ec.europa.eu/food/food/chemicalsafety/contaminants/out153_en.pdf. Accessed 18 August 2011.
8. BJØRSETH, A. and RAMDAHL, T. (1985). Source and emissions of PAH, in BJØRSETH and RAMDAHL (Eds.), *Handbook of Polycyclic Aromatic Hydrocarbons-emission Sources and Recent Progress in Analytical Chemistry*. Marcel Dekker, New York.
9. TÓTH, L. and BLAAS, W. (1972). Einfluß der Räuchertechnologie auf den Gehalt von geräucherten Fleischwaren an cancerogenen Kohlenwasserstoffen. *Fleischwirtschaft*. **11**: 1419–1422.
10. LAMBERT, O., VEYRAND, B., DURAND, S., MARCHAND, P., LE BIZEC, B., PIROUX, M., PUYO, S., THORIN, C., DELBAC, F. and POULIQUEN, H. (2012). Polycyclic aromatic hydrocarbons: Bees, honey and pollen as sentinels for environmental chemical contaminants. *Chemosphere*. **86**: 98–104.
11. VARANASI, U., COLLIER, T. K., KRONE, C. A., KRAHN, M. M., JOHNSON, L. L., MYERS, M. S. and CHAN, S. L. (1995). Assessment of oil spill impacts on fishery resources: Measurement of hydrocarbons and their metabolites, and their effects, in important species; Exxon Valdez Oil Spill State/Federal Natural Resource Damage Assessment Final Report (Subtidal Study Number 7); National Marine Fisheries Service, NOAA: Seattle, WA, 311 pp; NTIS No. PB96-194741.
12. GFRERERE, M. and LANKMAYR, E. (2003). Microwave-assisted saponification for the determination of 16 polycyclic aromatic hydrocarbons from pumpkin seed oils. *J. Sep. Sci.* **26**(14): 1230–1236.
13. MAIER, H. G. (1991). The carcinogen content of coffee beans. *Deutsche Lebensmittel-Rundschau*. **87**(3): 69–75.
14. CHEN, B. H. and LIN, Y. S. (1997). Formation of polycyclic aromatic hydrocarbons during processing of duck meat. *J. Agric. Food Chem.* **45**(4): 1394–1403.
15. MAGA, J. A. (1987). The flavor chemistry of wood smoke. *Food. Rev. Int.* **3**: 139–183.
16. DUEDAHL-OLESEN, L., CHRISTENSEN. J. H., HØJGÅRD, A., GRANBY, K. and TIMM-HEINRICH, M. (2010). Influence of smoking parameters on the concentration of polycyclic aromatic hydrocarbons (PAHs) in Danish smoked fish. *Food. Add. Contam. Part A*, **27**(9): 1294–1305.
17. SAINT-AUBERT, B., COOPER, J. F., ASTRE, C., SPILIOTIS, J. and JOYEUX, H. (1992). Evaluation of the induction of polycyclic aromatic hydrocarbons (PAH) by cooking on two geometrically different types of barbecue. *J. Food Comp. Anal.* **5**: 257–263.
18. GUILLÉN, M. D., SOPELANA, P. and PARTEARROYO, M. A. (2000). Polycyclic aromatic hydrocarbons in liquid smoke flavorings obtained from different types of wood.

Effect of storage in polyethylene flasks on their concentrations. *J. Agric. Food Chem.* **48**(10): 5083–5087.
19. MORET, S., DUDINE, A. and CONTE, L. S. (2000). Processing effects on the polyaromatic hydrocarbon content of grapeseed oil. *JAOCS*. **77**: 1289–1292.
20. TYS, J., RYBACKI, J. and MALCZYK, P. (2003). Sources for contamination of rapeseed with benzo(a)pyrene. *Int. Agrophys.* **17**: 131–135.
21. LIN, D. and ZHU, L. (2004). Polycyclic aromatic hydrocarbons: Pollution and source analysis of a black tea. *J. Agric. Food Chem.* **52**: 8268–8271.
22. LIJINSKY, W. and SHUBIK, P. (1964). Benzo(a)pyrene and other polynuclear hydrocarbons in charcoal-broiled meat. *Science.* **145**(3627): 53–55.
23. LIJINSKY, W. and SHUBIK, P. (1965). Polynuclear hydrocarbon carcinogens in cooked meat and smoked food. *Ind. Med. Surg.* **34**: 152.
24. LIJINSKY, W. and ROSS, A. E. (1967). Production of carcinogenic polynuclear hydrocarbons in the cooking of food. *Food Cosmet. Toxicol.* **5**: 343–347.
25. Codex Alimentarius. (2009). Code of Practice for the Reduction of Contamination of Food with Polycyclic Aromatic Hydrocarbons (PAH) from Smoking and Direct Drying Processes. CAC/RCP 68/2009.
26. HOPIA, A., PYVSALO, H. and WICKSTRØM, K. (1986). Margarines, butter and vegetable oils as sources of polycyclic aromatic hydrocarbons. *JAOCS*. **63**(7): 889–893.
27. CEJPEK, K., HAJŠLOVÁ, J., KOCOUREK, V., TOMANIOVÁ, M. and CMOLIK, J. (1998). Changes in PAH levels during production of rapeseed oil. *Food Add. Contamn.* **15**(5): 563–574.
28. LARSSON, B. K., ERIKSSON, A. T. and CERVENKA, M. (1987). Polycyclic aromatic hydrocarbons in crude and deodorized vegetable oil. *JAOCS*. **64**(3): 365–370.
29. VAESSEN, H. A. M. G., KAMP, C. G. AND JEKELL, A. A. (1988). Preparation and stability of ampouled polycyclic aromatic hydrocarbon solutions. *Z. Lebensmit. Unters Forsch.* **186**(4): 308–310.
30. MOTTIER, P., PARISOD, V. and TURESKY, R. J. (2000). Quantitative determination of polycyclic aromatic hydrocarbons in barbecued meat sausages by gas chromatography coupled to mass spectrometry. *J. Agric. Food Chem.* **48**(4): 1160–1166.
31. HOWARD, J. W. and FAZIO, T. (1980). Analytical methodology and reported findings of polycyclic aromatic hydrocarbons in foods. *JAOAC*. **63**(5): 1077–1104.
32. GUILLEN, M. D. (1994). Polycyclic aromatic compounds: Extraction and determination in food. *Food Addit. Contam.* **11**(6): 669–684.
33. HAMPIKYAN, H. and COLAK, H. (2010). Investigation of polycyclic aromatic hydrocarbons in foods. *Asian J. Chem.* **22**(8): 5797–5807.
34. MORET, S. and CONTE, L. S. (2000). Polycyclic aromatic hydrocarbons in edible fats and oils: Occurrence and analytical methods. *J. Chromatogr. A* **882**(1–2): 245–253.
35. PLAZA-BOLAÑOS, P., FRENICH, A. G. and VIDAL, J. L. M. (2010). Polycyclic aromatic hydrocarbons in food and beverages. Analytical methods and trends. *J. Chromatogr. A*, **1217**: 6303–6326.
36. MENICHINI, E., BOCCA, A., MERLI, F., IANNI, D. and MONFREDINI, F. (1991). Polycyclic aromatic hydrocarbons in olive oils on the Italian market. *Food Addit. Contam.* **8**(3): 363–369.
37. SWEETMAN, T. (1999). Contamination of coconut oil by PAH. *INF.* **10**(7): 706–712.
38. FROMBERG, A., HØJGÅRD, A. and DUEDAHL-OLESEN, L. (2007). Analysis of polycyclic aromatic hydrocarbons in vegetable oils combining gel permeation chromatography with solid-phase extraction clean-up. *Food Addit. Contam.* **24**(7): 758–767.
39. DANYI, S., BROSEA, F. O., BRASSEURA, C., SCHNEIDERB, Y.-J., LARONDELLEB, Y., PUSSEMIERC, L., ROBBENSD, J., DE SAEGERE, S., MAGHUIN-ROGISTER, G. and SCIPPOET

M.-L. (2009). Analysis of EU priority polycyclic aromatic hydrocarbons in food supplements using high performance liquid chromatography coupled to an ultraviolet, diode array or fluorescence detector. *Analytica Chimica Acta*. **633**: 293–299.
40. WINDAL, I., BOXUS, L. and HANOT, V. (2008). Validation of the analysis of the 15 + 1 European-priority polycyclic aromatic hydrocarbons by donnor–acceptor complex chromatography and high-performance liquid chromatography–ultraviolet/fluorescence detection. *J. Chromatogr. A*. **1212**: 16–22.
41. WELLING, P. and KANNDORP, B. (1986). Determination of polycyclic aromatic hydrocarbons (PAH) in edible vegetable oils by liquid chromatography and programmed fluorescence detection Comparison of caffeine complexation and XAD-2 chromatography sample clean-up. *Z. Lebensm. Unters.Forsch.* **183**(2): 111–115.
42. STIJVE, T. and HIRSCHENHUBER, C. (1987). Simplified determination of benzo[a]pyrene and other polycyclic aromatic hydrocarbons in various food materials by HPLC and TLC. *Dtsch. Lebensm. Rundsch.* **83**(9): 276–282.
43. ZHAO, X., LINGYUAN, F., JIA, H., JIANWANG, L., HUILI, W., CHANQJIANG, H. and XUEDONQ.W. (2009). Analysis of PAHs in water and fruit juice samples by DLLME combined with LC-Fluorescence Detection RID F-2644-2010. *Chromatographia* [0009–5893] Zhao, X yr:2009 vol:**69** iss:11–12 pg:1385.
44. ZHAO, X., XIUJUAN, L., ZHIXU, Z., CHANQJIANG, H., MINQHUA, Z., HUILI, W. and XUEDONQ.W. (2009). Homogeneous liquid-liquid extraction combined with high performance liquid chromatography-fluorescence detection for determination of polycyclic aromatic hydrocarbons in vegetables RID F-2644-2010. *J. Sep. Sci.* **32**(12): 2051–2057.
45. VAN STIJN, F., KERKHOFF, M. A. T. and VANDEGINSTE, B. G. M. (1996). Determination of polycyclic aromatic hydrocarbons in edible oil and fats by on-line donor-acceptor complex chromatography and high-performance liquid chromatography with fluorescence detection. *J. Chromatogr. A*. **750**: 263–273.
46. VAN DER WIELEN, J. C. A., JANSEN, J. T. A., MARTENA, M. J., DE GROOT, H. N. and IN'T VELD, P. H. (2006). Determination of the level of benzo[a]pyrene in fatty foods and food supplements. *Food Addit. Contam. A*, **23**(7): 709–714.
47. BARRANCO, A., ALONSO-SALCES, R., CRESPO, I., BERRUETA, L. A., GALLO, B., VINCENTE, F. and SAROBE M. (2004). Polycyclic aromatic hydrocarbon content in commercial Spanish fatty foods. *J. Food Prot.* **67**(12): 2786–2791.
48. SAGREDOS, A. N. (1979). A method for rapid determination of polycyclic aromatic hydrocarbons in fats and oils via coffee in complexes. *Deutsche Lebensmittel-Rundschau.* **75**(11): 350–352.
49. SAGREDOS, A. N., SINHAROY, D. and THOMAS, A. (1988). On the occurrence, determination and composition of polycyclic aromatic-hydrocarbons in crude oils and fats. *Fett Wissenschaft Technologie.* **90**(2): 76–81.
50. LAWRENCE, J. F. and DAS, B. S. (1986). Determination of nanogram/kilogram levels of polycyclic aromatic hydrocarbons in foods by HPLC with fluorescence detection. *Int. J. Environ. Anal. Chem.* **24**(2): 113–131.
51. HEDDEGHEM, A., HUYGHEBAERT, A. and DE MOOR, H. (1980). Determination of polycyclic aromatic hydrocarbons in fat products by high pressure liquid chromatography. *Z. Lenensm. Unters. Forsch.* **171**: 9–13.
52. GRIMMER, G. and BÖHNKE, H. (1975). Polycyclic aromatic hydrocarbon profile analysis of high-protein foods, oils, and fats by gas chromatography. *JAOAC.* **58**(4): 725–733.
53. WINKLER, E., BUCHELE, A. and MÜLLER, O. (1977). Method for the determination of polycyclic aromatic hydrocarbons in maize by capillary-column gas-liquid chromatography. *J. Chromatogr.* **138**: 151–164.

54. SPEER, K., HORSTMANN, P., STEEG, E., KÜHN, T. and MONTAG, A. (1990). PAH analysis in vegetable samples. *Z. Lebensmit. Unters. Forsch.* **191**: 442–448.
55. JANSKA, M., TOMANIOVA, M., HAJSLOVA, J. and KOCOUREK, V. (2006). Optimization of the procedure for the determination of polycyclic aromatic hydrocarbons and their derivatives in fish tissue: Estimation of measurements uncertainty. *Food Addit. Contamin.* **23**(3): 309–325.
56. TURRIO-BALDASSARRI, L., BAYARRI, S., DOMENICO, A., DI., IACOVELLA, N. and ROCCA, C. LA. (1999). Supercritical fluid extraction of bivalve samples for simultaneous GC-MS determination of polychlorobiphenyls and polycyclic aromatic hydrocarbons. *Int. J. Environ. Anal. Chem.* **75**(1/2): 217–227.
57. ALI, M. Y. and COLE, R. B. (1998). SFE-plus-C_{18} lipid cleanup method for selective extraction and GC/MS quantitation of polycyclic aromatic hydrocarbons in biological tissues. *Anal. Chem.* **70**: 3242–3248.
58. ALI, M. Y. and COLE, R. B. (2001). SFE-plus-C18 lipid cleanup and selective extraction method for GC/MS quantitation of polycyclic aromatic hydrocarbons in smoked meat. *J. Agric. Food Chem.* **49**(9): 4192–4198.
59. KAYALI-SAYADI, M. N., RUBIO-BARROSO, S., CUESTA-JIMENZ, M. P. and POLO-DIEZ, L. M. (2000). A new method for the determination of selected PAHs in coffee brew samples by HPLC with fluorimetric detection and solid-phase extraction. *J. Liquid Chromatogr. Relat. Technol.* **22**(4): 615–627.
60. WANG, G., LEE, A. S., LEWIS, M., KAMATH, B. and ARCHER, R. K. (1999). Accelerated solvent extraction and gas chromatography/mass spectrometry for determination of polycyclic aromatic hydrocarbons in smoked food samples. *J. Agric. Food Chem.* **47**(3):1062–1066.
61. JIRA, W. (2004). A GC/MS method for the determination of carcinogenic polycyclic aromatic hydrocarbons (PAH) in smoked meat products and liquid smokes. *Eur. Food Res. Technol.* **218**: 208–212.
62. JIRA, W., ZIEGENHALS, K. and SPEER, K. (2008). Gas chromatography-mass spectrometry (GC-MS) method for the determination of 16 European priority polycyclic aromatic hydrocarbons in smoked meat products and edible oils. *Food Addit. Contam. A.* **25**(6): 704–713.
63. VEYRAND, B., BROSSEAUD, B., SARCHER, L., VARLET, V., MONTEAU, F., MARCHAND, P. ANDRE, F. and LE BIZEC, B. (2007). Innovative method for determination of 19 polycyclic aromatic hydrocarbons in food and oil samples using gas chromatography coupled to tandem mass spectrometry based on an isotope dilution approach. *J. Chromatogr. A.* **1149**: 333–344.
64. LUND, M., DUEDAHL-OLESEN, L. and CHRISTENSEN, J. H. (2009). Extraction of polycyclic aromatic hydrocarbons from smoked fish using pressurized liquid extraction with integrated fat removal. *Talanta.* **79**: 10–15.
65. HERNANDEZ-POVEDA, G. F., MORALES-RUBIO, A., PASTOR-GARCIA, A. and DE LA GUARDIA, M. (2008). Extraction of polycyclic aromatic hydrocarbons from cookies: A comparative study of ultrasound and microwave-assisted procedures. *Food Addit. Contam. A.* **25**(3): 356–363.
66. RAMALHOSA, M. J., PAIGA, P., MORAIS, S., DELERUE-MATOS, C. and OLIVERIRA, M. B. P. P. (2009). Analysis of polycyclic aromatic hydrocarbons in fish: Evaluation of a quick, easy, cheap, effective, rugged, and safe extraction method. *J. Sep. Sci.* **32**(20): 3529–3538.
67. FORSBERG, N. D., WILSON, G. R. and ANDERSON, K. A. (2011). Determination of parent and substituted polycyclic aromatic hydrocarbons in high-fat salmon using a modified QuEChERS extraction, dispersive SPE and GC–MS. *J. Agric. Food Chem.* **59**(15): 8108–8116.
68. MORET, S., CONTE, L. and DEAN, D. (1999). Assessment of polycyclic aromatic hydrocarbon content of smoked fish by means of a fast HPLC/HPLC method. *J. Agric. Food Chem.* **47**(4): 1367–1371.

69. GÓMEZ-RUIZ, J. A. and WENZL, T. (2009). Evaluation of gas chromatography columns for the analysis of the 15 + 1 EU-priority polycyclic aromatic hydrocarbons (PAHs). *Anal. Bioanal. Chem.* **393**: 1697–1707.
70. PERRIN, J. L., POIROT, N., LISKA, P., THIENPONT, A. and FELIX, G. (1993). Trace enrichment and HPLC analysis of PAHs in edible oils and fat products, using liquid-chromatography on electron-acceptor stationary phases in connection with reverse phase and fluorescence detection. *Fett Wissen. Techn. Fat Sci. Technol.* **95**(2): 46–51.
71. ROBB, D., COVEY, T. R. and BRUINS, A. P. (2000). Atmospheric pressure photoionization: An ionization method for liquid chromatography–mass spectrometry. *Anal. Chem.* **72**(15): 3653–3659.
72. SHORT, L. C., CAI, S-S. and SYAGE, J. A. (2007). APPI-MS: Effects of mobile phases and VUV lamps on the detection of PAH compounds. *J. Am. Soc. Mass Spectrom.* **18**: 589–599.
73. AHMED A., CHOI, C. H., CHOI, M. C. and KIM, S. (2012). Mechanisms behind the generation of protonated ions for polyaromatic hydrocarbons by atmospheric pressure photoionization. *Anal. Chem.* **84**: 1146–1151.
74. SYAGER, J. A. (2004). Mechanism of $[M+H]^+$ formation in photoionization mass spectrometry. *J. Am. Soc. Mass Spectrom.* **15**: 1521–1533.
75. BROCKMANN, K. J., BENTER, T., LORENZ, M., SUAREZ, A. L. M., GAEB, S., SCHMITZ, O., SCHIEWEK, R. and MOENNIKES, R. (2008). Analytical innovations: Development and applications of atmospheric pressure laser ionization (APLI). In *Simulation and assessment of chemical processes in a multiphase environment*. NATO Science for Peace and Security Series C – Environmental Security. Barnes, I., Kharytonov, M. M. (Eds.) pp. 219–230.
76. ZHANG, J., DONG, X., CHENG, J., LI, J. and WANG, Y. (2011). Efficient analysis of non-polar environmental contaminants by MALDI-TOF MS with graphene as matrix. *J. Am. Soc. Mass Spectrom.* **22**: 1294–1298.
77. VREULS, J. J., JONG, G. J. and BRINKMAN, U. A. T. (1991). On-line coupling of liquid chromatography, capillary gas chromatography and mass spectrometry for the determination and identification of polycyclic aromatic hydrocarbons in vegetable oils. *Chromatographia*. **31**(3–4): 113–118.
78. European Commission (2007). Commission Regulation No 333/2007/EC of 28 March 2007 laying down the methods of sampling and analysis for the official control of the levels of lead, cadmium, mercury, inorganic tin, 3-MCPD and benzo(a)pyrene in foodstuffs. *OJEU.* **L88**: 29–38.
79. European Commission (2011). Commission Regulation (EU) No. 836/2011 of 19 August 2011 amending Regulation (EC) No. 333/2007 laying down the methods of sampling and analysis for the official control of the levels of lead, cadmium, mercury, inorganic tin, 3-MCPD, and benzo(a)pyrene in foodstuffs. *OJEU.* **L215**: 9–16.
80. DENNIS, M. J., MASSEY, R. C., MCWEENY, D. J., KNOWLES, M. E. and WATSON, D. (1983). Analysis of polycyclic aromatic hydrocarbons in UK total diets. *Fd. Chem. Toxic.* **21**(5): 569–574.
81. DE VOS, R. H., VAN DOKKUM, W., SCHOUTEN, A. and DE JONG-BERKHOUT, P. (1990). Polycyclic aromatic hydrocarbons in dutch total diet samples (1984–1986). *Fd. Chem. Toxic.* **28**(4): 263–268.
82. KAZEROUNI, N., SINHA, R., HSU, C-H., GREENBERG, A. and ROTHMAN, N. (2001). Analysis of 200 food items for benzo[a]pyrene and estimation of its intake in an epidemiologic study. *Fd. Chem. Toxic.* **39**: 423–436.
83. FALCO, G., DOMINGO, J. L., LLOBET, J. M., TEIXIDO, A., CASAS, C. and MÜLLER, L. (2003). Polycyclic aromatic hydrocarbons in foods: Human exposure through the diet in Catalonia, Spain. *J. Food. Prot.* **66**(12): 2325–2331.

84. MARTI-CID, R., LLOBET, J. M., CASTELL, V. and DOMINGO, J. L. (2008) Evolution of the dietary exposure to polycyclic aromatic hydrocarbons in Catalonia, Spain. *Food Chem. Toxic.* **46**: 3163–3171.
85. FAO/WHO (2006). Safety evaluation of certain contaminants in food. Prepared by the Sixty-fourth meeting of the Joint FAO/WHO Expert Committee on Food Additives (JECFA). WIIO Food Additives Series: 55; FAO Food and Nutrition Paper 82; World Health Organization WHO, 2006; Food and Agriculture Organization of the United Nations FAO,2006. http://whqlibdoc.who.int/publications/2006/9241660554_eng.pdf
86. European Food Safety Authority (2008). Polycyclic aromatic hydrocarbons in food. Scientific opinion of the panel on contaminants in the food chain. *EFSA J.* **724**: 1–114.
87. POTT, P. (1775). Chirurgical Observations Relative to the Cataract, Polypus of the Nose, the Cancer of the Scrotum, the Different Kinds of Ruptures and Mortification of the Toes and Feet. London, L., HAWES, W., CLARKE, and R., COLLINS (Eds.) Reprinted in Natl Cancer Inst Monogr **10**:7–13 (1963).
88. BUTLIN, H. T. (1892). Three lectures on cancer of the scrotum in chimney-sweeps and others. *Br. Med. J.* **2**: 1.
89. IARC (1987). Polynuclear aromatic compounds, part 1: Chemical, environmental and experimental data in IARC Monographs on the Evaluation of the Carcinogenic Risk of Chemicals to Humans, Vol. **32**, Suppl. 7. International Agency for Research on Cancer: Lyon, France.
90. WHO/IPCS (1998). Selected non-heterocyclic polycyclic aromatic hydrocarbons. Environmental Health Criteria. 202. International Programme on Chemical Safety, World Health Organization, Geneva.
91. CULP, S. J., GAYLOR, D. W., SHELDON, W. G., GOLDSTEIN, L. S. and BELAND, F. A. (1998). A comparison of the tumours induced by coal tar and benzo[a]pyrene in a 2-year bioassay. *Carcinogenesis.* **19**: 117–124.
92. EFSA (2005). Opinion of the Scientific Committee on a request from EFSA related to a harmonized approach for risk assessment of substances which are both genotoxic and carcinogenic. *EFSA J.* **282**: 1–31. http://www.efsa.europa.eu/EFSA/Scientific_Opinion/sc_op_ej282_gentox_en3,0.pdf
93. MORET, S., PIANI, B., BORTOLOMEAZZI, R. and CONTEL, R. S. (1997). HPLC determination of polycyclic aromatic hydrocarbons in olive oils. *Z Lebensm Unters Forsch A*. **205**: 116–120.
94. European Council (1993). Council Regulation No. 315/1993/EEC of 8 February 1993 laying down Community procedures for contaminants in food. *OJEU.* **L.37**: 1–3.
95. European Commission (2006). Commission Regulation No 1881/2006/EC of 19 December 2006 setting maximum levels for certain contaminants in foodstuffs. *OJEU.* **L364**: 5–24.
96. WENZEL, T., SIMON, R., KLEINER, J. and ANKLAM, E. (2006). Analytical methods for polycyclic aromatic hydrocarbons (PAHs) in food and the environment needed for new food legislation in the European Union. *Trends Anal. Chem.* **25**: 716–725.
97. European Commission (2005). Commission Recommendation of 4 February 2005 on further investigation into the levels of polycyclic aromatic hydrocarbons in certain foods. *OJEU.* **L34**: 43–45.
98. European Commission (2005). Commission Directive 2005/10/EC of 4 February 2005 laying down the sampling methods and methods of analysis for the official control of the levels of benzo(a)pyrene in foodstuffs. *OJEU.* **L34**: 15.
99. European Commission (2007). Commission Regulation No 333/2007/EC of 28 March 2007 laying down the methods of sampling and analysis for the official control of the levels of lead, cadmium, mercury, inorganic tin, 3-MCPD and benzo(a)pyrene in foodstuffs. *OJEU.* **L88**: 29–38.

100. European Council (2003). Council Regulation No. 2065/2003/EC of the European Parliament and of the council of 10 November 2003 on smoke flavourings used or intended for use in or on foods. *OJEU.* **L309**: 1–8.
101. EFSA (2009). List of opinions on smoking flavours. Available online: www.efsa.europa.eu/en. Accessed 17 August 2011.
102. European Commission (2006). Commission Regulation No 627/2006/EC of 21 April 2006 implementing Regulation (EC) No 2065/2003 of the European Parliament and of the Council as regards quality criteria for validated analytical methods for sampling, identification and characterization of primary smoke products. *OJEU.* **L109**: 3–6.
103. WILCKE, W. (2000). Polycyclic aromatic hydrocarbons (PAHs) in soil: A review. *J. Plant Nutr. Soil Sci.* **163**: 229–248.
104. ZIEGENHALS, K., SPEER, K. and JIRA, W. (2009). Polycyclic aromatic hydrocarbons (PAH) in chocolate on the German market. *J. Verbr. Lebensm.* **4**: 128–135.
105. WRETLING, S., ERIKSSON, A., ESKHULT, G. A. and LARSSON, B. (2010). Polycyclic aromatic hydrocarbons (PAHs) in Swedish smoked meat and fish. *J. Food Compos. Anal.* **23**: 264–272.

14
Phthalates in foods

T. Cirillo and R. Amodio Cocchieri,
University of Naples Federico II, Italy

DOI: 10.1533/9780857098917.2.334

Abstract: Phthalates are chemicals known mainly as plasticizers in the polymer industry, with worldwide extensive industrial application, that have caused widespread environmental pollution. This chapter describes the main sources of human exposure to phthalates; human health effects and the approach and methods to exposure assessment; the sources of food contamination during agricultural production, processing and packaging; data on occurrence in food and on dietary intake; the analytical methods applied in detecting phthalates in foods; and the implications for the food industry and policy makers for prevention and control of contamination and the future trends.

Key words: phthalates, phthalate acid esters (PAEs) food, toxicity, endocrine disruptors, dietary intake.

14.1 Introduction

Phthalates (phthalic acid esters (PAEs)) are the esters of 1, 2-benzenedicarboxylic acid. PAEs with shorter branching alkyl chain, such as dimethylphthalate (DMP), diethylphthalate (DEP) and dibutylphthalate (DBP), are typically used in cosmetics and personal care products. DBP is also used in epoxy resins, cellulose esters and special adhesive formulations. Longer branching alkyl chain PAEs, such as butylbenzylphthalate (BBP), dicyclohexylphthalate (DCHP), di-n-octylphthalate (DnOP), di-n-nonylphthalate (DnNP), di-isodecylphthalate (DiDP) and diethylhexylphthalate (DEHP) are widely used as plasticizers in the polymer industry to improve flexibility, workability and general handling properties. Their worldwide production is growing (from 1.8 million tons in 1975 to 4.3 million t in 2006), a quarter of which is represented by DEHP used in polyvinyl chloride (PVC) manufacturing.

Extensive industrial application of PAE esters has caused widespread environmental pollution because these products, which are not chemically bound in plastics or other products, are easily released into water, air and soil.

Humans are exposed by ingestion, inhalation, dermal contact and medical devices. For the general population, exposure by diet is considered a major route.

Overall, the adverse effects of PAEs are widely reported. The main health concerns regard their toxicity on development and the reproductive systems. Other effects (carcinogenicity; altered thyroid function; potential contribution to the overall population burden of insulin resistance; obesity; respiratory symptoms; skin sensitization and allergic contact dermatitis; and effects on dopamine system in the central nervous system) are suspected but are in need of further investigation.

In this contribution we summarize the sources of PAE exposure; the sources of food contamination, their occurrence in food and dietary estimates; human health effects; the methods of analysis and monitoring PAEs in foods; the implications for the food industry and policy makers for prevention and control of contamination and the future trends.

14.2 Human exposure to phthalates

Human exposure to PAEs is considered widespread as a consequence of their extensive industrial use in many products, for consumer and personal care requirement, and of diffuse environmental pollution. The ubiquitous exposure to PAEs among the general population was demonstrated by the finding of detectable concentrations of their metabolites in the urine of the majority of people aged over six years in USA (CDC, 2005). In Europe, Wormuth et al. (2006) studied the most important sources of PAEs in Europeans by applying a scenario-based approach in which realistic exposure situations were generated to calculate the age-specific range of daily consumer exposure to eight PAEs. The results provided a link between the knowledge on PAE sources and the concentrations of their metabolites found in human urine, and demonstrated that in many cases exposure in infants and adults may arise from different sources. Infants showed a significantly higher daily exposure to PAEs than older consumers in relation to their body weight. Consumer products and different indoor sources are the main sources of exposure to DMP, DEP, DiDP, BBP and DnNP, whereas food has a major influence on the exposure to DBP and DEHP. Recently, the findings of PAE metabolites analyses (9 compounds) in urine from infants of 2–36 months have shown overall exposure to one or more phthalates. In particular, 80% of samples, showed a contamination by at least seven of the nine compounds investigated. Levels of methylethylphthalate (MEP), monomethylphthalate (MMP) and methylbutylphthalate (MBP) were directly related to the number of baby lotions, creams and powders applied by their mothers (Sathyanarayana et al., 2008).

336 Persistent organic pollutants and toxic metals in foods

The main routes of exposure for the general population are considered ingestion, inhalation and skin contact (Adibi *et al.*, 2003; Rudel *et al.*, 2003). In hospitalized patients, parenteral and intravenous medical devices may represent major routes of exposure, principally to DEHP and DBP (ATSDR 2002; Green *et al.*, 2005; Weuve *et al.*, 2006).

14.2.1 Ingestion

PAE ingestion may occur via food and nutritional supplements. Children may be exposed to ingestion of PAEs by breast milk, infant formulas, cow's milk or plastic-packed food, indoor dust, and by sucking plastic teats and old toys and mouthing contaminated hands or other objects (Clark *et al.*, 2003a; Calafat *et al.*, 2004; Mortensen *et al.*, 2005; Sathyanarayana, 2008). Following the evaluation of the risks of PAEs (CSTEE, 1998) DnOP was banned from toys and child-care articles that can be mouthed by children, as were bis DEHP, DBP and BBP, DiNP and DiDP some years later (Commission Directive 2005/84/EC).

Dietary intake from contaminated food is considered the main source of PAE exposure in the general population (Spilmann *et al.*, 2006). Available data relate mainly to DEHP and DBP. Total daily intakes of DEHP and DBP via food in the adult population have been studied by Pfannhauser *et al.* (1995), the Ministry of Agriculture, Food and Fisheries (MAFF) (1996), Petersen and Breindahl (2000), DEHRM (2003), Fromme *et al.* (2007b) and Colacino *et al.* (2010), and in children by Müller *et al.* (2003). In adults, mean total daily dietary intakes (based on 70 kg b.w.) varied from 2.1 to 4.9 µg/kg b.w. for DEHP and from 0.2 to 4.2 µg/kg b.w. for DBP; in children aged 1–6 and 7–14 years, estimated mean total daily dietary intakes were, respectively, 11.0 and 26.0 µg/kg b.w. for DEHP, and 3.5 and 8.0 µg/kg b.w. for DBP.

Medical devices, mainly bags made of PVC softened by DEHP employed in enteral nutrition, may be an important source of PAE ingestion. The lipid content of the enteral/nutritional formulas may significantly condition the leaching of the plasticizer. Enteral formulas containing lipid emulsion prepared in PVC/DEHP bags and delivered through PVC/DEHP tubing were estimated to determine a daily DEHP exposure of 0.14 mg/kg/day in adults and 2.5 mg/kg/day in neonatal infants (FDA, 2001). Ingestion of commonly used drugs and medicines may be an important source of PAEs (Hauser *et al.*, 2004; Hernández-Díaz *et al.*, 2009) as many antibiotics, antihistamines, laxatives, herbal preparations and nutritional supplements are coated with films made of synthetic polymers containing PAEs that may leach into the gastrointestinal tract during drug release.

14.2.2 Inhalation

Inhalation of PAEs can occur in the general population, mainly through house dust and indoor air, including inside automobiles where PAE release from plasticized components can occur. PAE contamination of indoor air and dust may depend on atmospheric pollution and on specific indoor PAE sources, such as building materials, PVC flooring, furnishings, toys, clothing and PVC accessories. Since 1997 several studies have documented worldwide PAE indoor pollution. A mean total PAE content of 960 µg/g of dust (range 130–2920 µg/g dust) in 38 homes in Norway was reported by Oie et al. (1997), where the main compound was DEHP (mean 640, range 100–1610 µg/g dust). The mean DEHP inhalation exposure from this source in adults was estimated to be 0.76 µg/day. Pohner et al. (1997) reported a 95th percentile DEHP concentration of 2.0 mg/g dust for 272 German homes. Total PAE concentrations in indoor dust samples from 120 US homes ranged from 0.3 to 524 µg/g dust, and from 0.005 to 28 µg/m^3 in indoor air, as reported by Rudel et al. (2001), with DEHP and DBP levels ranging, respectively, from 20 to 114 ng/m^3 and from 101 to 431 ng/m^3. Butte et al. (2001) reported a 95th percentile DEHP concentration of 2.6 mg/g dust in 286 homes in Germany. PAE occurrence in air and dust from apartments and kindergartens in Berlin was documented by Fromme et al. (2004). In room air, DBP was the highest occurring compound, with median values of 1083 and 1188 ng/m^3, respectively, in apartments and kindergartens; in house dust, DEHP was the main PAE with a median value of 703 mg/kg dust. Becker et al. (2004) reported levels of DEHP of 508 µg/g dust in samples collected from different households in Germany. Levels of five PAEs in indoor air of houses in Tokyo are reported by Otake et al. (2004), who also estimated inhalation exposure levels in adults. Median concentrations of DEHP from 0.10 to 0.11 µg/m^3 were found, with results in inhalation exposures ranging between 2 and 22 µg/day, respectively. Bornehag et al. (2005) studied the association between persistent allergic symptoms in Swedish children and the concentrations of PAEs in dust taken from the children's bedrooms, finding DEHP in nearly all samples (median concentration 0.77 mg/g dust, max 40.459 mg/g dust). Kolarik et al. (2008) reported a mean concentration of 0.96 mg/g DEHP in Bulgarian children's bedroom dust. Thirty sampling sites representing three different indoor environments (private homes, day care centers and workplaces) in the Stockholm area in Sweden were analyzed by Bergh et al. (2011). Median values for total PAEs (6 compounds) were between 1500 and 2700 ng/m^3 (range of the concentrations 740–7400 ng/m^3); the main compounds were DEP and DBP. PAE esters (9 compounds) in indoor dust collected in houses from Albany, New York (USA) and several cities in China showed total concentrations ranging from 87.1 to 9670 µg/g, and from 24.4 to 8590 µg/g, respectively (Guo and Kanna, 2011). DBP, BzBP and DEHP were the prevalent compounds in dusts collected in Albany, while DiBP, DBP and DEHP were the major compounds in dusts collected in Chinese houses.

Some hobbies, such as clay modeling, may represent a considerable source of PAE inhalation for adults and children. Polymer clay is reported to be a major source of air-dispersed PAEs: this material is softened by various

PAEs, such as DnOP, BBP, DEHP and terephthalic acid, whose total content can vary from 3.5% to 14% by weight. They can be dispersed in the air during the firing of the modeled clay; ambient concentrations ranging from 32 to 2667 µg/m^3 for BBP, up to 6670 µg/m^3 for DnOP and from 6.05 to 4993 µg/m^3 for DEHP and/or chemically similar analogues, were measured in experimental conditions (Maas et al., 2004). Polymer clay can also be a source of dermal absorption of PAEs and, particularly in children, of ingestion from mouthing hands contaminated during modeling.

Inhalation of PAEs, principally DEP, DBP and BBP, can also occur via the use of perfumes and hair sprays (Blount et al., 2000).

In hospitalized patients subjected to respiratory therapy, DEHP may be transferred into respiratory gases passing through tubes made of PVC plasticized by PAEs. From direct measurements under experimental conditions Hill (1997) found that DEHP may be released into the air stream, leading to exposure in patients, estimated at 28.4–94.6 µg/day.

14.2.3 Intravenous exposure

DEHP is the main PAE involved in the release from PVC devices employed in medical care for invasive therapies, such as transfusion of blood and blood products, extracorporeal membrane oxygenation, and dialysis (Lee et al., 1999). The factors influencing DEHP leaching are lipid content, temperature, time of storage and characteristics of the contact (agitation). Exposure may vary greatly according to treatment conditions, equipment characteristics, infusion time, etc. (FDA, 2001). Premature babies undergoing intensive medical care in neonatal intensive care units are exposed to DEHP at higher concentrations than adults (Calafat et al., 2004; Green et al., 2005).

14.2.4 Skin absorption

Direct contact with clothing, personal care products such as cosmetics and sunscreens, synthetic modeling clay, toys, yoga pads, waxes, cleaning products, insecticides and denture material containing PAEs may lead to absorption via the skin (Munksgaard, 2004). Due to frequency of use, personal care products are considered the main exposure route to lower molecular weight PAEs for young women (Blount et al., 2000), for men using eau de cologne and aftershave (Duty et al., 2005), and for infants whose mothers use lotions, powders and shampoos (Sathyanarayana et al., 2008).

DMP and DBP may be used topically as insect repellants just as DEP and DnOP along with other ingredients in insecticides or topically used repellants, causing dermal or inhalation exposure (Ware and Whitacre, 2004).

14.2.5 Exposure assessment

Given that environmental epidemiological studies need proper exposure assessment, the approach and methods to measure exposure have to be appropriate to avoid measurement errors or misclassifications (Armstrong, 2003). The traditional methods of assessment, such as questionnaires and medical records, are considered inadequate to estimate individual exposure because many sources of PAE environmental exposure are unknown.

Use of metabolites as biomarkers is preferred, because the risk of accidental contamination during collection, storage and analysis is greatly reduced. PAE metabolites may be measured in urine, serum, saliva, seminal fluid, breast milk, amniotic fluid, meconium and placenta, but the most common approach is the detection of urinary metabolites (Calafat and McKee, 2006). Because PAEs have biological half-lives from hours to days, they are quickly excreted from the body. Hence urinary metabolite levels are higher than those found in other biological matrices, and their detection is consequently more precise.

Several studies have demonstrated a high temporal and within-individual variability of urinary PAE metabolites over several days (Hauser *et al.*, 2004; Fromme *et al.*, 2007a) suggesting that prolonging urinary metabolite measurements over longer exposure periods such as six months may be recommended (Teitelbaum *et al.*, 2008). Individual exposure levels may depend on several factors, such as changes in exposure sources (for instance diet and product use) as well as on changes in the metabolism of the considered chemicals. Therefore, studies based on single urine sampling can accurately measure the exposure at a single point in time, but may not reflect a long-term exposure level. If individual time-activity pattern and concentrations of PAEs in the environment and foods are stable, a person's exposure may lead to 'pseudo-steady state' metabolite concentrations over long periods of time and consequently the measure of urinary metabolite levels in a single sample at a single point in time may be considered effective (NRC, 2006). On the contrary, when substantial variability in exposure exists over time, the assessment of associations between exposure and health damage requires the monitoring of urinary levels of PAE metabolites over weeks or months (Meeker *et al.*, 2009).

14.3 Sources and occurrence in foods

PAEs can enter the environment by losses during manufacturing processes, by leaching or evaporating from final products, in which they are not chemically bound, and as well as those that are discarded to waste or recycling (Clara *et al.*, 2010).

Because of their relatively high vapor pressures, PAEs may volatilize to the atmosphere and get trapped in cloud masses. Both rainfall and runoff transfer these pollutants to surface water and soils. (Fromme *et al.*, 2002; Clark *et al.*, 2003b; EU-RAR, 2004, 2007, 2008; Gasperi *et al.*, 2009). Environmental pollution may also originate from treated and untreated domestic and industrial wastewater and sewage sludge (Fromme *et al.*, 2002; Fauser *et al.*, 2003; Marttinen *et al.*, 2003; EU-RAR, 2004, 2007, 2008; Roslev *et al.*, 2007;

Dargnat et al., 2009) in soils and sediments (Björklund et al., 2009; Zeng et al., 2009).

PAEs in particular DEHP, are consistently found in leachate from landfills receiving discarded consumer products and materials (Marttinen et al., 2003; Asakura et al., 2004). In Europe, the use of sludge in agriculture accounts for about 50% of its production, and the use of sludge for soil amendment may be an important input source of pollutants to the environment (Dargnat et al., 2009). Besides agricultural practices, the use of pesticides or insect repellants containing PAEs may represent a further source of contamination. PAEs present in agricultural soils are then available for crops and for uptake by grazing livestock (Rhind et al., 2002). BBP, DBP and DEHP have been found in raw agricultural products such as grains, nuts, beans, vegetables and fruits from the USA. The highest concentration was found for DBP (> 50 000 µg/kg) followed by DEHP at 2200 µg/kg and BBP at 0.005 µg/kg (Kavlock et al., 2006; HSDB DBP, 2008; HSDB BBP, 2009). Food may also be contaminated from processing and packaging.

In processed foods DEHP and DBP are particularly found in fatty foods, including dairy products. DEHP is reported as the predominant PAE in milk products. It was found at concentrations ranging from 20 000 to 120 000 µg/kg in retail milk from Norway, Spain and UK; in dairies in the UK, DEHP levels from 200 to 16 800 µg/kg were found in retail cheeses, from 200 to 2700 µg/kg in cream and from 2500 to 7400 µg/kg in butter (Sharman et al., 1994). High levels of BBP in butter (1600 µg/kg) and of BBP and DBP in Cheddar cheese (6400 and 1500 µg/kg, respectively) were found in Canada (Page and Lacroix, 1995).

Literature data show a decrease of PAE contamination levels in milk and dairy products over the years. Casajuana and Lacorte (2004) found DBP at levels of 7–50 µg/kg, DEHP at 15–27 µg/kg and DEP at 36–72 µg/kg in Spanish retail milk, while in Denmark Soerensen (2006) found DEHP levels of 15–37 µg/kg in yoghurt.

Regarding other foods, Casajuana and Lacorte (2004) evidenced in the muscle of livestock contents of PAEs ranging between 120 and 280 µg/kg, rarely exceeding 500 µg/kg; Wormuth et al. (2006) reported DEHP and DBP concentrations amounting up to about 10 mg/kg in cereals, bread, biscuits, cakes, nuts, spices, fat and oil from Germany, UK and Japan; Wang et al. (2007) detected in cucumbers and tomatoes purchased from the market 242–347 µg/kg of DBP and 311–517 µg/kg of DEHP, with a 100% detection frequency.

The presence of PAEs in foods extends well beyond common food products and includes baby food. In Germany, Bruns-Weller and Pfordt (2000) found DEHP and DBP in baby-food samples at concentrations ranging, respectively, from 10 to 20 µg/kg and from n.d. to 30 µg/kg; in Denmark, Petersen and Breindahl (2000) reported maximum concentrations of BBP, DBP and DEHP of 5, 40 and 630 µg/kg, respectively, in baby-food samples; in Spain, Casajuana and Lacorte (2004) evidenced DEP concentration at 76 µg/kg, DEHP at 20 µg/kg, DBP at 18 µg/kg, DMP, and BBP at levels lower than 1–2 µg/kg (all mean values) in powdered infant formulas packed in metal cans; in European, North American and Asian countries Yano et al. (2005) detected, in baby milk powders from supermarkets or local open markets, DBP and DEHP concentrations ranging from 30 to 280 ng/g and 20 to 80 ng/g, respectively. As recently reported, in addition to infant formulas and baby foods, infant exposure to PAEs may occur also through breast milk (Zhu et al., 2006).

Contact with equipment, surfaces, containers, etc., during processing and packaging are considered the major sources of PAE contamination of food. A number of studies have demonstrated that the migration of DEHP from PVC tubing in the milking machines used on dairy farms, which may contain up to 40% DEHP by weight, contributes significantly to the DEHP contamination of raw milk (Ruuska et al., 1987; Bluthgen, 2003; Feng et al., 2005; Lopez-Espinosa et al., 2007). Correlations between the DEHP leaching amount into raw milk and contact time, and between leaching rate and the percentage of DEHP in the PVC were evidenced (Bluthgen, 2003), but the same study also pointed out that, after DEHP plasticizers were no longer used in milking machines, retail milk from the UK still contained 35 µg/kg of DEHP. Tsumura et al. (2001b) evidenced an increase in DEHP levels in chicken from 80 µg/kg precooking to 13 100 µg/kg after frying in a Teflon coated pan, and further to 16 900 µg/kg after packing in polystyrene boxes.

Heating food in ready-to-eat packages greatly facilitates PAE migration from packaging materials to food. Additional processing, packaging and condensation can lead to 5–100-fold increases in DEHP concentrations in cream and cheese products (Petersen, 1991; Sharman et al., 1994; Casajuana and Lacorte, 2004; Mortensen et al., 2005). DEHP contamination of food may be attributed to leaching from PVC gloves used during food preparation, as demonstrated in a study conducted in 1999 by Tsumura et al. (2001a) on meals collected by the 'one week duplicate diet' sampling method in three Japanese hospitals, in which DEHP was detected at levels ranging from 10–4 400 ng/g wet weight (ww). In a follow-up study conducted in the same hospitals in 2001, after the banning of the use of DEHP in PVC gloves, DEHP concentrations in meals decreased significantly (6–675 ng/g ww) (Tsumura et al., 2003). Cirillo et al. (2013a) evaluated the levels of DEHP and DBP in ready-to eat packed meals for hospital patients, through "one week duplicate diet" method, over a two-week time period. The DEHP median concentration levels in total meals varied from 0.056 to 0.350 µg/g w.w, the DBP median levels varied from 0.020 to 0.166 µg/g w.w.

Direct contact with phthalate-containing packages is considered the major source of PAEs in fat foods. Over 1985–1989, Page and Lacroix (1992) measured PAEs and di-2-ethylhexyladipate (DEHA) plasticizers in Canadian food-packaging materials and in food samples. They showed widespread use of DEHP or DEHA-containing food-packaging materials and demonstrated the potential link between the presence of chemicals in packaging materials and in food. Polycoated aluminum/paper foil, too, may leach some PAEs, as demonstrated by the finding of DBP, BBP and DEHP at levels up to 10.6, 47.8 and 11.9 μg/g, respectively, in retail butter and margarine. In a study carried out by Cirillo *et al.* (2013b) on ready-to-eat meals packed in polyethylene terephthalate (PET) and aluminium dishes, supplied to patients in two hospitals, the probable migration of DBP and DEHP from the packaging into the food was evaluated. The meals were analysed according to three time ranges: before the packaging (T0) and 60 min (T1) and 120 min (T2) after the packaging, during storage in thermostatic delivery carts. Regarding meals packed in PET dishes, the significant increase of contaminant compounds levels during storage time, shows that migration of PAEs from packaging to food is possible.

In Europe, most of the food packaging has been reported to contain DEHP and DBP. Up to 11 mg/kg of DBP and up to 61 mg/kg of DEHP were reported in cardboard and paper used as food containers in four European countries (Aurela *et al.*, 1999). Measurements of PAEs in plastic foil, paper, cardboard and aluminum foil with color printing used for various foods (sweets, wafers, meat, sweets, milk products, frozen foods, vegetables, dry ready-to-cook products and potato chips) were carried out in the Czech Republic: DBP and DEHP were found in all the tested materials at concentrations up to 1 g/kg (Jarošová *et al.*, 2006). A study on Brazilian food-packaging materials acquired on the retail market showed the presence of DEHP and other plasticizers in PVC film manufactured for domestic use, squeeze bags for honey and wrapping for soft milk candy, at around 20–35% weight/weight (w/w). DEHP in packaging closure seals for fatty foods, such as palm oil, coconut milk and soft high-fat cheese, was 18–33% w/w (Freire *et al.*, 2006).

Although PAE contamination of foods has been widely demonstrated, there is currently not much data on dietary PAE intake under real-life conditions. In the UK, MAFF conducted a study using the market-basket approach, collecting retail food products, prepared as for consumption, then combined in amounts reflecting their relative importance in the average UK adult diet into 20 composite samples. DEHP and DBP concentrations in foods were combined with food consumption data for a 60 kg adult from the National Diet and Nutrition Study of British Adults. Mean intake was estimated to be 2.1 μg/kg body weight (b.w.)/day for DEHP and 0.2 μg/kg b.w./day for DBP (MAFF, 1996). In Switzerland during the period 1991–1996, Kuchen *et al.* (1999) analyzed a total of 36 ready-to-eat food samples representative of daily nutrition in the adult population, estimating mean intakes of 2.8 μg/kg b.w./day for DEHP and 4.2 μg/kg b.w./day for DBP.

Several studies have been carried out analyzing duplicate diets. In a study on Danish adults performed in 1994, Petersen and Breindahl (2000) analyzed meals collected by volunteers during 24 hours, in order to estimate total daily exposure, normalizing results to a daily diet of 10 MJ of energy and to a body weight of 70 kg. Mean daily DEHP and DBP intakes ranged from 2.7

to 4.3 µg/kg b.w. and 1.8 to 4.1 µg/kg b.w., respectively. In the same period in Austria, Pfannhauser et al. (1995) carried out a study on ten adults (18–50 years of age) over seven consecutive days, obtaining comparable mean daily intakes of 4.9 µg/kg b.w. for DEHP and 2.9 µg/kg b.w. for DBP. In a UK study, dietary DEHP and DBP intakes were studied by the 'one week duplicate diet' method with the contribution of two groups of volunteers, the former consuming a normal mixed diet and the latter a vegan diet (DEHRM, 2003). For the vegan consumers, the median dietary intake of DEHP was quite twice as high as that for the normal diet consumers (5.0 µg/kg b.w. vs 2.8 µg/kg b.w.), while negligible differences were found for DBP (0.39 µg/kg b.w. vs 0.33 µg/kg b.w.). In Germany, Fromme et al. (2007b) collected daily duplicate diet samples over seven consecutive days in 27 female and 23 male healthy subjects aged 14–60 years. The median daily intakes via food were 2.4 µg/kg b.w. for DEHP and 0.3 µg/kg b.w. for DBP.

Very few studies have focused on estimating DEHP and DBP intake in children. Danish estimates on PAE dietary exposure, based on a computer modeling program proposed by the European Union System for the Evaluation of Substances (EUSES), were 26.0 and 8.0 µg/kg b.w./day for DEHP and DBP respectively in children 1–6 years old and of 11.0 and 3.5 µg/kg b.w./day in those 7–14 years old (Müller et al., 2003).

Infant exposure to DEHP by milk formulas was estimated, on 3 kg children consuming 700 mL/day, to range from 2.5 to 16.1 µg/kg b.w. (Yano et al., 2005). Cirillo et al. (2011) in packed school meals for children 3–10 years old in Naples (Italy) evaluated the levels of DBP and DEHP to range from 18.3 to 775.0 ng/g ww, and from 21.3 to 1050.8 ng/g ww, respectively. DBP intake via school meals was found in nursery schoolchildren (ranging from 2.4–16.9 µg/kg b.w.) and in primary schoolchildren (ranging from 1.7–9.0 µg/kg b.w.); DEHP intake was found in children 3–5 years old (ranging from 4.2–17.7 µg/kg b.w.) and in those 6–10 years old (ranging from 2.8–12.9 µg/kg b.w.).

14.3.1 Toxicity of phthalates

Phthalates present a very low acute toxicity. Once incorporated, they are rapidly metabolized by hydrolysis and subsequent oxidation reactions. PAE metabolites are almost completely excreted via urine, partly as glucuronides. The simple monoesters are the major urinary metabolites of PAEs such as DnBP, DBP or BBzP, and their urinary excretion represents approximately. 70% of the oral dose (Anderson et al., 2001). In the case of PAEs such as DEHP, DiNP, DiDP and DPHP only between 2–7% of the dose is excreted as the simple monoester, because most of the simple monoesters are further metabolized to produce a number of oxidative metabolites such as alcohols, ketones and carboxylic acids (Koch et al., 2004; Silva et al., 2007a, 2007b; Wittassek et al., 2008).

With respect to toxicological mechanisms, significant differences exist between low molecular weight (LMW) and high molecular weight (HMW)

PAEs. LMW PAEs are those with alkyl side chains of C4–C8 total carbon number; they are irritants to the skin and mucoses in animals, but not usually in humans. Due to their chemical properties they do not tend to bioaccumulate. Thus, their most important toxicological effects are considered subchronic and chronic. PAEs are reported to be animal carcinogens and can cause fetal death, malformations, testicular injury, liver injury, anti-androgenic activity, teratogenicity, peroxisome proliferation and especially reproductive toxicity in laboratory animals (ATSDR 1995, 1997, 2001, 2002). Several PAEs particularly DEHP and DBP, are reported to reduce fetal testosterone and insulin-like growth factor-3 in animals, inducing abnormalities in male reproduction (Foster, 2006; Gray *et al*., 2006). Furthermore, a potential androgen deficiency during fetal development has been demonstrated in rodents by the short anogenital distance (AGD) and other abnormalities such as hypospadias (failure of the urethra to tubularize completely, resulting in a urethral opening on the ventral aspect of the penis), cryptorchidism (undescended testis) and malformations of the epididymis, vas deferens, seminal vesicles and prostate (Gray *et al*., 2006). Toxicity profiles and potency vary according to the specific PAE. The extent of these toxicities and their applicability to humans remain incompletely characterized and controversial. On the basis of results of experiments on animals and of various recent studies on humans, the main health concerns regard PAE toxicity on development and the reproductive systems (Duty *et al*., 2005; Swan *et al*., 2005; Marsee *et al*., 2006, Fromme *et al*., 2007a). On the basis of adverse effects observed in rodent studies, LMW PAEs are classified as toxic for reproductive and developmental systems.

HMW PAEs are those with carbon side chains of C9 and greater total carbon (typically to C13). These substances are not classified as reproductive and developmentally toxic and do not meet the international definitions for endocrine disrupters.

Mixture studies are designed to test interactions (e.g. dose-addition and/or response-addition, synergy, antagonism) for mixtures of chemicals. The US National Research Council (NRC) published a report on a cumulative risk assessment for mixtures of LMW and HMW PAEs to quantify the accumulation of risk from multiple chemical stressors that may interact (NRC, 2008). In initial PAE cumulative risk assessments, for adverse effects on the developing male reproductive tract leading to an adverse outcome of reduced fertility, HMW PAEs do not contribute substantially to overall risk. This conclusion further supports the differentiation between LMW and HMW PAEs and questions the need to include HMW PAEs in cumulative risk assessments.

14.4 Studies of the effects of phthalates on humans

In recent studies on the humans the main health concern is with regard to the effects of PAE exposure on development and reproductive systems; in particular, on Testicular Dysgenesis Syndrome (TDS), semen quality and adverse effects on the female reproductive system. Literature study, moreover, shows that exposure to PAEs is associated with cancer, alteration of thyroid function, metabolic syndrome, abnormal respiratory function or other effects such as urticaria syndrome, and decrease in children's intellectual function, as described below.

14.4.1 Testicular dysgenesis syndrome (TDS)

An association has recently been shown between prenatal exposure in humans, particularly to DEHP and its urinary metabolites, and a similar cluster of reproductive developmental outcomes in male infants (Swan et al., 2005; Swan, 2008). In addition, free serum testosterone in human male infants has been negatively correlated with levels of some PAE metabolites in breast milk (Main et al., 2006). The concentrations of PAE metabolites in breast milk were associated with an alteration of the pituitary–gonadal axis in three-month-old boys towards a net anti-androgenic effect (Main et al., 2006, 2010). This observation was in line with rat evidence, suggesting that prenatal exposure to some PAEs during a critical developmental window (Welsh et al., 2008) can cause testicular dysgenesis, which resembles TDS in humans. Another group of researchers also established a potential link between in utero DEHP exposure and hypospadias, showing that DEHP could up-regulate the expression of activating transcription factor 3 (ATF3), which is involved in the regulation of apoptosis in the mouse genital tubercle (Liu et al., 2009). These observed associations were weak, but were corroborated by an American baby cohort investigation (Swan et al., 2005). Prenatal exposure to PAEs, as assessed by maternal urinary excretion of metabolites during pregnancy, was correlated to a reduced AGD in infant boys. In turn, short AGD was associated with a high rate of cryptorchidism and smaller penis size (Swan et al., 2005).

PAEs also show anti-androgenic and oestrogeno-mimetic activities. Oestrogen receptor 1 (ESR1) was found to interact with both DEHP/MEHP and DBP/BBP/MBP in humans, but not in rodents. The widely used PAEs are known to interfere with endocrine systems, including the oestrogenic and thyroid hormones (Swan, 2008; Botelho et al., 2009).

However, the long-term consequences of these findings for humans are uncertain. In particular, the potential for these anti-androgens to influence the course of brain sexual differentiation has only recently been addressed (Engel et al., 2009). Data from Salazar-Martinez (2004), Swan et al. (2005) and Swan (2008) support the hypothesis that in humans, maternal exposure to PAEs particularly DEHP, lowers fetal testosterone production and results in incomplete masculinization of the genital tract, resulting in a shortened AGD as well as incomplete testicular descent and smaller penile size, suggesting that AGD is a marker for insufficient fetal androgenization and that low-dose PAE exposure may affect several markers of human male genital development. These data suggest that the same process might plausibly influence brain sexual differentiation and its expression in sexually dimorphic behaviors. Play behaviors may be used to test the hypothesis that PAE exposure during gestation may alter brain sexual differentiation and its behavioral outcomes (Swan *et al*., 2010).

The mechanism of PAE toxicity on fetal Leydig cell steroidogenesis has not been well characterized. Angiotensin and vasopressin have shown to inhibit testosterone production by fetal Leydig cells (Tahri-Joutei *et al*., 1991; Leung and Sernia, 2003). However, the induction of both angiotensin and vasopressin receptor (*Nalp6*), together with aminopeptidase A, an enzyme responsible for converting angiotensin II to angiotensin III, following PAE exposure, suggests a role for either abnormal angiotensin or vasopressin activity in the suppression of testosterone synthesis in fetal Leydig cells (Liu *et al*., 2005).

14.4.2 Semen quality

An inverse relationship between sperm concentration and DBP concentrations in the cellular fractions of ejaculates was reported by Murature *et al*. (1987). In another study carried out in India by Rozati *et al*. (2002) the concentration of PAEs was inversely correlated with sperm morphology but not correlated with ejaculate volume, sperm concentration or motility. A series of studies have examined urinary PAE levels and semen characteristics, sperm DNA damage and serum reproductive hormones in men attending an infertility clinic (Duty *et al*., 2003a, 2003b; Hauser *et al*., 2004; Duty *et al*., 2005). There were dose–response relationships of MBP with low sperm concentration and motility. There was suggestive evidence of an association between the highest MBP quartile and low sperm concentration. MEP, MMP and DEHP metabolites did not appear related with these semen parameters. On the contrary, Jonsson *et al*. (2005) did not observe significant associations between semen quality and PAE exposure in a study on young healthy military recruits in Sweden. However, several differences were noted between the US and Swedish studies (Hauser *et al*., 2006), including the age and fertility status of the study populations. DEHP, DBP and DEP were positively associated with semen liquefaction time, and DEHP was significantly correlated to its 'rate of malformation' (presumably impaired morphology) (Zhang *et al*., 2006). Sperm DNA damage, as assessed by the neutral comet assay, was associated with MEP and MEHP concentrations. PAE metabolites were associated with hormones. In particular, MBzP was associated with a decrease in FSH, and MBP with an increase of borderline significance in inhibin-B (Duty *et al*., 2003, 2005; Hauser *et al*., 2007). Besides, MBP concentration was associated with a decrease in free testosterone, and increased LH to free testosterone ratio in association in an occupationally exposed cohort (Pan *et al*., 2006).

14.4.3 Effects of phthalate exposure in women

Few adverse effects on the female reproductive system have been reported in humans. Colón *et al*. (2000) monitored levels of certain PAEs in the blood serum of young Puerto Rican girls aged 6 months to 8 years with a premature breast development (thelarche). The authors demonstrated significantly higher PAE levels in 68% of patients, which suggests possible negative effects of PAEs on the human reproduction and development system. Inverse relationship between PAE concentrations in the cord blood and pregnancy duration in humans has been reported (Latini *et al.*, 2003). In addition, increased serum DEHP concentrations have been found in women with endometriosis (Cobellis *et al.*, 2003; Reddy *et al.*, 2006).

14.4.4 Carcinogenic effects of phthalates

One of the best described carcinogenic effects of PAEs in rodents is hepatic cancer, although hepatic neoplasms are not observed in response to long-term exposure to all PAEs. Some PAE monoesters, including MEHP, MiNP, MBP, MBzP, MOP and MiDP, can activate peroxisome-proliferator-activated receptor-α (PPARα), as demonstrated by Bility *et al.* (2004) by an *in vitro* reporter assay. The ability of PAE monoesters to activate PPARα increases with increasing chain length. In addition to hepatic cancer, in rodents some PAEs can cause tumors in other cell types. For example, a 'tumor triad' – liver, testicular Leydig cell and pancreatic acinar-cell tumors – has been described for some PPARα ligands, such as DEHP (Klaunig *et al.* 2003). There is a known difference between rodents and humans in the ability of PPARα ligands to cause changes in the liver, including increases in cell growth and peroxisome proliferation (Peters *et al.*, 2005), which would suggest that the hepatocarcinogenic effects of DEHP and DiNP are unlikely to occur in humans (Klaunig *et al.*, 2003). Furthermore, PPARα-mediated events suggest that humans might not be susceptible to the nonhepatic tumors, even if the nonhepatic tumors may be mediated through mechanisms that are independent of PPARα.

In a recent study, López-Carrillo *et al.* (2010) in the northern states of Mexico show for the first time that exposure to DEP, as assessed by urinary MEP concentrations, may be associated with an increase in breast cancer (BC) risk, whereas exposure to other PAEs, measured by the urinary concentrations of MBzP (BBzP) and MCPP (DOP and other PAEs), was negatively associated with BC. The findings require confirmation to exclude the possibility that these parent/metabolite PAEs are surrogates of an unrecognized lifestyle or dietary BC risk factors.

14.4.5 Thyroid function

A limited number of studies suggest that exposure to some PAEs such as DBP, DEHP and DnOP may be associated with altered thyroid function, but human data remain insufficient (Jugan *et al.*, 2010). In particular, the association between environmental exposures to PAEs and serum thyroid hormone and TSH levels in humans was investigated within the Massachusetts (USA) male infertility clinic study (Meeker *et al.*, 2007) and among pregnant

Taiwanese women in their second trimester (Huang *et al.*, 2007), showing that urinary levels of PAE metabolites may be associated with altered free T4 and/or total T3 levels both in adult men and pregnant women, but the need for additional research to confirm these findings has been highlighted.

14.4.6 Metabolic syndrome

Owing to the documented anti-androgenic effects of certain PAEs in animal models, recent observations show that low testosterone in adult males may be associated with an increased prevalence of obesity and type 2 diabetes (Ding *et al.*, 2006; Selvin *et al.*, 2007). According to Stahlhut *et al.* (2007) MBzP, MEHHP, MEOHP and MEP are significantly associated with increased waist circumference and MBP, MBzP and MEP with increased HOMA (a measure of insulin resistance). While these findings provide preliminary evidence of a potential contributory role for PAEs in the overall population burden of insulin resistance, obesity and related clinical conditions, additional studies are required for corroboration.

14.4.7 Phthalates in relation to respiratory function, asthma and allergies

Epidemiologic occupational studies in adults have shown associations between heated PVC fumes and asthma and respiratory symptoms (Polakoff *et al.*, 1975; Falk and Portnoy, 1976; Brooks and Vandervort, 1977; Eisen *et al.*, 1985; Markowitz, 1989; Nielsen *et al.*, 2007). Furthermore, epidemiologic studies in adults provide some evidence of a relation between interior PVC surface materials and the risk of asthma. A variety of respiratory symptoms, such as coughing, work-related shortness of breath, wheezing and rhinitis, as well as a decline in forced expiratory volume in one second (FEV1), were found to increase in exposed workers compared with non-exposed reference groups. It should be noted, however, that the majority of these studies were not adjusted for confounders (Norback *et al.*, 2000; Tuomainen *et al.*, 2004; Jaakkola *et al.*, 2006, 2008).

In an American study on 240 adult participants in the Third National Health and Nutrition Examination Survey, Hoppin *et al.* (2004) observed an inverse association between MBP and three measures of pulmonary function, forced vital capacity (FVC), FEV1 and peak expiratory flow (PEF), and higher MEP was associated with lower FVC and FEV1 values in men. No consistent associations were seen in females or with other PAE metabolites (Hoppin *et al.*, 2004).

A number of studies have examined respiratory function, asthma and allergy in children in relation to the use of PVC-containing products in the home or PAEs in house dust (Jaakkola *et al.*, 1999, 2000; Bornehag *et al.*, 2004) although none used biomarkers to define exposure. Two cross-sectional studies found associations between allergic symptoms and house dust levels of BzBP and DEHP (Bornehag *et al.*, 2004; Kolarik *et al.*,

2008). Epidemiological data point to a possible correlation between PAE exposure and asthma and airway diseases in children. A limited amount of data also suggest PAE-induced enhancement of mast cell degranulation and eosinophilic infiltration. Thus, some of the early key mechanisms in the pathology of allergic asthma could be targeted by PAE exposure. However, the clinical relevance of real-life exposure and identification of molecular targets that can explain interactions remain largely to be solved (Bornehag and Nanberg, 2010). Moreover, even if some causal relationship between certain PAEs and asthma was found to exist, it would not necessarily imply that PAEs are able to enhance the development of allergic sensitization, the alternative argument being that exposure is associated with the exacerbation of respiratory symptoms in those who are already sensitized (Kimber and Dearman, 2010).

14.4.8 Other effects of phthalates

Although there are occasional case reports of contact urticaria syndrome to PAEs (Sugiura *et al.*, 2002), the available evidence in humans demonstrates clearly that they lack significant potential to cause skin sensitization and allergic contact dermatitis (Geier *et al.*, 2004). Moreover, PAEs are uniformly negative in assays where chemical respiratory allergens test are positive, including the guinea pig maximization test and the murine local lymph node assay (Kimber *et al.*, 2007). These data are consistent with the fact that, as a class, PAE esters have little or no ability to form stable associations with proteins.

Several animal studies have revealed that the dopamine system in the central nervous system is affected by PAEs. Low-dose PAEs can impair tyrosine hydroxylase immunoreactivity (Ishido *et al.*, 2004) and cause the loss of midbrain dopaminergic neurons, decreasing tyrosine hydroxylase – biosynthetic activity (Tanida *et al.*, 2009).

Kim *et al.* (2009) investigated the relationship between the urinary concentrations of PAE metabolites and children's intellectual functioning including general intelligence, attention and concentration. This study shows a negative association between IQ and urine PAEs in humans, but if adjusted for maternal IQ, inverse relationships between PAE metabolites and IQ scores were found; however, given the limitations in cross-sectional epidemiology, prospective studies are needed to fully explore these associations.

14.5 Methods of phthalate analysis and monitoring in foods

As discussed above, widespread use of PAEs has led to their occurrence as ubiquitous contaminants. It has been shown that most foodstuffs contain measurable amounts of PAEs some at low levels. Effective measurement of PAE content in food may be particularly difficult as contamination of food samples can easily take place within laboratory activities when analytical solvents, reagents and glassware containing traces of PAE may be used. Hence, in assessing levels of PAE contamination of food, suitable precautions have to be taken to avoid any contamination source and to ensure the quality of analytical results.

At present, official methods for determining the levels of PAE esters in food have not been established in the EU. However, the emerging problems connected with such contamination have raised the need to develop a standard system of measurements for these contaminants in foodstuffs. Before taking any action, the European Commission requested the Institute for Reference Materials and Measurements, which is part of the Joint Research Centre (JRC-IRMM), to conduct a survey among European food control laboratories in order to evaluate comparability and potential critical points of analytical method, applied to determine PAEs in food, while providing support for laboratories not experienced in the field in question (JRC, 2009).

The JRC report describes several analytical methods that are applied to simple matrices, such as beverages, or more complex samples, such as total diet, with regard to determining single PAEs or different congeners, including isomer mixtures. The most frequently determined congeners are as follows: bis(2-ethylhexyl) PAE (DEHP), which is the most commonly detected PAE in food, dibutylphthalate (DBP) and benzobutylphthalate (BBP).

Analytical procedures are developed through the traditional steps applied to determine organic chemical contaminants in food, i.e. extraction of contaminants from the food matrices, cleanup of extracts, and measurement of the contaminant concentrations in the extracts. The differences in the chemical structures of PAEs may considerably influence their polarity and hence the efficacy of the extraction system adopted. The characteristics of food matrices, particularly the fat content, also have to be considered in the choice of the method to apply. Furthermore, as pointed out above, due to the ubiquitous presence of PAEs in the environment, the laboratory air, the reagents, the glassware and the equipment may be contaminated, and therefore precautionary measures have to be taken to control contamination sources in the laboratory and quality control procedures have to be strictly applied. Since equipment generally used in collecting food, such as plastic containers, glass jars or aluminum foil, may itself be a source of PAE contamination, food specimens are best collected where possible in their original packaging, storing them at the appropriate temperatures until analysis.

Before analysis, liquid samples are usually sub-divided into aliquots (up to 300 mL) without any preliminary treatment, while solid foods are homogenized with the addition of distilled water or polar organic solvents for enhancing their pulping when necessary, and sub-divided into aliquots ranging from 0.1–10 g, according to the fat content. PAEs are usually extracted from nonfatty liquid samples, such as water, soft drinks and alcoholic beverages, with chloroform, n-hexane, n-heptane or isooctane, and frequently measured without any cleanup treatment. Acetonitrile or acetonitrile/water mixtures are mainly used in PAE extraction from nonfatty solid foods while, in the case of fatty foods, PAEs are extracted together with the fat by dichloromethane, alone or mixed with cyclohexane or n-hexane, or by mixtures of n-hexane with acetone or acetonitrile that allow a more selective extraction of PAEs from the food. Extraction may be accomplished by shaking the sample with the extraction solvent or mixture, or by the application of ultrasound or microwave systems.

Cleanup of extracts from nonfatty liquid foods are usually carried out by liquid/liquid (L/L) partition; L/L partition can also be applied to cleanup solid food extracts, such as total diet samples, removing co-extracted interferences by a further partition into n-hexane or dichloromethane (Pfannhauser *et al.*, 1995; Tsumura *et al.*, 2001a, 2003). In the case of fatty foods the JRC report (2009) refers that gel permeation chromatography (GPC) is the most widely used technique. Biobeads® S-X3 (Bio-Rad Laboratories, Hercules, CA, USA) and dichloromethane/cyclohexane (1:1 vv) or cyclohexane/ethyl acetate (1:1 vv) mixtures are, respectively, the stationary and mobile phases adopted. Preparative liquid chromatography on silica columns, Florisil® columns eluted by an n-hexane/diethyl ether (80:20 vv) mixture and sweep co-distillation on Florisil are also reported in the cleanup of extracts of fatty foods (JRC, 2009).

Measurement techniques comprise the application of gas chromatography with flame ionization (GC-FID) (Petersen, 1991; Page and Lacroix, 1995) or electron capture detection (GC-ECD), high performance liquid chromatography (HPLC) in combination with UV diode array detection (HPLC-DAD) or with tandem quadrupole mass spectrometry (HPLC-MS/MS) and gas chromatography with mass spectrometry (GC-MS) (JRC, 2009). The last is the technique most frequently applied. Low polarity columns and electron ionization and single ion monitoring mode are usually applied for GC-MS measurements, adopting different temperature programs depending on the kind and number of compounds to measure. In analyzing mixtures of different PAE isomers electron ionization in the scan mode, applying positive chemical ionization (PCI), covering a mass-to-charge range of 50 to 350 or higher, is considered better than electron ionization in identifying peaks and recognizing the different isomers (George and Prest, 2001).

Validation of analytical methods for determining PAEs in food shows various difficulties. Since reference certified matrices are not available, spiking experiments are carried out, correcting results for recoveries. Estimated recoveries are reported to be mostly between 80% and 110%. Participation in inter-laboratory comparison tests is considered of particular importance. Furthermore, since the results of analyses have to be corrected for blank levels, special attention has to be kept to maintain blanks under control as these levels impact adversely on method limits of detection. As indoor air contamination with PAEs may influence blank levels, specially designed laboratories, eliminating e.g. the use of PVC coating of floors, could reduce the air and particulate input to blank values; appropriate ventilation can further contribute significantly to avoid indoor air saturation with PAEs. Any plastic material has to be avoided in sample collection, preparation and handling. Although in order to minimize blank levels, glass is the preferred material, preventive solvent rinsing and thermal treatment of glassware are strongly suggested in any event.

Furthermore, solvents and chemicals have to be checked before use. Redistillation of organic solvents and the application of thermally cleaned aluminum oxide for cleaning up nonpolar solvents are considered efficient measures to reduce blank values. Analytical instruments have to be controlled by performing cleaning runs without injection. A charcoal filter installed in the gas supply of the gas chromatograph, frequent heating out of the injector and particular attention paid to the temperature of the injector head are also recommended.

14.6 Implications for the food industry and policy making for prevention and control of contamination

The prevention of PAE contamination of food depends on several factors, some of which entail international policy to reduce PAE environmental pollution. Appropriate waste and wastewater management may allow a significant reduction in PAE release into soil and surface waters, thereby limiting food contamination during agricultural production; the use of phthalate-free fertilizers and pesticides could contribute to reduce the background contamination of crops. Furthermore, the use in food production of plant, machinery and equipment with plastic parts or components containing PAEs and the use of packaging for foods containing PAEs as plasticizers or involved in the manufacture of adhesives and inks should be banned or strictly limited. Since the migration of PAEs resulting from contact between food matrices and plastic surfaces depends mainly on the lipid content of foods, time, temperature and characteristics of contact and on storage duration, the food industry should adopt appropriate measures to control every parameter influencing PAE migration to reduce the contamination of fatty foods, particularly when the end-users are infants and children.

Various pronouncements, opinions and laws about PAEs have been published by European Union. Since 2000 DEHP has been listed among the 33 hazardous substances to be controlled in surface water by the European Community (Commission Directive 2000/60/CE). Commission Directive 2008/105/EC defines a limit value for surface waters for DEHP of 1.3 μg/L.

At the present time, European legislation recommends reducing the use of PAEs in different products, namely cosmetics and food packaging. The Commission Directive 2007/19/EC imposed restrictions on the use of DBP, DEHP, BBP, DiNP and DiDP on plastic materials and articles intended to come into contact with food. The European Commission stated more specific migration limits (SMLs) for PAEs to control the safety of food contact materials and articles (Commission Regulation 10/2011).

Table 14.1 TDI levels for phthalates established by WHO/CICAD (2003) and by EFSA (2005)

Phthalate	TDI, EFSA 2005 (mg/kg b.w./day)	TDI, WHO CICAD 2003 (mg/kg b.w./day)
DBP	0.01	
DEHP	0.05	
BBP	0.50	
DiNP	0.15	
DiDP	0.15	
DEP	n/a	0.50

The Committee on Toxicity of Chemicals in Food Consumer Products and the Environment on the basis of information from a report by the Danish Environmental Protection Agency about exposure to PAEs and potential risks to health, suggests the selection of NOAELs/LOAELs for these compounds, that may be used for endocrine-disrupting effects in EU risk assessments (COT, 2011).

In 1995 the European Commission's Scientific Committee for Food set temporary tolerable daily intakes (t-TDIs) for DBP and BBP, a tolerable daily intake (TDI) for DEHP, and a group TDI for DiNP with DiDP. In 2005, the Scientific Panel on Food Additives, Flavourings, Processing Aids and Materials in Contact with Food (AFC) of the European Food Safety Authority (EFSA) was asked to re-evaluate DBP, BBP, DEHP, DiNP and DiDP for use in the manufacture of food contact materials (EFSA, 2005a, 2005b, 2005c, 2005d, 2005e). The AFC panel revised the previous t-TDIs and TDIs for the PAEs under consideration, concluding that a group TDI was inappropriate and set new TDI values (EFSA, 2005e) As a TDI for DEP was not set by the SCF or EFSA, the TDI of 0.5 mg/kg b.w./day proposed by the World Health Organization (WHO, 2003) may be considered (Table 14.1).

14.7 Future trends

The reality of the reproductive effects caused by PAEs at today's exposure levels highlights the urgent need to eliminate these plasticizers from products. Addressing two main types of products containing PAEs – PVC and cosmetics – would have a major impact in reducing exposure. A number of companies, hospitals and government agencies have taken steps to switch to alternative materials and phase out PVC use. Several approaches have been developed to reduce plasticizer migration from flexible PVC by crosslinking surfaces and leaving the additive behind 'bars'. Some strategies imply surface modification by peroxides, azides (Jayakrishnan *et al.*, 1995; Jayakrishnan and Sunny, 1996; Sacristan *et al.*, 2000), sulphides, or acrylates (Jayakrishnan and Lakshmi, 1998). Physical treatments of surfaces with γ-radiation or plasma exposure are also described (Duvis et al., 1991; Krishnan et al., 1991). Replacing classic plastics with biocompatible and/or oligomeric materials, as functionalized polyethylene oxide (PEO) or poly-ε-caprolactone or their combinations, is considered an alternative approach (Ferruti et al., 2003) because, according to the nature of these systems, the devices should be nontoxic. The covalent linkage of the additive to polymer chains in PVC, chemically modified by nucleophilic substitution of its chlorine atoms, represents the most recent strategy proposed to solve the PAE migration problem. Many studies in this respect have shown that aromatic para-substituted thiol compounds are the most appropriate nucleophiles to achieve high degrees of modification (up to 80 mol%) without any undesired side reactions (Navarro et al., 2008). Another interesting property of thiol groups is their ability to efficiently deactivate radicals. In a recent study Navarro et al. (2010) developed derivatives of phthalic and isophthalic acids that guarantee plasticization of PVC while avoiding any migration. Many ways may be proposed to reduce exposure to PAEs such as avoiding flexible PVC products, limiting the use of personal care products, cosmetics and fragrances, regular home vacuuming to minimize dust build-up and potential PAE exposure, etc.

Large-scale surveys of PAEs in multi-environmental compartments are urged so as to provide better risk assessment and establish a standard for future PAE control.

Future adoption of best management practices (BMP) is desirable to reduce the release of these substances into the environment and their consequent introduction into the food chain. Source control is important for reducing the emission of pollutants into urban watercourses, as is management of building materials and traffic density reduction (Björklund et al., 2008). However, to further decrease pollutant loads in storm water, structural BMPs, such as sedimentation facilities, permeable pavements and bioretention areas, need to be implemented.

Regarding toxicological scientific research in the field, the literature data suggest significant associations between exposure to some PAEs and altered endocrine function and reproductive or developmental effects, but the number of human studies is currently limited. Also, for some of the more studied associations, such as between PAEs and semen quality, the data across studies are not consistent. Owing to the complex nature of the endocrine system, studies should evaluate not only individual hormone levels, but also the ratios between relevant hormones (e.g. LH: testosterone ratio in males as a marker for Leydig cell function) that may help provide clues to the biological mechanisms of xenobiotic activity in humans. Finally, human research is needed on potential latent and transgenerational effects (e.g. epigenetic modifications) of exposure to plastic additives including the possibility of environmentally linked fetal origins of adult diseases, as well as genetic, metabolic, demographic or environmental characteristics resulting in increased individual susceptibility to adverse health effects following exposure (Koch and Calafat, 2009; Meeker *et al.*, 2009).

14.8 Sources of further information and advice

Websites

- http://www.atsdr.cdc.gov/toxprofiles/index.asp
- http://europe.eu.int/comm/food/food/chemicalsafety/foodcontact/legisl_list_en.htm
- http://www.foodstandards.gov.uk/safereating/phthalates/
- http://www.ourstolenfuture.com/
- Consumer Product Safety Commission (CPSC) has implemented a new ban on certain phthalates in children's toys and related products http://www.cpsc.gov/ABOUT/Cpsia/108rfc.pdf

Books and reports

- Charles Staples (2010), *Phthalate esters* (Springer-Verlag Berlin and Heidelberg GmbH & Co. K, Berlin). This book reports the state of scientific knowledge on phthalate esters in the environment. In this monograph the authors report information on analytical methodologies; a compilation of measurements of the concentration in water, sediment, soil, air, dust and food. In addition, key physical properties data and characteristics that control the exposure are described. Ecotoxicity data and relevant toxicity of mammals, human health and environmental risks are also exposed.
- Thomas Wenzl (2009), *Methods for the determination of phthalates in food* (Office for Official Publications of the European Communities Luxembourg).This report is carried out by the Institute for Reference Materials, which is part of the European Commission's Joint Research Centre (JRC-IRMM), and refers about a survey among European food control laboratories on analytical methods applied for the determination of phthalates in food. The aim of this survey is to evaluate comparability of the analysis protocols, to highlight potential pitfalls and as a follow-up to provide support to laboratories that are new in that field.
- Committee on the Health Risks of Phthalates, National Research Council Phthalates and Cumulative Risk Assessment (2008). *The Task Ahead* (The National Academies Press Washington, DC). The Committee on the Health Risks of Phthalates provides information regarding risk-assessment practices, describes their strengths and weaknesses, and provides recommendations for conducting the risk assessment.
- Consumer Product Safety Improvement Act (CPSIA) (2010). *Review of Exposure Data and Assessments for Select Dialkyl Ortho-Phthalates.*

Exposure and Risk Assessment Division (Springfield, VA). The study by CPSIA reported existing data of human exposure to phthalates, privileging the six dialkyl ortho-phthalates that were permanently or temporarily prohibited in children's toys or child-care articles. This report presents background information, such as the volumes of production and uses of phthalates, concentrations of these six phthalates, measured in environmental media (water, soil, sediment, sludge, solid waste, air, dust), in a variety of consumer products and in food and biota which humans may be exposed. Data on human biomonitoring studies to assess exposures to phthalates in the general population or in populations with high potential exposures to phthalates are reported too.
- Bizzari S, Blagoev M and Kishi A (2009), '*Plasticizers*', *Chemical Economics Handbook* (Palo Alto, CA, USA, SRI International). This report shows data about world utilization and forecasting about consumption growth of plasticizers. Consumption of some phthalates, such as linear phthalates, is forecast to decline during 2008–2013, while that of other compounds utilization is forecast to grow.

14.9 References

ADIBI J, PERERA F, JEDRYCHOWSKI W, CAMANN D, BARR D, JACEK R and WHYATT R (2003), 'Prenatal exposures to phthalates among women in New York City and Krakow, Poland', *Environ Health Persp*, **111**, 1719–1722.

ANDERSON W A C, CASTLE L, SCOTTER M J, MASSEY R C and SPRINGALL C (2001), 'A biomarker approach to measuring human dietary exposure to certain phthalate diesters', *Food Addit Contam*, **18**, 1068–1074.

ARMSTRONG B (2003), *Exposure measurement error: consequences and design issues*. In *Exposure assessment in occupational and environmental epidemiology*, New York, Oxford University Press.

ASAKURA H, MATSUTO T and TANAKA N (2004), 'Behavior of endocrine-disrupting chemicals in leachate from MSW landfill sites in Japan', *Waste Manage*, **24**, 613–622.

ATSDR (1995), *Toxicological profile for diethylphthalate*, Atlanta, GA, U.S. Department of Health and Human Services, Public Health Service. Available from: http://www.atsdr.cdc.gov/toxprofiles/index.asp [Accessed 12 March 2013].

ATSDR (1997), *Toxicological profile for di-n-octylphthalate*, Atlanta, GA, U.S. Department of Health and Human Services, Public Health Service. Available from: http://www.atsdr.cdc.gov/toxprofiles/index.asp [Accessed 12 March 2013].

ATSDR (2001), *Toxicological profile for di-n-butyl phthalate*, Atlanta, GA, U.S. Department of Health and Human Services, Public Health Service. Available from: http://www.atsdr.cdc.gov/toxprofiles/index.asp [Accessed 12 March 2013].

ATSDR (2002), *Toxicological profile of di(2-ethylhexyl)phthalate*, Atlanta, GA, U.S. Department of Health and Human Services, Public Health Service. Available from: http://www.atsdr.cdc.gov/toxprofiles/index.asp [Accessed 12 March 2013].

AURELA B, KULMALA H, and SODERHJELM L (1999), 'Phthalates in paper and board packaging and their migration into tenax and sugar', *Food Addit Contam*, **16**, 571–577.

BECKER K, SEIWERT M, ANGERER J, HEGER W, KOCH H, NAGORKA R, ROSSKAMP E, SCHLUTER C, SEIFERT B and ULLRICH D (2004), 'DEHP metabolites in urine of children and DEHP in house dust', *Int J Hyg Envir Heal*, **207**, 409–417.

BERGH C, TORGRIP R, EMENIUS G and ÖSTMAN C (2011), 'Organophosphate and phthalate esters in air and settled dust – a multi-location indoor study', *Indoor Air*, **21**(1), 67–76.

BILITY M T, THOMPSON J T, MCKEE R H, DAVID R M, BUTALA J H, VANDEN HEUVEL J P and PETERS J M (2004), 'Activation of mouse and human Peroxisome Proliferator-Activated Receptors (PPARs) by phthalate monoesters', *Toxicol Sci*, **82**(1), 170–182.

BIZZARI S, BLAGOEV M and KISHI A (2009), 'Plasticizers', *Chemical Economics Handbook*, Palo Alto, CA, USA, SRI International.

BJÖRKLUND K, ALMQVIST H, MALMQVIST P A and STRÖMVALL A M (2008), 'Best management practices to reduce phthalate and nonylphenol loads in urban runoff', 11th International Conference on Urban Drainage, Edinburgh, Scotland 31st August-5th September 2008, UK.

BJORKLUND K, COUSINS P, STROMVALL A and MALMQVIST P (2009), 'Phthalates and nonylphenols in urban runoff: occurrence, distribution and area emission factors', *Sci Total Environ*, **407**(16), 4665–4672.

BLOUNT B C, SILVA M J, CAUDILL S P, NEEDHAM L L, PIRKLE J L, SAMPSON E J, LUCIER G W, JACKSON R J and BROCK J W (2000), 'Levels of seven urinary phthalate metabolites in a human reference population', *Environ Health Persp*, **108**(10), 979–982.

BLUTHGEN A (2003), 'Organic migration agents into milk at farm level (illustrated with diethylhexylphthalate)', *Bull Int Dairy Fed*, **356**, 39–42.

BORNEHAG C G, SUNDELL J, WESCHLER C J, SIGSGAARD T, LUNDGREN B, HASSELGREN M and HÄGERHED-ENGMAN L (2004), 'The association between asthma and allergic symptoms in children and phthalates in house', *Environ Health Persp*, **112**(14), 1393–1397.

BORNEHAG C G, LUNDGREN B, WESCHLER C J, SIGSGAARD T, HÄGERHED-ENGMAN L and SUNDELL J (2005), 'Phthalates in indoor dust and their association with building characteristics', *Environ Health Persp*, **113**, 1399–1404.

BORNEHAG C G and NANBERG E (2010), 'Phthalate exposure and asthma in children', *Int J Androl*, **33**, 333–345.

BOTELHO G, GOLIN M, BUFALO A, MORAIS R, DALSENTER P and MARTINO-ANDRADE A (2009), 'Reproductive effects of di(2-ethylhexyl)phthalate in immature male rats and its relation to cholesterol, testosterone, and thyroxin levels', *Arch Environ Contam Toxicol*, **57**, 777–784.

BROOKS S M and VANDERVORT R (1977), 'Polyvinyl chloride film thermal decomposition products as an occupational illness: 2. clinical studies', *Jom-J Occup Med*, **19**(3), 192–196.

BRUNS-WELLER E and PFORDT J (2000), 'Determination of phthalic acid esters in foods, mother's milk, dust and textiles', *Umweltwiss Schadst Forsch*, **12**(3), 125–130.

BUTTE W, HOFFMANN W, HOSTRUP O, SCHMIDT A and WALKER G (2001), 'Endocrine disrupting chemicals in house dust : results of a representative monitoring', *Gefahrstoffe – Reinhalt Luft*, **61**(1–2), 19–23.

CALAFAT A, NEEDHAM L, SILVA M and LAMBERT G (2004), 'Exposure to di-(2-ethylhexyl) phthalate among premature neonates in a neonatal intensive care unit', *Pediatrics*, **113**, 429–434.

CALAFAT A M and MCKEE R H (2006), 'Integrating biomonitoring exposure data into the risk assessment process: phthalates as a case study', *Environ Health Persp*, **114**(11), 1783–1789.

CASAJUANA N and LACORTE S (2004), 'New methodology for the determination of phthalate esters, bisphenol A, bisphenol A diglycidyl ether, and nonylphenol in commercial whole milk samples', *J Agric Food Chem*, **52**(12), 3702–3707.

CDC (CENTERS FOR DISEASE CONTROL AND PREVENTION) (2005), *Third National Report on Human Exposure to Environmental Chemicals*, Atlanta, GA, Centers for Disease Control and Prevention, National Center for Environmental Health, Division of Laboratory Sciences.

CIRILLO F, FASANO E, ESPOSITO F, MONTUORI P and AMODIO COCCHIERI R (2013a), 'Di(2-ethylhexyl)phthalate (DEHP) and di-n-butylphthalate (DBP) exposure through diet in hospital patients', *Food Chem Toxicol*, **51**, 434–438.

CIRILLO F, FASANO E, ESPOSITO F, DEL PRETE E and AMODIO COCCHIERI R (2013b), 'Study on the influence of temperature, storage time and packaging type on di-n-butylphthalate and di(2-ethylhexyl)phthalate release into packed meals', *Food Addit Contam*, **30**, 403–411.

CLARA M, WINDHOFER G, HARTL W, BRAUN K, SIMON M, GANS O, SCHEFFKNECHT C and CHOVANEC A (2010), 'Occurrence of phthalates in surface runoff, untreated and treated wastewater and fate during wastewater treatment', *Chemosphere*, **78**(9), 1078–1084.

CLARK K, COUSINS I and MACKAY D (2003a), 'Assessment of critical exposure pathways', in STAPLES C, *The handbook of environmental chemistry. Phthalate esters,* New York, Springer-Verlag, 3(Q), 227–262.

CLARK K, COUSINS I, MACKAY D and YAMADA K (2003b), 'Observed concentrations in the environment', in STAPLES C, *The handbook of environmental chemistry. Phthalate esters*, New York, Springer-Verlag, 3(Q), 125–177.

COBELLIS L, LATINI G, DE FELICE C, RAZZI S, PARIS I, RUGGIERI F, MAZZEO P and PETRAGLIA F (2003), 'High plasma concentrations of di-(2-ethylhexyl)-phthalate in women with endometriosis', *Hum Reprod*, **18**(7), 1512–1515.

COLACINO J A, HARRIS T R and SCHECTER A (2010), 'Dietary intake is associated with phthalate body burden in a nationally representative sample', *Environ Health Persp*, **118**, 998–1003.

COLÓN I, CARO D, BOURDONY C J and ROSARIO O (2000), 'Identification of phthalate esters in the serum of young Puerto Rican girls with premature breast development', *Environ Health Persp*, **108**, 895–900.

Commission Directive 2000/60/EC of 23 October 2000 establishing a framework for Community action in the field of water policy, *Official Journal of the European Communities*, L327/1

Commission Directive 2005/84/EC of 14 December 2005 amending for the 22nd time Council Directive 76/769/EEC on the approximation of the laws, regulations and administrative provisions of the Member States relating to restrictions on the marketing and use of certain dangerous substances and preparations (phthalates in toys and childcare articles), *Official Journal of the European Union*, L344/40.

Commission Directive 2007/19/EC of 30 March 2007 amending Directive 2002/72/EC relating to plastic materials and articles intended to come into contact with food and Council Directive 85/572/EEC laying down the list of simulants to be used for testing migration of constituents of plastic materials and articles intended to come into contact with foodstuffs, *Official Journal of the European Union*, L91/17.

Commission Directive 2008/105/EC of 16 December 2008 on environmental quality standards in the field of water policy, amending and subsequently repealing Council Directives 82/176/EEC, 83/513/EEC, 84/156/EEC, 84/491/EEC, 86/280/EEC and amending Directive 2000/60/EC of the European Parliament and of the Council. *Official Journal of the European Union,* L348/84.

Commission Regulation (EU) 10/2011 of 14 January 2011 on plastic materials and articles intended to come into contact with food, *Official Journal of the European Union*, L12/1.

CSTEE (SCIENTIFIC COMMITTEE FOR TOXICITY, ECOTOXICITY AND THE ENVIRONMENT) (1998), *Opinion on Phthalate migration from soft PVC toys and child-care articles- opinion expressed at the 6th CSTEE plenary meeting Brussels.*

DARGNAT C, TEIL M J, CHEVREUIL M and BLANCHARD M (2009), 'Phthalate removal throughout wastewater treatment plant: case study of Marne Aval station (France)', *Sci Total Environ*, **407**, 1235–1244.

DEHRM (DIVISION OF ENVIRONMENTAL HEALTH & RISK MANAGEMENT) (2003), *Measuring the bioavailability of human dietary intake of PAH; phthalates and aromatic hydrocarbons*. Report prepared for the Food Standards Agency, London, UK.

DING E L, SONG Y, MALIK V S and LIU S (2006), 'Sex differences of endogenous sex hormones and risk of type 2 diabetes', *JAMA*, **295**(11), 1288–1299.

DUTY S M, SILVA M J, BARR D B, BROCK J W, RYAN L, CHEN Z, HERRICK R F, CHRISTIANI D C and HAUSER R (2003a), 'Phthalate exposure and human semen parameters', *Epidemiology*, **14**(3), 269–277.

DUTY S M, SINGH N P, SILVA M J, BARR D B, BROCK J W, RYAN L, HERRICK R F, CHRISTIANI D C and HAUSER R (2003b), 'The relationship between environmental exposures to phthalates and DNA damage in human sperm using the neutral comet assay', *Environ Health Persp*, **111**(9), 1164–1169.

DUTY S M, CALAFAT A M, SILVA M J, RYAN L and HAUSER R (2005), 'Phthalate exposure and reproductive hormones in adult men', *Hum Reprod*, **20**(3), 604–610.

DUVIS T, KARLES G and PAPASPYRIDES C D (1991), 'Plasticized PVC films/petroleum oils: the effect of ultraviolet irradiation on plasticizer migration', *J Appl Polym Sci*, **42**(1), 191–198.

EFSA (2005a), Opinion of the Scientific Panel on Food Additives, Flavourings, Processing Aids and Materials in Contact with Food (AFC) on a request from the Commission related to butylbenzylphthalate (BBP) for use in food contact materials, *The EFSA Journal*, **241**, 1–14.

EFSA (2005b), Opinion of the Scientific Panel on Food Additives, Flavourings, Processing Aids and Material in Contact with Food (AFC) on a request from the Commission related to di-butylphthalate (DBP) for use in food contact materials, *The EFSA Journal*, **242**, 1–17.

EFSA (2005c), Opinion of the Scientific Panel on Food Additives, Flavourings, Processing Aids and Materials in Contact with Food (AFC) on a request from the Commission related to bis(2-ethylhexyl)phthalate (DEHP) for use in food contact materials, *The EFSA Journal*, **243**, 1–20.

EFSA (2005d), Opinion of the Scientific Panel on Food Additives, Flavourings, Processing Aids and Materials in Contact with Food (AFC) on a request from the Commission related to di-isodecylphthalate (DIDP) for use in food contact materials, *The EFSA Journal*, **245**, 1–14.

EFSA (2005e), Opinion of the Scientific Panel on Food Additives, Flavourings, Processing Aids and Materials in Contact with Food (AFC) on a request from the Commission related to di-isononylphthalate (DINP) for use in food contact materials, *The EFSA Journal*, **244**, 1–18.

EISEN E A, WEGMAN D H and SMITH T J (1985), 'Across-shift changes in the pulmonary function of meat-wrappers and other workers in the retail food industry', *Scand J Work Environ Health*, **11**(1), 21–26.

ENGEL S M, ZHU C, BERKOWITZ G S, CALAFAT A M, SILVA J, MIODOVNIK A and WOLFF M S (2009), 'Prenatal phthalate exposure and performance on the Neonatal Behavioral Assessment Scale in a multiethnic birth cohort', *Neurotoxicology*, **30**(4), 522–528.

EU-RAR (2004), *Dibutyl Phthalate – Addendum to the Environmental Section*, 1st Priority List, 29, IHCP.

EU-RAR (2007), *Benzyl Butyl Phthalate (BBP)*, 3rd Priority List, 76, IHCP.

EU-RAR (2008), *Bis(2-ethylhexyl) Phthalate (DEHP)*, 2nd Priority List, 80, IHCP.

FALK H and PORTNOY B (1976), 'Respiratory tract illness in meat packers', *JAMA*, **235**, 915–917.

FAUSER P, VIKELSØE J, SØRENSEN P B and CARLSEN L (2003), 'Phthalates, nonylphenols and LAS in an alternately operated wastewater treatment plant – fate modelling based on measured concentrations in wastewater and sludge', *Water Res*, **37**, 1288–1295.

FDA (FOOD AND DRUG ADMINISTRATION) (2001), *Safety Assessment of di(2-ethylhexyl) Phthalate (DEHP) Released from Medical Devices*, Washington DC.

FENG Y L, ZHU J, and SENSENSTEIN R. (2005), 'Development of a headspace solid phase microextraction method combined with gas chromatography mass spectrometry for the determination of phthalate esters in cow milk', *Ana. Chim Acta*, **538**, 41–48.

FERRUTI P, MANCIN I, RANUCCI E, DE FELICE C, LATINI G and LAUS M (2003), 'Polycaprolactone-poly(ethylene glycol) multiblock copolymers as potential substitutes for di(ethylhexyl) phthalate in flexible poly(vinyl chloride) formulations', *Biomacromolecules*, **4**, 181–188.

FOSTER P (2006), 'Disruption of reproductive development in male rat offspring following in utero exposure to phthalate esters', *Int J Androl*, **29**(1), 140–147.

FREIRE M T, SANTANA I A and REYES F G R (2006), 'Plasticizers in brazilian food-packaging materials acquired on the retail market', *Food Addit Contam*, **23**(1), 93–99.

FROMME H, KÜCHLER T, OTTO T, PILZ K, MÜLLER J and WENZEL A (2002), 'Occurrence of phthalates and bisphenol A and F in the environment', *Water Res*, **36**, 1429–1438.

FROMME H, LAHRZ T, PILOTY M, GEBHART H, ODDOY A and RUDEN H (2004), 'Occurrence of phthalates and musk fragrances in indoor air and dust from apartments and kindergartens in Berlin (Germany)', *Indoor Air*, **14**, 188–195.

FROMME H, BOLTE G, KOCH H M, ANGERER B J, BOEHMER S, DREXLER H, MAYER R and LIEBL B (2007a), 'Occurrence and daily variation of phthalate metabolites in the urine of an adult population', *Int J Hyg Envir Heal*, **210**(1), 21–33.

FROMME H, GRUBER L, SCHLUMMER M, WOLZ G, BÖHMER S, ANGERER J, MAYER R, LIEBL B and BOLTE G (2007b), 'Intake of phthalates and di(2-ethylhexyl)adipate: results of the integrated exposure assessment survey based on duplicate diet samples and biomonitoring data', *Environ Int*, **33**(8), 1012–1020.

GASPERI J, GARNAUD S, ROCHER V and MOILLERON R (2009), 'Priority pollutants in surface waters and settleable particles within a densely urbanized area: case study of Paris (France)', *Sci Total Environ*, **407**, 2900–2908.

GEIER J, LESSMANN H, HELLRIEGEL S and FUCHS T (2008), 'Positive patch test reactions to formaldehyde releasers indicating contact allergy to formaldehyde', *Contact Dermatitis*, **58**, 175–177.

GRAY LE JR, LASKEY J and OSTBY J (2006), 'Chronic di-n-butyl phthalate exposure in rats reduces fertility and alters ovarian function during pregnancy in female long evans hooded rats', *Toxicol Sci*, **93**(1), 189–195.

GREEN R, HAUSER R, CALAFAT A M, WEUVE J, SCHETTLER T, RINGER S, HUTTNER K and HU H (2005), 'Use of di(2-ethylhexyl) phthalate-containing medical products and urinary levels of mono(2-ethylhexyl) phthalate in neonatal intensive care unit infants', *Environ Health Persp*, **113**(9), 1222–1225.

GUO Y and KANNAN K (2011), 'Comparative assessment of human exposure to phthalate esters from house dust in China and the United States', *Environ Sci Technol*, **45**, 3788–3794.

HAUSER R, DUTY S, GODFREY-BAILEY L and CALAFAT A (2004), 'Medications as a source of human exposure to phthalates', *Environ Health Persp*, **112**, 751–753.

HAUSER R, MEEKER J D, DUTY S, SILVA M J and CALAFAT A M (2006), 'Altered semen quality in relation to urinary concentrations of phthalate monoester and oxidative metabolites', *Epidemiology*, **17**(6), 682–691.

HAUSER R, BARTHOLD J S and MEEKER J D (2007), 'Epidemiologic evidence on the relationship between environmental endocrine disruptors and male reproductive and developmental health', *Cont Endocrinol*, **II**, 225–251.

Hazardous Substances Data Bank (HSDB), Hazardous substances data bank query for Butyl Benzyl Phthalate. National Library of Medicine's TOXNET system. Available from: http://toxnet.nlm.nih.gov [accessed on 12 March 2013].

Hazardous Substances Data Bank (HSDB), Hazardous substances data bank query for Dibutyl Phthalate. National Library of Medicine's TOXNET system. Available from: http://toxnet.nlm.nih.gov [accessed on 12 March 2013].

HERNÁNDEZ-DÍAZ S, MITCHELL A A, KELLEY K E, CALAFAT A M and HAUSER R (2009), 'Medications as a potential source of exposure to phthalates in the U.S. population', *Environ Health Persp*, **117**(2), 185–189.

HILL S (1997), 'Analysis of contaminants in oxygen from PVC tubing in respiratory therapy, chromatic components in electrochemical sensors, and a model for the degradation of electrical cable insulation', *PhD Thesis*, University of Connecticut, CT, USA.

HOPPIN J A, ULMER R and LONDON S J (2004), 'Phthalate exposure and pulmonary function', *Environ Health Persp*, **112**(5), 571–574.

ISHIDO M, MASUO Y, SAYATO-SUZUKI J, OKA S, NIKI E and MORITA M (2004), 'Dicyclohexylphthalate causes hyperactivity in the rat concomitantly with impairment of tyrosine hydroxylase immunoreactivity', *J Neurochem*, **91**, 69–76.

JAAKKOLA J J, OIE L, NAFSTAD P, BOTTEN G, SAMUELSEN S O and MAGNUS P (1999), 'Interior surface materials in the home and the development of bronchial obstruction in young children in Oslo, Norway', *Am J Public Health*, **89**, 188–192.

JAAKKOLA J J, VERKASALO P K and JAAKKOLA N (2000), 'Plastic wall materials in the home and respiratory health in young children', *Am J Public Health*, **90**, 797–799.

JAAKKOLA J J, IEROMNIMON A and JAAKKOLA M S (2006), 'Interior surface materials and asthma in adults: a population-based incident case-control study', *Am J Epidemiol*, **164**, 742–749.

JAAKKOLA J J and KNIGHT T L (2008), 'The role of exposure to phthalates from polyvinyl chloride products in the development of asthma and allergies: a systematic review and meta-analysis', *Environ Health Persp*, **116**, 845–853.

JAROŠOVÁ A (2006), 'Phthalic Acid Esters (PAEs)' in the Food Chain', *Czech J Food Sci*, **24**, 223–231.

JAYAKRISHNAN A, SUNNY M C AND RAJAN M N (1995), 'Photocrosslinking of azidated poly(vinyl chloride) coated onto plasticized PVC surface-route to containing plasticizer migration', *J Appl Polym Sci*, **56**(10), 1187–1195.

JAYAKRISHNAN A and SUNNY M C (1996), 'Phase transfer catalysed surface modification of plasticized poly(vinyl chloride) in aqueous media to retard plasticizer migration', *Polymer*, **37**(23), 5213–5218.

JAYAKRISHNAN A and LAKSHMI S (1998), 'Immobile plasticizer in flexible PVC', *Nature* **396**, 638.

JÖNSSON B A, RICHTHOFF J, RYLANDER L, GIWERCMAN A and HAGMAR L (2005), 'Urinary phthalate metabolites and biomarkers of reproductive function in young men', *Epidemiology*, **16**, 487–493.

JRC (JOINT RESEARCH CENTRE) (2009), *Annual report 2009 EUR 24246 EN of The Insitute for Reference Materials and Measurements*, Belgium.

JUGAN M L, LEVI Y and BLONDEAU J P (2010), 'Endocrine disruptors and thyroid hormone physiology', *Biochem Pharmacol*, **79**, 939–947.

KAVLOCK R, BARR D, BOEKELHEIDE K, BRESLIN W, BREYSSE P, CHAPIN R, GAIDO K, HODGSON E, MARCUS M, SHEA K and WILLIAMS P (2006), 'NTP-CERHR expert panel update on the reproductive and developmental toxicity of di(2-ethylhexyl) phthalate', *Reprod Toxicol*, **22**, 291–399.

KIM B, CHO S, KIM Y, SHIN M, YOO H, KIM J, YANG Y H, KIM H, BHANG S and HONG Y (2009), 'Phthalates exposure and attention-deficit/hyperactivity disorder in school-age children', *Biol Psychiat*, **66**(10), 958–963.

KIMBER I, AGIUS R, BASKETTER D A, CORSINI E, CULLINAN P, DEARMAN R J, GIMENEZ-ARNAU E, GREENWELL L, HARTUNG T, KUPER F, MAESTRELLI P, ROGGEN E and ROVIDA C

(2007), 'Chemical respiratory allergy: opportunities for hazard identification and characterization', *ATLA*, **35**, 243–265.

KIMBER I and DEARMAN R J (2010), 'An assessment of the ability of phthalates to influence immune and allergic responses', *Toxicology*, **271**, 73–82.

KLAUNIG J E, BABICH M A, BAETCKE K P, COOK J and FANKHAUSER C (2003), 'PPAR[alpha] agonist-induced rodent tumors: modes of action and human relevance[double dagger]', *Cr Rev Toxicol*, **33**(6), 655–780.

KOCH H M, BOLT H M and ANGERER J (2004), 'Di(2-ethylhexyl)-phthalate (DEHP) metabolites in human urine and serum after a single oral dose of deuterium-labelled DEHP', *Arch Toxicol*, **78**, 123–130.

KOCH H M and CALAFAT A M (2009), 'Human body burdens of chemicals used in plastic manufacture', *Philos Trans R Soc Lond B Biol Sci*, **364**, 2063–2078.

KOLARIK B, NAYDENOV K, LARSSON M, BORNEHAG C and SUNDELL J (2008), 'The association between phthalates in dust and allergic diseases among Bulgarian children', *Environ Health Persp*, **116**, 98–103.

KRISHNAN V K, JAYAKRISHNAN A and FRANCIS J D (1991), 'Radiation grafting of hydrophilic monomers on to plasticized poly(vinyl chloride) sheets. II. migration behaviour of the plasticizer from N-vinyl pyrrolidone grafted sheets', *Biomaterials*, **12**(5), 489–492.

KUCHEN A, MÜLLER F, FARINE M, ZIMMERMANN H, BLASER O and WÜTHRICH C (1999), 'Die mittlere tägliche Aufnahme von Pestiziden und anderen Fremdstoffen über die Nahrung in der Schweiz', *Mitt Lebensm Hyg*, **90**, 78–107.

LATINI G, DE FELICE C, PRESTA G, DEL VECCHIO A, PARIS I, RUGGIERI F and MAZZEO P (2003), 'Exposure to di-(2-Ethylhexyl)-phthalate in humans during pregnancy: a preliminary report', *Biol Neonate*, **83**, 22–24.

LEE J H, KIM K O AND JU Y M (1999), 'Polyethylene oxide additive entrapped polyvinyl chloride as a new blood bag material', *J Biomed Mater Res*, **48**, 328–334.

LEUNG P S and SERNIA C (2003), 'The renin-angiotensin system and male reproduction: new functions for old hormones', *J Mol Endocrinol*, **30**, 263–270.

LIU K, LEHMANN K P, SAR M, YOUNG S S and GAIDO K W (2005), 'Gene expression profiling following in utero exposure to phthalate esters reveals new gene targets in the etiology of testicular dysgenesis', *Biol Reprod*, **73**, 180–192.

LIU X, ZHANG D Y, LI Y S, JING XIONG, HE D W, LIN T, LI X L and WEI G H (2009), 'Di-(2-ethylhexyl) phthalate up regulates ATF3 expression and suppresses apoptosis in mouse genital tubercle', *J Occup Health*, **51**, 57–63.

LOPEZ-CARRILLO L, HERNANDEZ-RAMIREZ R U, CALAFAT A M, TORRES-SANCHEZ L, GALVAN-PORTILLO M, NEEDHAM L L, RUIZ-RAMOS R and CEBRIAN M E (2010), 'Exposure to phthalates and breast cancer risk in northern Mexico', *Environ Health Persp*, **118**, 539–544.

LOPEZ-ESPINOSA M J, GRANADA A, ARAQUE P, MOLINA-MOLINA J M, PUERTOLLANO M C, RIVAS A, FERNÁNDEZ M, CERRILLO I, OLEA-SERRANO MF, LÓPEZ C and OLEA N (2007), 'Oestrogenicity of paper and cardboard extracts used as food containers', *Food Addit Contam*, **24**, 95–102.

MAAS R, PATCH S and PANDOLFO T (2004), 'Inhalation and ingestion of phthalate compounds from use in synthetic modeling clays', *Bull Environ Contam Toxicol*, **73**, 227–234.

MAFF (MINISTRY OF AGRICULTURAL, FISHIERIES AND FOOD) (1996), *Food Surveillance Information Sheet – Phthalates in Food*, UK, Joint Food Safety and Standards Group.

MAIN K M, MORTENSEN G K, KALEVA M, BOISEN K A, DAMGAARD I N, CHELLAKOOTY M, SCHMIDT I M, SUOMI A, VIRTANEN H E, PETERSEN J H, ANDERSSON A, TOPPARI J and SKAKKEBAEK N E (2006), 'Human breast milk contamination with phthalates and alteration of endogenous reproductive hormones in three months old infants', *Environ Health Perspec*, **114**, 270–276.

MAIN K M, SKAKKEBÆK N E, VIRTANEN H E and TOPPARI J (2010), 'Genital anomalies in boys and the environment', *Best Pract Res Cl En*, **24**(2), 279–289.

MARKOWITZ J S, GUTTERMAN E M, SCHWARTZ S, LINK B and GORMAN S M (1989), 'Acute health effects among firefighters exposed to a polyvinyl chloride (PVC) fire', *Am J Epidemiol*, **129**(5), 1023–1031.

MARSEE K, WOODRUFF T J, AXELRAD D A, CALAFAT A M and SWAN S H (2006), 'Estimated Daily phthalate exposures in a population of mothers of male infants exhibiting reduced anogenital distance', *Environ Health Persp*, **114**(6), 805–809.

MARTTINEN S K, KETTUNEN R H, SORMUNEN K M and RINTALA J K (2003), 'Removal of bis-(2- ethylhexyl) phthalate at a sewage treatment plant', *Water Res*, **37**, 1385–1393.

MEEKER J D, CALAFAT A M and HAUSER R, (2007), 'Di(2-ethylhexyl) phthalate metabolites may alter thyroid hormone levels in men', *Environ Health Persp*, **115**, 1029–1034.

MEEKER J D, SATHYANARAYANA S and SHANNA H S (2009), 'Phthalates and other additives in plastics: human exposure and associated health outcomes', *Phil Trans R Soc B*, **364**, 2097–2113.

MORTENSEN G K, MAIN K M, ANDERSSON A, LEFFERS H and SKAKKEBÆK N E (2005), 'Determination of phthalate monoesters in human milk, consumer milk, and infant formula by tandem mass spectrometry (LC-MS-MS)', *Anal Bioanal Chem*, **82**(4), 1084–1092.

MÜLLER A K, NIELSEN E and LADEFOGED O (2003), 'Institute of Food Safety and Nutrition (2003). *Human exposure to selected phthalates in Denmark*. FødevareRapport. ISBN: 87–91399–20–3.

MUNKSGAARD E (2004), 'Leaching of plasticizers from temporary denture soft lining materials', *Eur J Oral Sci*, **112**, 101–104.

MURATURE D A, TANG S Y, STEINHARDT G and DOUGHERTY R C (1987) 'Phthalate esters and semen quality parameters', *Biomed Environ Mass Spectrom*, **14**, 473–477.

NAVARRO R, BIERBRAUER K, MIJANGOS C, GOITI E and REINECKE H (2008), 'Modification of PVC with new aromatic thiol compounds. Synthesis and characterization', *Polym Degrad Stab*, **93**(3), 585–591.

NAVARRO R, PEREZ M, TARDAJOS M and REINECKE H (2010), 'Phthalate plasticizers covalently bound to PVC: plasticization with suppressed migration', *Macromol*, **43**(5), 2377–2381.

NIELSEN G D, LARSEN S T, OLSEN O, LOVIK M, POULSEN S K, GLUE C and WOLKOFF P (2007) 'Do indoor chemicals promote development of airway allergy?', *Indoor Air*, **17**, 236–255.

NORBACK D, WIESLANDER G, NORDSTROM K and WALINDER R (2000), 'Asthma symptoms in relation to measured building dampness in upper concrete floor construction, and 2-ethyl-1-hexanol in indoor air', *Int J Tuber Lung Dis*, **4**, 1016–1025.

NRC (NATIONAL RESEARCH COUNCIL) (2006), *Human biomonitoring for environmental chemicals*, Washington, DC, National Academies Press.

NRC (NATIONAL RESEARCH COUNCIL) (2008), *Phthalates and Cumulative Risk Assessment*, Washington, DC, National Academies Press.

OIE L, HERSOUG L and MADSEN J (1997), 'Residential exposure to plasticizers and its possible role in the pathogenesis of asthma', *Environ Health Persp*, **105**, 972–978.

OTAKE T, YOSHINAGA J and YANAGISAWA Y (2004), 'Exposure to phthalate esters from indoor environment', *J Expo Anal Environ Epidemiol*, **14**, 524–528.

PAGE B D and LACROIX G M (1992), 'Studies into the transfer and migration of phthalate esters from aluminium foil paper laminates to butter and margarine', *Food Addit Contam*, **9**, 197–212.

PAGE D B and LACROIX G M (1995), 'The occurrence of phthalate ester and di-2-ethylhexyl adipate plasticizers in Canadian packaging and food sampled in 1985–1989: A survey', *Food Addit Contam*, **2**, 129–151.

PAN G, HANAOKA T, YOSHIMURA M, ZHANG S, WANG P, TSUKINO H, INOUE K, NAKAZAWA H, TSUGANE S and TAKAHASHI K (2006), 'Decreased serum free testosterone in workers exposed to high levels of Di-n-butyl Phthalate (DBP) and Di-2-ethylhexyl Phthalate (DEHP): A cross-sectional study in China', *Environ Health Persp*, **114**(11), 1643–1648.

PETERS J M, CHEUNG C and GONZALEZ F J (2005), 'Peroxisome proliferator-activated receptor-α and liver cancer: where do we stand?', *J Mol Med*, **83**, 774–785.

PETERSEN J H (1991), 'Survey of di-(2-ethylhexyl)phthalate plasticizer contamination of retail Danish milks', *Food Addit Contam*, **8**(6), 701–705.

PETERSEN J H and BREINDAHL T (2000), 'Plasticizers in total diet samples, baby food and infant formulae', *Food Addit Contam*, **17**, 133–141.

PFANNHAUSER W, LEITNER E and SIEGL H (1995), 'Phthalate in Lebensmitteln', *Forschungsbericht*, Sektion III, in German. Österreichisches Bundesministerium für Gesundheit und Konsumentenschutz. Wien.

PÖHNER A, SIMROCK S, THUMULLA J, WEBER S and WIRNER T (1997), 'Background contamination in household dust in private homes with semi- and nonvolatile organic compounds', *Analyse und Bewertung von Umweltschadstoffen*, in German. AnBUS (Eds) Germany.

POLAKOFF P L, LAPP N L and REGER R (1975), 'Polyvinyl chloride pyrolysis products. A potential cause for respiratory impairment', *Arch Environ Health*, **30**(6), 269–271.

REDDY B S, ROZATI R, REDDY B V and RAMAN N V (2006), 'Association of phthalate esters with endometriosis in Indian women', *BJOG*, **113**, 515–520.

RHIND S M, KYLE C E, TELFER G, MARTIN G, DUFF E and SMITH A (2002), 'Phthalate and alkyl phenol concentrations in soil following applications of inorganic fertilizer or sewage sludge to pasture and potential rates of ingestion by grazing ruminants', *J Environ Monit*, **4**,142–148.

ROSLEV P, VORKAMP K, AARUP J, FREDERIKSEN K and NIELSEN P H (2007), 'Degradation of phthalate esters in an activated sludge wastewater treatment plant', *Water Res*, **41**(5), 969–976.

ROZATI R, REDDY P P, REDDANNA P and MUJTABA R (2002), 'Role of environmental estrogens in the deterioration of male factor fertility', *Fertil Steril*, **78**(6), 1187–1194.

RUDEL R, BRODY J, SPENGLER J, VALLARINO J, GENO P, SUN G and YAU A (2001), 'Identification of selected hormonally active agents and animal mammary carcinogens in commercial and residential air and dust samples', *J Air Waste Manag Assoc*, **51**, 499–513.

RUDEL R A, CAMANN DE, SPENGLER J D, KORN L R and BRODY J G (2003), 'Phthalates, Alkylphenols, Pesticides, Polybrominated Diphenyl Ethers, and Other Endocrine-Disrupting Compounds in Indoor Air and Dust', *Environ Sci Technol*, **37**, 4543–4553.

RUUSKA R M, KORKEALA H, LIUKKONEN-LILJA H, SUORTTI T and SALMINEN K (1987), 'Migration of contaminants from milk tubes and teat liners', *J Food Protect*, **50**, 316–320.

SACRISTAN J, MIJANGOS C, REINECKE H, SPELLS S and YARWOOD J (2000), 'Selective surface modification of PVC films as revealed by confocal Raman microspectroscopy', *Macromolecules*, **33**(16), 6134–6139.

SALAZAR-MARTINEZ E, ROMANO-RIQUER P, YANEZ-MARQUEZ E, LONGNECKER M P and HERNANDEZ-AVILA M (2004), 'Anogenital distance in human male and female newborns: a descriptive, cross-sectional study', *Environ Health*, **3**(1), 8.

SATHYANARAYANA S (2008), 'Phthalates and children's health', *Curr Probl Pediatr Adolesc Health Care*, **38**(2), 34–49.

SELVIN E, FEINLEIB M, ZHANG L, ROHRMANN S, RIFAI N, NELSON W G, DOBS A, BASARIA S, GOLDEN S H and PLATZ E A (2007), 'Androgens and diabetes in men: results from the Third National Health and Nutrition Examination Survey (NHANES III)', *Diabetes Care*, **30**, 234–238.

SHARMAN M, READ W A, CASTLE L and GILBERT J (1994), 'Levels of di-(2 ethylhexyl) phthalate and total phthalate esters in milk, cream, butter and cheese', *Food Addit Contam*, **11**, 375–385.

SILVA M J, REIDY J A, KATO K, PREAU J L JR, NEEDHAM LL and CALAFAT AM (2007a), 'Assessment of human exposure to di-isodecyl phthalate using oxidative metabolites as biomarkers', *Biomarkers*, **12**(2), 133–144.

SILVA M J, SAMANDAR E, REIDY J A, HAUSER R, NEEDHAM L L and CALAFAT A M (2007b), 'Metabolite profiles of di-n-butyl phthalate in humans and rats', *Environ Sci Technol*, **41**, 7576–7580.

SOERENSEN L (2006), 'Determination of phthalate in milk and milk products by liquid chromatography/tandem mass spectrometry', *Rapid Commun Mass Spectrom*, **20**(7), 1135–1143.

STAHLHUT R W, VAN WIJNGAARDEN E, DYE T D, COOK S and SWAN S H (2007), 'Concentrations of urinary phthalate metabolites are associated with increased waist circumference and insulin resistance in adult U.S. males', *Environ Health Persp*, **115**(6), 876–882.

SUGIURA K, SUGIURA M, HAYAKAWA R, SHAMOTO M and SASAKI K (2002), 'A case of contact urticaria syndrome due to di(2-ethylhexyl) phthalate (DOP) in work clothes', *Contact Dermatitis*, **46**(1), 13–16.

SWAN S H, MAIN K M, LIU F, STEWART S L, KRUSE R L, CALAFAT A M, MAO C S, BRUCE REDMON J, TERNAND C L, SULLIVAN S and LYNN TEAGUE J (2005), 'Decrease in anogenital distance among male infants with prenatal phthalate exposure', *Environ Health Persp*, **113**, 1056–1061.

SWAN S H (2008), 'Environmental phthalate exposure in relation to reproductive outcomes and other health endpoints in humans', *Environ Res*, **108**, 177–184.

SWAN S H, LIU F, HINES M, KRUSE R L, WANG C, REDMON J B, SPARKS A and WEISS B (2010), 'Prenatal phthalate exposure and reduced masculine play in boys', *Int J Androl*, **33**, 259–269.

TAHRI-JOUTEI A, FILLION C, BEDIN M, HUGUES J N and POINTIS G (1991), 'Local control of Leydig cell arginine vasopressin receptor by naloxone', *Mol Cell Endocrinol*, **79**, 21–24.

TANIDA T, WARITA K, ISHIHARA K, FUKUI S, MITSUHASHI T, SUGAWARA T, TABUCHI Y, NANMORI T, QI W M, INAMOTO T, YOKOYAMA T, KITAGAWA H and HOSHI N (2009), 'Fetal and neonatal exposure to three typical environmental chemicals with different mechanisms of action: mixed exposure to phenol, phthalate, and dioxin cancels the effects of sole exposure on mouse midbrain dopaminergic nuclei', *Toxicol Lett*, **189**(1), 40–47.

TEITELBAUM S L, BRITTON J A, CALAFAT A M, YEB X, SILVA M J, REIDY J A, GALVEZA M P, BRENNER B L and WOLFF M S (2008), 'Temporal variability in urinary concentrations of phthalate metabolites, phytoestrogens and phenols among minority children in the United States', *Environ Res*, **106**, 257–269.

TSUMURA Y, ISHIMITSU S, SAITO I, SAKAI H, KOBAYASHI Y and TONOGAI Y (2001a), 'Eleven phthalate esters and di(2-ethylhexyl) adipate in one-week duplicate diet samples obtained from hospitals and their estimated daily intake', *Food Addit Contam*, **18**(5), 449–460.

TSUMURA Y, ISHIMITSU S, KAIHARA A, YOSHII K, NAKAMURA Y and TONOGAI Y (2001b), 'Di(2-ethylhexyl) phthalate contamination of retail packed lunches caused by PVC gloves used in the preparation of foods', *Food Addit Contam*, **18**(6), 569–579.

TSUMURA Y, ISHIMITSU S, SAITO I, SAKAI H, TSUCHIDA Y and TONOGAI Y (2003), 'Estimated daily intake of plasticizers in 1-week duplicate samples following regulation of DEHP-contaminated PVC gloves in Japan', *Food Addit Contam*, **20**(4), 317–324.

TUOMAINEN A, SEURI M and SIEPPI A (2004), 'Indoor air quality and health problems associated with damp floor coverings', *Int Arch Occup Environ Health*, **77**, 222–226.

WANG M, KOU L, ZHANG Y and SHI Y (2007), 'Matrix solid-phase dispersion and gas chromatography/mass spectrometry for the determination of phthalic acid esters in vegetables', *Chin J Chromatogr*, **25**(4), 577–580.

WARE G and WHITACRE D (2004), 'An introduction to insecticides', in WILLOUGHBY O, *The Pesticide Book*, Willoughty, OH, USA, MeisterPro Information Resources, pp. 496.

WELSH M, SAUNDERS P T, FISKEN M, SCOTT H M, HUTCHISON G R, SMITH L B and SHARPE R M (2008), 'Identification in rats of a programming window for reproductive tract masculinization, disruption of which leads to hypospadias and cryptorchidism', *J Clin Invest*, **118**, 1479–1490.

WENZL T (2009), 'Methods for the determination of phthalates in food. Outcome of a survey conducted among European food control laboratories', *European Commission. Joint Research Center, Institute for Reference Materials and Measurement*, Belgium.

WEUVE, J, SANCHEZ B N, CALAFAT A M, SCHETTLER T, GREEN R A, HU H and HAUSER R (2006), 'Exposure to phthalates in neonatal intensive care unit infants: Urinary concentrations of monoesters and oxidative metabolites', *Environ Health Persp*, **114**(9), 1424–1431.

WITTASSEK M and ANGERER J (2008), 'Phthalates: metabolism and exposure', *Int J Androl*, **31**, 131–138.

WORMUTH M, ACHERINGER M, VOLLENWEIDER M and HUNGERBUHLER M (2006), 'What are frequently used phthalic acid esters in Europeans?' *Risk Anal*, **26**(3), 803–824.

YANO K, HIROSAWA N, SAKAMOTO Y, KATAYAMA H, MORIGUCHI T and ASAOKA K (2005), 'Phthalate levels in baby milk powders sold in several countries', *Bull Environ Contam Toxicol*, **74**, 373–379.

ZENG F, CUI K C, XIE Z Y, WU L N, LUO D L, CHEN L X, LIN Y J, LIU M and SUN G X (2009), 'Distribution of phthalate esters in urban soils of subtropical city, Guangzhou, China', *J Hazard Mater*, **164**, 1171–1178.

ZHANG Y H, ZHENG L X and CHEN B H (2006), 'Phthalate exposure and human semen quality in Shanghai: a cross sectional study', *Biomed Environ Sci*, **19**, 205–209.

ZHU J, PHILLIPS S, FENG Y and YANG X (2006), 'Phthalates esters in human milk: concentration variations over a 6-month postpartum time', *Environ Sci Technol*, **40**, 5276–5281.

15
Polychlorinated naphthalenes (PCNs) in foods: sources, analytical methodology, occurrence and human exposure

A. Fernandes, The Food and Environment Research Agency (FERA), UK

DOI: 10.1533/9780857098917.2.367

Abstract: Polychlorinated naphthalenes (PCNs) have long been recognised as legacy contaminants originating through releases from electrical equipment and industrial chemicals and processes. In recent years, the advances in measurement techniques have allowed characterisation of total PCN occurrence, yielding more specific concentration data including individual congeners, some of which are known to elicit aryl hydrocarbon receptor (AhR) mediated responses. The main path to human exposure is generally through the diet and this chapter investigates the most recent information on PCN occurrence in food and in human tissues.

Key words: congener-specific, dietary intake, dioxin-like, fish, shellfish, toxic equivalent (TEQ).

15.1 Introduction

Polychlorinated napthalenes (PCNs) are members of the class of chlorinated polycyclic aromatic hydrocarbons – a potentially vast group of little-known environmental contaminants of anthropogenic origin. PCNs are the most studied subgroup of these compounds and some of the 75 congeners have recognised toxic, bio-accumulative and persistence properties which characterise these compounds as environmental contaminants that have been reported to elicit severe or even fatal biological effects in humans and animals depending on the severity of exposure. Nine fatalities arising through occupational exposure to chlorinated naphthalene have been reported, mostly involving cases of intoxication among cable workers, assemblers and labourers ('Trichloronaphthalenes' – US library of medicine-hazmap database;

CICAD, 2001). In two similar occurrences, the Yusho incident in Japan in 1968, and the Yu-Cheng incident in Taiwan in 1979, rice oil was contaminated with PCNs, among other similar contaminants (Haglund *et al.*, 1995), which were then detected in the blood and tissues of the victims (Ryan and Masuda, 1994). Inadvertent contamination of feed by PCNs caused cattle poisoning incidents in the 1940s and 1950s in the USA. Initially called disease X, the poisoning was characterised by latterly recognised symptoms of bovine hyperkeratosis, weakness and emaciation, followed by death (Bell, 1953; CICAD, 2001). The majority of the most severe cases of poisoning occurred before the widespread recognition of the toxic potential of these compounds, in particular aryl hydrocarbon receptor (AhR) mediated responses. Some of the most studied toxic effects of PCNs are thought to occur through interaction with the AhR, a dioxin-like property of some PCN congeners that almost certainly arises from the structural similarity in the planar configuration of these compounds to polychlorinated dibenzo-p-dioxins and dibenzofurans (PCDD/F) and some laterally substituted polychlorinated biphenyls (PCB). In common with other persistent lipophilic contaminants, PCNs are ingested by humans and farm animals during feeding, and this transfer pathway is widely recognised as being the main route to human exposure, unless there is specific occupational exposure. PCNs are commonly compared to PCBs, with which they share a number of properties and industrial applications. In general, their levels of occurrence are lower than PCBs, but their toxicity is comparable to the most potent PCB congeners.

15.2 Sources of PCNs

The pathways leading to the occurrence of PCNs in food have their origins in a number of sources, the most important of which is direct production. There are, however, other, mostly inadvertent, sources of PCNs, such as occurrence as by-products in other mass produced chemicals and industrial processes, and formation during combustion processes.

15.2.1 Direct production

PCNs are an industrial chemical, produced over most of the last century, principally from the 1930s to the 1980s. An estimate of the total global production was reported as approximately 1 50 000 metric tonnes (Falandysz, 1998). They were sold as technical mixtures (e.g. Halowax in the US, Nibren in Germany, Clonacire in France, Seekay in the UK, etc.) of the commercial PCN product dissolved in mineral oil. The mixtures differed in the degree of substitution of chlorine present on the naphthalene molecule, ranging from mono-chloro- to octa-chloro-substituted naphthalene. Apart from widespread commercial use as dielectrics, PCNs were also used as lubricants, electric cable insulation, wood, paper and fabric preservatives, cutting and grinding fluids, and plasticisers. Tens of thousands of tonnes were produced in many industrial countries, until manufacture ceased (as in the USA, in 1980) or their use was prohibited (as in Germany, in 1983). The production and use of PCNs are now severely restricted or banned in many countries.

15.2.2 Occurrence in other chemicals and processes

While inadvertent release from the production, usage and recycling/disposal of PCN-containing material is likely to be the largest source of environmental burdens, another important source of PCNs arises from their occurrence as contaminants in other similar industrial chemicals, such as PCBs, which were manufactured on a large scale. It is widely known, for example, that PCNs occur in commercial PCB mixtures (Haglund et al., 1995; Falandysz, 1998; Helm et al., 2000; Kannan et al., 2000a) such as Aroclors, Kanechlors, Phenoclors, Sovol, etc. Naphthalene can occur as a contaminant in commercially produced biphenyl and undergoes chlorination during the production of PCBs. The differences in reactivity (naphthalene is more reactive) afford a higher degree of chlorination than biphenyl, and the exact characteristics of co-formation vary depending on the specific commercial PCB formulations. Yamashita et al. (2000), conducted a comparative study of PCN concentrations (tri- to octa-chlorinated congeners) and profiles in 18 different commercial PCB mixtures. The report estimated a global by-production of approximately 169 tonnes of PCNs occurring through this inadvertent mechanism. The study suggested that the lowest concentrations occurred in Aroclors, ranging from 5.2 to 67 µg/g. In Sovol, a product from the former USSR, however, the PCN concentration was estimated at ~ 731 µg/g. The study also reported that the total PCN concentration in Aroclors and Kanechlor was proportional to increasing chlorine content, a trend that was not observed in Phenoclors. In comparison to other similar studies conducted earlier, the PCN concentrations detected were lower, a finding that was attributed to selective volatilisation of the lower chlorinated congeners (mono-, di- and tri-CNs). This property was also cited as a reason for the differences in concentration and congener profiles in three different samples of the same PCB mixture – Aroclor 1254 (Yamashita et al., 2000).

Other legacy sources of PCNs result from the covert movement of commercial mixtures from areas and countries where production and use is banned, or from historical unintentional emissions associated with the chloro-alkali industry, where they were used as lubricants or impregnants for the graphite electrodes used in the chemical processes, and from releases from the refining of metals such as aluminium, magnesium and copper (Vogelgesang, 1986; Jarnberg et al., 1993; Theisen et al., 1993; Aittola et al., 1994; Takasuga et al., 2004). Kannan et al. (2000a) reported concentrations of total PCNs as high as 23 µg/g dry wt, in sediments from coastal Georgia, USA, contaminated by the disposal of wastes from the chloro-alkali processes. Considerably lower (3–5 orders of magnitude) concentrations were observed in associated

370 Persistent organic pollutants and toxic metals in foods

biota (blue crab, fish and birds), with biota profiles predominated by tetra- or penta-chloronaphthalenes, while hepta- and octa-chloronaphthalenes were dominant in the sediments. (Kannan *et al.*, 2000a).

15.2.3 Combustion sources

It is recognised that PCNs can also be formed through industrial thermodynamic processes such as incineration. Falandysz (1998), providing an environmental update on earlier literature, reported similar patterns of PCN occurrence observed in common incineration products such as fly ash and flue gas (Imagawa *et al.*, 1993; Imagawa and Yamashita, 1994; Jarnberg *et al.*, 1997). Formation pathways resulting from *de novo* synthesis during combustion have been documented (Iino *et al.*, 1999; Imagawa and Lee, 2001), suggesting that the chlorination of naphthalene and other aromatic fused-ring structures occurs beyond the combustion stages, on the active surfaces of fly ash. The full range of mono- to octa-chlorinated PCNs has been observed (Takasuga *et al.*, 1994; Abad *et al.*, 1999), although the relative proportions have varied depending on the conditions of combustion. In a study of emission samples from municipal incinerators, Abad *et al.* (1999), reported levels of total PCNs from 1.08 up to 21.36 ng/Nm^3 for mono- to octa-CNs and 0.33 to 5.72 ng/Nm^3 for tetra- to octa-CNs. Under the thermodynamic conditions of combustion, thermal degradation of the more highly chlorinated PCNs (hepta- and octa-CNs), has also been reported (Schneider *et al.*, 1996), which could lead to the formation of the more toxic, lower chlorinated congeners. Emissions from combustion-related processes input directly into ambient air and can influence local atmospheric concentrations, as reported in more recent studies on PCN formation (Helm and Bidleman, 2003; Takasuga *et al.*, 2004).

15.3 Toxicology

A number of studies have reported on the biological effects of PCNs in test animals and humans.

15.3.1 Metabolism and half-lives

The limited data on metabolism of PCNs have generally been derived from acute dosing studies on test animals e.g. rats and rabbits (Chu *et al.*, 1976, 1977a, 1977b) and pigs (Ruzo *et al.*, 1975, 1976a, 1976b), followed by urine analysis. The most common metabolites found were chlorinated naphthols which were either partially dechlorinated or which had undergone rearrangement of the substituted chlorines. These observations indicate that metabolic pathways proceed mainly through hydroxylation of the parent PCN molecule, sometimes with accompanying dechlorination and rearrangement of the chlorine substituents.

Given the structural similarities it is unsurprising that half-lives of PCNs in mammalian systems should be similar to chlorinated dioxins and PCBs. In a study on selective retention of hexachloronaphthalenes in rat liver and adipose tissue, Asplund et al. (1994) reported half-lives of 36 days in the liver and 41 days in adipose tissue. However, much longer half-lives have been reported in humans, with analysis of blood samples taken from three victims of the Yu-Cheng rice oil contamination incident in 1979, indicating a half-life of 1.5 to 2.4 years (Ryan and Masuda, 1994). A later study on human milk samples from Sweden collected between 1972 and 1992 (Noren and Meironyte, 2000) showed a 16% decrease in PCN concentrations between these two periods and the half-life (given as t_{dec} ½, the time given to halve the concentration) was reported as 8 years. While the results of the studies vary, it is clear that PCNs are strongly retained in human tissues and are persistent.

15.3.2 Ah receptor-mediated toxicity

The halogenated fused diaromatic structure provides strong chemical stability and the molecule is resistant to attack by strong acids. All chloronaphthalene congeners are planar and lipophilic compounds. The structural similarity of the molecules to the highly toxic 2,3,7,8-tetrachlorodibenzo-p-dioxin molecule suggests an AhR-mediated mechanism of toxicity. The induction of aryl hydrocarbon hydroxylase (AHH) and ethoxyresorufin O-de-ethylase (EROD) enzymes is a short term biological response that is specifically indicative of planar diaromatic halogenated hydrocarbons, such as dioxin and dioxin-like compounds. The few available data indicate that several PCNs are potent inducers of H4IIE-EROD, AHH and H4IIE-luciferase (Hanberg et al., 1990; Engwall et al., 1994; Blankenship et al., 2000). Among the PCN congeners tested, the most potent EROD, AHH and luciferase inducers were hexa- and hepta-chloronaphthalenes and, to a lesser extent, penta-chloronaphthalenes.

A number of studies over the years have sought to determine the Ah receptor binding potencies of PCNs, and these have utilised either commercial mixtures or individual congeners. In situations where direct exposure to commercial PCNs occurs, such as accidental discharge or inadvertent contamination of food, the data from commercial mixture studies of induction potencies may prove useful. However, most exposure (apart from occupational) is likely to be through the diet, where removal mechanisms, initially in the environment, such as weathering and environmental distillation, followed by metabolism and selective uptake in animals used for food, would modify the profiles that originally characterised the sources. Thus, in order to evaluate the risk from exposure of general populations, induction data derived from individual or selected congener data are likely to be more useful. In assays such as EROD, AHH and luciferase, the relative induction potencies of individual PCN congeners appear to be related to the degree and position of chlorine substitution. This specificity is common to other similar classes of similar compounds, such as PCDD/Fs and PCBs. A number of studies (Hanberg et al.,

1990, 1991; Norrgren *et al.*, 1993; Engwall *et al.*, 1994; Brunström *et al.*, 1995) concur on the relatively high potencies shown in EROD and AHH induction of three PCN congeners PCNs 66/67 and 73 (1,2,3,4,6,7/1,2,3,5,6,7-hexa-CN and 1,2,3,4,5,6,7-hepta-CN). PCN 75 (octachloronaphthalene) also showed a dose-dependent induction of AHH in rat microsomes (Campbell *et al.*, 1981). These findings are generally confirmed by later *in vitro* studies (Hanberg *et al.*, 1990; Villeneuve *et al.*, 2000; Behnisch *et al.*, 2003) on individual congeners, or quantitative structure–activity relationship predictions of relative *in vitro* potencies (Falandysz and Puzyn, 2004, Puzyn *et al* 2007).The more potent responses were seen for congeners with 4 or more chlorines, with a maximum response for the hexa- and hepta-chlorinated compounds, as observed earlier.

A number of other biological effects have been reported, including a combination of toxic responses such as mortality, embryotoxicity, hepatotoxicity, immunotoxicity, dermal lesions, teratogenicity and carcinogenicity (Hanberg *et al.*, 1990; Engwall *et al.*, 1994; Blankenship *et al.*, 1999; Blankenship *et al.*, 2000; Villeneuve *et al.*, 2000). In humans, severe skin reactions (chloracne) and liver disease have been reported after occupational exposure to PCNs. Other symptoms found in occupationally exposed workers include irritation of the eyes, cirrhosis of the liver, fatigue, headache, anaemia, haematuria and nausea. Workers exposed to PCNs also had a slightly higher risk of all cancers combined (CICAD, 2001).

15.4 Methods of analysis of PCNs in foods

PCNs are comprised of 8 homologue groups (mono- to octa-chlorinated) and 75 possible structural configurations. Based on the chemical kinetics of addition/substitution reactions during the manufacture of commercial products and the thermodynamics of formation during combustion processes, varying amounts of the different congeners are formed. Thus, in common with other organic environmental contaminants, most of the methodologies that have been developed for PCN analysis have drawn on the experience gained from the analysis of similar contaminants, such as PCBs. Like PCBs, PCNs are amenable to capillary column gas chromatography (GC), which provides a powerful tool for the separation of individual congeners. Although historically, GC was commonly coupled with electron capture detection, the trend has moved to using the greater specificity of mass spectrometry, with the higher sensitivity that is provided in selected ion monitoring mode. In the most recent reports of PCN determinations in food, these highly selective and sensitive measurement techniques have been complemented with internal standardisation using 13C labelled PCN analogues of the main congeners of interest. Used in conjunction with selective extraction and purification techniques which exclude potential interferants such as PCBs, these methodologies are capable of producing targeted, reliable data which can then be combined with relevant toxic equivalence factor (TEF) or relative potency (REP) values, to produce realistic estimates of the PCN toxicity associated with food samples.

In general, the chromatographic profiles of PCN congeners in environmental samples and some biota still bear a resemblance to some commercial mixtures (Fernandes et al., 2010). Some researchers have reported findings in biota and food on the basis of this similarity and have reported the sum total of occurring PCNs, or the sums of homologue groups, usually tetra- to octa-chlorinated, as these comprise the majority of congeners that are not environmentally degraded. Historically, studies have tended to characterise commercial PCN mixtures using in-house synthesised standards or the small number of commercially available individual congeners, which can then be used as secondary standards for quantification. Other studies have used limited numbers of congeners e.g. Wiedmann and Ballschmiter, 1993 have suggested a method for quantification of PCNs based on only one or two congeners. A drawback of using commercial mixtures as analytical standards is the issue of matching profiles observed in samples with those of the standard mixtures. The differences may arise as a result of environmental weathering mechanisms, metabolism and selective uptake in species, and input of selected congeners from combustion sources. Consequently, more recent investigations of the occurrence of these chemicals in biota and food and the resulting dietary exposure have targeted individual congeners, the selection of which is based on the levels and patterns of occurrence as well as the toxicological characteristics of the individual congeners (Fernandes et al., 2010, 2011). Although it is currently possible to procure some individual congener standards commercially, the poor availability of a large range of relevant individual congener standards continues to hinder a full and accurate measurement of, in particular, the toxicologically significant congeners – an important factor for meaningful risk assessment. More recent methodologies have used individual native congener standards (Isosaari et al., 2006; Fernandes et al., 2010, 2011), complemented by the use of ^{13}Carbon labelled PCNs.

15.4.1 Analytical methodologies

The similarities in molecular structure and physical and chemical properties of PCNs to other environmental contaminants that are more routinely measured, such as polychlorinated dibenzo-p-dioxins, polychlorinated dibenzofurans (PCDD/F) and PCBs, suggest that a similar sample extraction scheme to that used for PCDD/F analysis would prove equally effective. This is borne out by analytical methodologies reported by different researchers working on PCNs who have also used similar extraction solvents (Kannan et al., 2000b; Domingo et al., 2003; Isosaari et al., 2006; Marti-cid et al., 2008; Fernandes et al., 2010, 2011), usually hexane/dichloromethane, hexane/acetone or toluene.

As most foods contain varying amounts of lipids that are co-extracted, most methodologies have used exclusion techniques to remove these undesirable compounds, either gel permeation (Haglund et al., 1993; Domingo 2003) or, more commonly, acid hydrolysis (Kannan et al., 2000b; Guruge et al., 2004; Jiang et al., 2007; Fernandes et al., 2010), usually in the form of sulphuric acid coated on particulate silica. An important consideration in the purification of PCN extracts is the need to exclude other similar lipophilic contaminants that may have been co-extracted but are not removed or degraded with the lipids. Some compounds, such as the PCBs in particular, have the potential to compromise the subsequent measurement stages, as they are generally present at greater concentrations than the PCNs. Some methodologies incorporate a fractionation stage using either alumina (Jiang et al., 2007; Marti-cid et al., 2008), porous graphitic carbon (Jiang et al., 2007) or activated carbon (Kuehl et al., 1994; Isosaari et al., 2006; Fernandes et al., 2010). These adsorbents exploit the planar structure of the PCNs, in particular the strong retention on activated carbon which requires reverse elution with toluene to desorb, as opposed to the more abundant PCBs which are, relatively, poorly retained on carbon. Some methodologies effectively combine extraction, purification and fractionation in a single stage (Fernandes et al., 2010).

Due to the relatively low concentrations of PCNs found in foods, coupled with the desire for specificity of measurement, most recently reported methodologies for the analysis of PCNs have used high resolution GC coupled to high resolution mass spectrometry (HRGC-HRMS), usually operating at 10 000 resolution in the electron ionisation mode, for measurement. Although non-polar gas chromatographic columns, typically 5% phenylmethly polysiloxane, are commonly used (Guruge et al., 2004; Llobet et al., 2007; Marti-Cid et al., 2008; Kunisue et al., 2009; Fernandes et al., 2010; Park et al., 2010), some congeners remain unresolved. Examples of this are PCNs 66 and 67 which co-elute, (but fortunately have similar toxicities), but can be separated using other stationary phases, such as a Rt-βDEXcst column (Helm et al., 1999), if complete resolution is required, as in studies on environmental transport.

As mentioned earlier, historical reporting of data has sometimes used an approximation to the composition of commercial mixtures, but more recent and sophisticated analyses are able to characterise PCNs as homologue groups, or more specifically as individual congeners. This allows presentation of data to be made either as concentration values or as a more integrative representation of the toxicological content of a sample. This can be done by reference to a relative potency value, or a TEF value such as that currently in use for dioxins and PCBs (Van den Berg et al., 2006). This latter approach would be reasonable because, as per the rules of additivity of response, the candidate compounds should demonstrate the ability to bind to the Ah receptor, be structurally similar to dioxins (chlorinated planar aromatics), be bio-accumulative and persistent in food webs, and elicit AhR-mediated biochemical and toxic responses. As already seen, a number of PCN congeners are able to fulfil these criteria.

15.5 Occurrence in foods

In a manner similar to PCDD/Fs and PCBs, the main pathway of human exposure to PCNs is likely to be through dietary intake. This is also expected to be the most likely pathway in aquatic biota, particularly fish species and their predators such as black cormorants, white-tailed eagles and porpoises (Falandysz, 1997). Further, as the log K_{ow} values for the higher chlorinated (and more persistent) PCNs are greater than 5 (Environment Canada, 2011), it is predicted that for aquatic fish species, dietary uptake is likely to be much more significant than uptake from water (Arnot and Gobas, 2003).

However, despite the recognition of their toxic potential, very few data exist on the occurrence of these contaminants in foods or food chains, which provides poor support to risk assessments based on dietary exposure. The dearth of data (particularly individual compounds) may be, at least in part, due to analytical inaccessibility, caused by the lack of a selective and sensitive analytical methodology, and a lack of reliable individual congener standards. Recently, some standards have become commercially available and, given the rising interest, it is expected that more will follow.

A review carried out by Falandysz, 2003, on PCNs as food chain contaminants adequately covers earlier findings (prior to 2000) relating to toxicology, environmental occurrence and persistency, the fate of PCNs in environmental compartments, and incorporation into marine and terrestrial sources of human food. The following section collates more recent findings (over the last decade) on PCNs in species of fish and marine mammals that are consumed by humans, and the recent investigations into PCN concentrations in retail foods.

15.5.1 Marine and aquatic biota

PCNs have been detected in several environmental compartments, including biota that are used, or have the potential to be used, as food. They have been measured in fish collected in 1996–1997, from the Great Lakes, and lakes and rivers from inland Michigan, in various species such as trout, carp, bass and pike, with detection of PCNs ranging from 19 to 31 400 ng of summed PCNs/ kg on a wet (whole) weight basis (Kannan *et al.*, 2000b). The toxic equivalence associated with these concentrations was reported as 0.007 to 11 ng/kg wet weight, with the major contribution to TEQ arising from hexa-chlorinated congeners – PCN 66, 67 and 69. Fish from the Detroit river showed the highest concentrations. In the same year, Ishaq *et al.*, 2000, reported PCN contamination in harbour porpoises from the west coast of Sweden with concentrations of up to 730 ng/kg wet weight (ww) in blubber nuchal (posterior of the neck) fat and liver. Hexa-chlorinated congeners detected in the liver were the biggest PCN contributors to the TEQ, providing about half the total amount. Helm *et al.* (2000), reported concentrations of 35.4 to 383 ng/kg lipid weight in blubber from ringed seal and beluga whale. A range of fish species from the Baltic sea and three Finnish lakes were measured, with levels ranging from 1–170 ng/kg ww for Baltic sea samples and 2–66 ng/kg ww for the lakes, reflecting much higher contamination levels for the same species in the Baltic sea compared to the lakes (Isosaari *et al.*, 2006). In a study investigating PCN concentrations in edible fish and seafood collected from retail outlets in six Catalonian cities in 2005, Llobet *et al.* (2007) reported on 14 of the most consumed fish and shellfish species in Spain, ranging from sardines, tuna, salmon and mackerel to mussels, clams and shrimp. PCN concentrations reported as the totals of tetra-octa-chlorinated homologues ranged from 3 ng/kg ww for cuttlefish to 227 ng/kg ww for salmon.

A study of seafood from China (Jiang *et al.*, 2007) reported levels of 93.8 ng/kg lipid weight in crabs to 545 ng/kg lipid weight in fish, and 1300 ng/kg lipid weight in cephalopods, for the sum of PCN homologue groups. The composition of PCN occurrence favoured the lower chlorinated PCNs with tri-, tetra- and penta-chlorinated homologues generally dominating, perhaps reflecting different patterns of pollution and commercial mixtures used in the region. Nine pooled samples, belonging to two food categories (fish and duck), were collected in 2006 from coastal waters of Qingdao and Chongming Island, off Shanghai (Yang *et al.*, 2009). Concentrations of coplanar polychlorinated biphenyls (co-PCB) and PCNs were measured by HRGC-HRMS, and potential health risks for consumption of fish and duck were preliminarily discussed. In fish, the mean concentrations of co-PCBs and PCNs respectively were about 4041 and 225 ng/kg lipid from Qingdao, and 3318 and 640 ng/kg lipid from Chongming Island; in duck meat, the mean concentrations were about 966 and 43.8 ng/kg lipid, respectively. The homologue profiles of PCNs in fish from the two locations were quite different. The total co-PCB-TEQ and PCN-TEQ ranged from 0.68 to 11.40 ng (WHO-TEQ)/kg lipid and from 0.001 to 0.210 ng (WHO-TEQ)/kg lipid for fish and duck meat, respectively. The study examined the average daily intake of dioxin-like compounds from the consumption of fish and duck meat, and based on local population consumption patterns, estimated that the tolerable daily intake (TDI), 1–4 pg (WHO-TEQ)/kg body weight(b.w.) would not be exceeded (Yang *et al.*, 2009).

A study on temporal trends reported that the concentration of PCNs in archived Lake Ontario trout had declined eight-fold from 1979 to 2004 (Gewurtz *et al.*, 2009). However, the decline was congener-specific, being most marked for the higher chlorinated congeners. The authors suggested that PCN concentrations in whole lake trout may still be relatively high. In a later study on Scottish marine and aquatic fish species (n=52) sampled in 2008, Fernandes *et al.* (2009), reported 3-fold higher concentrations (as a sum of 12 individual PCN congeners) in aquatic species (2.5 to 103 ng/kg ww; mean 22 ng/kg) compared to the marine (deep-sea) species (0.3–63 ng/kg ww; mean 7.6 ng/kg). Marine shellfish (mussels) showed lower concentrations, with a mean value of 2.5 ng/kg ww. Some potential species-selective bio-accumulation was noted, as the highest levels (35 to 63 ng/kg ww) of PCNs in the deep-sea marine fish were all observed in three samples of spurdog.

15.5.2 Occurrence in commonly consumed foods

Most of the studies discussed above, have investigated marine/aquatic foods. However there are fewer studies that have investigated the full range of retail food items that could be used to derive a general dietary intake. A Spanish study reports the measurement of PCNs in a range of foods, randomly acquired in seven cities in Catalonia, Spain (Domingo *et al.*, 2003). The highest (wet weight) concentration of total PCNs was found in oils and fats (447 ng/kg), followed by cereals (71 ng/kg), fish and shellfish (39 ng/kg) and dairy products (36 ng/kg). In general, tetra-CN was the predominant homologue group in all foods except for fruits and pulses, which had greater proportions of hexa-CNs. The high levels observed in oils and fats were not reflected in vegetable or meat based products. Foods that were subject to some measure of processing (oils and fats, cereals, dairy products, meat and meat products), and fish showed the highest levels. An exposure assessment based on the data indicated that children showed higher exposure (1.65 ng/kg b.w./day) than the oldest adults (0.54 ng/kg b.w./day). This was largely due (as is the normal case) to the higher total food consumption by children than adults, on a body-weight basis, rather than due to particular contamination of food groups of especial importance to children. Extending this study with a view to establishing temporal trends, Marti-Cid *et al.* (2008), reported concentrations of PCNs in a similar set of retail foods from 12 representative cities in Spain. The foods covered a selection of commonly consumed items, including milk, fish, vegetables, fruits, eggs, meat, cereals, oils and processed foods. In contrast to the earlier study, the follow-up found unexpectedly lower PCN concentrations in oils and fats (average 21.5 ng/kg ww), with the highest concentrations found in fish and seafood (average 47.1 ng/kg ww) The lowest concentrations were, as observed in the earlier study, in milk, with an average concentration of 0.8 ng/kg ww. These concentrations corresponded to a total dietary intake of 7.25 ng/day (0.1 ng kg/b.w./day) (sum tetra- to octa-chloro PCN homologues rather than individual congeners) for an adult man (70 kg standard b.w.).

In a more recent investigation of individual PCN congeners in UK retail foods (Fernandes *et al.*, 2010), specific PCN congeners (PCNs 52/60, 53, 66/67, 68, 69, 71/72, 73, 74 and 75) were selected for analysis, based on the available literature on current occurrence and toxicology, and limited by the commercial availability of reference standards. The investigation showed PCN occurrence in all 44 of the studied foods – meat, milk, fish, dairy and meat products, eggs, poultry, vegetables, fruits, etc. The most frequently detected congeners were PCN 52, PCNs 66/67 and PCN 73. The highest concentrations were observed in fish (highest value of 37 ng/kg ww for the sum of the measured congeners). The dioxin-like toxicity (PCN-TEQ) associated with

378 Persistent organic pollutants and toxic metals in foods

these concentrations, (reported range 0.0001 ng TEQ/kg for fruits and vegetables to an average of 0.01 ng TEQ/kg for fish) was 1–2 orders of magnitude lower than those reported for chlorinated dioxins or PCBs in food. These concentrations recorded for individual foods represent most, but not all, of commonly consumed foods, and were used to provide a preliminary indication of UK dietary exposure. On a worst case basis, assuming that all foods contained the maximum concentrations found – 0.02 ng TEQ/kg, (as found in eel, farmed salmon, sprats, lambs' liver, lamb shoulder and butter, and that solid food eaten by adults and young children (4–6 years) was 1.5 kg/day and 1.0 kg/day respectively for 97.5 percentile consumers), estimated dietary intakes would be 0.39 pg TEQ/kg b.w./day and 0.98 pg TEQ/kg b.w./day, respectively (Fernandes et al., 2010). This compares to the tolerable daily intake of 2 pg WHO-TEQ/kg b.w./day for dioxin-like compounds (PCDD/Fs and PCBs).

More recently, a larger study on the PCN content of 100 commonly consumed foods collected in the Republic of Ireland between 2007 and 2008, has been reported (Food Safety Authority of Ireland, 2010; Fernandes et al., 2011). As with the earlier studies in Spain and the UK, the results showed PCN occurrence in the majority of studied foods – milk, fish, dairy and meat products, eggs, animal fat, shellfish, offal, vegetables, cereal products, etc. Concentrations ranged from 0.09 ng/kg ww for milk to 59.3 ng/kg ww for fish, for the sum of the measured PCNs. The highest concentrations were observed in fish, which generally showed congener profiles that reflected some commercial mixtures, followed by animal fat and dairy products. The calculated PCN-TEQ values associated with the reported concentrations, ranged from 0.0001 ng/kg TEQ mainly for vegetable based foods and some shellfish to 0.03 ng/kg TEQ for fish.

It is important to note that in some studies that have estimated PCN toxicity as TEQs, not all PCN congeners with reported relative potency values have been included in the estimates of PCN-TEQ. Thus, these are likely to be underestimates of the full PCN-TEQ. A number of studies, most of them recent (Kannan et al., 2000b; Jiang et al., 2007; Fernandes et al., 2010; Park et al., 2010; Fernandes et al., 2011) have started quoting TEQ values for PCNs based on the relative potency studies discussed earlier. These reports have used different potency values to compute TEQs, and it is clear that comparative studies would be considerably aided if the same, harmonised set of TEF values were used for those PCN congeners that were identified as being toxicologically significant. This would require a meta-analysis of reported REP values and possibly, confirmatory studies which would allow harmonisation of the values and a consensus on their use. TEF values for PCNs are clearly under consideration, but have not yet been adopted by regulatory bodies.

These data are summarised in Table 15.1 which presents an overview of the concentrations of PCNs in foods, reported over the last decade or so. The majority of the data have been reported on a whole (wet) weight basis to allow comparison. An aspect of data presentation that the table highlights

Table 15.1 A summary of recent PCN analyses in foods

Food type	No. of samples	Sampling period	Geographical origin	Report/quantitation type	PCN concentration sum – ng/kg whole (wet) weight	Notes	Reference
Fish	22	1996–97	Great Lakes and inland Michigan, N America	Homologue totals (tri–octa)/individual congeners	19–31 400	9 species, carp, trout, pike, etc.	Kannan et al., 2000a
	16	2000	Catalonia, Spain	Homologue totals tetra–octa	39 (average)	Retail (including shellfish)	Domingo et al., 2003
	52	2001–2003	Baltic Sea	Sum of 16, tetra–octa congeners	20–450	Baltic herring and salmon	Isosaari et al., 2006
	45	2001–2003	Remote lakes, Finland	Sum of 16, tetra–octa congeners	2–66	9 freshwater species	Isosaari et al., 2006
	53	2001–2003	Baltic Sea	Sum of 16, tetra–octa congeners	1–190	Baltic fish species (various)	Isosaari et al., 2006
	8	2003–2004	East and South China Sea	Homologue totals tri–octa	137–545 (lipid wt., average)	4 species, retail	Jiang et al., 2007
	27	2005	Catalonia, Spain	Homologue totals tetra–octa	13–227	Retail, composites	Llobet et al., 2007
	9	2006	Catalonia, Spain	Homologue totals tetra–octa	12.8–226.9	9 species, retail, composites	Marti-cid et al., 2008
	9 (max)	2006	China	Homologue totals/ congeners	225–640 (lipid wt., average)	Pooled samples coastal waters	Yang et al., 2009
	12	2007	UK	Sum of 12, penta–octa congeners	0.73–37.3	Wild and farmed, retail	Fernandes et al., 2010
	12	2007–2008	Ireland	Sum of 12, penta–octa congeners	2.08–59.3	Wild and farmed species	Fernandes et al., 2011

(*Continued*)

Table 15.1 Continued

Food type	No. of samples	Sampling period	Geographical origin	Report/quantitation type	PCN concentration sum – ng/kg whole (wet) weight	Notes	Reference
	16	2008	Scotland	Sum of 12, penta–octa congeners	2.5–103	5 freshwater species	Fernandes et al., 2009
	32	2008	Scotland	Sum of 12, penta–octa congeners	0.3–63	18 marine (including deep-sea) species	Fernandes et al., 2009
Shellfish	15	2003–2004	East and South China Sea	Homologue totals tri–octa	94–1300 (lipid wt., average)	7 shellfish, 2 cephalopod species	Jiang et al., 2007
	15	2005	Catalonia, Spain	Homologue totals tetra–octa	3–22	Retail, composites	Llobet et al., 2007
	5	2006	Catalonia, Spain	Homologue totals tetra–octa	2.7–21.8	Retail, composites	Marti-cid et al., 2008
	5	2007–2008	Ireland	Sum of 12, penta–octa congeners	0.18–2.34	Oysters	Fernandes et al., 2011
	5	2008	Scotland	Sum of 12, penta–octa congeners	0.8–6.5	Mussels	Fernandes et al., 2009
Eggs and poultry	4	2000	Catalonia, Spain	Homologue totals tetra–octa	23 (average)	Retail eggs	Domingo et al., 2003
	1	2006	Catalonia, Spain	Homologue totals tetra–octa	1.7–4.3	Chicken, retail eggs, composites	Marti-cid et al., 2008
	9 (max)	2006	China	Homologue totals/congeners	43.8 (lipid wt., average)	Pooled samples coastal waters	Yang et al., 2009
	9	2007	UK	Sum of 12, penta–octa congeners	0.48–8.29	Retail	Fernandes et al., 2010
	15	2007–2008	Ireland	Sum of 12, penta–octa congeners	0.23–2.22	Composite eggs, including barn, free range, etc.	Fernandes et al., 2011

Milk	4	2000	Catalonia, Spain	Homologue totals tetra–octa	0.4 (average)	Retail, composites	Domingo et al., 2003
	2	2006	Catalonia, Spain	Homologue totals tetra–octa	0.5–1.2	Retail, composites	Marti-cid et al., 2008
	2	2007	UK	Sum of 12, penta–octa congeners	0.19–0.22	Retail	Fernandes et al., 2010
	15	2007–2008	Ireland	Sum of 12, penta–octa congeners	0.09–0.38	Composite from farms	Fernandes et al., 2011
Dairy products	4	2000	Catalonia, Spain	Homologue totals tetra–octa	36 (average)	Retail, composites	Domingo et al., 2003
	2	2006	Catalonia, Spain	Homologue totals tetra–octa	0.8–22.7	Retail, composites	Marti-cid et al., 2008
	7	2007	UK	Sum of 12, penta–octa congeners	0.52–6.09	Retail	Fernandes et al., 2010
	6	2007–2008	Ireland	Sum of 12, penta–octa congeners	0.68–3.13	Retail	Fernandes et al., 2011
Meat	30	2000	Catalonia, Spain	Homologue totals tetra–octa	18 (average)	Retail, composites, including meat products	Domingo et al., 2003
	9	2006	Catalonia, Spain	Homologue totals tetra–octa	1.8–5.8	Retail, composites, including meat products	Marti-cid et al., 2008
	4	2007	UK	Sum of 12, penta–octa congeners	0.19–5.69	Retail meat	Fernandes et al., 2010
Animal fat	8	2001	Japan	Tri– to octa individual congeners	125–308	Fats – pork and chicken	Guruge et al., 2004
	21	2007–2008	Ireland	Sum of 12, penta–octa congeners	1.6–13.4	Fats – beef, pork, lamb, avian	Fernandes et al., 2011

(*Continued*)

Table 15.1 Continued

Food type	No. of samples	Sampling period	Geographical origin	Report/quantitation type	PCN concentration sum – ng/kg whole (wet) weight	Notes	Reference
Meat products and offal	5	2007	UK	Sum of 12, penta–octa congeners	0.34–4.88	Retail	Fernandes et al., 2010
	12	2007–2008	Ireland	Sum of 12, penta–octa congeners	0.18–4.26	Retail liver, 5 species	Fernandes et al., 2011
	2	2007–2008	Ireland	Sum of 12, penta–octa congeners	0.13–0.54	Processed meat	Fernandes et al., 2011
Fats and oils	6	2000	Catalonia, Spain	Homologue totals tetra–octa	447 (average)	Retail, composites	Domingo et al., 2003
	4	2006	Catalonia, Spain	Homologue totals tetra–octa	20.4–22.3	Retail, composites	Marti-cid et al., 2008
Cereals and bread	12	2000	Catalonia, Spain	Homologue totals tetra–octa	3–71 (average)	Retail pulses, cereals, composites	Domingo et al., 2003
	6	2006	Catalonia, Spain	Homologue totals tetra–octa	5.3–15.1	Retail, composites, including pulses	Marti-cid et al., 2008
	1	2007	UK	Sum of 12, penta–octa congeners	0.28	Retail	Fernandes et al., 2010
	4	2007–2008	Ireland	Sum of 12, penta–octa congeners	0.25–0.77	Retail	Fernandes et al., 2011
Fruits and vegetables	32	2000	Catalonia, Spain	Homologue totals tetra–octa	0.7–4 (average)	Retail, composites	Domingo et al., 2003
	9	2006	Catalonia, Spain	Homologue totals tetra–octa	0.9–3.9	Retail, composites	Marti-cid et al., 2008
	2	2007	UK	Sum of 12, penta–octa congeners	0.15–0.16	Retail	Fernandes et al., 2010
	8	2007–2008	Ireland	Sum of 12, penta–octa congeners	0.16–2.84	Retail	Fernandes et al., 2011

is the different ways in which PCN data have been condensed and reported (either as a sum of different congener sets or homologue totals), a factor that can make direct comparisons difficult.

15.6 PCN occurrence in humans

Although human exposure to chlorinated naphthalenes can occur via oral, inhalative and dermal routes, the dietary pathway is expected to be the most significant for general populations except, of course, for those individuals who have been occupationally exposed. Although data on the occurrence of PCNs in human tissues and body burdens are scarce, a few studies have reported on concentrations in humans with detection in blood, breast milk, adipose tissue and liver. The chlorinated naphthalene congener profiles found in humans appear to show significantly fewer congeners than those observed in commercial mixtures or indeed environmental matrices or biota. Figure 15.1 demonstrates this difference, showing the hexa-chlorinated naphthalene isomers that occur in river fish compared to a human milk sample. The observed differences are unsurprising, given the omnivorous nature of the human diet, which will reflect environmental degradation or modification for plant based foods, and selective uptake and metabolism for marine and animal based foods. The latter two processes will also be repeated in humans, further modifying

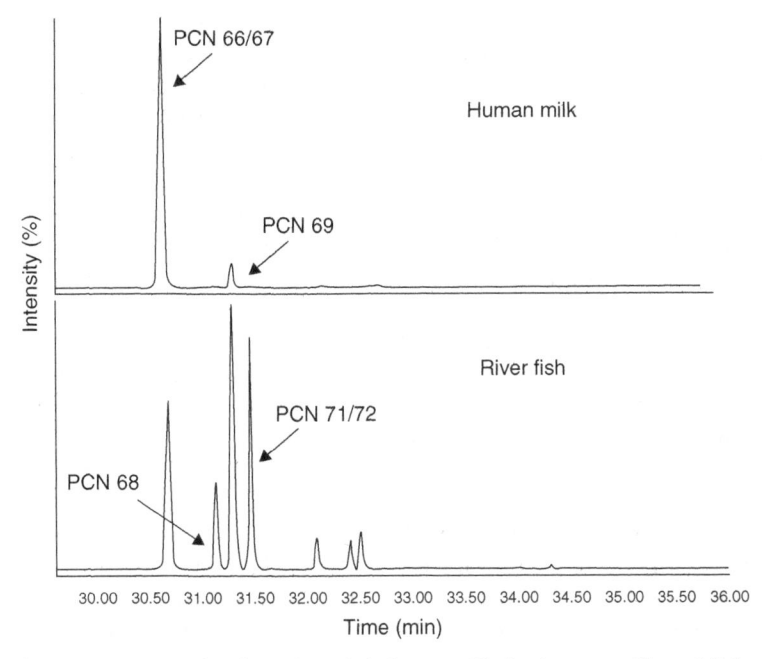

Fig. 15.1 Hexa-chlorinated naphthalene profile for human milk and fish.

384 Persistent organic pollutants and toxic metals in foods

the PCN congener profile. Thus, in the limited studies carried out on human samples, the PCN congener profiles can be expected to be simplified, showing relatively fewer congeners.

15.6.1 Concentrations in blood

The most prominent congeners observed in human samples are two penta-chlorinated and two hexa-chlorinated congeners, namely 1,2,3,5,7/1,2,4,6,7-pentachloronaphthalene (PCNs 52/60) and 1,2,3,4,6,7/1,2,3,5,6,7-hexachloronaphthalene (PCNs 66/67), and to a lesser extent some tetra-chlorinated congeners (CICAD, 2001; Kunisue et al., 2009; Schiavone et al., 2010). The predominance of these hexachloronaphthalene congeners is consistent with the profiles observed in animal tissues (Fernandes et al., 2010, 2011). The observed profiles may also be expected to reflect the input from other pathways, such as dermal contact and inhalation, although the magnitude of these is expected to be relatively small. A targeted study investigating a small number ($n = 5$) of electronics workers who had possibly been occupationally exposed (Weistrand et al., 1997), failed to find a significant difference in PCN blood plasma concentrations from those of controls ($n = 6$). The sum of the concentrations for three penta-/hexa-chlorinated congeners ranged from 152 to 684 ng/kg lipid. The corresponding range for controls was 255 to 654 ng/kg lipid. In contrast, another study (Asplund et al., 1994) on 37 males consuming varying amounts of Baltic fish, reported plasma concentrations of PCNs 66/67 of 9400 ng/kg in high fish consumers, 3800 ng/kg in moderate fish consumers, reducing to 2400 ng/kg in those who did not consume fish. The association between blood plasma levels and the estimated fish intake was reported as being statistically significant for PCNs 66/67, but not for a range of other congeners. An earlier study on accidental exposure (Ryan and Masuda, 1994) arising from the consumption of PCB contaminated rice oil during the Yu-Cheng incident (as observed earlier PCNs are present at significant levels in commercial PCB mixtures) found blood lipid concentrations for PCNs 66/67 which peaked at 30 400 ng/kg at 171 days after the first sampling. In a recent report on PCN concentrations in human serum derived from 61 healthy human volunteers from Seoul, Korea in 2007, Park et al. (2010), reported a mean concentration of 2170 ng/kg lipid. It is noteworthy that PCDD/Fs and PCBs were also analysed in the same samples, and the TEQ (computed using TEF values from Puzyn et al., 2007) arising from PCNs was 26.8% of the combined TEQ.

15.6.2 Concentrations in breast milk

The trend observed in a study on breast milk from Sweden, carried out between 1972 and 1992, was a reduction over time in PCN concentrations. Milk from several hundred mothers was analysed and the average PCN concentrations declined from 3081 to 483 ng/kg lipid (Lunden and Noren, 1998). A more recent study from the Swedish national screening programme (Haglund *et al.*, 2011) reported similar PCN profiles in three out of four human milk samples. The fourth contained the highest total concentrations and a larger percentage of less halogenated, and less persistent, PCNs, which might indicate that the subject was recently exposed. The report concluded that the levels were similar to those of PCDD/F-TEQs if expressed on a REP-basis. The weekly PCN-REP intake for a 1-month old baby was estimated to be 22 pg/kg b.w., which is above the tolerable weekly intake of 14 pg TEQ/kg b.w. recommended by the EU Scientific Committee on Food (Haglund *et al.*, 2011). PCN concentrations measured in 11, pooled breast milk samples from 109 first-time mothers in the Republic of Ireland in 2010 (Pratt, 2011, personal communication), ranged from 59 to 168 ng/kg lipid for the sum of 12 congeners (PCNs 52/60, 53, 66/67, 68, 69, 71/72, 73, 74 and 75). As observed earlier, PCN 52/60 and 66/67 occurred at the highest concentrations.

15.6.3 Concentrations in adipose tissue and liver

In common with other lipophilic contaminants such as PCDD/F and PCBs, PCNs are expected to accumulate in adipose tissue (Weistrand and Noren, 1998; Witt and Niessen, 2000; Kunisue *et al.*, 2009). In an early study, on adipose tissue from Canada, the concentrations of PCN 66/67 and a penta-chlorinated naphthalene congener (not identified), ranged from 100 to 2400 ng/kg lipid and 199–25 000 ng/kg lipid, respectively (Williams *et al.*, 1993). The corresponding average levels occurred between 1000 and 6200 ng/kg lipid for total penta-chloronaphthalenes and hexachloronaphthalenes. Reported concentrations of PCNs in the adipose tissue of Swedish subjects ranged from 1000 to 3900 ng/kg fat for the sum of tetra- to hexa-PCNs (Weistrand and Norén, 1998). Witt and Niessen (2000), measured PCN concentrations (six tetra- to hexa-chlorinated congeners) in the adipose tissue of 48 children (median age 6–10 yrs), collected during non-elective surgery, from Germany, Kazakhstan and Russia, which ranged from 900 to 34 600 ng/kg fat. The measured distribution varied between adipose tissue from different locations with the lowest median value of 1700 ng/kg fat from Stralsund, Germany, ranging up to the highest of 8500 ng/kg lipid from Saratov, Russia.

Individual PCN congeners (tri- to hexa-chloro) were measured in adipose tissue from 14 male and 29 female subjects in New York, USA collected by liposuction, between 2003 and 2005, and ranged from 60 to 2500 ng/kg fat (males) and 20 to 900 ng/kg fat (females) (Kunisue *et al.*, 2009). In common with other studies, PCNs 52/60 and PCNs 66/67 dominated the profiles and collectively accounted for 66% of the total PCN concentration.

Adipose tissue samples collected from 12 subjects in Siena, Italy returned summed PCN concentrations ranging from 500 to 14 000 ng/kg fat (mean 2600 ng/kg fat). The most abundant PCN congeners detected were PCNs

66/67 (1400 ng/kg fat), followed by PCNs 50/51 (1,2,3,4,6/1,2,3,5,6-pentachloronaphthalene) (370 ng/kg fat) and PCNs 52/60 (260 ng/kg fat), accounting for an average of 56%, 14% and 10%, respectively, of the total PCNs measured (Schiavone et al., 2010).

Concentration of summed tetra- to hexa-chlorinated PCNs in liver samples taken from seven subjects during autopsies, in Sweden, ranged from 1375 to 26 113 ng/kg fat (Weistrand and Noren, 1998).

These reports of PCN occurrence in human tissue (blood, milk adipose and liver) confirm human exposure to these contaminants. The more recent studies shed light on the selectivity of bio-accumulation of PCN congeners in humans and provide the basis for a choice of PCN congeners to be targeted for analysis and exposure studies.

15.7 Conclusions and future trends

Overall, the data that are currently available on PCN occurrence in foods suggest a widespread current distribution of these contaminants in foods and food webs. This is remarkable, given the time that has elapsed since the unrestricted use of these compounds, and those other commercial chemicals such as PCBs which are known sources of PCNs, ceased (particularly in Western Europe and North America), and underlines the persistence and ubiquity of PCNs. The more recent reports suggest a renewed interest in this class of contaminants, possibly associated with reports on the contribution of specific PCN congeners to an Ah receptor-mediated, (dioxin-like) mode of toxicity. The availability of commercial standards has also helped identify the more potent of these compounds in foods and human tissues, although more standards would certainly be instrumental in quantitatively targeting a wider range of the more relevant compounds in toxicological, occurrence and exposure studies. Although the reported contribution is smaller than PCDD/Fs and PCBs (the limited studies suggest a ten-fold lower level, but do not take into account all the relevant congeners), PCN toxicity is likely to add to the cumulative toxicity of other dioxin-like compounds.

More studies are required, both, in terms of occurrence and toxicology, in order to elaborate a clearer picture of human dietary exposure to PCNs particularly in those parts of the world where data, and details on current PCN usage, is scarce, such as the emerging economies of Brazil, Mexico, Argentina, Turkey, India and South Africa. Additional follow-up work is also required in those countries where baseline studies have been carried out, in order to establish trends. Given that most of the recent literature puts more emphasis on the measurement of individual congeners that are of greater toxicological significance, and the focus on AhR-mediated toxicity, it is clear that a set of consensus potency factors or TEFs would be instrumental in allowing comparisons, as well as simplifying the presentation of PCN toxicity in food and tissue samples. This is in line with current views on the inclusion of PCNs in the PCDD/F/PCB TEF scheme (Van den Berg et al., 2006).

Although this chapter focuses on PCNs in food and the resulting human exposure, further insight into environmental pathways, occurrence in different geographical regions and toxicology would be useful in order to understand the behaviour of these compounds as environmental and food contaminants, and human toxins. Therefore attention is drawn to further reading in these areas, references for which, are included in the next section:

- Environmental and Biota – Falandysz, 1998; Falandysz, 2003; Environment Canada, 2011.
- Toxicology and Foods – CICAD 37, 2001; COT, 2009.

15.8 References

ABAD, E., CAIXACH, J. and RIVERA, J., (1999). Dioxin like compounds from municipal waste incinerator emissions: Assessment of the presence of polychlorinated naphthalenes. *Chemosphere*, **38**(1): 109–120.

ARNOT, J. and GOBAS, F., (2003). A generic QSAR for assessing the bioaccumulation potential of organic chemicals in aquatic food webs. *Quant Struct Act Relat*, **22**: 1–9.

ASPLUND, L., JAKOBSSON, E., HAGLUND, P. and BERGMAN, A., (1994). 1,2,3,5,6,7-Hexachloronaphthalene and 1,2,3,4,6,7-hexachloronaphthalene selective retention in rat liver and appearance in wildlife. *Chemosphere*, **28**(12): 2075–2086.

AITTOLA, J-P., PAASIVIRTA, J., VATTULAINEN, A., SINKKONEN, S., KOISTINEN, J. and TARHANEN, J., (1994). Formation of chloroaromatics at a metal reclamation plant and efficiency of stack filter in their removal from emission. *Organohalogen Compd*, **19**: 321–324.

BEHNISCH, P.A., HOSOE, K. and SAKAI, S., (2003). Brominated dioxin-like compounds: in vitro assessment in comparison to classical dioxin-like compounds and other polyaromatic compounds. *Environ Int*, **29**: 861–877.

BELL, W. D., (1953). The relative toxicity of the chlorinated naphthalenes in experimentally produced bovine hyperkeratosis (X-disease). *Vet Med*, **48**: 135–146.

BLANKENSHIP, A., KANNAN, K., VILLALOBOS, S., VILLENEUVE, D., FALANDYSZ, J., IMAGAWA, T., JAKOBSSON, E. and GIESY, J.P., (1999). Relative potencies of halowax mixtures and individual polychlorinated naphthalenes (PCNs) to induce Ah receptor-mediated responses in the rat hepatoma H4IIE-luc cell bioassay. *Organohalogen Compd*, **42**: 217–220.

BLANKENSHIP, A., KANNAN, K., VILLALOBOS, S., VILLENEUVE, D., FALANDYSZ, J., IMAGAWA, T., JAKOBSSON, E. and GIESY, J.P., (2000). Relative potencies of Halowax mixtures and individual polychlorinated naphthalenes (PCNs) to induce Ah receptor-mediated responses in the rat hepatoma H4IIE-Luc cell bioassay. *Environ Sci Technol*, **34**(15): 3153–3158.

BRUNSTR, Ö. M. B., ENGWALL, M., HJELM, K., LINDQVIST, L. and ZEBUHR, Y., (1995). EROD induction in cultured chick embryo liver: A sensitive bioassay for dioxin-like environmental pollutants. *Environ Toxicol Chem*, **14**(5): 837–842.

CAMPBELL, M. A., BANDIERA, S., ROBERTSON, L., PARKINSON, A. and SAFE, S., (1981). Octachloronaphthalene induction of hepatic microsomal aryl hydrocarbon hydroxylase activity in the immature male rat. *Toxicology*, **22**: 123–132.

CHU, I., SECOURS, V. and VIAU, A., (1976). Metabolites of chloronaphthalene. *Chemosphere*, **6**: 439–444.

CHU, I., VILLENEUVE, D.C., SECOURS, V. and VIAU, A., (1977a). Metabolism of chloronaphthalenes. *J Agric Food Chem*, **25**: 881–883.

CHU, I., SECOURS, V., VILLENEUVE, D.C. and VIAU, A., (1977b). Metabolism and tissue distribution of (1,4,5,8–14C) – 1,2-dichloronaphthalene in rats. *Bull Environ Contam Toxicol*, **18**: 177–183.

CICAD (Concise International Chemical Assessment Document) No. 34. Polychlorinated Naphthalenes. International Programme on Chemical Safety (2001). available at: www.who.int/ipcs/publications/cicad/en/cicad34.pdf

COT (Committee on Toxicity of Chemicals in Food, Consumer Products and the Environment), 2009. Polychlorinated Naphthalenes in Food. Available at: http://cot.food.gov.uk/pdfs/tox200902pcnscp.pdf

DOMINGO, J., FALCO, G., LLOBET, J., CASAS, C., TEIXIDO, A. and MULLER, L., (2003). PCNs in foods: estimated dietary intake by the population of Catalonia, Spain. *Environ Sci Technol*, **37**(11): 2332–2335.

ENGWALL, M., BRUNSTROM, B. and JAKOBSSON, E., (1994). Ethoxyresorufin O-de-ethylase (EROD) and arylhydrocarbon hydroxylase (AHH) – inducing potency and lethality of chlorinated naphthalenes in chicken (Gallus domesticus) and eider duck (Somateria mollissima) embryos. *Arch Toxicol*, **68**: 37–42.

ENVIRONMENT CANADA (2011). Ecological risk assessment. Polychlorinated Naphthalenes. Available at: http://www.ec.gc.ca/lcpe-cepa/default.asp?lang=En&n=DB21D3D2-1

FALANDYSZ, J., (1997). Bioaccumulation and biomagnification features of polychlorinated naphthalenes. *Organohalogen Compd*, **32**: 374–379.

FALANDYSZ, J., (1998). Polychlorinated naphthalenes: an environmental update. *Env Poll*, **101**(1): 77–90.

FALANDYSZ, J., (2003). Chloronaphthalenes as food chain contaminants: a review. *Food Add Contam*, **20**: 995–1014.

FALANDYSZ, J. and PUZYN, T., (2004). Computational prediction of 7-ethoxyresorufin-O-diethylase (EROD) and luciferase (luc) inducing potency for 75 congeners of chloronaphthalene. *J Environ Sci Health A Tox Hazard Subst Environ Eng*, **39**: 1505–1523.

FERNANDES, A., SMITH, F., PETCH, R., BRERETON, N., BRADLEY, E., PANTON, S., CARR, M. and ROSE, M., (2009). Investigation into the levels of environmental contaminants in Scottish marine and freshwater fin fish and shellfish. Report FD 09/01 to the Food Standards Agency, UK London. Available at: http://www.foodbase.org.uk/results.php?f_category_id=&f_report_id=572

FERNANDES, A., MORTIMER, D., GEM, M., SMITH, F., ROSE, M., PANTON, S. and CARR, M., (2010). Polychlorinated Naphthalenes (PCNs): Congener specific analysis, occurrence in food and dietary exposure in the UK. *Environ Sci Technol*, **44**: 3533–3538.

FERNANDES A., TLUSTOS C., ROSE M., SMITH F., CARR M. and PANTON S., (2011). Polychlorinated Naphthalenes (PCNs) in Irish foods: Occurrence and human exposure. *Chemosphere*, **85**: 322–328.

FOOD SAFETY AUTHORITY OF IRELAND, (2010). Investigation into Levels of Polychlorinated Naphthalenes (PCNs) in Carcass Fat, Offal, Fish, Eggs, Milk and Processed Products. Available at: http://www.fsai.ie/search-results.html?searchString=PCNs%20dietary%20exposure

GEWURTZ S., LEGA R., CROZIER P., FAYEZ L., REINER E., HELM P., MARVIN C., TOMY G., (2009). Factors influencing trends of polychlorinated naphthalenes and other dioxin-like compounds in Lake Trout (Salvelinus Namaycush) from Lake Ontario, North America (1979–2004). *Environ Toxicol Chem*, **28**(5): 921–930.

GURUGE, K.S., SEIKE, YAMANAKA, N. and MIYAZAKI, S., (2004). Accumulation of PCNs in domestic animal related samples. *J Env Monit*, **6**: 753–757.

HAGLUND, P., JAKOBSSON, E., ASPLUND, L., ATHANASIADOU, M. and BERGMAN, A., (1993). Determination of polychlorinated naphthalenes in polychlorinated biphenyl products via capillary gas chromatography-mass spectrometry after separation by gel permeation chromatography. *J Chromatogr*, **634**: 79–86.

HAGLUND, P., JAKOBSSON, E. and MASUDA, Y., (1995). Isomer-specific analysis of polychlorinated naphthalenes in Kanechlor KC 400, Yusho rice oil, and adipose tissue of a Yusho victim. *Organohalogen Compd*, **26**: 405–410.

HAGLUND, P., KAJ, L. and BRORSTROM-LUNDEN, E., (2011). Results from the Swedish National Screening programme 2010: Subreport 1 (PCNs). Available at: http://www.naturvardsverket.se/upload/02_tillstandet_i_miljon/Miljoovervakning/rapporter/miljogift/B2002_NV_Screen_2010_PCN.pdf

HANBERG, A., WAERN, F., ASPLUND, L., HAGLUND, E. and SAFE, S., (1990). Swedish dioxin survey: determination of 2,3,7,8-TCDD toxic equivalent factors for some polychlorinated biphenyls and naphthalenes using biological tests. *Chemosphere*, **20**: 1161–1164.

HANBERG A., STAHLBERG M., GEORGELLIS A., DE WIT C. and AHLBORG U.G., (1991). Swedish dioxin survey: Evaluation of the H-4-II E bioassay for screening environmental samples for dioxin-like enzyme induction. *Pharmacol Toxicol*, **69**(6): 442–449.

HELM P., JANTUNEN L., BIDLEMAN T. and DORMAN F., (1999). Complete separation of isomeric penta- and hexachloronaphthalenes by capillary gas chromatography. *J. High Res Chromatogr*, **11**(11): 639–643.

HELM P., BIDLEMAN T.F., JANTUNEN L.M.M. and RIDAL J., (2000). Polychlorinated naphthalenes in Great Lakes air: sources and ambient air profiles. *Organohalogen Compd*, **47**: 17–20.

HELM, P. and BIDLEMAN, T.F., (2003). Current combustion-related sources contribute to polychlorinated naphthalene and dioxin-like polychlorinated biphenyl levels and profiles in air in Toronto, Canada. *Environ Sci Technol*, **37**: 1075.

IINO, F., IMAGAWA, T., TAKEUCHI, M. and SADAKATA, M., (1999). De novo synthesis mechanism of polychlorinated dibenzofurans from polycyclic aromatic hydrocarbons and the characteristic isomers of polychlorinated naphthalenes. *Environ Sci Technol*, **33**(7): 1038–1043.

IMAGAWA, T. and YAMASHITA, N., (1994). Isomer specific analysis of polychlorinated naphthalenes in halowax and flyash. *Organohalogen Compd*, **19**: 215–218.

IMAGAWA, T. and LEE, C., (2001). Correlation of polychlorinated naphthalenes with polychlorinated dibenzofurans formed from waste incineration. *Chemosphere*, **44**: 1511–1520.

ISHAQ R., KARLSSON K. and NÄF C., (2000). Tissue distribution of polychlorinated naphthalenes (PCNs) and non-ortho chlorinated biphenyls (non-ortho PCBs) in harbour porpoises (Phocoena phocoena) from Swedish waters. *Chemosphere*, **41**(2): 1913–1925.

ISOSAARI, P., HALLIKAINEN, A., KIVIRANTA, H., VUORINEN, P., PARMANNE, R., KOISTINEN, J. and VARTAINEN, T., (2006). Dioxins, PCBs, PCNs and PBDEs in edible fish caught from the Baltic sea and lakes in Finland. *Environ Pollut*, **141**: 213–225.

JARNBERG U., ASPLUND L., DE WITT C., GRAFSTROM A., HAGLUND P., JANSSON B., LEXEN K., STRANDELL M., OLSSON M. and JONSON B., (1993). Polychlorinated biphenyls and polychlorinated naphthalenes in Swedish sediment and biota: levels, patterns and time trends. *Environ Sci Tech*, **27**: 1364–1474.

JARNBERG, U., ASPLUND, L., DEWIT, C., EGEBACK, A.L., WIDEQVIST, U. and JAKOBSSON, E., (1997). Distribution of polychlorinated naphthalene congeners in environmental and source-related samples. *Arch Environ Contam Toxicol*, **32**(3): 232–245.

JIANG, Q., HANARI, N., MIYAKE, Y., OKAZAWA, T., LAU, R., CHEN K., WYRZYKOWSKA B., SO M., YAMASHTA N. and LAM P., (2007). Health risk assessment for PCBs, PCDD/Fs and PCNs in seafood from Guangzhou and Zhoushan, China. *Environ Poll*, **148**: 31–39.

KANNAN K., BLANKENSHIP A.L., GIESY J.P. and IMAGAWA T., (2000a). Polychlorinated naphthalenes in soil, sediment, and biota collected near a former chloralkali plant in coastal Georgia, USA. *Cent Eur J Pub Health*, **8** (supplement): 10–12.

KANNAN, K., YAMASHITA, N., IMAGAWA, T., DECOEN, W., KHIM, J., DAY, R., SUMMER, C. and GIESY, J., (2000b). PCNs and PCBs in fishes from Michigan waters including the Great Lakes. *Environ Sci Technol*, **34**: 566–572.

KUEHL, D.W., HAEBLER, R. and POTTER, C., (1994). Coplanar PCB and metal residues in dolphins from the US Atlantic coast including Atlantic bottlenose obtained during the 1987/88 mass mortality. *Chemosphere*, **28**: 1245–1253.

KUNISUE T., JOHNSON-RESTREPO B., HILKER D., ALDOUS K. and KANNAN K., (2009). Polychlorinated naphthalenes in human adipose tissue from New York, USA. *Environ Poll*, **157**: 910–915.

LLOBET, J.M., FALCO, G., BOCIO, A. and DOMINGO, J.L., (2007). Human exposure to polychlorinated naphthalenes through the consumption of edible marine species. *Chemosphere*, **66**: 1107–1113.

LUNDEN, A. and NOREN, K., (1998). Polychlorinated naphthalenes and other organochlorine contaminants in Swedish human milk, 1972–1992. *Arch Environ Contam Toxicol*, **34**: 414–423.

MARTI-CID, R., LLOBET, J., CASTELL, V. and DOMINGO, J., (2008). Human exposure to PCNs and PCDEs from foods in Catalonia, Spain: Temporal trends. *Environ Sci Technol*, **42**: 4195–4201.

NOREN, K. and MEIRONYTE, D., (2000). Certain organochlorine and organobromine contaminants on Swedish human milk in perspective of the past 20–30 years. *Chemosphere*, **40**: 1111–1123.

NORRGREN L., ANDERSSON T. and BJÖRK M., (1993). Liver morphology and cytochrome P450 activity in fry of rainbow trout after microinjection of lipid-soluble xenobiotics in the yolk-sac embryos. *Aquat toxicol*, **26**: 307–316.

PARK H., KANG J-H., BAEK S-Y. and CHANG S-Y., (2010). Relative importance of polychlorinated naphthalenes compared to dioxins and polychlorinated biphenyls in human serum, from Korea: Contribution to TEQ and potential sources. *Environ Poll*, **158**: 1420–1427.

PUZYN, T., FALANDYSZ, J., JONES, P.D. and GIESY, J.P., (2007). Quantitative structure-activity relationships for the prediction of relative in vitro potencies (REPs) for chloronaphthalenes. *J Environ Sci Health A Tox Hazard Subst Environ Eng*, **42**: 573–590.

RUZO, L. O., SAFE, S., HUTZINGER, O., PLATONOW, N. and JONES, D., (1975). Hydroxylated metabolites of chloronaphthalenes (Halowax 1031) in pig urine. *Chemosphere*, **3**: 121–123.

RUZO, L., JONES, D., SAFE, S. and HUTZINGER, O., (1976a). Metabolism of chlorinated naphthalenes. *J Agric Food Chem*, **24**: 581–583.

RUZO, L.O., JONES, D. and PLATONOW, N., (1976b). Uptake and distribution of chloronaphthalenes and their metabolites in pigs. *Bull Environ Contam Toxicol*, **16**: 233–239.

RYAN J.J. and MASUDA Y., (1994). Polychlorinated naphthalenes (PCNs) in the rice oil poisonings. *Organohalogen Compd*, **21**: 251–254.

SCHIAVONE A., KANNAN K., HORII Y., FOCARDI S. and CORSOLINI S., (2009). Occurrence of brominated flame retardants, polycyclic musks, and chlorinated naphthalenes in seal blubber from Antarctica: Comparison to organochlorines. *Mar Pollut Bull*, **58**(9): 1415–1419.

SCHNEIDER, M., STIEGLITZ, L., WILL, R. and ZWICK, G., (1996). Formation of polychlorinated naphthalenes on fly ash. *Organohalogen Compd*, **27**: 192–195.

TAKASUGA T., INOUE, T., OHI, E. and IRELAND P., (1994). Development of an all congener specific HRGC/HRMS analytical method for polychlorinated naphthalenes in environmental samples. *Organohalogen Compd*, **19**: 41–44.

TAKASUGA T., TSUYOSHI I., OHI E. and KUMAR K. S., (2004). Formation of polychlorinated naphthalenes, dibenzo-p-dioxins, dibenzofurans, biphenyls and organochlorine

pesticides in thermal processes and their occurrence in ambient air. *Arch Environ Contam Toxicol*, **46**: 419–431.
THEISEN J., MAULSHAGEN A. and FUCHS J., (1993). Organic and inorganic substances in copper slag 'kieselrot'. *Chemosphere*, **26**: 881–896.
US National Library of Medicine. "Trichloronaphthalenes" Available at: http://hazmap.nlm.nih.gov/category-details?id=662&table=copytblagents
VAN DEN BERG, M., BIRNBAUM, L.S., DENISON, M., DE VITO, M., FARLAND, W., FEELEY, M., FIEDLER, H., HAKANSSON, H., HANBERG, A., HAWS, L., ROSE, M., SAFE, S., SCHRENK, D., TOHYAMA, C., TRITSCHER, A., TUOMISTO, J., TYSKLIND, M., WALKER, N. and PETERSON, R. E., (2006). The 2005 World Health Organization re-evaluation of human and mammalian toxic equivalency factors for dioxins and dioxin-like compounds. *Toxicol Sci*, **93**: 223–241.
VILLENEUVE, D.L., KHIM J.S., KANNAN K., FALANDYSZ J., BLANKENSHIP A.L., NIKIFOROV V. and GIESY J.P., (2000). Relative potencies of individual polychlorinated naphthalenes to induce dioxin-like response in fish and mammalian in vitro bioassays. *Arch Environ Contam Toxicol*, **39**(3): 273–281.
VOGELGESANG J., (1986). Hexachlorobenzene, octachlorostyrene and other organochlorine compounds in waste water from industrial high-temperature processes involving chlorine. *Wasser u. Abwasserforsch*, **19**: 140–144.
WEISTRAND, C. and NOREN, K., (1998). Polychlorinated naphthalenes and other organochlorine contaminants in human adipose and liver tissue. *J. Toxicol. Environ. Health Part A*, **53**(4): 293–311.
WEISTRAND, C., NOREN, K. and NILSSON, A., (1997). Occupational exposure – Organochlorine compounds in blood plasma from potentially exposed workers. *Environ Sci Poll Res*, **4**(1): 2–9.
WIEDMANN T. and BALLSCHMITER K., (1993). Quantification of chlorinated naphthalenes with GC-MS using the molar response of electron impact ionization. *Fresen J Anal Chem*, **346**: 800–804.
WILLIAMS, D., KENNEDY, B. and LEBEL, G., (1993). Chlorinated naphthalenes in human adipose tissue from Ontario municipalities. *Chemosphere*, **27**: 795–806.
WITT, K. and NIESSEN, K.H., (2000). Toxaphenes and chlorinated naphthalenes in adipose tissue of children. *J. Pediatr. Gastroenterol. Nutr*, **30**: 164–169.
YAMASHITA, N., KANNAN, K., IMAGAWA, T., MIYAZAKI, A. and GIESY, J.P. (2000). Concentrations and profiles of polychlorinated naphthalene congeners in eighteen technical polychlorinated biphenyl preparations. *Environ. Sci. Technol*, **34**(19): 4236–4241.
YANG, Y., PAN, J., ZHU, X., LIU, X., LU, G., LI, Q., LIU, X., 2009. Studies on coplanar-PCBs and PCNs in edible fish and duck in Qingdao and Chongming Island. *Res. Environ. Sci*. **22**(2), 187–193.

16
Mercury in foods

E. M. Sunderland and M. Tumpney, Harvard
University School of Public Health, USA

DOI: 10.1533/9780857098917.2.392

Abstract: Mercury (Hg) in foods can elicit a variety of toxic effects in humans. High levels of exposure to inorganic mercury are known to cause kidney and liver failure, while much lower levels of exposure to methylmercury (MeHg) are associated with a variety of long-term neurodevelopmental deficits in children and may impair cardiovascular health in adults. Low absorption of inorganic mercury from foods by humans means that MeHg is the primary form of concern. Virtually all (~99%) of the MeHg exposure in the typical North American diet is from fish and shellfish (for consumers of these products) because most of the mercury found in other foods is present as inorganic Hg at relatively low concentrations (< 10 ng/g). Recently, a number of studies have documented MeHg production in anaerobic rice paddies in mining contaminated areas of China and subsequent uptake in the rice grain. MeHg concetrations are slighlty lower in most rice varieties than seafood. However, rice is the dominant source of MeHg exposure in many areas of China because it is consumed in large quantities as part of the staple Aisan diet. Exposure of Chinese individuals from seafood is also diminished by the relatively lower concentrations of MeHg in farmed fish that comprise 75% of the seafood consumed. A variety of studies show that seafood consumed with other foods, such as tropical fruit, tea and coffee, can substantially lower MeHg uptake.

Key words: fish, shellfish, rice, exposure, methylmercury.

16.1 Introduction

Mercury (Hg) is a naturally occurring heavy metal that has been released in large quantities by human activities (Nriagu 1993; Streets *et al.*, 2011). Liberating Hg from deep mineral reservoirs through mining and fossil fuel combustion allows it to cycle continuously through the atmosphere, terrestrial ecosystems and the oceans for hundreds to thousands of years (Gill and Fitzgerald 1988;

Sunderland and Mason, 2007; Selin *et al.*, 2008; Smith-Downey *et al.*, 2010). Human activity has enriched the global atmospheric mercury reservoir and associated deposition to terrestrial and aquatic ecosystems by at least three to five times preindustrial levels (Fitzgerald *et al.*, 1998; Mason *et al.*, 1994; Mason, 2002; Pirrone *et al.*, 2010).

In aquatic ecosystems, some of the inorganic mercury present in the environment is converted to methylmercury (MeHg), the only species that bioaccumulates in food webs (Benoit *et al.*, 2003). Concentrations of MeHg in predatory fish can reach more than a million times those found in water, and as a result seafood is the primary food source of human exposure (Wiener *et al.*, 2003; Sunderland, 2007). MeHg production also occurs in flooded rice paddies and accumulates in rice (Horvat *et al.*, 2003; Li *et al.*, 2010). Due to the large quantities consumed in many Asian countries and cultivation in contaminated regions, rice is becoming an increasingly important source of MeHg exposure (Appleton *et al.*, 2006; Li *et al.*, 2010; Rothenberg *et al.*, 2011).

This chapter reviews mercury concentrations in various foods. We use illustrative examples of food choices and resulting Hg exposures for diets typical of the US general population and the Chinese population, and calculate the exposures attributable to major food categories. Mediating effects of other foods on MeHg absorption in the human body and toxicological outcomes are also discussed.

16.1.1 Health effects of mercury exposure

Food sources containing Hg and MeHg pose potential health risks because of the toxic properties of both species (Clarkson and Magos, 2006; Mergler *et al.*, 2007). MeHg is the form of mercury of greatest concern because it easily crosses the blood–brain and placental barriers and has an extremely high (> 90%) absorption efficiency when ingested in foods (Mahaffey *et al.*, 2011). This partially explains the greater relative hazard associated with MeHg exposure compared to inorganic Hg species. Less than 10% of inorganic Hg in foods is absorbed by the body, while the rest is rapidly excreted within 24 hours (Clarkson and Magos, 2006). Some forms of inorganic Hg, such as the quicksilver found in thermometers, have negligible absorption efficiency (~0.01%) and rapidly pass through the human gastrointestinal tract. MeHg has a half-life of 50–70 days in the human body and there is no reliable method for eliminating it more quickly.

Effects of MeHg exposure are particularly pronounced on the developing brain in the third trimester of pregnancy because pathways for nervous system functions are being formed at this time (Mahaffey *et al.*, 2011). Although the toxicity of MeHg exposure for adults was documented as far back as the nineteenth century (Julvez *et al.*, 2012), some of the earliest evidence of developmental toxicity of MeHg was provided by food-related poisonings that occurred in Japan and Iraq in the 1950s and 1970s (Bakir *et al.*, 1973; Harada, 1995).

High exposure to MeHg has also been associated with heart disease risk (Roman et al., 2011). Seven epidemiologic studies, conducted in five different populations, have examined the relationship between MeHg exposure and the risk of heart attack in humans (Salonen et al., 1995; Rissanen et al., 2000; Guallar et al., 2002; Yoshizawa et al., 2002; Virtanen et al., 2005; Hallgren et al., 2007; Mozaffarian et al., 2011). Studies in three of these epidemiological populations showed increased risks of heart attack due to MeHg exposure, while two did not. In those studies that showed effects, increased risk of myocardial infarction occurred at an exposure level corresponding to approximately 2 ppm in hair for men (Salonen et al., 1995; Rissanen et al., 2000). Although more research is needed to confirm these findings and better develop dose–response relationships for MeHg exposures and cardiovascular health impacts, a recent review panel convened by the US EPA concluded that evidence for the cardiovascular impact of MeHg exposures was sufficient to be included in regulatory determinations (Roman et al., 2011).

16.1.2 Major food-related poisonings

The first documented examples of widespread environmental Hg toxicity occurred in Minamata, Japan in the 1950s, where a plastic factory had discharged MeHg directly into a local estuary over many years (Harada, 1995). The discharged MeHg had accumulated to extremely high concentrations in fish, which were subsequently consumed by the local population. Symptoms of what is now known as 'Minamata disease' include a variety of effects on the central nervous system, exhibited by numbness at the extremities, tremors, hair loss, loss of basic motor function and, at the highest levels of exposure, death (Harada, 1995). Children born to victims of the poisoning in Minamata Bay exhibited a variety of effects depending on the exposed dose, including long-term developmental delays, weakness, seizure and hyperactivity (Harada, 1978). Studies of patients from Minamata showed an increased incidence of neurological impairment, especially perioral paraesthesias, among residents with hair mercury content below 50 µg/g, which was previously thought to be a safe limit (Yorifuji et al., 2009). Another study showed that paraesthesias remained even after 30 years cessation of mercury exposure (Ninomiya et al., 2005). This poisoning event was one of the first to document the susceptibility of the developing brain to MeHg toxicity (Choi and Grandjean, 2008). The extent of mercury poisoning in this example prompted widespread, international concern about anthropogenic sources of mercury and their impacts on human and ecological health.

Another major poisoning event occurred in Iraq where individuals consumed bread made from grain coated in a methylmercurial fungicide that was intended for planting (Bakir et al., 1973). Exposures in this case were constrained to a period of several months and were more acute in nature relative to the poisoning that had occurred in Minamata, Japan, which lasted for many years. For adult toxicity, the lowest level observable damage to the nervous system was characterized by loss of sensation at the extremities of the hands and feet and in areas around the mouth (paraesthesia) between 240 and 480 µg/L in blood. These symptoms were followed at higher levels of exposure by loss of coordination in gait (ataxia), slurred speech (dysarthria), and visual and hearing defects (Bakir *et al.*, 1973).

16.1.3 Safety thresholds for mercury exposure

The World Health Organization (WHO) originally established a no-observed-adverse-effects-level (NOAEL) for Hg based on the Iraqi food contamination incident of 200–500 µg/L blood Hg and 50 µg/g hair mercury. Many epidemiological studies have since documented much more sensitive neurocognitive effects below initial safety standards set based on epidemiological data from the Iraqi poisoning. (Julvez *et al.*, 2012). In 2000, the US National Research Council (NRC) reviewed the major epidemiological evidence for MeHg toxicity (NRC, 2000) that led to the US EPA's present reference dose (RfD = acceptable daily intake of MeHg that is not expected to appreciably increase risks over a lifetime) of 0.1 µg/kg body weight per day. The RfD is equivalent to a blood-mercury concentration of about 5.8 µg/L in adults and ~1 µg/g in hair. However, effects below this threshold on the developing brain are an on-going area of research (Grandjean and Budtz-Jorgensen, 2007).

Other governmental agencies representing a number of countries (such as Japan, New Zealand, Australia, UK, Canada and those relying on the WHO) have developed MeHg intake levels to protect the public that range from 0.1 to 0.47 µg/kg/day. Most of the variability in the RfD for MeHg across these agencies stems from differences in the uncertainty factors used to extrapolate from the lowest observed effects on children's neurodevelopmental indicators (Mahaffey *et al.*, 2011). For example, the US EPA applied an uncertainty factor of 10 to epidemiological data representing the lowest level quantifiable response from a long-term cohort in the Faroe Islands population to account for variability and uncertainty within the human population (Rice *et al.*, 2003).

Various studies suggest that many populations and/or subpopulations exceed body burden Hg levels as a result of fish consumption rates that are generally considered acceptable. For example, approximately 10% of women in a study of pregnant women in Massachusetts, USA (Project Viva) had hair mercury concentrations equivalent to or greater than the US EPA RfD (Oken *et al.*, 2005). These results are similar to those observed in the US National Health and Nutrition Examination Survey (NHANES), which showed 4–10% of women of childbearing age exceeded the US EPA's RfD for MeHg between 1999 and 2004 (Mahaffey *et al.*, 2004; McDowell *et al.*, 2004; Mahaffey *et al.*, 2009). Another study in New York City adults showed blood Hg levels were three times those of the national average (McKelvey *et al.*, 2007). Foreign-born Chinese in this study had a geometric mean blood Hg of 7.26 µg/L, with > 50% exceeding the EPA's RfD (McKelvey *et al.*, 2007). Similar results have been observed within communities of Korean and Japanese Americans in the Arsenic Mercury Intake Biometric Study (AMIBS), where many women within both communities exceeded the MeHg RfD (Tsuchiya *et al.*, 2008a, 2008b, 2009). In a statistically representative survey of the Korean population from the Korean National Human Exposure and Biomonitoring Examination, one in three women and one in six children had hair mercury levels exceeding 1 μg/g and > 25% of all individuals exceeded the US EPA RfD (Son *et al.*, 2009; Kim and Lee, 2010).

16.1.4 Analytical techniques for measuring mercury in foods

Mercury is a naturally occurring element, commonly associated with many sulfidic ores and minerals. Its ubiquitous presence in the environment at trace levels makes contamination free sampling and measurement extremely difficult. As a result, prior to 1980, many efforts to quantify trace concentrations of Hg in the environment were hampered by large measurement errors resulting from contamination of samples and poor analytical reliability (Fitzgerald *et al.*, 2007). These methods have been refined and new methods for quantifying trace levels of Hg and MeHg in surface waters have been developed (Gill and Fitzgerald 1985; Bloom and Fitzgerald, 1988). In foods these improvements in analytical capability mean that Hg is detectable at much lower levels today than several decades ago, not necessarily that concentrations have risen in samples where Hg was previously reported as 'below detection'.

16.2 Concentrations of mercury in foods

The following sections review Hg concentrations in many commonly consumed foods.

16.2.1 Fish and shellfish

Figure 16.1 shows Hg concentrations reported by the US Food and Drug Administration (US FDA) for the top 30 ranked fish and shellfish in the commercial market. Safety thresholds for maximum allowable MeHg in fish tissue range between 0.3 and 1.0 μg/g. Highest concentrations that exceed these values are observed in predatory, long-lived fish. For example, highest Hg concentrations in the US commercial market are generally found in Gulf of Mexico tilefish, king mackerel, swordfish, shark and orange roughy (mean ~0.5–1.0 μg/g). Accordingly, many consumption advisories recommend women of childbearing age and children, limit or avoid these species.

Fish are the main source of MeHg exposure in most human populations (Mergler *et al.*, 2007). Unlike most persistent organic contaminants, MeHg accumulates in the thiol group of the cysteine residues in fish muscle tissue and is not removed/reduced by cooking (Harris *et al.*, 2003). High levels of

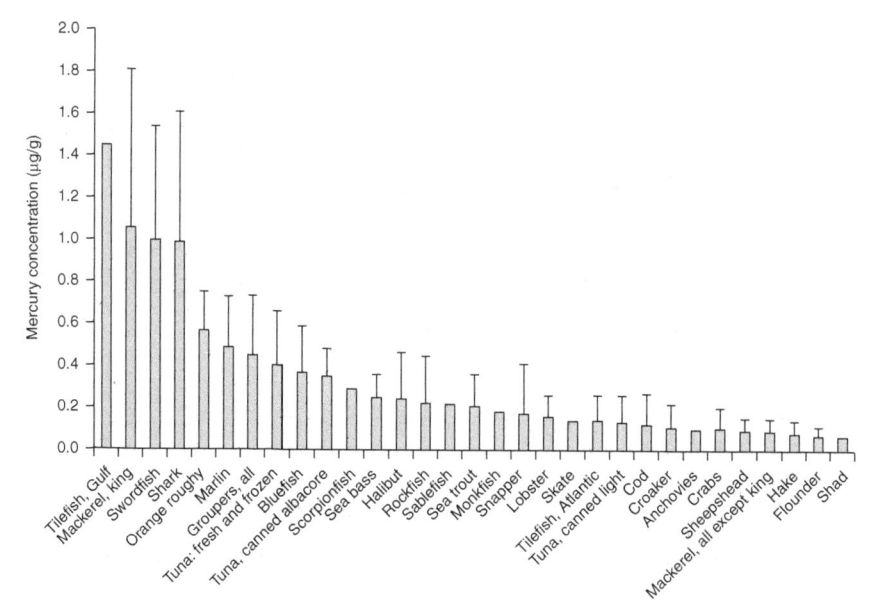

Fig. 16.1 Top 30 ranked seafood species by Hg concentration from the US FDA. Means and standard deviations are shown.

MeHg in fish tissue result from the combination of efficient MeHg assimilation from food and slow elimination rates (Trudel and Rasmussen 2006). Concentrations of mercury tend to be highest in long-lived predatory species that inhabit both freshwater and marine ecosystems (Wiener et al., 2003; Trudel and Rasmussen 2006; Storelli et al., 2007). Rapid and efficient fish growth can have a diluting effect on total mercury concentrations (Karimi et al., 2007), resulting in lower concentrations in fast growing species compared to slow growing fish of the same age (Doyon et al., 1998).

Most monitoring programs measure Hg rather than MeHg in fish because virtually all (> 95%) of the Hg is usually assumed to be MeHg in higher trophic level organisms (Bloom, 1992). However, selected studies show unexpectedly high variability in this fraction for certain marine and estuarine species, suggesting this hypothesis may need to be revisited (Table 16.1). In lower trophic level organisms, such as shellfish, the fraction of MeHg is highly variable and thus needs to be measured directly (Table 16.2).

Karimi et al., (2012) recently developed a large database of Hg concentrations in commercial market fish and showed that within species (seafood categories), Hg concentrations are highly variable spanning 0.3–2.4 orders of magnitude. Large variability in fish Hg concentration occurs both within and across species. Karimi et al., (2012) found large differences between Hg concentrations reported by the US FDA and other studies, where the FDA concentrations were generally lower (Fig. 16.2). For example, imported shrimp

Table 16.1 Selected studies of marine fish with < 85% total mercury as methylmercury in tissue. Most studies typically assume > 95% MeHg in fish tissue

Species	Study location	% MeHg	Range	n	Source
Cod (*Gadus morhua*)	Bay of Fundy	77 ± 14	61–88	3	Harding et al., 2005
Haddock (*Melanogrammus aeglefinus*)	Bay of Fundy	71 ± 22	54–97	8	Harding et al., 2005
Haddock (*Melanogrammus aeglefinus*)	Barents Sea	61	11–100	26	Joiris et al., 1995
Hake (*Urophycis tenuis*)	Bay of Fundy	80 ± 22	45–100	8	Harding et al., 2005
Herring (muscle, commercial)	Scheldt estuary	45	n/a	5	Baeyens et al., 2003
Herring (*Clupea harengus*)	Bay of Fundy	83	37–103	15	Harding et al., 2005
Flounder (muscle, commercial)	Scheldt estuary	67	52–83	14	Baeyens et al., 2003
Flounder (muscle, commercial)	Belgian coast	82	61–102	24	Baeyens et al., 2003
Flounder (*Platichthys flesus*)	French coast	81	77–85	90	Cossa et al., 2002
Mackerel (*Clupea harengus*)	Bay of Fundy	81 ± 16	57–114	14	Harding et al., 2005
Mackerel (*Rastrelliger kanagurta*)	Bangladesh	77	n/a	34	Joiris et al., 2000
Marlin (*Makaira nigricans*)	Hawaii	21 ± 13	2–62	19	Schultz et al., 1976
Marlin (*Makaira nigricans*)	Hawaii	52 ± 47	13–103	19	Brooks, 2004
Marlin (edible flesh)	Canada Market	66	51–84	3	Forsyth et al., 2004
Ocean perch (*Sebastes* spp.)	Barents Sea	53	22–80	50	Joiris et al., 1995
Shark (edible flesh)	Canada Market	65	48–94	12	Forsyth et al., 2004
Sole (common)	Scheldt estuary	32	18–46	16	Baeyens et al., 2003
Swordfish (*Xiphias gladius*)	Bay of Fundy	58 ± 18	37–92	11	Harding et al., 2005
Swordfish (*Xiphias gladius*)	New Zealand	77	n/a	10	Vleig et al., 1993
Tuna (*Thunnus alalunga*)	New Zealand	78	n/a	6	Vlieg et al., 1993
Tuna (canned albacore)	Canada Market	65	52–79	16	Forsyth et al., 2004

Mercury in foods 399

Table 16.2 Reported fraction of total mercury as methylmercury in selected shellfish species

Species	Study location	% MeHg	Range	n	Source
Crabs (*Cancer pagurus*)	Azores	66	38–123	6	Andersen and Depledge, 1997
Crabs (*Eriocheir sinensis*)	San Francisco Bay	27	n/a	27	Hui *et al.*, 2005
Lobster (*Nephrops norvegicus*)	Ligurian Sea	58 ± 14	31–88	37	Minganti *et al.*, 1990
Mussels (*Perna perna*)	Ghana	36	12–100	69	Joiris *et al.*, 2000
Mussels (commercial)	Sheldt estuary	24	n/a	24	Baeyens *et al.*, 2003
Oysters (*Crassotrea tulipa*)	Ghana	41	17–100	65	Joiris *et al.*, 2000
Oysters (*Crassotrea tulipa*)	Ghana	61	17–100	98	Joiris *et al.*, 2000
Shrimp (*Pasiphaea sivado*)	Ligurian Sea	85 ± 7	68–98	17	Minganti *et al.*, 1996
Shrimp (*Aristeus antennatus*)	Ligurian Sea	85	70–100	18	Minganti *et al.*, 1996

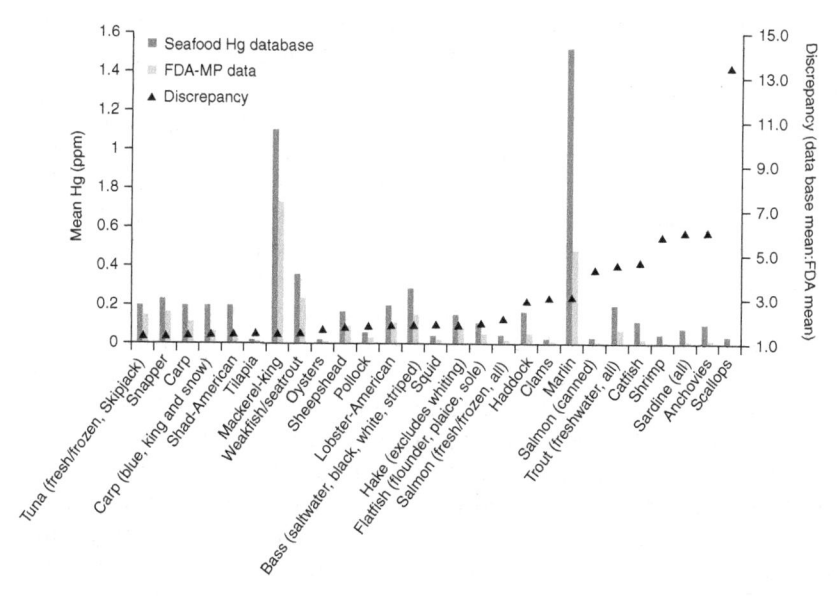

Fig. 16.2 Comparison of Hg concentrations reported by the US FDA and Seafood Hg database representing 300 independent data sources. Figure from Karimi *et al.* (2012).

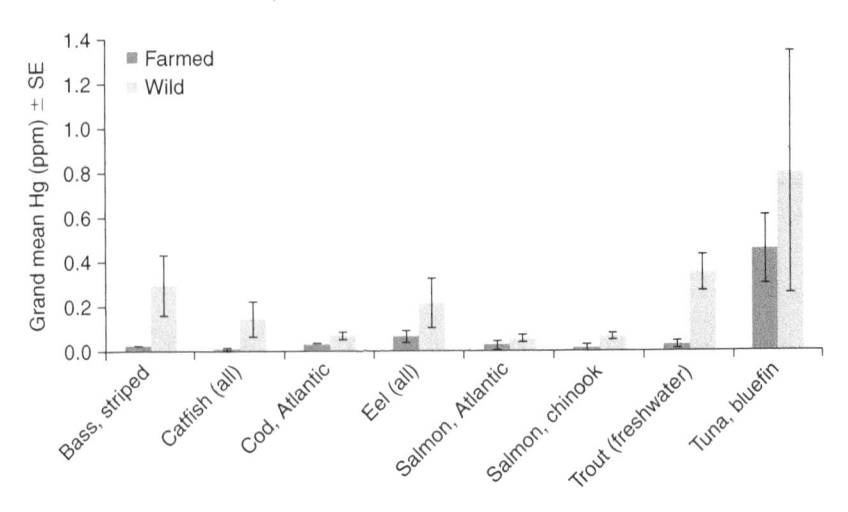

Fig. 16.3 Hg concentrations in farmed and wild fish in the US commercial market. Figure from Karimi et al. (2012).

make up a large fraction of US domestic seafood consumption, and measured mercury concentrations in shrimp caught in a variety of countries vary by an order of magnitude (Minganti et al., 1996; Plessi et al., 2001; Ruelas-Inzunza et al., 2004). Since fish Hg concentrations reflect exposures from ecosystems, understanding the harvesting regions and geographic variability in fish Hg concentrations is key to understanding trends in human exposures from seafood (Sunderland 2007).

Available data suggest farmed fish generally have lower Hg concentrations than the same species of wild fish. Karimi et al., (2012) showed that Hg concentrations in wild fish were 2–12 times higher than in farmed species (Fig. 16.3). However, measured Hg concentrations in farmed fish were comparatively limited. This is extremely important for exposure estimates for many Asian countries where over 90% of global aquaculture takes place (FAO, 2006).

In China, more than 74% of the fish supply comes from aquaculture (Alder and Pauly, 2006). Concentrations of Hg in farmed fish from China are generally low. For example, a survey of carp and snakehead in a Beijing commercial market reported concentrations between 2 and 77 ng/g total Hg, and 76–82% MeHg (Jin et al., 2006). Concentrations of Hg in farmed fish are a function of their diet, which can include a combination of fish meal and vegetable protein. Traditionally, farmed fish are fed fish meal from small forage fish. However, some experimentation with substitute foods such as soy protein are now taking place (Hardy et al., 1987; Berntssen et al., 2010). Alternative (non-fish) food for farmed fish may be effective for growth and contaminant reduction, but also seems to change the desired fatty acid profile that is associated with many health benefits (Berntssen et al., 2010).

Many studies advocate that nutrients in seafood, particularly long-chained polyunsaturated fatty acids (omega-3 fatty acids) offset the neurological and cardiovascular health risks associated with MeHg in seafood (Mozaffarian and Rimm, 2006; Domingo et al., 2007; Mahaffey et al., 2011). Accordingly, negative associations with prenatal MeHg exposure increase in strength when these benefits from omega-3 fatty acids are taken into account (Davidson et al., 2008; Davidson et al., 2011; Lynch et al., 2011).

In epidemiological studies, omega-3 fatty acids are generally positively associated with improvements in child IQ (Oken et al., 2005) and decreased risk of adult myocardial infarction (Mozaffarian, 2009). Fish that are highest in mercury are often not high in omega-3 fatty acids (Table 16.3). Thus, it is possible to choose a seafood diet that optimizes nutrient content with species like salmon, herring and sardine that are all low in Hg (Mahaffey et al., 2008). Unfortunately, fatty acids tend to be highly correlated with organic, lipid soluble contaminants, and the best fish choices for consumers become more complex is when multiple contaminants, fisheries sustainability, costs, and nutrients are all simultaneously considered (Oken et al., 2012).

16.2.2 MeHg in rice

In addition to fish consumption, a number of studies have found rice grown at contaminated sites also contributes substantially to MeHg exposure in countries where it is a dietary staple consumed in large quantities (Li et al., 2010). In contaminated regions of China, rice concentrations up to 174 ng MeHg/g and up to 66% MeHg as a fraction of total Hg have been reported (Table 16.4). Mercury-contaminated rice is a growing concern in other areas, such as the Philippines. For example, rice paddy fields along the Naboc River, Philippines have been irrigated by water from a contaminated river, and MeHg intake from rice for local residents exceeds one-third of their total exposure (Appleton et al., 2006).

Most foods other than seafood have MeHg concentrations below 20 ng/g. Although MeHg concentrations in rice are still lower than many fish, large quantities of rice consumed (100–400 g/day) in many Asian countries cause it to be a dominant MeHg exposure source (Table 16.4). In the Wanshan mining area of China, where fish are not part of the staple diet, rice accounted for between 93% and 98% of MeHg exposure of the population (Feng et al., 2008). A recent study found that rice grown in aerobic conditions had substantially lower MeHg concentrations, providing one potential mitigation measure for this exposure source (Peng et al., 2012). In addition, certain varieties of rice appear to accumulate much less MeHg than others with comparable yields (Rothenberg et al., 2012). Thus, advice to local farmers could also be an effective means for reducing this exposure source.

Table 16.3 Hg concentrations in selected fish and shellfish based on frequency of consumption reported in the US NHANES

Rank	Seafood category	EPA + DHA (mg/100 µg of fish)	Hg (µg/100 g fish)
1	Shrimp	390	0.03–0.04
2	Tuna (all, average)	630	0.24–0.48
	Canned, Light	128–270	0.11–0.12
	Canned, White	862	0.35–0.37
	Fresh, Bluefin (7 kg)	1173–1504	0.13
	Fresh, Skipjack (3 kg)	256–328	0.17
	Fresh, Yellowfin (5–20 kg)	100–120	0.06–0.31
3	Breaded fish products	0.26	0.135
4	Salmon	1590	0.04–0.13
5	Crabs	36	0.06–0.26
6	Catfish	280	0.16
7	Other fish	54	0.223
8	Scallops	270	0.05
9	Lobster	360	0.10–0.28
10	Clams	240	0.01–0.06
11	Cod	240	0.06–0.11
12	Oysters	350	0.01–0.07
13	Other shellfish	310	0.061
14	Flatfish	15	0.092
15	Unknown fish	53	0.223
16	Pollock	260	0.02–0.06
17	Mussels	350	0.03–0.08
18	Trout	580	0.14–0.15
19	Haddock	180	0.03–0.06
20	Crayfish	380	0.03
21	Perch	300	0.09–0.11
22	Sardines	980	0.02–0.03
23	Swordfish	580	0.98–1.03
24	Bass (freshwater)	640	0.38
25	Sea bass	490	0.14–0.22
26	Pike	140	0.31
27	Mackerel (except King)	1790	0.09–0.220
	King Mackerel	401	0.73–1.06
28	Shark	220	0.75–0.99
29	Walleye	530	0.52
30	Porgy	210	0.08

Table from Mahaffey et al. (2011)
EPA, eicosapentaenoic acid; DHA, docosahexaenoic acid
Data are from Sunderland (2007), Mahaffey et al. (2004) and references therein. Ranges in Hg concentrations represent variability in sample means across different harvesting regions.

16.2.3 Other foods

Tables 16.5 and 16.6 show a summary of literature values on Hg and MeHg concentrations in fruit, vegetables, various meats and dairy products. Concentrations are very low in all other foods, with maximum inorganic Hg levels of < 150 ng/g in extremely contaminated regions of China.

Table 16.4 Hg concentrations in rice from various regions of China

Location	MeHg (ng/g)	Total Hg (ng/g)	% MeHg/ total Hg	Intake MeHg (µg/g/day)	Reference
Qingzhen Hg polluted area	0.71–28	2.53–33.5	28.1–83.7	0.005–0.19	Horvat et al. (2003)
Wanshan Hg mining area	8.03–144	11.1–569	5.46–72.6	0.05–0.96	Horvat et al. (2003)
	1.9–27.6	4.9–214.7	2.4–75.1	0.01–0.21	Feng et al. (2008)
	1.61–174	10.3–1120	1.4–93	0.016–1.74	Qiu et al. (2008)
Wuchaun Hg mining area	4.2–18	9.1–570	2–66	0.04–0.18	Qiu et al. (2008)
	3.1–13.4	6.0–113	6.0–83.6	0.03–0.12	Li et al. (2008)
15 Chinese Provinces	1.9–10.5	6.3–39.3	7–44	0.02–0.105	Shi et al. (2005)

From Li et al. (2010)

Outside contaminated sites (Tables 16.5 and 16.6), all fruit, vegetable and meat products have total Hg levels < 10 ng/g. In addition, almost all of the mercury is present as inorganic Hg rather than MeHg (Feng et al., 2008), which means absorption will be extemely low (Clarkson and Magos, 2006). One relatively unexplored area of food contamination is the MeHg content of livestock and poultry fed fish meal as part of their diet (Alder et al., 2008). One study of Hg biomarkers in Sweden of individuals who did not consume seafood suggested that non-fish sources must be the dominant exposure source (Lindberg et al., 2004). Interestingly, unlike fish, chicken appear to have lower concentrations of Hg in their tissues than in their diet (Shah et al., 2010), perhaps suggesting that demethylation or some other elimination mechanism occurs in their gut.

16.3 Mercury exposures and risks from major food categories

Extremely high levels of inorganic Hg in the diet are required to elicit a toxicological response compared to levels present in most foods due to the poor absorption efficiency of inorganic Hg and its inability to cross biological membranes (Clarkson and Magos, 2006). Thus, MeHg exposure is a much greater concern in terms of health risks. Fish and shellfish are the primary sources of MeHg exposure for most populations. Exceptions are apparent in northern populations, that consume marine mammals as part of their traditional diets (Johansen et al., 2004), and in Asian populations consuming rice grown anaerobically in contaminated areas.

Table 16.7 contrasts MeHg exposure from typical diets in China and the US. Fish and shellfish are the overwhelming majority of MeHg exposure in

Table 16.5 Summary of concentrations of mercury in fruit and vegetables from various countries

Food	Country	Total Hg (ng/g) mean (min–max)	MeHg (ng/g) mean (min–max)	Reference
Vegetables	China (mining area)	87 (4–266)	0.10 (0.04–0.51)	Feng et al. (2008)
Vegetables	China (coastal city)	1 (ND–30)	n/a	Chen et al. (2011)
Vegetables	China (mining area)	2–130	n/a	Wang et al. (2011)
Vegetables	China (national monitoring)	35.2 (81.4 max)	n/a	Maoqi et al. (2003)
Vegetables	China (contaminated area)	46–132	n/a	Qian et al. (2009)
Vegetables	China (mining area)	4 (5–15)	n/a	Zhen et al. (2007)
Fruit	China (coastal city)	10 (1–58)	n/a	Chen et al. (2011)
Grain	China (national monitoring)	7.3 (13.5 max)	n/a	Maoqi et al. (2003)
Fruit	China (national monitoring)	0.9 (1.4 max)	n/a	Maoqi et al. (2003)
Juice	China (national monitoring)	2.4 (3.4 max)	n/a	Maoqi et al. (2003)
Snacks	China (national monitoring)	3.7 (7.9 max)	n/a	Maoqi et al. (2003)
Bread and cereal	UK total diet study	2–4	n/a	Ysart et al. (2000)
Vegetables	UK total diet study	0.4–1	n/a	Ysart et al. (2000)
Fruit	UK total diet study	0.6–0.8	n/a	Ysart et al. (2000)
Vegetables	Belgium	1.1–37.3	n/a	De Temmerman et al. (2009)
Mushrooms, raw	US total diet survey	3 (max 20)	n/a	US FDA (2005)
Fruit	Two Canadian cities	<1	n/a	Dabeka et al. (2003)
Vegetables	Two Canadian cities	<1	n/a	Dabeka et al. (2003)

ND = not detectable. n/a = not available.

Table 16.6 Summary of Hg concentrations in meat and dairy products from various countries

Food	Country	Total Hg (ng/g) mean (min–max)	MeHg (ng/g) mean (min–max)	Reference
Pork meat	China (mining area)	215.8 (7.5–564.6)	0.85 (0.05–3.43)	Feng et al. (2008)
Chicken	China (coastal city)	5 (ND–21)	n/a	Chen et al. (2011)
Pork	China (coastal city)	4 (ND–32)	n/a	Chen et al. (2011)
Meat	China (national monitoring)	2.9 (10.8 max)	n/a	Maoqi et al. (2003)
Egg (preserved)	China (national monitoring)	18.5 (36.7 max)	n/a	Maoqi et al. (2003)
Milk	China (national monitoring)	3.4 (6.3 max)	n/a	Maoqi et al. (2003)
Canned meat and fish	China (national monitoring)	5.7 (12.5 max)	n/a	Maoqi et al. (2003)
Meat products	UK total diet study	1–3	n/a	Ysart et al. (2000)
Chicken	UK total diet study	2	n/a	Ysart et al. (2000)
Eggs	UK total diet study	1.3	n/a	Ysart et al. (2000)
Eggs	Belgium	3.15–4.44	n/a	Waegeneers et al (2009)
Chicken	US Total Diet Study	1 (37 max)	n/a	US FDA (2005)
Beef	Egypt	5 (1–27)	n/a	Khalafalla et al. (2011)
Swine	Czech Republic	3 (1–22)	n/a	Ulrich et al. (2001)
Chicken	Korea, national survey	<1	n/a	Kwon et al. (2008)
Pork	Korea, national survey	1.5	n/a	Kwon et al. (2008)
Beef	Korea, national survey	<1	n/a	Kwon et al. (2008)
Meat and meat products	Two Canadian cities	0.3–2.3	n/a	Dabeka et al. (2003)
Dairy products	Two Canadian cities	0.11–1.8	n/a	Dabeka et al. (2003)
Chicken and turkey	Two Canadian cities	1.8	n/a	Dabeka et al. (2003)

ND = not detectable. n/a = not available.

Table 16.7 Summary of MeHg exposures from various food sources

Food	MeHg (ng/g)	Consumption (g/day)	Exposure (ng MeHg/day)	% Total
United States[1]				
Fish/shellfish	110	18.9	2079	98.6
Meat	0.04	145	6	0.3
Vegetables, grains, nuts, oil	0.01	592	7	0.3
Fruit, juice	0.01	162	2	0.1
Eggs, dairy	0.04	374	15	0.7
China[2]				
Fish/shellfish	50	34	1700	28.3
Rice	15	273	4095	68.2
Meat	1	95.3	95	1.6
Vegetables + grains	0.1	566	57	0.9
Soy, beans, oil	0.1	66	7	0.1
Fruit	0.1	36	4	0.1
Eggs, dairy	1	45	45	0.7

[1] US population-wide consumption of fish from NMFS, 2003 and market basket fish MeHg from Sunderland (2007). US Food consumption information from Bowman *et al.* (2011). Concentrations of inorganic Hg in foods are < 10 ng/g for all foods other than fish. Fish Hg is from Sunderland (2007). Fraction of MeHg is inferred from Feng *et al.* (2008) using 0.1% MeHg in vegetables and fruits and 0.4% MeHg in meat and meat products.
[2] Fish MeHg assumed to be ~50 ng/g based on Jin *et al.* (2006), assume vegetable and meat MeHg concentrations (0.1 ng/g and 1 ng/g, respectively) based on Feng *et al.* (2008).

the US (> 98.6%), due to the extremely low concentrations of MeHg in other foods. In contrast, most of MeHg exposure in the Chinese diet can easily be rice if concentrations are in the mid range of those reported in Table 16.4. Lower MeHg concentration in Chinese farmed fish also helps to explain the relatively lower fish and shellfish contribution to dietary MeHg exposure of the average individual in China.

A variety of studies have noted mediating effects of other foods on MeHg absorption. For example, tea and black coffee consumed with seafood appear to lower the bioaccessibility of MeHg by up to 60% (Canuel *et al.*, 2006a; Ouédraogo and Amyot 2011). Tropical fruit has also been observed to lower MeHg absorption by individuals, potentially through the mediating effects of phytochemicals on excretion, transport and/or binding (Passos *et al.*, 2003). Finally, individual assimilation efficiencies may also naturally vary across and within populations, adding an additional dimension to interpretation of Hg biomarker data (Canuel *et al.*, 2006b). These factors together may help to explain why, in certain populations, body burdens are not strongly correlated with MeHg intakes from seafood, despite no other obvious sources of MeHg exposures (Airaksinen *et al.*, 2010).

16.4 References

AIRAKSINEN, R., TURUNEN, A.W., RANTAKOKKO, P., MANNISTO, S., VARTIAINEN, T. and VERKASALO, P.K., 2010. Blood concentration of methylmercury in relation to food consumption. *Public Health Nutrition*, **14**(03), pp. 480–489.

ALDER, J. and PAULY, D., 2006. *On the multiple uses of forage fish: from ecosystems to markets*, Fisheries Centre, University of British Columbia, Canada.

ALDER, J., CAMPBELL, B., KARPOUZI, V., KASCHNER, K. and PAULY, D., 2008. Forage Fish: From Ecosystems to Markets. *Annual Review of Environment and Resources*, **33**(1), pp. 153–166.

ANDERSEN, J.L. and DEPLEDGE, M.H., 1997. A survey of total mercury and methylmercury in edible fish and invertebrates from Azorean waters. *Marine Environmental Research*, **44**(3), pp. 331–350.

APPLETON, J.D., APPLETON, J.D., WEEKS, J. M., CALVEZ, J.P.S. and BEINHOFF, C., 2006. Impacts of mercury contaminated mining waste on soil quality, crops, bivalves, and fish in the Naboc River area, Mindanao, Philippines. *Science of The Total Environment*, **354**(2–3), pp. 198–211.

BAEYENS, W., LEERMAKERS, M., PAPINA, T., SAPRYKIN, A., BRION, N., NOYEN, J., DEGIETER, M., ELSKENS, M. and GOEYENS, L., 2003. Bioconcentration and biomagnification of mercury and methylmercury in North Sea and Scheldt estuary fish. *Archives of Environmental Contamination and Toxicology*, **45**(4), pp. 498–505.

BAKIR, F., DAMLUJI, S.F., AMINZAKI, L., MURTADHA, M., KHALIDI, A., ALRAWI, N.Y., TIKRITI, S., DHAHIR, H.I., CLARKSON, T.W., SMITH, J.C. and DOHERTY, R.A., 1973. Methylmercury poisoning in Iraq. *Science*, **181**(4096), pp. 230–241.

BENOIT, J.M. et al., 2003. Geochemical and Biological Controls Over Methylmercury Production and Degradation in Aquatic Ecosystems. *ACS Symposium Series*, **835**, pp. 262–297.

BERNTSSEN, M.H.G., JULSHAMN, K. and LUNDEBYE, A.-K., 2010. Chemical contaminants in aquafeeds and Atlantic salmon (Salmo salar) following the use of traditional-versus alternative feed ingredients. *Chemosphere*, **78**(6), pp. 637–646.

BLOOM, N.S., 1992. On the chemical form of mercury in edible fish and marine invertebrate tissue. *Can J Fish Aquat Sci*, **49**, pp. 1010–1017.

BLOOM, N.S. and FITZGERALD, W.F., 1988. Determination of volatile mercury species at the picogram level by low-temperature gas chromatography with cold-vapour atomic fluorescence detection. *Analytica Chimica Acta*, **208**, pp. 151–161.

BOWMAN S., MARTIN C., FRIDAY J., CLEMENS J., MOSHFEGH A., LING B. and WELLS H., 2011. *Retail food commodity intakes: Mean amount of retail commodities per individual, 2001-2002*. U.S. Department of Agriculture, Agricultural Research Service and Economic Research Service.

BROOKS B., 2004. Mercury levels in Hawaiian commercial fish. In: *Proceedings of the U.S. EPA National Forum on Contaminants in Fish*, 25–28 January 2004, San Diego, CA. EPA-823-R-04-006. Washington, DC: U.S. Environmental Protection Agency, pp. 24–25.

CANUEL, R., DE GROSBOIS, S.B., LUCOTTE, M., ATIKESSÉ, L., LAROSE, C. and RHEAULT, I. 2006a. New evidence on the effects of tea on mercury metabolism in humans. *Arch Environmental and Occupational Health*, **61**(5), 232–238.

CANUEL, R., DE GROSBOIS, S.B., ATIKESSÉ, L., LUCOTTE, M., ARP, P., RITCHIE, C., MERGLER, D., HING MAN, C., AMYOT, M. and ANDERSON, R., 2006b. New Evidence on Variations of Human Body Burden of Methylmercury from Fish Consumption. *Environmental Health Perspectives*, **114**(2), pp. 302–306.

CHEN, C., QIAN, Y., CHEN, Q. and LI, C., 2011. Assessment of Daily Intake of Toxic Elements Due to Consumption of Vegetables, Fruits, Meat, and Seafood by Inhabitants of Xiamen, China. *Journal of Food Science*, **76**(8), pp. T181–T188.

CHOI, A.L. and GRANDJEAN, P., 2008. Methylmercury exposure and health effects in humans. *Environmental Chemistry*, **5**(2), p. 112.

CLARKSON, T.W. and MAGOS, L., 2006. The Toxicology of Mercury and Its Chemical Compounds. *Critical Reviews in Toxicology*, **36**(8), pp. 609–662.

COSSA, D., AUGER, D., AVERTY, B., LUCON, M., MASSELIN, P., and NOEL, J., 1992. Flounder (Platichthys flesus) muscle as an indicator of metal and organochlorine contamination of French Atlantic coastal waters. *Ambio*, **21**(2), pp. 176–182.

DABEKA, R.W., MCKENZIE, A.D. and BRADLEY, P., 2003. Survey of total mercury in total diet food composites and an estimation of the dietary intake of mercury by adults and children from two Canadian cities, 1998-2000. *Food Additives and Contaminants*, **20**(7), pp. 629–638.

DAVIDSON, P.W., CORY-SLECHTA, D.A., THURSTON, S.W., HUANG, L., SHAMLAYE, C.F., GUNZLER, D., WATSON, G., WIJNGAARDEN, E., ZAREBA, G., KLEIN, J.D., CLARKSON, T.W., STRAIN, J.J. and MYERS, G.J., 2011. Fish consumption and prenatal methylmercury exposure: Cognitive and behavioral outcomes in the main cohort at 17 years from the Seychelles child development study. *NeuroToxicology*, **32**(6),pp. 711–717.

DAVIDSON, P.W., STRAIN, J.J., MYERS, G.J., THURSTON, S.W., BONHAM, M.P., SHAMLAYE, C.F., ABBIE, S-R., WALLACE, J.M.W., ROBSON, P.J., DUFFY, E.M., GEORGER, L.A., SLOANE-REEVES, J., CERNICHIARI, E., CANFIELD, R.L., COX, C., HUANG, L.S., JANCIURAS, J. and CLARKSON, T.W., 2008. Neurodevelopmental effects of maternal nutritional status and exposure to methylmercury from eating fish during pregnancy. *NeuroToxicology*, **29**(5), pp. 767–775.

DE TEMMERMAN, L., WAEGENEERS, N., CLAEYS, N. and ROEKENS, E., 2009. Comparison of concentrations of mercury in ambient air to its accumulation by leafy vegetables: An important step in terrestrial food chain analysis. *Environmental Pollution*, **157**(4), pp.1337–1341.

DOMINGO, J.L., BOCIO, A., FALCO, G. and LLOBET, J.M., 2007. Benefits and risks of fish consumption. *Toxicology*, **230**(2-3), pp.219–226.

DOYON, J.-F., SHETAGNE, R. and VERDON, R., 1998. Different mercury bioaccumulation rates between sympatric populations of dwarf and normal lake whitefish (Coregonus clupeaformis) in the La Grande complex watershed, James Bay, Quebec. *Biogeochemistry*, **40**, pp.203–216.

FENG, X., LI, P., GUANGLE, Q., WANG, S., LI, G., SHANG, L., MENG, B., JIANG, H., BAI, W., LI, Z. and FU, X., 2008. Human Exposure To Methylmercury through Rice Intake in Mercury Mining Areas, Guizhou Province, China. *Environmental Science and Technology*, **42**(1), pp.326–332.

FITZGERALD, W.F., ENGSTROM, D.R., MASON, R.P. and NATER, E.A., 1998. The Case for Atmospheric Mercury Contamination in Remote Areas. *Environmental Science and Technology*, **32**(1), pp.1–7.

FITZGERALD, W.F., LAMBORG, C.H. and HAMMERSCHMIDT, C.R., 2007. Marine biogeochemical cycling of mercury. *Chemical Reviews*, **107**(2), pp.641–662.

FORSYTH, D.S., CASEY, V., DABEKA, R.W. and MCKENZIE, A., 2004. Methylmercury levels in predatory fish species marketed in Canada. *Food Additives and Contaminants*, **21**(9), pp.849–856.

GILL, G.A. and FITZGERALD, W.F., 1985. Mercury sampling of open ocean waters at the picomolar level. *Deep-Sea Research*, **32**(3), pp.287–297.

GILL, G.A. and FITZGERALD, W.F., 1988. Vertical mercury distributions in the oceans. *Geochimica et Cosmochimica Acta*, **52**, pp.1719–1728.

GUALLAR, E., SANZ-GALLARDO, I., BODE, P., ARO, A., GOMEZ-ARACENA, J., KARK, J.D., RIEMERSMA, R.A., MARTIN-MORENO, J.M. and KOK, F.J., 2002. Mercury, fish oils, and myocardial infarction. *The New England Journal of Medicine*, **347**(22), pp.1747–1754.

HALLGREN, C.G., HALLMANS, G., JANSSON, J.H., MARKLUND, S.L., HUHTASAARI, F., SCHUTZ, A., STROMBERG, U., VESSBY, B. and SKERFVING, S., 2007. Markers of high fish intake are associated with decreased risk of a first myocardial infarction. *British Journal of Nutrition*, **86**(3), p.397.

HARADA, M., 1995. Minamata disease: Methylmercury poisoning in Japan caused by environmental pollution. *Critical Reviews in Toxicology*, **25**(1), pp.1–24.

HARADA, M., 1978. Congenital minamata disease: Intrauterine methylmercury poisoning. *Teratology*, **18**, pp.285–288.

HARDING G., DALZIEL J. and VASS P., 2005. *Prevalence and bioaccumulation of methylmercury in the food web of the Bay of Fundy, Gulf of Maine*. In PERCY J.A., EVANS A.J., WELLS P.G. and ROLSTON S.J. (Eds.), Proceedings of the 6th Bay of Fundy Workshop, September 29–2 October 2004, Cornwallis, Nova Scotia, Canada. Dartmouth, Nova Scotia, Canada: Environment Canada, Atlantic Region, 76–77.

HARDY, R.W., SCOTT, T.M. and HARRELL, L.W., 1987. Replacement of herring oil with mehaden oil, soybean oil, or tallow in the diets of Atlantic salmon raised in marine net-pens. *Aquaculture*, **65**, pp.267–277.

HARRIS, H.H., PICKERING, I.J. and GEORGE, G.N., 2003. The chemical form of mercury in fish. *Science*, **301**(5637), p.1203.

HORVAT, M., NOLDE, N., FAJON, V., JEREB, V., LOGAR, M., LOJEN, S., JACIMOVIC, R., FALNOGA, I., LIYA, Q., FAGANELI, J. and DROBNE, D., 2003. Total mercury, methylmercury and selenium in mercury polluted areas in the province Guizhou, China. *The Science of the Total Environment*, **304**, pp.231–256.

HUI, C., RUDNICK, D. and WILLIAMS, E., 2005. Mercury burdens in Chinese mitten crabs (Eriocheir sinensis) in three tributaries of southern San Francisco Bay, California, USA. *Environmental Pollution*, **133**(3), pp.481–487.

JIN, S., CHUN-YING, C., BAI, L. and YURFENG, L., 2006. Analysis of total mercury and methylmercury concentrations in four commercially important freshwater fish species obtained from Beijing markets. *Journal of Hygiene Research*, **35**(6), pp. 722–725.

JOHANSEN, P., MUIR, D., ASMUND, G. and RIGET, F., 2004. Human exposure to contaminants in the traditional Greenland diet. *Science of The Total Environment*, **331**(1–3), pp.189–206.

JOIRIS, C.R., ALI, I.B., HOLSBEEK, L., BOSSICART, M. and TAPIA, G., 1995. Total and organic mercury in Barents Sea pelagic fish. *Bulletin of Environmental Contamination and Toxicology*, **55**, pp.674–681.

JOIRIS, C.R., HOLSBEEK, L. and OTCHERE, F.A., 2000. PII: S0025-326X(00)00014-X bivalves Crassotrea tulipa and Perna perna from Ghana. *Marine Pollution Bulletin*, **40**(5), pp.457–460.

JULVEZ, J., YORIFUJI, T., CHOI, A.L. and GRANDJEAN, P., 2012. *Methylmercury and Neurotoxicity* S. CECCATELLI and M. ASCHNER, eds., Boston, MA: Springer US.

KARIMI, R., CHEN, C.Y., PICKHARDT, P.C., FISHER, N.S. and FOLT, C.L., 2007. Stoichiometric controls of mercury dilution by growth. *Proceedings of the National Academy of Sciences*, **104**(18), pp.7477–7482.

KARIMI, R., FITZGERALD, T.P. and FISHER, N.S., 2012. A quantitative synthesis of mercury in commercial seafood and implications for exposure in the U.S. *Environmental Health Perspectives*, **22**(11), 1512–1519.

KHALAFALLA, F.A., ALI, F.H., SCHWAGELE, F and ABD-EL-WAHAB, M.A., 2011. Heavy metal residues in beef carcasses in Beni – Suef abattoir, Egypt. *Veterinaria Italiana*, **47**(3), pp.351–361.

KIM, N.-S. and LEE, B.-K., 2010. Blood total mercury and fish consumption in the Korean general population in KNHANES III, 2005. *The Science of the Total Environment*, **408**(20), pp.4841–4847.

KWON, Y.-M., LEE, K-H., LEE, H-S., PARK, S-O., PARK, J-M., KIM, J-M., KANG, K-M., NO, K-M., KIM, D-S., LEE, J-O., HONG, M-K and CHOI, D-W., 2008. Risk assessment for heavy metals in Korean foods and livestock foodstuffs. *Korean Journal of Food Science and Animal Resources*, **28**(3), pp.373–389.

LI, P., FENG, X. and QIU, G., 2010. Methylmercury exposure and health effects from rice and fish consumption: A review. *International Journal of Environmental Research and Public Health*, **7**(6), pp.2666–2691.

LINDBERG, A., BJÖRNBERG, K.A., VAHTER, M. and BERGLUND, M., 2004. Exposure to methylmercury in non-fish-eating people in Sweden. *Environmental Research*, **96**(1), pp.28–33.

LYNCH, M.L. HUANG, L.S., COX, C., STRAIN, J.J., MYERS, G.J., BONHAM, M.P., SHAMLAYE, C.F., STOKES-RINER, A., WALLACE, J.M.W., DUFFY, E.M., CLARKSON, T.W. and DAVIDSON, P.W., 2011. Varying coefficient function models to explore interactions between maternal nutritional status and prenatal methylmercury toxicity in the Seychelles Child Development Nutrition Study. *Environmental Research*, **111**(1), pp.75–80.

MAHAFFEY, K.R., SUNDERLAND, E.M., CHAN, H.M., CHOI, A.L., GRANDJEAN, P., MARIËN, K., OKEN, E., SAKAMOTO, M., SCHOENY, R., WEIHE, P., YAN, C-H and YASUTAKE, A., 2011. Balancing the benefits of n-3 polyunsaturated fatty acids and the risks of methylmercury exposure from fish consumption. *Nutrition Reviews*, **69**(9), pp.493–508.

MAHAFFEY, K.R., CLICKNER, R.P. and BODUROW, C.C., 2004. Blood organic mercury and dietary mercury intake: National health and nutrition examination survey, 1999 and 2000. *Environmental Health Perspectives*, **112**(5), pp.562–570.

MAHAFFEY, K.R., CLICKNER, R.P. and JEFFRIES, R.A., 2009. Adult women's blood mercury concentrations vary regionally in the United States: Association with patterns of fish consumption (NHANES 1999–2004). *Environmental Health Perspectives*, **117**(1), pp.47–53.

MAHAFFEY, K.R., CLICKNER, R.P. and JEFFRIES, R.A., 2008. Methylmercury and omega-3 fatty acids: Co-occurrence of dietary sources with emphasis on fish and shellfish. *Environmental Research*, **107**(1), pp.20–29.

MASON, R.P., 2002. Role of the ocean in the global mercury cycle. *Global Biogeochemical Cycles*, **16**(4).

MASON, R.P., FITZGERALD, W.F. and MOREL, F.M.M., 1994. The biogeochemical cycling of elemental mercury: Anthropogenic influences. *Geochimica et Cosmochimica Acta*, **58**(15), pp.3191–3198.

MCDOWELL, M.A., DILLON, C.F., OSTERLOH, J., BOLGER, P.M., PELLIZZARI, E., FERNANDO, R., DE OCA, R.M., SCHOBER, S.E., SINKS, T., JONES, R.L. and MAHAFFEY, K.R., 2004. Hair mercury levels in U.S. children and women of childbearing age: Reference range data from NHANES 1999-2000. *Environmental Health Perspectives*, **112**(11), pp.1165–1171.

MCKELVEY, W., GWYNN, C., JEFFERY, N., KASS, D., THORPE, L., GARG, R., PALMER, C.D. and PARSONS, P.J., 2007. A biomonitoring study of lead, cadmium and mercury in the blood of New York City adults. *Environmental Health Perspectives*, **115**, pp.1435-1441.

MERGLER, D., ANDERSON, H.A., CHAN, H.M., MAHAFFEY, K.R., MURRAY, M., SAKAMOTO, M. and STERN, A.H., 2007. Methylmercury exposure and health effects in humans: A worldwide concern. *Ambio*, **36**(1), pp. 3–11.

MINGANTI, V., CAPELLI, R., DE PELLEGRINI, R., RELINI, L.O. and RELINI, G., 1996. Total and organic mercury concentrations in offshore crustaceans of the Ligurian Sea and their relations to the trophic levels. *Science of The Total Environment*, **184**, pp.149–162.

MOZAFFARIAN, D., 2009. Fish, mercury, selenium and cardiovascular risk: Current evidence and unanswered questions. *International Journal of Environmental Research and Public Health*, **6**(6), pp.1894–1916.

MOZAFFARIAN, D. and RIMM, E.B., 2006. Fish intake, contaminants, and human health. *JAMA*, **296**(15), pp.1885–1899.

MOZAFFARIAN, D., SHI, P., MORRIS, J.S., SPIEGELMAN, D., GRANDJEAN, P., SISCOVICK, D.S., WILLETT, W.C. and RIMM, E.B., 2011. Mercury exposure and risk of cardiovascular disease in two U.S. cohorts. *The New England Journal of Medicine*, **364**(12), pp.1116–1125.

NINOMIYA, T., IMAMURA, K., KUWAHATA, M., KINDAICHI, M., SUSA, M. and EKINO, S., 2005. Reappraisal of somatosensory disorders in methylmercury poisoning. *Neurotoxicology and Teratology*, **27**(4), pp.643–653.

NRIAGU, J., 1993. Legacy of mercury pollution. *Nature*, **363**, p.589.

OKEN, E., WRIGHT, R.O., KLEINMAN, K.P., BELLINGER, D., AMARASIRIWARDENA, C.J., HU, H., RICH-EDWARDS, J.W. and GILLMAN, M.W., 2005. Maternal fish consumption, hair mercury, and infant cognition in a U.S. cohort. *Environmental Health Perspectives*, **113**(10), pp.1376–1380.

OKEN, E., CHOI, A.L., KARAGAS, M.R., MARIËN, K., RHEINBERGER, C.M., SCHOENY, R., SUNDERLAND, E. and KORRICK, S., 2012. Which fish should I eat? Perspectives influencing fish consumption choices. *Environmental Health Perspectives*, **120**(6), pp.790–798.

OUÉDRAOGO, O. and AMYOT, M., 2011. Effects of various cooking methods and food components on bioaccessibility of mercury from fish. *Environmental Research*, **111**(8), pp.1064–1069.

PASSOS, C.J., MERGLER, D., GASPAR, E., MORAIS, S., LUCOTTE, M., LARRIBE, F., DAVIDSON, R. and DE GROSBOIS, S., 2003. Eating tropical fruit reduces mercury exposure from fish consumption in the Brazilian Amazon. *Environmental Research*, **93**(2), pp.123–130.

PENG, X., LIU, F., WANG, W-X. and YE, Z., 2012. Reducing total mercury and methylmercury accumulation in rice grains through water management and deliberate selection of rice cultivars. *Environmental Pollution*, **162**(C), pp.202–208.

PIRRONE, N., CINNIRELLA, S., FENG, X., FINKELMAN, R.B., FRIEDLI, H.R., LEANER, J., MASON, R., MUKHERJEE, A.B., STRACHER, G.B., STREETS, D.G. and TELMER, K., 2010. Global mercury emissions to the atmosphere from anthropogenic and natural sources. *Atmospheric Chemistry and Physics*, **10**(13), pp.5951–5964.

PLESSI, M., BERTELLI, D. and MONZANI, A., 2001. Mercury and selenium content in selected seafood. *Journal of Food Composition and Analysis*, **14**, pp.461–467.

QIAN, J., ZHANG, L., CHEN, H., HOU, M., NIU, Y., XU, Z. and LIU, H., 2009. Distribution of Mercury Pollution and Its Source in the Soils and Vegetables in Guilin Area, China. *Bulletin of Environmental Contamination and Toxicology*, **83**(6), pp.920–925.

QIU, G., FENG, X., LI, P., WANG, S., LI, G., SHANG, L. and FU, X., 2008. Methylmercury accumulation in rice (Oryza *sativa L)* grown at abandoned mercury mines in Guizhou, China. *Journal of Agricultural and Food Chemistry*, **56**, pp. 2465–2468.

RICE, D.C., SCHOENY, R. and MAHAFFEY, K.R., 2003. Methods and rationale for derivation of a reference dose for methylmercury by the U.S. EPA. *Risk Analysis*, **23**(1), pp.107–115.

RISSANEN, T., VOUTILAINEN, S., NYYSSONEN, K., LAKKA, T.A. and SALONEN, J.T., 2000. Fish oil-derived fatty acids, docosahexaenoic acid and docosapentaenoic acid, and the risk of acute coronary events : The kuopio ischaemic heart disease risk factor study. *Circulation*, **102**(22), pp.2677–2679.

ROMAN, H.A., WALSH, T.L., COULL, B.A., DEWAILLY, E., GUALLAR, E., HATTIS, D., MARIËN, K., SCHWARTZ, J., STERN, A.H., VIRTANEN, J.K. and RICE, G., 2011. Evaluation of the cardiovascular effects of methylmercury exposures: Current evidence supports development of a dose–response function for regulatory benefits analysis. *Environmental Health Perspectives*, **119**(5), pp.607–614.

ROTHENBERG, S.E., FENG, X., ZHOU, W., TU, M., JIN, B. and YOU, J., 2012. Environment and genotype controls on mercury accumulation in rice (Oryza sativa L.) cultivated along a contamination gradient in Guizhou, China. *Science of the Total Environment*, **426**(C), pp.272–280.

ROTHENBERG, S.E., FENG, X. and LI, P., 2011. Low-level maternal methylmercury exposure through rice ingestion and potential implications for offspring health. *Environmental Pollution*, **159**(4), pp.1017–1022.

RUELAS-INZUNZA, J., GARCÍA-ROSALES, S.B. and PÁEZ-OSUNA, F., 2004. Distribution of mercury in adult penaeid shrimps from Altata-Ensenada del Pabellón lagoon (SE Gulf of California). *Chemosphere*, **57**(11), pp.1657–1661.

SALONEN, J.T., SEPPANEN, K., NYYSSONEN, K., KORPELA, H., KAUHANEN, J., KANTOLA, M., TUOMILEHTO, J., ESTERBAUER, H., TATZBER, F. and SALONEN, R., 1995. Intake of mercury from fish, lipid peroxidation, and the risk of myocardial infarction and coronary, cardiovascular, and any death in eastern finnish men. *Circulation*, **91**, pp.645–655.

SCHULTZ, C.D., CREAR, D., PEARSON, J.E., RIVERS, J.E. and HYLIN, J.W., 1976. Total and organic mercury in the Pacific blue marlin. *Bulletin of Environmental Contamination and Toxicology*, **15**(2), pp. 230–234.

SELIN, N.E. *et al.*, 2008. Global 3-D land-ocean-atmosphere model for mercury: Present-day versus preindustrial cycles and anthropogenic enrichment factors for deposition. *Global Biogeochemical Cycles*, **22**(2).

SHAH, A.Q., KAZI, T.G., BAIG, J.A., AFRIDI, H.I., KANDHRO, G.A., KHAN, S., KOLACHI, N.F. and WADHWA, S.K., 2010. Determination of total mercury in chicken feed, its translocation to different tissues of chicken and their manure using cold vapour atomic absorption spectrometer. *Food and Chemical Toxicology*, **48**(6), pp.1550–1554.

SHI, J., LIANG, L., JIANG, G., 2005. Simultaneous determination of methylmercury and ethylmercury in rice by capillary gas chromatography coupled on-line with atomic fluorescence spectrometry. *Journal of AOAC International*, **88**, pp. 665-669.

SMITH-DOWNEY, N.V., SUNDERLAND, E.M. and JACOB, D.J., 2010. Anthropogenic impacts on global storage and emissions of mercury from terrestrial soils: Insights from a new global model. *Journal of Geophysical Research*, **115**(G3).

SON, J.-Y., LEE, J., PAEK, D. and LEE, J-T., 2009. Blood levels of lead, cadmium, and mercury in the Korean population Results from the Second Korean National Human Exposure and Bio-monitoring Examination. *Environmental Research*, **109**(6), pp.738–744.

STORELLI, M.M., BARONE, G., PISCITELLI, G. and MARCOTRIGIANO, G.O., 2007. Mercury in fish: Concentration vs. fish size and estimates of mercury intake. *Food Additives and Contaminants*, **24**(12), pp.1353–1357.

STREETS, D.G., DEVANE, M.K., LU, Z., BOND, T.C., SUNDERLAND, E.M. and JACOB, D.J., 2011. All-time releases of mercury to the atmosphere from human activities. *Environmental Science and Technology*, **45**(24), pp.10485–10491.

SUNDERLAND, E.M., 2007. Mercury exposure from domestic and imported estuarine and marine fish in the U.S. seafood market. *Environmental Health Perspectives*, **115**(2), pp.235–242.

SUNDERLAND, E.M. and MASON, R.P., 2007. Human impacts on open ocean mercury concentrations. *Global Biogeochemical Cycles*, **21**(4).

TRUDEL, M. and RASMUSSEN, J.B., 2006. Bioenergetics and mercury dynamics in fish: a modelling perspective. *Canadian Journal of Fisheries and Aquatic Sciences*, **63**(8), pp.1890–1902.

TSUCHIYA, A., HINNERS, T., KROGSTAD, F., WHITE, J., BURBACHER, T., FAUSTMAN, E. and MARIËN, K., 2009. Longitudinal mercury monitoring within the Japanese And Korean communities (U.S.); Implications for exposure determination and public health protection. *Environmental Health Perspectives*, **117**(11), 1760–1766.

TSUCHIYA, A., HINNERS, T.A., BURBACHER, T.M., FAUSTMAN, E.M. and MARIËN, K., 2008a. Mercury exposure from fish consumption within the Japanese and Korean communities. *Journal of Toxicology and Environmental Health, Part A*, **71**(15), pp.1019–1031.

TSUCHIYA, A., BURBACHER, T.M., FAUSTMAN, E.M. and MARIËN, K., 2008b. Fish intake guidelines: incorporating n-3 fatty acid intake and contaminant exposure in the Korean and Japanese communities. *American Journal of Clinical Nutrition*, **87**(6), pp.1867–1875.

ULRICH, R., RASZYK, J. and NAPRAVNIK, A., 2001. Variations in contamination by mercury, cadmium and lead on swine farms in the district of Hodonín in 1994 to 1999. *Veterinarni Medicina*, **46**(5), pp.132–139.

U.S. FDA (U.S. Food and Drug Administration), 2005. Total Diet Study. Silver Spring, MD. Available:http://www.fda.gov/Food/FoodSafety/FoodContaminants Adulteration/TotalDietStudy/default.htm

VIRTANEN, J.K., VOUTILAINEN, S., RISSANEN, T.H., MURSU, J., TUOMAINEN, T-P., KORHONEN, M.J., VALKONEN, V-P., SEPPANEN, K., LAUKKANEN, J.A. and SALONEN, J.T., 2005. Mercury, fish oils, and risk of acute coronary events and cardiovascular disease, coronary heart disease, and all-cause mortality in men in eastern Finland. *Arteriosclerosis, thrombosis, and vascular biology*, **25**(1), pp.228–233.

VLIEG, P., MURRAY, T. and BODY, D.R., 1993. Nutritional data on six oceanic pelagic fish species from New Zealand waters. *Journal of Food Composition and Analysis*, **6**, pp.45–54.

WAEGENEERS, N., HOENIG, M., GOEYENS, L. and DE TEMMERMAN, L., 2009. Trace elements in home-produced eggs in Belgium: Levels and spatiotemporal distribution. *Science of the Total Environment*, **407**(15), pp.4397–4402.

WANG, X., LI, Y-F., LI, B., DONG, Z., QU, L., GAO, Y., CHAI, Z. and CHEN, C., 2011. Multielemental contents of foodstuffs from the Wanshan (China) mercury mining area and the potential health risks. *Applied Geochemistry*, **26**(2), pp.182–187.

WIENER, J.G., KRABBENHOFT, D.P., HEINZ, G.H. and SHEUHAMMER, A.M., 2003. Ecotoxicology of mercury. In HOFFMAN, D.J., RATTNER, B.A., BURTON, B.A. JR., CAIRNS, J. JR., (Eds.), *Handbook of Ecotoxicology*. BOCA RATON, Florida: CRC Press, pp. 409–463.

YORIFUJI, T., TSUDA, T., TAKAO, S., SUZUKI, E. and HARADA, M., 2009. Total mercury content in hair and neurologic signs. *Epidemiology*, **20**(2), pp.188–193.

YOSHIZAWA, K., RIMM, E.B., MORRIS, J.S., SPATE, V.L., HSIEH, C-C., SPIEGLEMAN, D., STAMPFER, M.J. and WILLETT, W.C., 2002. Mercury and the risk of coronary heart disease in men. *The New England Journal of Medicine*, **347**(22), pp.1755–1760.

YSART, G., MILLER, P., CROASDALE, M., CREWS, H., ROBB, P., BAXTER, M. and DE L'ARGY, C., 2000. 1997 UK Total Diet Study – dietary exposures to aluminium, arsenic, cadmium, chromium, copper, lead, mercury, nickel, selenium, tin and zinc. *Food Additives and Contaminants*, **17**(9), pp.1–12.

17

Arsenic in foods: current issues related to analysis, toxicity and metabolism

K. A. Francesconi and G. Raber, University of Graz, Austria

DOI: 10.1533/9780857098917.2.414

Abstract: We present a brief overview of the major arsenic species found in the environment and in our food, before describing some common methods for determining total arsenic content and arsenic species in food. Particular attention is placed on the measurement of inorganic arsenic since future food legislation for arsenic is likely to be concerned with this known toxic species. We point out the current weakness in methods based on high performance liquid chromatography (HPLC) in terms of the extraction efficiency, and note the recent harsher conditions employed to simultaneously extract inorganic arsenic and convert it to arsenate before the measurement. Finally, we discuss those aspects of arsenic toxicity relevant to health authorities and legislators of food safety.

Keywords: arsenic, arsenic species, toxicity, food regulations, speciation analysis, high performance liquid chromatography (HPLC), inductively coupled plasma mass spectrometry (ICPMS).

17.1 Introduction

Ask people to name a toxic element, and the answer would very probably be arsenic. This toxic metalloid, in the form of arsenic trioxide or 'white arsenic', was in previous times a favoured homicidal agent. Not only was white arsenic highly toxic but it was also odourless, tasteless and colourless, so it could be easily slipped into food meant for an unsuspecting victim. The cause of death was difficult to establish because analytical methods for measuring arsenic were unreliable. This all changed in 1835, when Dr Marsh, a forensic chemist, developed a sensitive method for determining arsenic based on the conversion of inorganic arsenic to gaseous arsine, which could be easily separated from the sample matrix and detected (in Dr Marsh's case, by thermal decomposition to elemental arsenic, which formed a highly visible 'mirror' on the inside of a heated glass tube). The chemistry of Marsh is still used today in many determinations of the arsenic content of food, albeit we detect arsenic in a different way and at much lower concentrations.

Our knowledge of arsenic in food has advanced considerably since the time of Dr Marsh. In particular, we now know that the assessment of arsenic in foods is often based on data from a two-step analytical process. A first analysis of the sample for its total arsenic content is essential, and when the arsenic content is sufficiently high an analytical method for determining the various species of arsenic is needed. These aspects will be addressed in this chapter, which deals with arsenic in foods (including water and other liquids) from an analytical perspective – what we need to measure, why and how – within the framework of the human health issues related to metabolism and toxicity of arsenic species. The arsenic species will be considered in terms of inorganic and organic species; we will use the abbreviation iAs to refer to inorganic arsenic reported as the sum of arsenite and arsenate.

17.2 Sources and occurrence in foods

Although there have been isolated incidents in the past of acute arsenic poisoning from arsenic-contaminated foods, for example the famous arsenic-beer episode in Manchester in 1900 (Reynolds 1901), such events are now rare and will not be discussed here. Rather, the focus will be on the main natural sources of exposure to arsenic for general populations, taking into account environmental factors and diet.

17.2.1 Origin and distribution of inorganic arsenic and organic arsenic in the environment

Arsenic occurs in the earth's crust primarily as oxides or sulfides at concentrations typically within the range 1–20 mg/kg (Mandal and Suzuki, 2002). This arsenic is essentially iAs, although simple methylated forms, such as dimethylarsinate (DMA), may be present, often as a consequence of its use in agriculture (Huang *et al.*, 2011). Although seawater arsenic content is remarkably constant globally (about 1–2 µg/L), arsenic levels in fresh water can vary widely as a consequence of localised geological conditions. Concentrations of arsenic in groundwater are commonly less than 10 µg/L although values up to 5000 µg/L have been reported (Ryker 2001; Smedley and Kinniburgh, 2002). Surface waters generally contain lower arsenic concentrations than groundwaters. In terms of drinking water supplies, high arsenic levels can have catastrophic consequences for human health, as seen in Bangladesh and various other regions of the world (Rahman *et al.*, 2009; Vahter *et al.*, 2010; Bhattacharya *et al.* 2011). Arsenic in drinking water is present essentially totally as iAs. This iAs exists mainly as arsenate in oxygenated waters, such as surface waters, whereas under reducing conditions, which can exist in some groundwaters, arsenite can be the dominant species (Postma et al., 2007). Organisms from the terrestrial environment generally have low levels of total arsenic, and a major proportion of this is iAs.

In contrast to the situation with terrestrial systems, organisms from marine ecosystems have a particular facility to convert iAs into organic arsenic compounds, and to accumulate arsenic in these organic forms. More than 70 organic arsenic compounds have been identified in the marine environment (Fig. 17.1). Arsenobetaine is by far the predominant arsenical (arsenic-containing compound) in marine animals, whereas arsenosugars dominate the arsenic content of marine algae. The reasons why arsenic occurs at such high levels in marine organisms is still unknown, but it is very likely a result of the similar chemistry arsenic shares with phosphorus and nitrogen (all group 15 elements). Marine algae possibly take up arsenate from seawater using transport processes designed for the essential phosphate ion, and the potentially toxic arsenate is then detoxified by transformation into organic arsenic compounds, primarily arsenosugars. The biosynthesis of arsenobetaine is more uncertain; its accumulation by marine animals, however, appears to be related to its structural similarity to glycine betaine, a nitrogen betaine which is an important osmolyte for aquatic organisms. The level of glycine betaine in an organism increases with salinity; were arsenobetaine to serve as an adventitiously acquired osmolyte it would also show this behaviour, and an explanation for why arsenobetaine levels are so very much higher in marine animals than in freshwater animals, would be forthcoming.

17.2.2 Major arsenic species in food

The following summarises information contained in a report by the European Food Safety Authority published in 2009 (EFSA Panel on Contaminants in the Food Chain (CONTAM), 2009). This report contains total arsenic data for 77 000 food samples. Most food categories contained low levels of arsenic (< 0.02 mg/kg): some mean values include water (0.003 mg/kg), rice (0.14 mg/kg) and seafood/shellfish (5 mg/kg). The highest value reported was for a species of algae (30.9 mg/kg). Unfortunately, there were very few speciation data (ca 1% of total arsenic data set). Nevertheless, the speciation data reported were consistent with iAs being the major form of arsenic in most terrestrial foods, and with organic arsenic predominating in seafoods.

It is well established that arsenobetaine is the dominant arsenic species in marine fish and most other seafoods, usually constituting more than 70% of the total arsenic content (typical levels are 2–50 mg arsenic/kg dry mass). Although arsenobetaine occurs in some terrestrial foods, for example in some types of mushrooms, it is generally a minor species

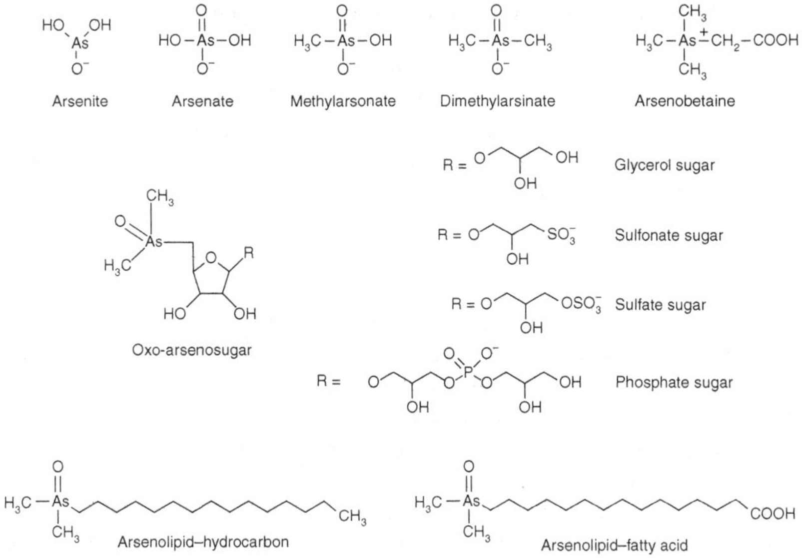

Fig. 17.1 Some arsenic species in the marine environment.

(Francesconi and Kuehnelt, 2002). There have also been several reports of arsenobetaine in freshwater organisms (Slejkovec *et al.*, 2004; Schaeffer *et al.*, 2006), although the levels (< 0.1 mg arsenic/kg dry mass) are generally much lower than those found in marine samples. Farmed freshwater fish, however, might contain higher arsenobetaine concentrations because they are usually provided with feed containing marine ingredients (Soeroes *et al.*, 2005). The differences in arsenobetaine content between marine and freshwater fish appear to be related to salinity (Larsen and Francesconi, 2003).

Arsenosugars are the major arsenic species in marine algae (typically 2–50 mg arsenic/kg dry mass). The edible alga hijiki, however, is a notable exception because it contains mainly iAs (up to 60 mg/kg) (Rose *et al.*, 2007). Arsenosugars are also found at significant levels in mussels and oysters (typically 0.2–5 mg/kg dry mass), and in many other marine organisms as well, albeit at lower concentrations. In terrestrial foods, arsenosugars occur generally at trace levels only.

Lipid-soluble (fat-soluble) organic arsenic species have also been found in fish oils (Rumpler *et al.*, 2008; Taleshi *et al.*, 2008; Arroyo-Abad *et al.*, 2010) and in sashimi tuna fish (Taleshi *et al.*, 2010). Quantitative data are not yet available; the estimated content of arsenolipids in sashimi tuna, a fatty fish, was ca 2.4 mg arsenic/kg, and hence the values for arsenolipids in fish are generally less than 1 mg arsenic/kg dry mass. The origin of these compounds in fish is probably algae (Amayo *et al.*, 2011).

17.3 Methods for determining arsenic in foods

Analytical aspects relevant to the analysis of arsenic in food are covered in three parts. First, the methods commonly used for determining the total arsenic content of foods are discussed, followed by an overview of arsenic speciation methods. Finally, we present information on appropriate reference materials certified for total arsenic and arsenic species.

17.3.1 Determining total arsenic content of foods

Although solid sample introduction methods can be used for arsenic measurements, the vast majority of the analyses of food for arsenic are performed on samples that have undergone an acid mineralisation process. The usual procedure is to subject the sample to a microwave-assisted acid digestion (mineralisation) step, thereby producing a sample solution that can be analysed by an instrumental method.

The two most commonly used instrumental methods for determining total arsenic content in food samples are atomic absorption spectrometry (AAS) and inductively coupled plasma mass spectrometry (ICPMS) (EFSA Panel on Contaminants in the Food Chain (CONTAM), 2009). Although AAS analyses can be performed directly on sample digests by using electrothermal AAS, application of that method is restricted because of severe sample matrix effects. The AAS method more commonly used for arsenic measurements is vapour generation AAS (also called hydride generation AAS). Following the classic sample preparation technique of Marsh, the vapour generation step converts inorganic arsenic to arsine (AsH_3, boiling point $-55°C$), which is efficiently transported to the AAS resulting in greatly improved sensitivity. The sample mineralisation procedure is absolutely critical for total arsenic measurement with vapour generation AAS, because arsenic species give different responses by this method and all arsenic must be completely decomposed to arsenate (and then pre-reduced with potassium iodide/ascorbic acid to arsenite) before the arsine formation step. Some arsenic species, for example arsenobetaine, are particularly resistant to acid mineralisation and require very forcing conditions to be converted to arsenate (Goessler and Pavkov, 2003). Vapour generation AAS is a simple inexpensive method that can deliver reliable quantitative data for arsenic in food down to levels of about 0.05 mg/kg (dry mass).

Inductively coupled plasma mass spectrometry (ICPMS) is the other main instrumental technique for determining arsenic in food. Although ICPMS is a robust technique it does not handle sample solutions with high total dissolved solids ($> 0.1\%$ mass/volume) very well. Interferences are generally not a problem; samples high in chloride can cause spectral interference ($ArCl^+$ formed in the argon plasma has the same nominal mass as As^+), although this is readily overcome with collision/reaction cell technology incorporated into all modern ICPMS instruments or by using high-resolution ICPMS. ICPMS is a very sensitive instrumental technique, easily providing quantitative data for arsenic in foods at levels down to 1 µg/kg (dry mass). Hydride generation may also be used in combination with ICPMS, resulting in improved instrumental sensitivity and, depending on reagent blanks, lower limits of quantification. However, the improvement is not great and the application of vapour generation ICPMS, rather than using conventional ICPMS, is generally not warranted.

17.3.2 Determining arsenic species in foods

Several methods are available for the quantitative measurement of arsenic species. The first commonly used arsenic speciation method was based on vapour generation followed by cold-trapping or gas chromatography and arsenic-selective detection. Although the method gained wide acceptance in environmental (water) and clinical analysis, it was generally not appropriate for food analysis (Francesconi and Kuehnelt, 2004). The most widely used method for arsenic species in food is high performance liquid chromatography/inductively coupled plasma mass spectrometry (HPLC/ICPMS); it has the ability to separate the many arsenicals often present in food samples and to quantify them at low levels. Before discussing the recent advances in arsenic speciation by HPLC/ICPMS, however, it is worth considering a much simpler method that uses solvent-partitioning to separate the iAs component of food samples.

Solvent-partitioning method
The so-called solvent-partitioning method is based on the conversion of iAs to arsenic trichloride, by treating the sample with strong HCl, and selective extraction of this analyte into chloroform. The method was first developed in the 1970s, and is relevant today because of the current focus on methods for determining iAs in foodstuffs. A positive aspect is that essentially all the sample is brought into solution, so that the arsenic extraction efficiency is high. The early results, however, seemed to overestimate the iAs content, which suggests a lack of selectivity (Edmonds and Francesconi, 1993). Nevertheless, there are still applications of this solvent-partitioning method (Munoz *et al.*, 2000; Almela *et al.*, 2002), although its selectivity for all types of samples still remains to be proven.

HPLC/ICPMS
The many advantages of using HPLC/ICPMS for arsenic speciation analysis have been discussed in an earlier review published in 2004 (Francesconi and Kuehnelt, 2004) and substantiated by the work over the time since then. A critical step, and perhaps the Achilles' heel in the method, is the extraction; we discuss those difficulties, before briefly discussing the issues related to chromatography and detection.

The extraction conditions should be sufficiently forcing to solubilise most of the arsenic in the sample, but they should not be so strong that the native arsenic species are degraded. Furthermore, because food samples can contain many arsenic species with a range of physical properties, a single extraction method for all arsenic species in all foodstuffs is not feasible (Francesconi, 2003). For food samples high in organic arsenic species, methanol/water mixtures with gentle mixing or sonication are often used with typical extraction efficiencies of > 80 %. The non-extractable arsenic is often referred to as lipid-soluble or protein-bound arsenic, although few data have been reported to support this assignment.

The recent focus on iAs content of foods has encouraged investigation of more forcing extraction methods employing acids and bases. For example, a mixture of sodium hydroxide and ethanol with microwave-assisted heating was reported by Larsen *et al.* (2005), and subsequently successfully applied to provide data on the iAs content in a range of marine samples (Sloth *et al.*, 2005). This mixture solubilised arsenite and arsenate from the sample, while simultaneously converting arsenite to arsenate which was then measured by anion-exchange HPLC/ICPMS (see below). Difficulties were encountered, however, when carbohydrate-rich samples were tested, because an intractable jelly was formed during the extraction procedure.

Trifluoroacetic acid has also been investigated as an extraction medium to extract inorganic arsenic species from rice (Heitkemper *et al.*, 2001; Williams *et al.*, 2005; Williams *et al.*, 2006). Partial reduction of arsenate to arsenite can occur under these conditions, so those studies reported a total inorganic arsenic value. More recently, solutions of trifluoroacetic acid/H_2O_2 have been used to extract rice, wheat and fish samples (Raber *et al.*, 2012). Here, the use of an oxidising extraction mixture ensured that all iAs was present as arsenate, which simplified the chromatography in the following step (see below).

Narukawa *et al.* (2008) showed that good extraction efficiency could be achieved even with water alone. They investigated several mixtures of solvents and found that water at 80ºC gave good extraction efficiencies for iAs (arsenite and arsenate were determined separately) and DMA. The other solvent mixtures investigated in this study were aqueous methanol solutions from 25% to 100% methanol. It was shown that the efficiency mainly depends on the temperature during the extraction process. At room temperature the extraction yields were low with all solvents investigated. At elevated temperatures, generally higher amounts of arsenic could be extracted without conversion of the species.

Because most of the arsenic species in food are water-soluble and charged, ion-pairing or ion-exchange HPLC are the most commonly used separation methods for arsenic speciation analysis of foods (Francesconi and Kuehnelt, 2004). However, some arsenic species, for example arsenolipids and some of the sulfur-containing arsenicals, require reversed-phase HPLC (Raml *et al.*, 2006; Taleshi *et al.*, 2010). The recovery of the various arsenic species from the HPLC columns is usually good (> 80%) but mass balance calculations should always be performed to identify those samples where recovery is unacceptably low, and hence the speciation analysis no longer gives a true indication of the main arsenic species present. A recently reported separation has been optimised for the separation of MA, DMA and arsenate, which served as the analyte for iAs (Fig. 17.2) (Raber et al., 2012).

ICPMS is a very sensitive and selective arsenic detector with relatively few interferences. The ICPMS response depends only on the amount of arsenic, and is independent of the way arsenic is bound in the species. This enables unknown arsenic species to be quantified against known arsenic standards. Provided extraction efficiencies and column recoveries are high; HPLC/ICPMS can determine arsenic species in food at levels down to 1 μg As/kg (dry mass).

17.3.3 Certified reference materials relevant for measuring total arsenic and arsenic species in food

There are several reference materials based on foodstuffs that have certified arsenic concentrations and hence can be used to validate analytical methods for determining arsenic content of foods (Table 17.1). Unfortunately, most of these reference materials are of marine origin and hence their arsenic content is high. There is a need for reference materials with low arsenic concentrations.

Although there are some food-relevant reference materials certified for arsenic species, the dominant certified species is arsenobetaine, a known non-toxic arsenical (Table 17.2). There is an urgent need for certification of iAs content in a range of foodstuffs.

17.4 Toxicity of arsenic

Arsenic and food are discussed in terms of the relative toxicity of the various arsenic species, their mode of toxic action and the consequences for identifying and assessing the major toxic concerns.

17.4.1 Dependence of toxicity on species

Most of the data relating to arsenic toxicity relates to inorganic arsenic, almost always as arsenite or arsenate. There are good practical experimental reasons for this in terms of solubility and concomitant ease of administration and controlling exposure. It is also most relevant in terms of bioavailable fractions. Studies based on (initially) insoluble arsenic compounds are more difficult to perform but can provide important information (Cui et al., 2011). Toxicity testing of organoarsenicals generally shows no observable toxicity or very low toxicity. Exceptions are the trivalent methyl arsenic species, which show high toxicity in various cytotoxicity tests (Sakurai et al., 2006; Wang et al., 2007). These species are not usually found in food, although methylarsonite

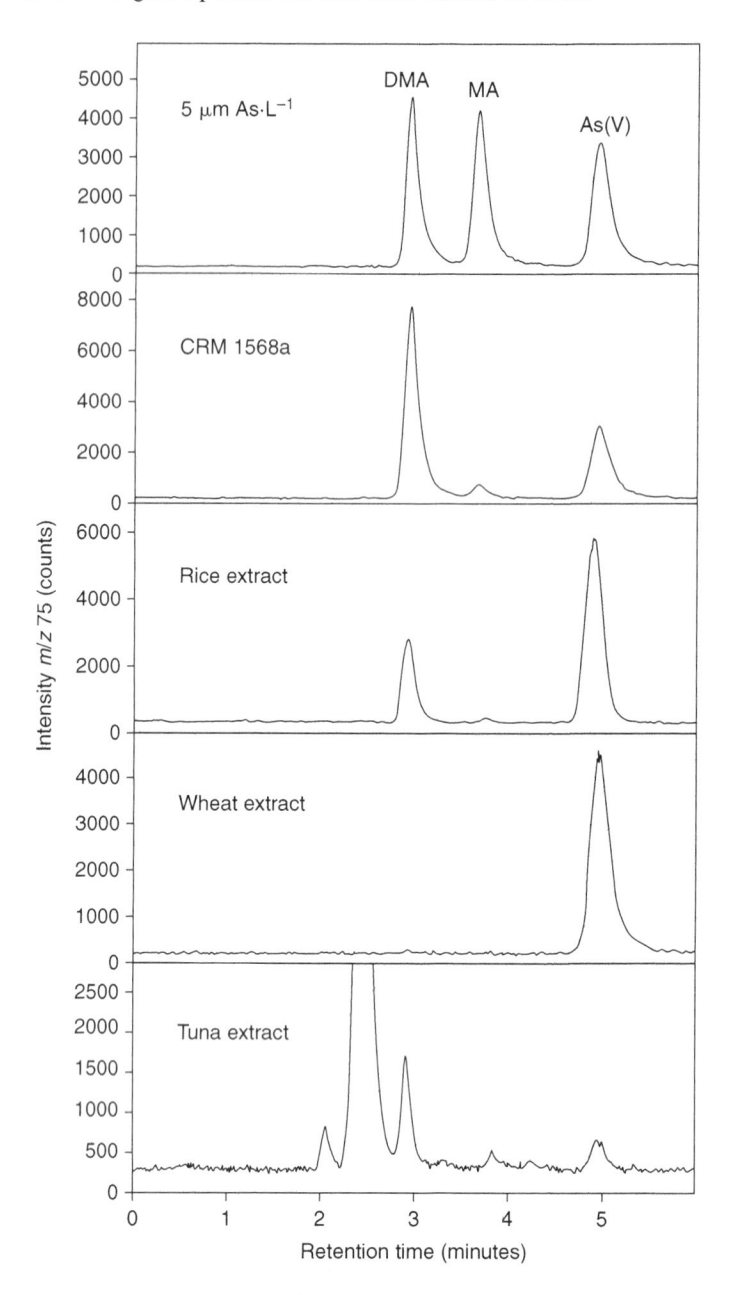

Fig. 17.2 Anion-exchange HPLC separation of arsenic species in extracts of rice, wheat and tuna fish using malonic acid buffer. Conditions: Hamilton PRP-X100 (4.1 × 250 mm, 5 μm particle size) at 40°C and a mobile phase of 10 mM malonic acid buffer pH 5.6 (adjusted with aqueous NH_3) at 1.2 mL·min^{-1}; 20 μL injection volume except for tuna extract (10 μL). The large peak for tuna at retention time of 2.5 minutes is arsenobetaine. m/z means mass to charge ratio.

Table 17.1 Reference materials relevant to food analysis certified for total arsenic content

Food type	Description[a]	Certified total arsenic content (mg arsenic/kg dry mass)
Tomato leaves	SRM 1573 (NIST)	0.27 ± 0.05
Rice flour	SRM 1568a (NIST)	0.290 ± 0.030[b]
Tuna fish	CRM BCR-627 (IRMM)	4.8 ± 0.3
Mussel tissue	ERM-CE278 (IRMM)	6.07 ± 0.13
Dogfish muscle	CRM Dorm-3 (NRCC)	6.88 ± 0.30
Oyster tissue	SRM 1566b (NIST)	7.65 ± 0.65
Dogfish liver	CRM Dolt-4 (NRCC)	9.66 ± 0.62
Cod muscle	CRM BCR-422 (IRMM)	21.1 ± 0.5
Lobster hepatopancreas	CRM TORT-2 (NRCC)	21.6 ± 1.8

Adapted from EFSA (EFSA Panel on Contaminants in the Food Chain (CONTAM) 2009).
SRM: standard reference material; CRM: certified reference material; ERM: European reference material.
[a] NIST: National Institute of Standards and Technology (USA); NRCC: National Research Council of Canada (Canada); IRMM: Institute for Reference Materials and Measurements (Belgium)
[b] The uncertainty usually given as 95 % confidence interval.

Table 17.2 Reference materials relevant to food analysis certified for arsenic species content

Food type	Description[a]	Certified arsenic species content (mg arsenic/kg dry mass)
Dogfish muscle	DORM-2 (NRCC)	Arsenobetaine (16.4 ± 1.1)[b]
		Tetramethylarsonium ion (0.248 ± 0.054)
Tuna fish	CRM BCR-627 (IRMM)	Arsenobetaine (3.9 ± 0.2)
		Dimethylarsinate (0.15 ± 0.01)

Adapted from EFSA (EFSA Panel on Contaminants in the Food Chain (CONTAM) 2009).
CRM: certified reference material;
[a] NRCC: National Research Council of Canada (Canada); IRMM: Institute for Reference Materials and Measurements (Belgium).
[b] The uncertainty usually given as 95 % confidence interval.

was reported in some old samples of carrots that had been stored for 25 years (Yathavakilla *et al.*, 2008). These reduced methylated species, however, are highly relevant as possible intermediates in pathways of arsenic biotransformation in humans.

It is worth mentioning the toxicity data available on the most common arsenic species in foods, i.e. arsenobetaine and arsenosugars. In acute toxicity experiments with rats fed up to 10 g of compound/kg no deaths were recorded (Kaise *et al.*, 1985). This is a remarkable tolerance to arsenobetaine,

424   Persistent organic pollutants and toxic metals in foods

at least by the rat, equivalent to the ingestion of about 1.5 kg of compound by an average-sized man. Cytotoxicity tests with mammalian cells exposed to arsenobetaine also produced no effects, indicating that arsenobetaine was at least 14 000-fold less toxic than arsenite (Kaise *et al.*, 1998). Cytotoxicity tests with arsenosugars have shown some small effects indicating low toxicity (at least 1000-fold less toxic than arsenite) (Sakurai *et al.*, 1997). DMA has been associated with various toxic effects but only at high exposures (Kenyon and Hughes, 2001; Arnold *et al.*, 2006; Irvine *et al.*, 2006).

17.4.2 Mode of toxic action for the major arsenic species and major toxicity concerns

The human biotransformation of iAs is a well-studied field, and for very good reasons. In many parts of the world, human populations are exposed to iAs in their drinking water, often at levels > 100 µg/L. This is not a result of man-induced pollution, but rather reflects naturally high arsenic environments. Many studies have demonstrated that long-term exposure from iAs at these levels results in increased incidence of various human maladies including cancers and cardiovascular disease (e.g. Chen *et al.*, 1992). More recent studies, however, have indicated that detrimental health effects, including cardiovascular disease (Navas-Acien *et al.*, 2005) and diabetes (Navas-Acien *et al.*, 2008), can also occur in populations exposed to iAs in drinking water at levels < 100 µg As/L.

Arsenic biotransformation primarily involves methylation, to produce first MA and then DMA as the major product. These two methylated metabolites are then excreted in the urine together with 'unchanged' iAs. The mechanism of methylation was long thought to follow the Challenger pathway, first proposed to explain the formation of methylated arsenic products from iAs by fungi (Challenger, 1945). An alternative pathway, however, has more recently been proposed (Hayakawa *et al.*, 2005). Key intermediates in both pathways are the reduced methylated species MA(III) and dimethylarsinite. However, the order in which they are formed relative to MA and DMA differs; this has consequences for determining the fate and effects of other organoarsenic compounds following their ingestion and metabolism in the body.

Arsenobetaine appears to be metabolically unavailable to man. It is rapidly excreted in the urine with no significant metabolites. There is evidence, however, of a biphasic elimination, which has consequences for studies estimating arsenic exposure (Newcombe *et al.*, 2010). Arsenosugars, on the other hand, are bioavailable and are biotransformed by humans primarily to DMA, although as many as ten significant metabolites have been identified (Francesconi *et al.*, 2002; Raml *et al.*, 2005). Preliminary data on the human metabolism of arsenolipids indicates that they also are bioavailable and are biotransformed primarily to DMA (Schmeisser *et al.*, 2005). Thus, ingestion by humans of both arsenosugars and arsenolipids results in the formation of DMA, the same major metabolite produced from ingested iAs. If DMA(III) is critical to the mode of toxic action of arsenic, it is then essential that we understand the mechanism for its formation. DMA(III) formation by the Challenger pathway (i.e. before the production of DMA) would suggest that iAs and arsenosugars/arsenolipids can be treated separately in terms of toxic effects. Should the Hayakawa pathway be correct, however, the toxic DMA(III) would be formed after DMA, and hence the contribution of DMA from arsenosugars/arsenolipids would need to be given equal consideration to that from iAs.

17.5 Implications for the food industry and policy makers

The approach taken in forming food regulations for organic contaminants is quite different from that taken for metals. Organic contaminants are, in the main, artificial, and their presence in food is a direct result of human activity. Metals and metalloids, on the other hand, are natural environmental constituents, and organisms have evolved to either use them (e.g. Cu) or, in the case of toxic elements, to handle them in a way that reduces their toxicity (e.g. Cd). Thus, in many pristine environments around the world, far from human contamination, organisms can naturally contain high levels of certain metals.

The task for food regulators then becomes: how to impose maximum permissible levels for metals in food products that protect both the consumer and the food industry? So far a pragmatic approach has been taken, with the assumption that naturally acquired metal is not (or at least is less) harmful compared with the same metal accumulated from a pollution source (Francesconi, 2007). This has resulted in maximum permissible levels being set that partly takes into account the natural levels. For example, allowable Cd levels are higher for molluscs than for fish, because molluscs (filter-feeders) are known to naturally accumulate Cd. Food regulators will no doubt point out that the levels set also reflect the contributions of particular food items to the average diet, but this seems to be outweighed by natural occurrence factors.

Consider then how much more difficult the situation is to establish food regulations for arsenic. Arsenic is a natural 'contaminant' present in our food as iAs and a whole range of organoarsenic forms. The levels of arsenic found in uncontaminated food can vary by a factor of 1000 depending on the food type. The toxicities of the arsenic species vary from innocuous (arsenobetaine) to highly toxic (iAs), with a host of other organoarsenicals for which toxicity information is limited. Therefore, information on the type of arsenic species is also necessary to perform a risk assessment of arsenic in food.

Against this background, it is a daunting task to set maximum permissible concentrations for arsenic. Nevertheless, a start needs to be made, and a logical first step is to focus on iAs, the most toxic form of arsenic commonly found in foods. Early attempts were made to estimate iAs exposure to food from the total arsenic data by estimating the average proportion of total arsenic present as iAs for a particular food (EFSA Panel on Contaminants in the Food Chain (CONTAM), 2009). This approach seems reasonable for terrestrial foods, where iAs constitutes the major part, but falls down when seafoods are considered. This is particularly so for fish, where the relationship between iAs and total arsenic (essentially arsenobetaine) is poor. A pragmatic economic-driven response may be to measure total As and then to measure iAs in samples where the total As exceeds a certain pre-determined threshold.

Although the methods based on bulk separation of the iAs and then analysis of total arsenic has advantages, it is unlikely to be sufficiently robust to handle the various types of food samples. HPLC/ICPMS methods can suffer from varying food-dependent extraction efficiencies. Nevertheless, HPLC/ICPMS methods are robust and sensitive, and they have the potential to provide reliable quantitative data for a range of food products. The methods, however, are far from being routine in food laboratories. Steps should be taken now to identify those laboratories capable of doing these analyses, and to co-ordinate their work, with the aim of providing standardised methods for iAs determinations in food.

17.6 References

ALMELA, C., ALGORA, S., BENITO, V., CLEMENTE, M.J., DEVESA, V., SUNER, M.A., VELEZ, D. and MONTORO, R. 2002, "Heavy metal, total arsenic, and inorganic arsenic contents of algae food products", *Journal of Agricultural and Food Chemistry*, vol. **50**, no. 4, pp. 918–923.

AMAYO, K.O., PETURSDOTTIR, A., NEWCOMBE, C., GUNNLAUGSDOTTIR, H., RAAB, A., KRUPP, E.M. and FELDMANN, J. 2011, "Identification and Quantification of Arsenolipids Using Reversed-Phase HPLC Coupled Simultaneously to High-Resolution ICPMS and High-Resolution Electrospray MS without Species-Specific Standards", *Analytical Chemistry*, vol. **83**, no. 9, pp. 3589–3595.

ARNOLD, L.L., ELDAN, M., NYSKA, A., VAN GEMERT, M. and COHEN, S.M. 2006, "Dimethylarsinic acid: Results of chronic toxicity/oncogenicity studies in F344 rats and in B6C3F1 mice", *Toxicology*, vol. **223**, no. 1–2, pp. 82–100.

ARROYO-ABAD, U., MATTUSCH, J., MOTHES, S., MOEDER, M., WENNRICH, R., ELIZALDE-GONZALEZ, M.P. and MATYSIK, F. 2010, "Detection of arsenic-containing hydrocarbons in canned cod liver tissue", *Talanta*, vol. **82**, no. 1, pp. 38–43.

BHATTACHARYA, P., HOSSAIN, M., RAHMAN, S.N., ROBINSON, C., NATH, B., RAHMAN, M., ISLAM, M.M., VON BROMSSEN, M., AHMED, K.M., JACKS, G., CHOWDHURY, D., RAHMAN, M., JAKARIYA, M., PERSSON, L.A. and VAHTER, M. 2011, "Temporal and seasonal variability of arsenic in drinking water wells in Matlab, southeastern Bangladesh: a preliminary evaluation on the basis of a 4 year study.", *Journal of environmental science and health. Part A, Toxic/hazardous substances and environmental engineering*, vol. **46**, no. 11, pp. 1177–1184.

CHALLENGER, F. 1945, "Biological Methylation", *Chemical reviews*, vol. **36**, no. 3, pp. 315–361.

CHEN, C.J., CHEN, C.W., WU, M.M. and KUO, T.L. 1992, "Cancer Potential in Liver, Lung, Bladder and Kidney due to Ingested Inorganic Arsenic in Drinking-Water", *British journal of cancer*, vol. **66**, no. 5, pp. 888–892.

CUI, L., NEWCOMBE, C., URGAST, D.S., RAAB, A., KRUPP, E.M. and FELDMANN, J. 2011, "Assessing the toxicity of arsenic-bearing sulfide minerals with the bio-indicator Corophium volutator", *Environmental Chemistry*, vol. **8**, no. 1, pp. 52–61.

EDMONDS, J.S. and FRANCESCONI, K.A. 1993, "Arsenic in Seafoods – Human Health-Aspects and Regulations", *Marine pollution bulletin*, vol. **26**, no. 12, pp. 665–674.

EFSA Panel on Contaminants in the Food Chain (CONTAM) 2009, 27.09.2010-last update, *Scientific Opinion on Arsenic in Food*. Available: http://www.efsa.europa.eu/en/efsajournal/pub/1351.htm

FRANCESCONI, K.A. 2003, "Complete extraction of arsenic species: a worthwhile goal?", *Applied Organometallic Chemistry*, vol. **17**, no. 9, pp. 682–683.

FRANCESCONI, K.A. 2007, "Toxic metal species and food regulations making a healthy choice", *Analyst*, vol. **132**, no. 1, pp. 17–20.

FRANCESCONI, K.A. and KUEHNELT, D. 2004, "Determination of arsenic species: A critical review of methods and applications, 2000–2003", *Analyst*, vol. **129**, no. 5, pp. 373–395.

FRANCESCONI, K.A. and KUEHNELT, D. 2002, "Arsenic compounds in the environment", *Environmental Chemistry of Arsenic*, pp. 51–94.

FRANCESCONI, K.A., TANGGAARD, R., MCKENZIE, C.J. and GOESSLER, W. 2002, "Arsenic metabolites in human urine after ingestion of an arsenosugar", *Clinical chemistry*, vol. **48**, no. 1, pp. 92–101.

GOESSLER, W. and PAVKOV, M. 2003, "Accurate quantification and transformation of arsenic compounds during wet ashing with nitric acid and microwave assisted heating", *Analyst*, vol. **128**, no. 6, pp. 796–802.

HAYAKAWA, T., KOBAYASHI, Y., CUI, X. and HIRANO, S. 2005, "A new metabolic pathway of arsenite: arsenic-glutathione complexes are substrates for human arsenic methyltransferase Cyt19", *Archives of Toxicology*, vol. **79**, no. 4, pp. 183–191.

HEITKEMPER, D.T., VELA, N.P., STEWART, K.R. and WESTPHAL, C.S. 2001, "Determination of total and speciated arsenic in rice by ion chromatography and inductively coupled plasma mass spectrometry", *Journal of Analytical Atomic Spectrometry*, vol. **16**, no. 4, pp. 299–306.

HUANG, J., HU, K. and DECKER, B. 2011, "Organic Arsenic in the Soil Environment: Speciation, Occurrence, Transformation, and Adsorption Behavior", *Water Air and Soil Pollution*, vol. **219**, no. 1–4, pp. 401–415.

IRVINE, L., BOYER, I.J. and DESESSO, J.M. 2006, "Monomethylarsonic acid and dimethylarsinic acid: Developmental toxicity studies with risk assessment", *Birth Defects Research Part B-Developmental and Reproductive Toxicology*, vol. **77**, no. 1, pp. 53–68.

KAISE, T., OCHI, T., OYA-OHTA, Y., HANAOKA, K., SAKURAI, T., SAITOH, T. and MATSUBARA, C. 1998, "Cytotoxicological aspects of organic arsenic compounds contained in marine products using the mammalian cell culture technique", *Applied Organometallic Chemistry*, vol. **12**, no. 2, pp. 137–143.

KAISE, T., WATANABE, S. and ITOH, K. 1985, "The Acute Toxicity of Arsenobetaine", *Chemosphere*, vol. **14**, no. 9, pp. 1327–1332.

KENYON, E.M. and HUGHES, M.F. 2001, "A concise review of the toxicity and carcinogenicity of dimethylarsinic acid", *Toxicology*, vol. **160**, no. 1–3, pp. 227–236.

LARSEN, E.H., ENGMAN, J., SLOTH, J.J., HANSEN, M. and JORHEM, L. 2005, "Determination of inorganic arsenic in white fish using microwave-assisted alkaline alcoholic sample dissolution and HPLC-ICP-MS", *Analytical and Bioanalytical Chemistry*, vol. **381**, no. 2, pp. 339–346.

LARSEN, E.H. and FRANCESCONI, K.A. 2003, "Arsenic concentrations correlate with salinity for fish taken from the North Sea and Baltic waters", *Journal of the Marine Biological Association of the United Kingdom*, vol. **83**, no. 2, pp. 283–284.

MANDAL, B.K. and SUZUKI, K.T. 2002, "Arsenic round the world: a review", *Talanta*, vol. 58, no. 1, pp. 201–235.

MUNOZ, O., DEVESA, V., SUNER, M.A., VELEZ, D., MONTORO, R., URIETA, I., MACHO, M.L. and JALON, M. 2000, "Total and inorganic arsenic in fresh and processed fish products", *Journal of Agricultural and Food Chemistry*, vol. 48, no. 9, pp. 4369–4376.

NARUKAWA, T., INAGAKI, K., KUROIWA, T. and CHIBA, K. 2008, "The extraction and speciation of arsenic in rice flour by HPLC-ICP-MS", *Talanta*, vol. 77, no. 1, pp. 427–432.

NAVAS-ACIEN, A., SHARRETT, A.R., SILBERGELD, E.K., SCHWARTZ, B.S., NACHMAN, K.E., BURKE, T.A. and GUALLAR, E. 2005, "Arsenic exposure and cardiovascular disease: A systematic review of the epidemiologic evidence", *American Journal of Epidemiology*, vol. 162, no. 11, pp. 1037–1049.

NAVAS-ACIEN, A., SILBERGELD, E.K., PASTOR-BARRIUSO, R. and GUALLAR, E. 2008, "Arsenic exposure and prevalence of type 2 diabetes in US adults", *Jama-Journal of the American Medical Association*, vol. 300, no. 7, pp. 814–822.

NEWCOMBE, C., RAAB, A., WILLIAMS, P.N., DEACON, C., HARIS, P.I., MEHARG, A.A. and FELDMANN, J. 2010, "Accumulation or production of arsenobetaine in humans?", *Journal of Environmental Monitoring*, vol. 12, no. 4, pp. 832–837.

POSTMA, D., LARSEN, F., HUE, N.T.M., DUC, M.T., VIET, P.H., NHAN, P.Q. and JESSEN, S. 2007, "Arsenic in groundwater of the Red River floodplain, Vietnam: Controlling geochemical processes and reactive transport modeling", *Geochimica et Cosmochimica Acta*, vol. 71, no. 21, pp. 5054–5071.

RABER, G., STOCK, N., HANEL, P., MURKO, M., NAVRATILOVA, J. and FRANCESCONI, K.A. 2012, "An improved HPLC-ICPMS method for determining inorganic arsenic in food: Application to rice, wheat and tuna fish", *Food Chemistry*, vol. 134, no. 1, pp 524–532.

RAHMAN, A., VAHTER, M., SMITH, A.H., NERMELL, B., YUNUS, M., EL ARIFEEN, S., PERSSON, L. and EKSTROM, E. 2009, "Arsenic Exposure During Pregnancy and Size at Birth: A Prospective Cohort Study in Bangladesh", *American Journal of Epidemiology*, vol. 169, no. 3, pp. 304–312.

RAML, R., GOESSLER, W., TRAAR, P., OCHI, T. and FRANCESCONI, K.A. 2005, "Novel thioarsenic metabolites in human urine after ingestion of an arsenosugar, 2 ',3 '-dihydroxypropyl 5-deoxy-5-dimethylarsinoyl-beta-D-riboside", *Chemical research in toxicology*, vol. 18, no. 9, pp. 1444–1450.

RAML, R., GOESSLER, W. and FRANCESCONI, K.A. 2006, "Improved chromatographic separation of thio-arsenic compounds by reversed-phase high performance liquid chromatography-inductively coupled plasma mass spectrometry", *Journal of Chromatography a*, vol. 1128, no. 1–2, pp. 164–170.

REYNOLDS, E.S. 1901, "An account of the epidemic outbreak of arsenical poisoning occurring in beer-drinkers in the north of England and the midland counties-in 1900", *Lancet*, vol. 1, pp. 166–170.

ROSE, M., LEWIS, J., LANGFORD, N., BAXTER, M., ORIGGI, S., BARBER, M., MACBAIN, H., and THOMAS, K. 2007, "Arsenic in seaweed-Forms, concentration and dietary exposure", *Food and Chemical Toxicology*, vol 45, no 7, pp 1263–1267

RUMPLER, A., EDMONDS, J.S., KATSU, M., JENSEN, K.B., GOESSLER, W., RABER, G., GUNNLAUGSDOTTIR, H. and FRANCESCONI, K.A. 2008, "Arsenic-containing long-chain fatty acids in cod-liver oil: A result of biosynthetic infidelity?", *Angewandte Chemie-International Edition*, vol. 47, no. 14, pp. 2665–2667.

RYKER, S.J. 2001, "Mapping arsenic in groundwater: A real need, but a hard problem – Why was the map created?", *GeoTimes*, vol. 46, no. 11, pp. 34–36.

SAKURAI, T., KAISE, T., OCHI, T., SAITOH, T. and MATSUBARA, C. 1997, "Study of in vitro cytotoxicity of a water soluble organic arsenic compound, arsenosugar, in seaweed", *Toxicology*, vol. 122, no. 3, pp. 205–212.

SAKURAI, T., KOJIMA, C., KOBAYASHI, Y., HIRANO, S., SAKURAI, M.H., WAALKES, M.P. and HIMENO, S. 2006, "Toxicity of a trivalent organic arsenic compound, dimethylarsinous glutathione in a rat liver cell line (TRL 1215)", *British journal of pharmacology*, vol. **149**, no. 7, pp. 888–897.

SCHAEFFER, R., FRANCESCONI, K.A., KIENZL, N., SOEROES, C., FODOR, P., VARADI, L., RAML, R., GOESSLER, W. and KUEHNELT, D. 2006, "Arsenic speciation in freshwater organisms from the river Danube in Hungary", *Talanta*, vol. **69**, no. 4, pp. 856–865.

SCHMEISSER, E., RUMPLER, A., KOLLROSER, M., RECHBERGER, G., GOESSLER, W. and FRANCESCONI, K.A. 2005, "Arsenic fatty acids are human urinary metabolites of arsenolipids present in cod liver.", *Angewandte Chemie (International ed. in English)*, vol. **45**, no. 1, pp. 150–4.

SLEJKOVEC, Z., BAJC, Z. and DOGANOC, D.Z. 2004, "Arsenic speciation patterns in freshwater fish", *Talanta*, vol. **62**, no. 5, pp. 931–936.

SLOTH, J.J., LARSEN, E.H. and JULSHAMN, Y. 2005, "Survey of inorganic arsenic in marine animals and marine certified reference materials by anion exchange high-performance liquid chromatography-inductively coupled plasma mass spectrometry", *Journal of Agricultural and Food Chemistry*, vol. **53**, no. 15, pp. 6011–6018.

SMEDLEY, P.L. and KINNIBURGH, D.G. 2002, "A review of the source, behaviour and distribution of arsenic in natural waters", *Applied Geochemistry*, vol. **17**, no. 5, pp. 517–568.

SOEROES, C., GOESSLER, W., FRANCESCONI, K.A., KIENZL, N., SCHAEFFER, R., FODOR, P. and KUEHNELT, D. 2005, "Arsenic speciation in farmed Hungarian freshwater fish", *Journal of Agricultural and Food Chemistry*, vol. **53**, no. 23, pp. 9238–9243.

TALESHI, M.S., EDMONDS, J.S., GOESSLER, W., RUIZ-CHANCHO, M.J., RABER, G., JENSEN, K.B. and FRANCESCONI, K.A. 2010, "Arsenic-Containing lipids Are Natural Constituents of Sashimi Tuna", *Environmental science and technology*, vol. **44**, no. 4, pp. 1478–1483.

TALESHI, M.S., JENSEN, K.B., RABER, G., EDMONDS, J.S., GUNNLAUGSDOTTIR, H. and FRANCESCONI, K.A. 2008, "Arsenic-containing hydrocarbons: natural compounds in oil from the fish capelin, Mallotus villosus", *Chemical Communications*, no. **39**, pp. 4706–4707.

VAHTER, M., SOHEL, N., STREATFIELD, K. and PERSSON, L.A. 2010, "Arsenic exposure from drinking water and mortality in Bangladesh", *Lancet*, vol. **376**, no. 9753, pp. 1641.

WANG, T., JAN, K., WANG, A.S.S. and GURR, J. 2007, "Trivalent arsenicals induce lipid peroxidation, protein carbonylation, and oxidative DNA damage in human urothelial cells", *Mutation Research-Fundamental and Molecular Mechanisms of Mutagenesis*, vol. **615**, no. 1–2, pp. 75–86.

WILLIAMS, P.N., ISLAM, M.R., ADOMAKO, E.E., RAAB, A., HOSSAIN, S.A., ZHU, Y.G., FELDMANN, J. and MEHARG, A.A. 2006, "Increase in rice grain arsenic for regions of Bangladesh irrigating paddies with elevated arsenic in groundwaters", *Environmental science and technology*, vol. **40**, no. 16, pp. 4903–4908.

WILLIAMS, P.N., PRICE, A.H., RAAB, A., HOSSAIN, S.A., FELDMANN, J. and MEHARG, A.A. 2005, "Variation in arsenic speciation and concentration in paddy rice related to dietary exposure", *Environmental science and technology*, vol. **39**, no. 15, pp. 5531–5540.

YATHAVAKILLA, S.K.V., FRICKE, M., CREED, P.A., HEITKEMPER, D.T., SHOCKEY, N.V., SCHWEGEL, C., CARUSO, J.A. and CREED, J.T. 2008, "Arsenic speciation and identification of monomethylarsonous acid and monomethylthioarsonic acid in a complex matrix", *Analytical Chemistry*, vol. **80**, no. 3, pp. 775–782.

18

Organotin compounds in foods

E. Rosenberg, Vienna University of Technology, Austria

DOI: 10.1533/9780857098917.2.430

Abstract: This chapter discusses the sources and distribution of organotin compounds (OTCs) in the environment. Due to the intensive use of OTCs as stabilizers for polyvinyl chloride (PVC), as antifouling paint for ship hulls, as wood preservatives, and as fungicides in plant protection, OTCs have entered the hydrosphere. Being present in the water column at low and sub-ppb concentrations, they are strongly enriched in the trophic web and can reach significant ppb concentrations in fish and shellfish (up to 100 ng Sn/g dw in fish and up to 500–1000 ng Sn/g dw in shellfish). Depending on the actual consumption of fish and shellfish in different countries, the individual populations may be at lower or higher risk from the exposure to OTCs through food. While it can be concluded, based on the existing data, that the uptake of OTCs does not exceed the tolerable daily intake (TDI) level for OTCs for the average consumer, particular groups (e.g. children, high consumers of fish and fishery products) may still be at risk. The European Food Safety Authority has adopted a group TDI of 0.25 µg/kg body weight/day for the sum of tributyltin (TBT), dibutyltin (DBT), triphenyltin (TPhT) and dioctyltin (DOcT), based on their similar toxicity and mechanism of action. In the comparison to this, the daily intake of OTCs was estimated, on the basis of various national studies, to be 18 ng/kg bw/day if median concentration values were taken into consideration, or 83 ng/kg bw/day for the group of normal consumers. In the case of high consumers of fish and fishery products, these values approximately doubled to 37 ng/kg bw/day when using the median for the calculation, and 171 ng/kg bw/day when using the mean. In addition to a detailed discussion of the technological uses of OTCs, a short summary of the toxicology and biological effects of OTCs is also given.

Key words: butyltin compounds, phenyltin compounds, octyltin compounds, organotin compounds, PVC stabilizer, antifouling paint, wood preservative, tolerable daily intake (TDI), tolerable average residue level (TARL), food contact material (FCM).

18.1 Introduction

Organotin compounds (OTCs) are organic compounds containing the element tin with the general formula R_nSnX_{4-n} in which at least one direct tin–carbon bond is established, and where R is an alkyl or aryl moiety, and X is a halogen or an anionic moiety, such as chloride, acetate, hydroxide or other (Davies, 2004). The most relevant organotin compounds in this context belong to the classes of butyltin compounds (monobutyltin (MBT), dibutyltin (DBT) and tributyltin (TBT) and phenyltin compounds (monophenyltin (MPhT), diphenyltin (DPhT) and triphenyltin (TPhT)). Tetra-substituted butyltin (TeBT) and phenyltin (TePhT) compounds also exist, but do not have any practical relevance. Among the other organotin compounds, only octyltins such as monooctyltin (MOcT) and dioctyltin (DOcT) have technical relevance. With few exceptions, which will be discussed further below, the other organotin compounds do not have any technical or environmental relevance. Table 18.1 gives an overview of the major applications of organotin compounds and the quantities used in the European Union (RPA, 2007).

Table 18.1 Major applications of organotin compounds and quantities used in the EU in 2002 and in 2007. (RPA, 2007)

Applications	Quantity (tons/year) 2002[1]	Quantity (tons/year) 2007[2]
PVC stabilizers	15 000	>16 000
Catalysts	1300–1650	~2,000
• Plasticizers	150–350[3]	
• Silicones	50–100	
• Electrodeposition coatings	700–800	950
• Polyurethanes	400	750
Other uses:		
• Glass coating	760–800	Same
• Biocide in antifouling paints	1,250	Phased out globally
• Synthesis	< 150	~500
• Biocide (other)[4]	< 100	Reduced
• Pesticide	100	Unknown
• Intermediate in synthesis (tetra-substituted)[5]	N/A	Unknown
All uses	**Approx. 19 000**	**Approx. 21 000**

[1] RPA (2005).
[2] From consultation for the RPA study (2007).
[3] Derived by subtracting sub-totals for silicones, EDC and polyurethanes from total for 'catalysts'.
[4] Use of tributyltin compounds for these applications is now prohibited within the EU as they have not been notified under the Biocidal Products Directive.
[5] ETINSA (European Tin Stabilisers Manufacturers Association) has advised that the total quantity of tri-substituted tins for use as an intermediate in 2004 was substantially higher than the estimate for 2002. Although not clarified, this could perhaps be because the quantities present in the tetra-substituted tins had been excluded.

432 Persistent organic pollutants and toxic metals in foods

The first organotin compound was described more than one and a half centuries ago by Löwig (1852), while Sir Edward Frankland can be credited for the first systematic study on organotin compounds, in the frame of which he synthesized and characterized diethyltin diiodide (Frankland, 1854) and tetraethyltin in 1859. For about 80 years, however, organotin compounds had no particular application or any industrial use, until a patent was issued to Standard Oil Development Co. in 1932 for the use of organotin compounds as stabilizers in transformer oils. The credit for the original developments of organotin stabilizers for vinyl polymers in the early 1930s goes to Yngve and coworkers, who at that time worked at Union Carbide Co. (Champ and Seligman, 1996). Yngve filed a patent application in 1936 in which he proposed that tetraphenyltin and diphenyldipropyltin functioned as effective heat stabilizers in the fabrication of PVC plastic materials (Yngve, 1940). He and others extended this patent further over the next few years, to show that many other organotin compounds, such as DBT oxide, and various DBT salts of short chain carboxylic acids also provide excellent stabilization of PVC plastics against heat and weathering. It was shown that, of the many types of organotin compounds, the best stabilizers resulted from the $(C_4H_9)_2SnX_2$ structure (or $(C_4H_9)_2SnX$, where X is divalent.)

18.2 Technical, agricultural and industrial uses of organotin compounds

Among the various technical, agricultural and industrial uses of organotin compounds, only the four most significant are highlighted here:

(i) as stabilizers for PVC,
(ii) as catalysts in the production of polymers,
(iii) as preservatives to protect wood, ceramics, plastics and fabrics from fungal attack, and
(iv) as biocides to protect plants from insects, and ship hulls from fouling due to marine organisms.

Directly or indirectly, they all can contribute to the exposure of humans to organotin compounds, for which reason they are discussed in the following in more detail.

18.2.1 Organotin compound use as a stabilizer for polyvinylchloride
The use of organotin compounds as a stabilizer for PVC was mainly driven by the wish to produce a PVC material with crystal clarity. This was a property that could not be achieved with the then-available heat stabilizers, which were based on lead or other metals.

After the pioneering work of Yngve in the early 1930s, the first organotin stabilizer of significant commercial importance was DBT dilaureate, patented by Quattlebaum and Rugely in 1939 (Wilkes et al., 1996). The lubricity of dibutyl dilaureate, however, limited its use levels. Later, DBT was modified to produce dibutyl maleate and dibutyl maleate half esters. These not only had reduced lubricity, but also markedly improved heat stabilizing performance. This favorable behavior – resulting in a reduction of the yellowing of colorless PVC material – is believed to be an effect of the ability of the maleate moiety to add to long polyene sequences by a Diels–Alder reaction, thereby breaking the color-forming sequence of cumulated double bonds. Consequently, DBT laureate-maleate products set the industry standard for PVC stabilization with regard to clarity and performance in the 1940s (Wilkes et al., 1996).

The next step forward in the development of organotin stabilizers was the discovery of the effectiveness of organotin mercaptides as PVC stabilizers. It was Rowland and Reid who in 1947 first claimed the improved heat stability of mixtures of mercaptans with organotin carboxylates (Rowland and Reid, 1949). Following a series of inventions and improvements filed by other authors for patent, organotin compounds of mercaptoacids and their esters were introduced by Weinberg and Johnson in 1951 (Weinberg and Johnson, 1954). For more than one and a half decades, one of these compounds, dibutyl-S,S'-bis-(isooctyl mercaptoacetate) was to become the most important stabilizer for rigid PVC, due to its useful properties. After the end of patent protection for these products in 1968, a wide variety of compounds entered the market to replace the dominant dibutyl-S,S'-bis-(isooctyl mercaptoacetate) (Bacloglu et al., 2001). Concurrent with the market entry of other organotin stabilizers, prices fell when the basic organotin mercaptide patents expired.

In response to the demand for low-cost tin stabilizers in the North American rigid pipe and PVC profile market, low-cost methyltin-based stabilizers were developed. Methyltin mercaptoacetate esters achieved approval in indirect food contact in 1975, joining octyltin mercaptoacetates, which had gained approval already in 1961 (Wilkes et al., 1996). Although continued price pressure led the manufacturers of organotin compounds to develop (cheaper) alternatives with reduced levels of tin, for example by diluting organotin stabilizers with active barium and calcium complexes, or by completely eliminating tin by its homologue antimony compounds, none of these approaches gained a wider popularity, due to their inferior performance.

In the early 1980s, lubricating low-cost mixed sulphur-bridged methyltin stabilizers were developed with unmatched performance for rigid PVC pipes and profiles, thus securing its market position against all competitive non-tin products.

Organotin compounds most commonly used as PVC stabilizers are mono and/or di-alkyltin compounds of the general formula $RSnX_3$ and R_2SnX_2, where the group R is methyl, n-butyl, or n-octyl, and the ligand X can be mercaptoester, carboxylate, or sulphide (see Fig. 18.1).

434 Persistent organic pollutants and toxic metals in foods

$$\begin{array}{c}R^1\\ \diagdown\\ R^1\end{array}\!\!Sn\!\!\begin{array}{c}S-R^2\\ \diagup\\ \diagdown\\ S-R^2\end{array}\qquad R^1\!-\!Sn\!\!\begin{array}{c}S-R^2\\ -S-R^2\\ S-R^2\end{array}$$

Sn-R¹:
Methyltin: CH_3-
Butyltin: $n\text{-}C_4H_9-$
Octyltin: $n\text{-}C_8H_{17}-$

S-R²:
-S-CH$_2$-CO-O-alkyl
-S-CH$_2$-CH$_2$-CO-O-alkyl
-S-alkyl
-S-

S-R²:
Thioglycolates (alkyl is mostly ethyl hexyl or *iso*-octyl)
Mercaptopropionates
Mercaptoethanol esters (so-called reverse esters)
Alkylmercaptides
Sulfides

Fig. 18.1 Scheme of the most common organotin stabilizers for PVC.

Table 18.2 Acute toxicity of some organotin chlorides

Compound	LD$_{50}$ value
Monomethyltin trichloride	575–1370
Dimethyltin dichloride	74–237
Trimethyltin chloride	9–20
Mono-*n*-butyltin trichloride	2200–2300
Di-*n*-butyltin dichloride	112–219
Tri-*n*-butyltin chloride	122–349
Mono-*n*-octyltin trichloride	2400–3800
Di-*n*-octyltin dichloride	4000–7000
Tri-*n*-octyltin chloride	4000–29 000

From Wilkes *et al.*, 1996.

A consideration of greatest importance in this context is the toxicity of the different organotin compounds: certain monoalkyltins are less toxic than di-alkyltins, and generally toxicity decreases (for humans) as the chain length of the alkyl group increases. Since trialkyltins, R_3SnX are typically more toxic than the mono- and di-alkyltins, great care must be taken during the production process of organotin stabilizers that the concentration of triorganotins is kept to the absolute minimum. To illustrate this, Table 18.2 reports the acute toxicity of some organotin chlorides (Jennings and Starnes, 1996).

Organotin stabilizers migrate from rigid PVC only very slightly. This fact, together with favorable toxicological properties, is the basis for the worldwide approval of certain types of organotin compounds (particular methyl- and octyl-tin isooctylthioglycolates) for use in food packaging and potable water pipe (Bacaloglu *et al.*, 2001; Muncke, 2011).

In addition to negligible mammalian toxicity, dioctyltin compounds have lower odor levels and better lubrication properties than the corresponding butyltin compounds. In turn, the stabilization efficiency of the octylthiotin compounds does not completely match that of DBT derivatives. Still, the best results among octyltins are obtained using di-*n*-octyltin derivatives of thioglycolic acid (Midwest Research Institute, 1976).

In 2001, it was estimated that between 12 000 to 13 000 tons of tin are used annually in tin stabilizers worldwide (Batt, 2001). This figure increased to more than 16 000 tons of tin in 2007, showing a steady and significant increase with a growth rate of about 4% per year (RPA, 2007).

18.2.2 Organotin compound use as a catalyst in the production of polymers

Catalysts are widely used in industrial processes to speed up chemical reactions, such as polymerization. Mono- and di-organotin catalysts are commonly used as catalysts in chemical synthesis and in the curing of coatings.

Taking advantage of their basic character, the organotins catalyze the esterification and transesterification of mono- and poly-esters in chemical synthesis. These products are then used for plasticizers, synthetic lubricant manufacturing and polyester polyol production, as well as for some coating applications.

As curing catalysts, one of the largest uses of organotins is in electrocoat (Ecoat) coatings. These electrocoating products have a wide range of different applications, of which probably the most important is in the automotive industry, where they provide excellent rust resistance. The catalysts are also used in urethane coatings, as well as polyurethane foam production. Other applications include curing silicones (from which they are phased out by now from the use in food contact materials (FCMs)) and silanes.

Of the large number of organotin compounds the following appear to have been used more commonly as catalysts:

- Hydrated MBT oxide
- Butyl chlorotin dihydroxide
- Butyltin-*tris*-(2-ethylhexoate)
- DBT diacetate
- DBT oxide
- DBT dilaureate
- Butyl stannoic acid
- Dioctyltin dilaurate
- Dioctyltin maleate

Other tin compounds, such as stannous oxide, stannous oxalate and stannous *bis*-(2-ethylhexoate) are used extensively in chemical synthesis as catalysts, although they cannot be counted as organotins, since they do not have a tin–carbon bond.

As is true also for antifouling paints, there are substitute products that are not based on organotin chemistry (but beryllium and bismuth instead); however, they do not offer a viable alternative in terms of toxicity and a favorable cost/performance ratio (Batt, 2001).

18.2.3 Organotin compounds as preservatives to protect wood from fungal attack

From the 1950s, TBTO and tributyltin naphthenate (TBTN) were also used for industrial wood treatment and preservatives. This use declined over the decades and was mainly concentrated in tropical areas (Richardson, 1993). It can be assumed that, after 2000, OTCs had completely disappeared from these applications in industrialized nations.

18.2.4 Organotin compounds as biocides in marine antifoulant paints and pesticides for crop protection

Marine antifoulant paints

Prior to its ban in Europe in 2003 (and from 2008 when the ban came into force worldwide), TBT had been extensively used for decades as a biocide in paint formulations to provide protection against the fouling of ship hulls of both large ships and smaller vessels (Champ, 2000). TBT is unique among the organotins in that it exhibits a very high and targeted toxicity to target organisms. The MBTs and DBTs hardly exhibit these properties, but are typically found to accompany TBT either as a by-product from the synthesis, or as a degradation product.

OTC-containing paints had been formerly used to protect the underwater surface area of a ship's hull against barnacles, algae, etc., in order to avoid increased fuel consumption and to minimize the need for premature dry-docking. Triorganotins were introduced for this application during the 1950s. In the earliest applications of TBT-containing antifouling paints, tributyltin oxide (TBTO) was freely dispersed in what were called Free Association Paints (FAP). These paints had uncontrolled, rapid biocide leaching rates.

In response to the negative performance (short effective protection time) and environmental effects of FAPs, copolymer systems of TBT and methacrylate were developed which had 'self-polishing' behavior. These revolutionary new systems, called self-polishing copolymers (SPC) had controlled, uniform biocide leach rates by incorporating the TBT biocide into the polymeric binding system of the paint formulation. This polymeric system gradually dissolved, thus releasing TBT at a controlled rate over a period of up to 5 years.

Throughout the 1980s, many countries worldwide began restricting the use of TBT paints because of their environmental impact, resulting in considerable part also from the misuse of the product in pleasure craft and other small coastal vessels (Champ, 2000). In the late 1980s, the Organotin Antifoulant Paint Control Act was passed in the United States, restricting the use of TBT paints to vessels greater than 25 meters in length. It also specified the allowable leach rates of MAF paints sold in the USA, as well as restrictions on applications and waste disposal. Taking the initiative, the International Maritime Organization (IMO) proposed a worldwide ban on TBT MAF paints, commencing with a ban on its application to vessels from 1 January 2003, and a total ban on its presence on vessel hulls from 1 January 2008. This ban – the IMO's Antifouling System Convention – entered into force on 17 September 2008, some seven years after its adoption. The Convention, which bans the use of TBT-based antifouling paints globally, and contains a regime for restricting the use of other harmful antifouling paint technologies, met the requirements of ratification by 25 states, representing 25% of the world's shipping tonnage, only in 2007. The Convention was adopted in London in October 2001 but the ratification process has been slow, with the first of the large flag states (Panama) only ratifying in 2007. The Convention has till 2008 been ratified by 34 states, representing 53% of the world's merchant shipping tonnage (IMO, 2008).

The provisions of the Convention in respect of TBT were incorporated into EU law in 2003 (Regulation (EC) No. 782/2003) and it has been an offence since 1 January 2008 for any ship visiting an EU port to have TBT on its hull. The entry into force of the Convention brings the rest of the world into line with this. Table 18.3 presents an overview of the regulations that control the use of TBT in antifouling paints in Europe, North America, the southern hemisphere, Japan and worldwide.

The reason for the increasing awareness raised by the use of TBT in antifouling paints was due to the fact that triorganotins were found to be not only highly toxic even at very low levels and to target specific groups of organisms, but in addition to the desired effect of preventing the growth of unwanted organisms on the ship hulls, TBT was also found to demonstrate unwanted and severe side effects. The two most-studied effects were the thickening of the shells of oysters, and growth anomalies and imposex in the common dog-whelk *Nucella lapillus*, a marine snail.

In the early 1970s, organotin compounds were marketed as excellent biocides. They were known to be highly toxic and specific to molluscs, and also effective biocides against a wide range of marine fouling organisms (Champ and Seligman, 1996). Early OTC-containing paints for ship hulls were merely physical mixtures of organotin compounds and paint, often at very high concentrations, which explains the enormous amounts of OTC emitted into the maritime environment. This was common practice while OTCs were known to be effective at very low concentrations, which, as analytical capabilities improved, turned out to be at the ppt (ng per liter) level, and thus required the development of analytical methods and protocols of greater sensitivity and accuracy. The toxicity of TBT to certain marine organisms occurred at concentration levels that, in the 1970s and first half of the 1980s, were below the detection limits for most analytical laboratories (Hoch, 2001; Rosenberg, 2005).

With the improvement of analytical methods and the ability to detect organotin compounds at environmentally relevant concentrations, public awareness arose about the environmental effects of organotin compounds used as antifouling paints. Studies were performed both in the United States

Table 18.3 Regulations controlling the use of TBT paints worldwide by country up to 1992 and IMO proposals for 2003 and 2008 (Bray, 2006)

	Year	Regulations
Europe		
Austria	NA	Banned the use of TBT antifouling paint in freshwater lakes.
Europe (EC States)	1991	Prohibited the use of TBT-based paints on vessels less than 25 m length overall (LOA). TBT antifoulants available only in 20 L containers.
Europe (non-EC)	Various	Prohibited use of TBT-based paints on vessels less than 25 m (most States).
Finland	1991	Prohibited the use of TBT-based paints on boats less than 25 m LOA.
France	1982	Prohibited the use of TBT-based paints on vessels less than 25 m LOA, except for aluminum-hulled vessels.
Germany	1990	Prohibited the use of TBT-based paints on vessels less than 25 m LOA. Ban on retail sale. Ban on its use on structures for mariculture. Regulation for the safe disposal of antifouling paints after removal.
The Netherlands	1990	Prohibited the use of TBT-based paints on vessels less than 25 m LOA. Washing/blasting slurry used to prepare TBT antifoulants may be treated as hazardous waste. TBT antifoulants available only in 20 L containers. All antifoulants must be registered.
Norway	1989	Prohibited the use of TBT-based paints on vessels less than 25 m LOA.
Sweden	1989	Prohibited the use of TBT-based paints on vessels less than 25 m LOA.
	1992	Maximum leaching rate of 4 $\mu g/cm^2/day$ for vessels greater than 25 m LOA. All antifoulants must be registered.
Switzerland	1987	The use of TBT-based antifouling paints is banned in fresh water lakes. All antifoulants must be registered.
United Kingdom	1985	Sale of TBT-based paints restricted, effective bar on TBTO FAPs.
	1987	Prohibited the use of TBT-based paints on vessels less than 25 m LOA and on fish-farming equipment. TBT antifoulants available only in 20 L containers. All antifoulants to be registered as pesticides. Advisory Pesticides Committee must approve sale and use. Washing/blasting slurry treated as hazardous.
North America		
Canada	1989	Prohibited the use of TBT-based paints on vessels less than 25 m LOA, except for aluminum-hulled vessels. Maximum leaching rate set for vessels greater than 25 m in length. All antifoulants must be registered.
United States	1988	Prohibited the use of TBT-based paints on vessels less than 25 m LOA, except for aluminum-hulled vessels.

(*Continued*)

Table 18.3 Continued

	Year	Regulations
	1990	Maximum leaching rate of 4 µg/cm^2/day for vessels greater than 25 m in length. All antifoulants must be registered. TBT-based antifouling paints can only be applied by certified applicators.
Southern hemisphere		
Australia	1989	Prohibited the use of TBT-based paints on vessels less than 25 m LOA. Maximum leaching rate of 5 µg/cm^2/day for vessels greater than 25m LOA. All dry-docks must be registered with the Environmental Protection Agency because of discharges. All antifoulants must be registered.
New Zealand	1989	The application of TBT copolymer antifouling paint is banned with three exceptions: hulls of aluminum vessels, the aluminum out-drive or any vessel greater than 25 m LOA.
	1993	The application of TBTO FAPs is banned. Maximum leaching rate of 5µg/cm^2/day for vessels greater than 25 m LOA. All antifoulants must be registered. Use of any organotin containing antifouling paint prohibited.
South Africa	1991	Prohibited the use of TBT-based paints on vessels less than 25 m LOA. TBT antifoulants available only in 20 L containers All antifoulants to be registered.
Hong Kong	NA	All TBT antifoulants must have a valid permit for import/supply. All antifoulants must be registered.
Japan		
Japan	1990	TBT banned for all new vessels.
	1992	TBT banned for all vessels.
IMO measures		
Worldwide ban proposal	2003	Proposed ban for 1 Jan 2003 – no reapplication of TBT.
	2008	1 Jan 2008 No ships or structures shall bear TBT. Ratified on 17 Sept 2008 when > 25% of shipping tonnage or 25 of the world's shipping nations signed in.

(e.g. by the US Navy (NAVSEA 1984, 1986)) and in Europe (e.g. Alzieu *et al.*, 1986; Bryan *et al.*, 1986; ten Hallers-Tjabbes, 1994; Waldock *et al.*, 1983) and revealed the detrimental effects of organotin compounds on certain nontarget marine organisms. Still, the use of organotin-based antifouling paints was considered a technically and, particularly, economically viable approach, notably as the early free-association OTC-containing paints were substituted by TBT-based copolymer paints that exhibited significantly lower release rates.

The detrimental effects of organotin compounds being used as antifouling agents only eventually surfaced during the late 1970s and early 1980s, mainly as a consequence of two isolated observations, which were attributed to the extensive use of OTCs as antifouling paints for ship hulls: these were the collapse of the oyster-farming industry in Arcachon Bay in southern France (Alzieu, 2000; Alzieu et al., 1986), and the decline of the population of certain marine snail species in Northern England (Bryan et al., 1986; Evans et al., 1996; OSPAR, 2009).

In the former case, widespread failures in the reproduction ability of the commercially important Pacific oyster (*Crassostrea gigas*) were observed by the oyster farmers in Arcachon Bay in the South of France in the late 1970s. It would take, however, several years until high levels of organotin compounds, and in particular TBT, in the waters of the bay were identified as the cause of this ecological catastrophe. Being a very popular holiday resort, the Bay of Arcachon was home to many private yacht owners who anchored their boats there. There was thus a permanent and high input of organotin compounds into the water of the bay, as at that time OTC-containing paints were in common use as antifouling paints, even for leisure boats. The high concentrations of organotin compounds impaired reproduction of the Pacific oyster by inhibiting embryogenesis and larval development, and causing the deformation of their shells (Alzieu et al., 1986; Alzieu, 2000; Ruiz et al., 1996). Similar effects to oysters were also observed along the coasts of the United Kingdom (Thain and Waldock, 1986; Waldock and Thain, 1983; Waldock et al., 1987). After restrictions of the use of TBT-containing antifouling paints for ship hulls starting from 1983, a gradual improvement of the situation occurred. OTC concentrations decreased relatively quickly, while the concentrations of OTCs in biota and sediments have decreased only very slowly, supporting the fact that particularly (anoxic) sediments have a large storage capacity for OTCs and will act as sources for the reemission of OTCs even many years after their ban (Alzieu, 2000; Ruiz et al., 1996).

The second drastic example was the observation of the development of male sexual characteristics (i.e. penises) on female dogwhelks (*Nucella lapillus*) in Plymouth Sound, UK. Both the percentage and the severity of this phenomenon increased with proximity to the harbor of Plymouth. Triggered by this initial alarming observation, further research revealed that large parts of the population of *Nucella lapillus* along the coastline of the southern UK were affected by this phenomenon called *imposex*. Also, this effect was linked to exposure to high concentrations of TBT, which exhibits strong endocrine activity, mimicking the effect of a male hormone and therefore leading to the formation of male sexual characteristics imposed on female specimens, further leading to infertility and a rapid decline of the population (Bryan et al., 1986). Also with the phenomenon of imposex, a gradual recovery was seen after the ban of OTC compounds for biocidal use (Evans et al., 1996). An adverse effect on reproductive ability was seen not only with the common

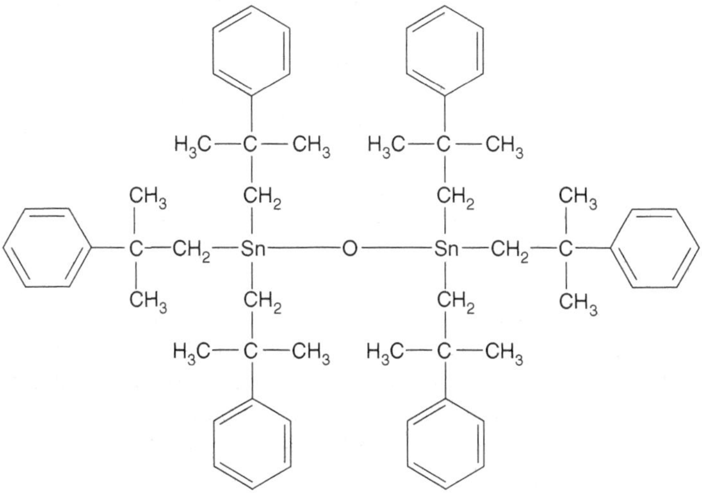

Fig. 18.2 Chemical structure of fenbutatin oxide (*bis*-[tris-(2-methyl-2-phenylpropyl) tin] oxide).

dogwhelk (*Nucleus lapillus*), but subsequent investigations demonstrated that more than 150 species of molluscs and gastropods were affected by the phenomenon of imposex caused by TBT worldwide (Matthiessen and Gibbs, 1998; ten Hallers-Tjabbes, 2003).

TBT also affects higher animals. In this case, the effects are not directed towards their reproductive abilities; however, higher levels of TBT have been shown to affect the immune system. TBT has been detected in a number of birds, fish, whales and mammals (Kannan and Falandysz, 1997; Kannan et al., 1996).

Pesticides for crop protection
There are five main triorganotin ingredients approved and used as pesticides for crop protection (Batt, 2001):

- TPT hydroxide (TPhTOH or fentin hydroxide)
- Tricyclohexyltin hydroxide (TCyTOH or cyhexatin)
- Tricyclohexyltin triazole (TCyTT or azocyclotin)
- Trineophenyltin oxide (TNTO or fenbutatin oxide) (see structure in Fig. 18.2)
- TPhT acetate (TPhTAc or fentin acetate)

The main use of these products is as fungicides and acaracides. Fungicides are pesticides that kill or inhibit the growth of fungi, while acaracides kill mites and ticks (acarides).

TPhTAc and TPhTOH are used primarily for high value crops as fungicides. Tin-containing fungicides are used when the possibility of disease is very high, which justifies the added costs. They are used on potatoes, sugar beets and pecans (Stäb et al., 1994).

TCyTOH, TCyTT and TNTO possess excellent efficiency as acaracides. What makes them attractive is that they are not considered susceptible to resistance development. They are used on citrus, top fruit, vines, vegetables and hops.

Depending on the particular application, there are nowadays substitute products available. For example, on potatoes that have been mostly treated with TPhTOH, alternative substances such as propamocarb hydrochloride/ chlorothalonil and dimethomorph/mancozeb may be used. On sugar beets, tetraconazole can be used. TCyTOH may be substituted by dicofol, hexythiazox, propargite, pyribaden and tebufenpyrad. Following the trend of generally reevaluating the risks associated with the use of organotin compounds, fenbutatin is currently on the priority list of chemicals scheduled for evaluation and reevaluation by the JMPR (FAO/WHO Joint Meeting on Pesticides Residues for the period of 2011 to 2016 – JMPR, 2009).

When comparing the organotin products and their substitutes that are devoid of tin, the former typically have a cost advantage for growers when one considers the rate at which the product is applied, the cost of the product itself and the number of days between sprayings. However, a single treatment method is not normally used. Growers typically rotate two or three treatment types in order to avoid the build-up of resistance to any one fungicide.

The world production of organotin compounds was ca 50 t/a in 1950 – that is, before the onset of its large-scale use as biocide – 35 000 t/a in 1981, and before the IMO's ban on the use of organotin compounds in 2003, it was estimated at 40 000 t/a. As the tin content of these materials is ca 25%, the production of organotin compounds represented thus approximately 7–10% of the usage of Sn metal and about 20% of the total annual production of organotins (Bennett, 1996; Donard and Pinel, 1989).

In the main producing and consuming areas – the United States, Western Europe and Japan – 76% of the organotin compounds are used as stabilizers for PVC, 10% as antifouling biocides, 8% as agricultural biocides, and 5% as catalysts for the production of polyurethanes and silicones. In less industrialized countries, the pattern of use is more biased towards agricultural applications.

18.2.5 Regulation of exposure from consumer products

On 4 June 2009, a new European Commission Decision 2009/425/EC was published (EC, 2009), which amended Council Directive 76/769/EEC (EC, 1976) as regards restrictions on the marketing and use of organostannic compounds for the purpose of adapting its Annex I to technical progress. The usage of tri-substituted organotin compounds such as TBT compounds and TPhT compounds, DBT compounds and DOcT compounds in consumer products are governed by this Decision.

Table 18.4 Summary of the new threshold limits for organotin compounds in consumer products as imposed by the European Council's Decision 2009/425/EC (SGS, 2009)

Substance	Scope	Requirement	Effective
Tri-substituted organostannic compounds such as tributyltin (TBT) compounds and triphenyltin (TPhT) compounds	Article or part of an article	≤ 0.1 %	1 July 2010
Dibutyltin (DBT) compounds	1. Mixture 2. Article or part of an article (except FCMs)	≤ 0.1 %	1 January 2012
	1. One-component and two-component room temperature vulcanization sealants (RTV-1 and RTV-2 sealants) and adhesives 2. Paints and coatings containing DBT compounds as catalysts when applied on articles 3. Soft PVC profiles whether by themselves or coextruded with hard PVC 4. Fabrics coated with PVC containing DBT compounds as stabilizers when intended for outdoor applications 5. Outdoor rainwater pipes, gutters and fittings, as well as covering material for roofing and façades	≤ 0.1 %	1 January 2015
Dioctyltin (DOcT) compounds	1. Textile articles intended to come into contact with the skin 2. Gloves 3. Footwear or part of footwear intended to come into contact with the skin 4. Wall and floor coverings 5. Childcare articles 6. Female hygiene products 7. Nappies 8. Two-component room temperature vulcanization moulding kits (RTV-2 moulding kits)	≤ 0.1 %	1 January 2012

444 Persistent organic pollutants and toxic metals in foods

From 1 July 2010 any article, or part of an article, that contains tri-substituted organotin compounds such as TBT and TPhT in a concentration of greater than 0.1% by weight of tin, is prohibited from being placed on the market.

From 1 January 2012, DBT compounds with a concentration greater than 0.1% by weight of tin shall not be used in mixtures and articles for supply to the general public. DOcT compounds in an article shall not be in a concentration greater than 0.1% by weight of tin (Table 18.4).

18.3 Physical and chemical properties of organotin compounds

Organotin compounds are organic compounds in which at least one carbon–tin bond is established. The vast majority of these compounds contain tin in the oxidation state +4. Although tin is a homologue of carbon and a member of main group 14 of the periodic table of elements, it has a lower electronegativity than carbon (1.96 vs 2.55 according to the Pauling scale), which causes the carbon–tin bond to be more susceptible to cleavage. This does not mean, however, that the C–Sn bond is inherently instable. Yet, under the influence of light, atmospheric oxygen or certain microorganisms, organotin compounds are degraded in a relatively short time (Fig. 18.3). When the hydrocarbon groups are split off, they eventually leave behind nontoxic inorganic products after a step-wise degradation.

The symmetrical tetraalkyltin compounds are colorless, form monomolecular solutions, are fairly stable towards water and air, and can be distilled without decomposition at < 200°C. This demonstrates that thermal decomposition is not a significant route of degradation in the environment. With longer alkyl chain length, the tetraalkyltins become increasingly waxy substances.

The symmetrical tetraaryltin compounds are stable in air and water and are also colorless. They melt at temperatures above 150°C.

The tin–carbon bond is readily cleaved by strong acids, halogens and other electrophilic agents. Tin establishes preferentially covalent bonds with other

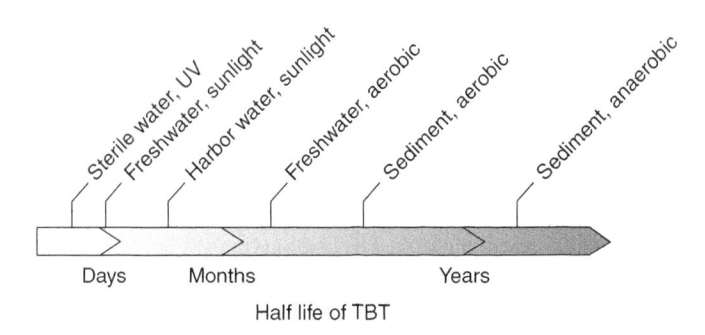

Fig. 18.3 Half-life of organotin compounds under different conditions.

elements, with these bonds possessing a strong ionic character. Tin is usually the electropositive partner in these compounds. The triorganotin hydroxides act more like inorganic bases than as alcohols, due to the amphoteric character of tin. The *bis*-(triorganotin) compounds, $(R_3Sn)_2O$, are thus strong bases and react with both inorganic and organic acids, forming salt-like compounds which, however, are nonconducting and insoluble in water. Tin does not form double bonds with oxygen, and di-organotin oxides, R_2SnO, are polymers of usually high degree of polymerization via intermolecular tin–oxygen bonds.

It is evident that the number of organic substituents and, particularly, their nature has a significant effect on the physicochemical properties of organotin compounds, especially their solubility. Generally speaking, the water solubility decreases with the degree of substitution and the length of the substituent chain (if aliphatic). However, experimental solubility data vary greatly as a result of different experimental conditions (Table 18.5).

18.4 Analysis of organotin compounds in foods

The very different physicochemical, and also ecotoxicological, properties of organotin compounds calls for a detailed analysis that is capable of differentiating the individual organotin compounds (species). This type of analysis is typically referred to as 'speciation analysis' and is mostly performed using hyphenated techniques, which consist of the combination of a powerful separation technique and a sensitive and selective detection technique (Lobinski and Szpunar, 2003).

Separation is mostly achieved by gas or liquid chromatography. Considering the ionic nature of most organotin compounds, separation in liquid phase has the advantage of not requiring derivatization of the analytes to increase their volatility. This is important, as this step may be associated with inadequate recoveries, particularly in the presence of a complex interfering matrix (Chau *et al.*, 1997). Gas chromatographic techniques, in contrast, do require that the analytes be transformed into a more volatile form. This is typically done by one of the following reactions:

(a) $R_n SnX_{4-n} + (4-n) NaBH_4 \rightarrow R_n SnH_{4-n} + (4-n) NaX + (4-n) BH_3$

(b) $R_n SnX_{4-n} + (4-n) NaBEt_4 \rightarrow R_n SnEt_{4-n} + (4-n) NaX + (4-n) BEt_3$

(c) $R_n SnX_{4-n} + (4-n) R'MgX \rightarrow R_n SnR'_{4-n} + (4-n)/2 \, MgX_2$

Reaction (a) is the hydridization of organotin compounds with sodium tetrahydroborate as a reagent to form the corresponding alkyl- (or aryl-)tin hydride. It is a very versatile reaction, in that it can be performed in aqueous solution, and that the reagent is commonly available at low cost (Donard

Table 18.5 Physicochemical properties of di- and tri-substituted organotins (butyltin compounds, octyltin compounds and phenyltin compounds) (RPA, 2007)

Property	DBTC	DBTO	DOcTC	DOcTO	TBTC	TBTO	TPhTC	TPhTH
CAS No.	683–18–1	818-08-6	3542–36–7	870–08–6	1461–22–9	56–35–9	639–58–7	76–89–9
EINECS No.	211–670–0	212–449–1	222–583–2	212–791–1	215–958–7	200–268–0	211–358–4	200–990–6
Molecular formula	$(C_4H_9)_2Cl_2Sn$	$(C_4H_9)_2OSn$	$(C_8H_{17})_2Cl_2Sn$	$(C_8H_{17})_2OSn$	$(C_4H_9)_3ClSn$	$(C_4H_9)_6OSn_2$	$(C_6H_5)_3ClSn$	$(C_6H_5)_3OHSn$
Mol. weight (g/mol)	303.8	248.9	416.0	361.1	325.5	596.1	385.5	367.0
% Tin (w/w)	39.1%	47.7%	28.5%	32.9%	36.5%	19.9%	30.8%	32.3%
Melting point (°C)	42	105	47	230	–19	–45	106	123
Boiling point (°C)*	>250	>250	>250	>250	>250	>250	>250	>250
Vapor pressure (25°C) (Pa)	0.15	4.2E-06	2.63E-04	9.5E-02	1.00	1.0E-03	2.10E-05	4.7E-03
Water solubility (mg/L)	33	4.0	1.6	0.23	10	35	40	1
Source in RAR (RPA, 2005)	Parametrix, 2004	Parametrix, 2004	Parametrix, 2004	Parametrix, 2004	Parametrix, 2005	BUA, 2003	CICAD 13 (WHO, 1999)	CICAD 13 (WHO, 1999)

* As most organotins decompose, boiling points of >250°C are reported in the absence of a 'true' boiling point (at atmospheric pressure).

et al., 1986). Its disadvantages are that highly volatile organotin hydrides are formed, which may suffer from increased losses during the subsequent steps of sample preparation, clean-up or preconcentration. Also, the organotin hydrides are not particularly thermally stable and may thus undergo degradation when in contact with hot parts of the analytical system, such as the injector of the gas chromatograph, and the reaction may be interfered by metal ions concurrently present in the samples.

Reaction (b) is the alkylation (or arylation) of organotin compounds with sodium tetraalkyl- (or aryl-)borates, which are the alkyl/aryl homologues of the sodium tetrahydroborate (Rapsomanikis, 1994). The reaction products formed are fully alkylated (arylated) organotin compounds, where the moiety R' has to differ from the residue R in order to preserve the original species information. The reaction has the same advantages and disadvantages as listed above for the hydridization reaction. Only the volatility of the derivatives is lower in comparison to the organotin hydrides and can be, depending on the alkyl moiety R' of the particular tetraalkyl- (or aryl-) borate reagent, adjusted to the particular analysis conditions. Common residues R' for this derivatization reaction are ethyl and propyl; tetrabutyl and tetraphenyl reagents also exist but are rarely used, due to the loss of species information in the case of butyltin analysis and the relatively high molecular weight of phenylated organotin derivatives. As the reaction yield depends on the pH, derivatization is often performed in a sodium acetate buffer, which provides a pH value (pH 4–5) at which the reaction yield is close to the optimum.

Reaction scheme (c) denotes derivatization with Grignard reagents, that is, organometallic compounds of the general structure R'MgX with R' being an alkyl or aryl moiety and X a halogen (typically Cl or Br) which historically has been the most widely used reaction scheme. It must be performed in a water-free reaction medium. Since this requires a solvent exchange in most cases, the application of Grignard derivatization is more tedious than other derivatization schemes. In turn it is very flexible, as a wide range of Grignard reagents are commercially available, and provide high reaction yields and stable derivatives (Morabito *et al.*, 2000).

The derivatives are readily soluble in organic solvents such as hexane, isooctane or toluene, and are chromatographed with apolar or low-polarity capillary GC columns (typical stationary phase choices are 100%) polydimethylsiloxane columns (DB-1 or equivalent) or 95% poyldimethylsiloxane/5% polydiphenylsiloxane (DB-5 or equivalent).

Detection can be either performed in an element- or compound-specific mode. While in the former case the identification must be achieved by comparison of retention times with those of authentic standards, in the latter case a positive identification is possible, based on the analyte's (or its derivative's) spectrum. A variety of element-specific detectors are available for qualitative and quantitative analysis of organotin compound (speciation analysis). These include the atomic emission detector (AED), the continuously operated flame photometric detector (FPD) and its pulsed version (P-FPD) and, of course,

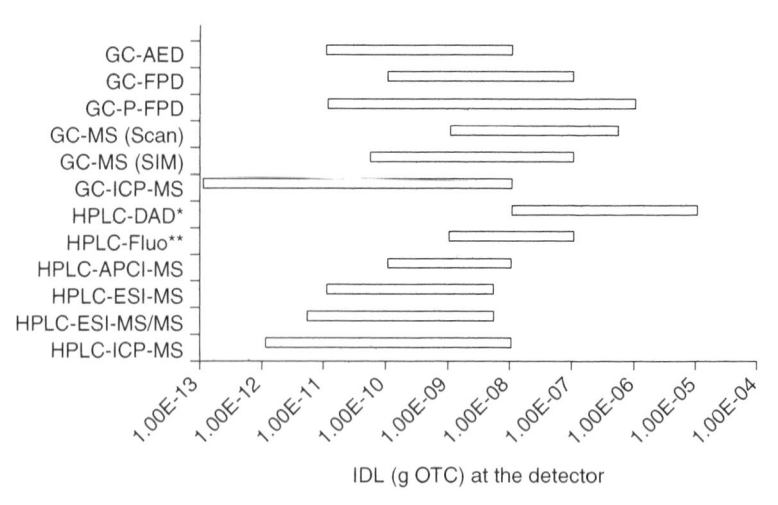

448 Persistent organic pollutants and toxic metals in foods

Fig. 18.4 Comparison of instrumental detection limits (IDLs) for the determination of organotin compounds.

inductively coupled plasma mass spectrometry (ICP-MS). The merits and applicability of these detectors for organotin speciation have been compared by Aguerre *et al.* (2001). Of the above mentioned element-specific detectors, the two types of FPD are operationally the simplest, while still providing excellent selectivity and sensitivity. Also, the AED is a most versatile detector for organotin compound speciation, both in terms of selectivity and sensitivity, although at significantly higher instrument costs. Being commercially available only from one producer, it nowadays mostly has been supplanted by ICP-MS detection, particularly in those cases where ultimate sensitivity is required and operational costs are not of primary concern. The relative sensitivities of various gas chromatographic detectors for organotin compounds (expressed as instrumental detection limits, IDL) are summarized in Fig. 18.4. ICP-MS is not only the currently most sensitive (commercially available) detector for the determination of organotin compounds by gas chromatography, but provides also the possibility of using isotopically labeled organotin compounds as internal standards. This is very important from the point of view of accuracy, precision and analytical quality assurance.

Mass spectrometry (with electron impact ionization) provides also molecule-specific detection of OTC for gas chromatography. Latest generation GC/MS systems provide a sensitivity that is comparable with that of element-specific detectors and, similarly to GC with ICP-MS detection, the possibility of using isotopically labeled internal standards.

Although the use of liquid phase separation (HPLC) appears an attractive choice for organotin compound speciation, in that it does not require derivatization of the ionic organotin compounds, it has several disadvantages. The first relates to the poorer chromatographic separation which, when the reversed phase separation mode is used, hardly allows baseline separation between all relevant OTCs. In addition, chromatographic separation requires highly deactivated stationary phases and/or the use of high concentrations of acetic acid (up to 10% v/v) or trifluoroacetic acid (up to 1% v/v) in the mobile phase, which very negatively affects column life. Moreover, the absence of chromophoric groups for the butyltin compounds either necessitates post column (fluorescence) derivatization, or calls for mass spectrometric detection (González-Toledo *et al.*, 2003). In the latter case, both common types of atmospheric pressure ionization interfaces, electrospray ionization (ESI, e.g. by González-Toledo *et al.*, 2002; White *et al.*, 1998) and – more rarely –atmospheric pressure chemical ionization (APCI) have been used (Rosenberg *et al.*, 2000; Siu *et al.*, 1988). For both interfaces, the use of high acid or salt concentrations in the mobile phase is unfavorable, as it leads to signal suppression.

Successful speciation analysis of organotin compounds requires an appropriate procedure for sample preparation. This sample preparation procedure must fulfil various functions, namely (a) isolation (extraction) of the analytes from the matrix, (b) clean-up, if necessary, (c) preconcentration, and (d) derivatization (Lee and Chen, 2011).

Organotin compounds can be extracted from lipid-free samples either with aqueous acetic, trifluoroacetic or hydrochloric acid, or with an organic solvent with the addition of a complexing agent, e.g. tropolone in toluene (Abalos *et al.*, 1997). The function of the anion of the acid is to neutralize the charge of the alkyl- (or aryl) tin cation, and at the same time to render it more hydrophobic, thus enabling a more efficient extraction.

In the case of samples containing a significant fraction of fat tissue, digestion/hydrolysis of the matrix is required. This is typically done by hydrolyzing and thereby dissolving the matrix, either by methanolic sodium hydroxide or, more efficiently, by aqueous tetramethyl ammonium hydroxide (TMAH) (Nóbrega *et al.*, 2006). This leads to a complete dissolution and hydrolysis of fat tissue. Once complete dissolution of the matrix has been achieved, the organotin compounds can be extracted into an organic phase (hexane or *iso*-octane) after *in situ* derivatization (with $NaBEt_4$ or $NaBPr_4$) (Szpunar *et al.*, 1996). Since OTCs in fish and shellfish are incorporated in the fat tissue, classical extraction procedures will not provide adequate recoveries of organotin compounds, and complete digestion of the matrix is essential. It is necessary to carefully optimize the exact conditions of this solubilization step, since too rigid reaction conditions (e.g. by the application of heat, ultrasound or microwaves) or an extended duration of the solubilization step may lead to degradation of the organotin compounds (Pereiro *et al.*, 1996). The alkaline digestion/solubilization procedure is effective in terms of liberating the organotin compounds and making them accessible for derivatization and subsequent analysis. However, the lipid and tissue matrix constituents may not be completely digested, which will lead to serious interferences in the chromatographic determination (Szpunar-Lobinska *et al.*, 1994). Even though the commonly used hyphenated techniques may be sufficiently selective to cope

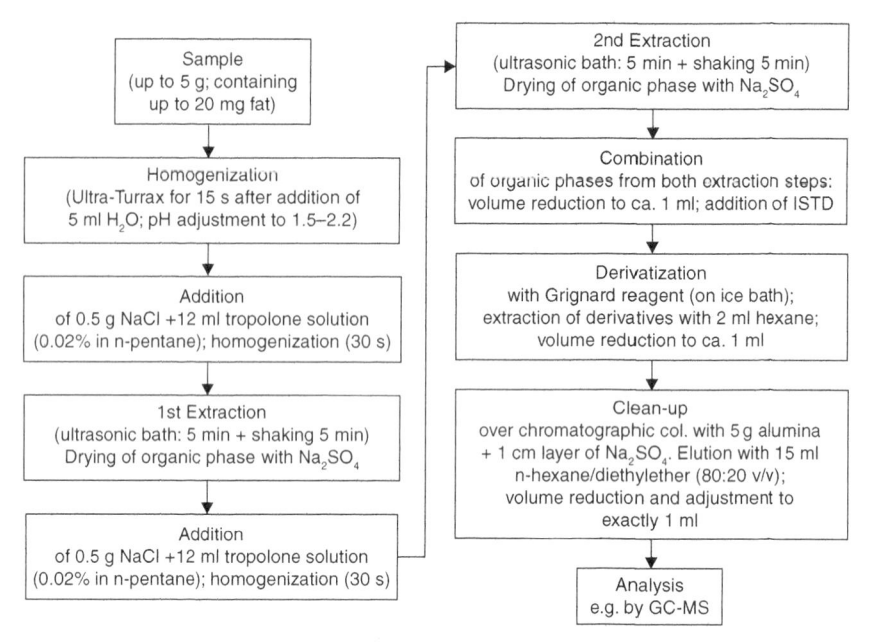

450 Persistent organic pollutants and toxic metals in foods

Fig. 18.5 General scheme for the determination of organotin compounds from environmental and food matrices (after Thomaidis, 2007).

with the co-eluting lipid matrix, it is recommended to perform a clean-up of the sample extract prior to analysis. This clean-up either in the form of a gel permeation chromatography where only the relevant fractions containing OTCs are pooled and analyzed by GC (Følsvik et al., 2002)), or by column chromatographic clean-up using a small alumina or silica column to remove the lipids – is important to maintain reproducible chromatographic conditions and to minimize instrument downtime (Ceulemans et al., 1994).

For the analysis of ionic OTCs from water samples, solid-phase extraction (SPE) has been proposed from as early as the late 1980s (Junk and Richard, 1987). However, SPE with 'classical' reversed phase C18 cartridges no longer finds much use, most probably due to the difficulties of quantitatively retaining analytes of very different polarity at the same time.

In contrast, two modern variations of SPE do find frequent application in OTC analysis, namely: solid-phase microextraction (SPME) and stir-bar sorptive extraction (SBSE). The common feature of these particular forms of SPE is that extraction of the in situ-derivatized analytes (e.g., by using $NaBEt_4$) takes place into a polymeric membrane, which in the case of SPME is coated onto the fused silica or metal tip of a syringe-like device (length: 1 cm, polymer film thickness: up to 100 µm with total diameter up to 500 µm). This fiber is either exposed to the headspace of the sample, or is directly immersed into the sample to extract the in situ-derivatized organotin compounds (Crnoja et al., 2001). Thermal desorption of the analytes takes place directly in the

Organotin compounds in foods 451

injection port of the GC. Since thermal desorption allows transferring the total amount of analytes extracted to the gas chromatographic system, the achievable sensitivity is much higher than with solvent desorption, in which only an aliquot of the extracted analytes is subjected to detection.

In the case of SBSE, the polymeric film (film thickness ca 0.5 mm) is coated around a magnetic stir bar of dimensions 2 cm × 4 mm (Baltussen *et al.*, 1999). The significantly larger volume of extractant phase in SBSE, as compared to SPME, provides nearly quantitative extraction yields even for analytes that are not extremely hydrophobic, i.e., that have log K_{ow} values in the range of 3–5 (for log K_{ow} values > 5, both techniques achieve quantitative extraction). The transfer of the extracted and *in situ*-derivatized analytes from the SBSE device onto the GC system again is achieved by thermal desorption. In this case, however, a dedicated thermal desorption unit must be used. It has been claimed that with SBSE and GC-ICP-MS detection limits for OTCs down to the ppq-level can be reached (Vercauteren *et al.*, 2001). A general scheme for the determination of organotin compounds from environmental and food matrices is shown in Fig.18.5.

18.5 Human dietary exposure to organotin compounds from foods

Human exposure to organotin compounds has four possible sources (Fig. 18.6): (i) food consumption (including drinking water), (ii) ingestion of contaminated soil/sediments, (iii) dermal absorption, and (iv) inhalation. These routes of exposure differ strongly both in intensity, as well as in probability of occurrence. The most important source of exposure of the general population is through food (in particular fish and other seafood). Uptake of organotin compounds by humans through their diet is more important the larger the fraction of fish or seafood other than fish is in the particular diet. Strong traditional and geographic differences exist between populations that consume large amounts of fish (e.g., Japan and the Mediterranean countries) and the inland countries.

Food contamination is primarily caused by the use of tri-substituted OTCs as biocides, components of antifouling paints and as agricultural pesticides. OTCs may accumulate in water bodies – particularly if these have a low water exchange rate – due to boating activities and agricultural runoff, with subsequent accumulation in the food chain. As seawater is normally not used for drinking-water abstraction, organotin compound contamination of seawater does not directly represent a significant source of intake of OTCs via drinking water. In few cases, contamination of drinking water has been reported, caused by the organotin stabilizers contained in new PVC tubing used for drinking-water supply (Sadiki and Williams, 1999; Sadiki *et al.*, 1996). Considering that even elevated levels of OTC release into drinking water may only – if at all – be observed for a short period of time until a new piece

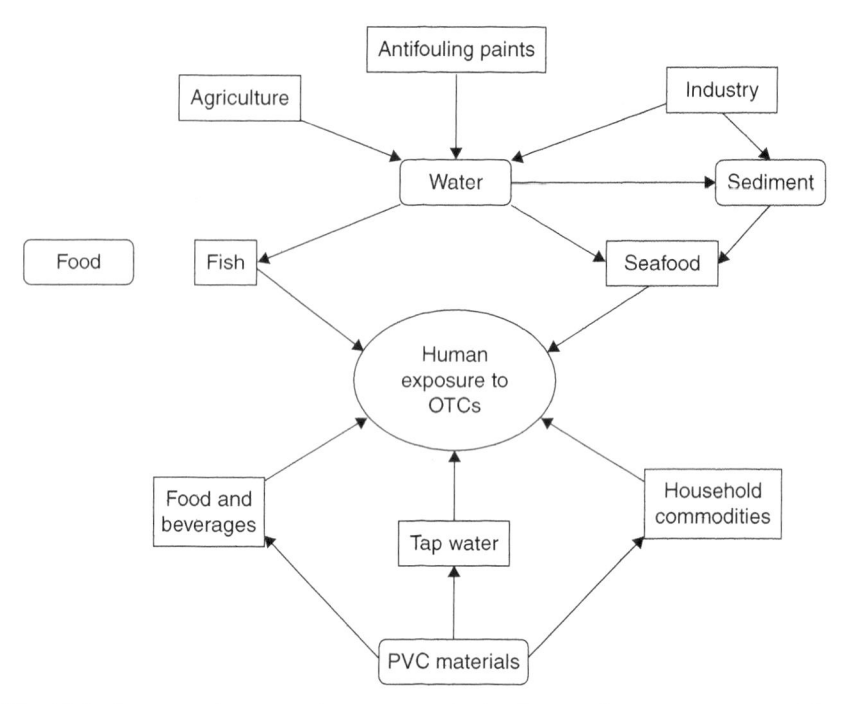

Fig 18.6 Sources of organotin compounds leading to human exposure (after Thomaidis, 2007).

of PVC pipe is equilibrated, both practical measurements (Jones-Lepp et al., 2001) and theoretical simulations (Fristachi et al., 2009) arrived at the conclusion that no significant risk to human health arises form OTC-contaminated drinking water.

18.5.1 Initial studies of organotin compound uptake from food

Japan is probably the country in which uptake of organotin compounds through food has reached attention first[1]. Surveys were conducted there to assess human exposure to TBT and TPhT in the early 1990s. The average TBT intake was estimated at 4.7 ± 7.0 and 2.2 ± 2.2 µg/day/person, respectively, in these studies from 1991 and 1992. Using the market-basket method, the national average daily intake of TBT in Japan was studied from 1990 through 1997, with intake-levels between 1.5 and 9.9 µg/day/person (WHO-IPCS, 1999b). More recent data exist for the estimate of TBT intake in Asia, Australia, Europe and the USA from the analysis of seafood species purchased from markets in eight cities (Keithly et al., 1999). Based on national diets and geometric means of seafood contamination, TBT doses

---

[1] Sections 18.5.1 and 18.5.2 are based on EFSA (2004).

were estimated to range from 0.18 (United Kingdom) to 2.6 (Korea) µg/day/person (Keithly *et al.*, 1999). Similar results were given in another study on seafood collected in the USA (Cardwell *et al.*, 1999) and in an overview prepared by Belfroid *et al.* (2000).

Between 1990 and 1997, TPhT intake was estimated in Japan by the market-basket method in the range of 0.7–10.4 µg/day/person (WHO-IPCS, 1999a).

18.5.2 More recent data: the database of the EU SCOOP Task 3.2.13 report

In October 2003, the final report of Scientific Co-operation (SCOOP) Task 3.2.13, entitled '*Assessment of Dietary Exposure to Organotin Compounds of the Population of the EU Member States*', was released (EC, 2003b). This project had been initiated by the European Commission and was coordinated by Italy. Its intention was to provide the Commission with information on dietary exposure to OTC in European countries. Eight countries – Belgium, Denmark, France, Germany, Greece, Italy, Norway and the Netherlands, which in the following will be referred to as 'participating countries' – delivered the available data on the occurrence of organotin compounds in food products, i.e. fish and seafood products.

The data that have been compiled in this report derive from country-specific samples that have been collected at various sites, including also heavily polluted areas during the period 1993–2002. The analytical data that was included in the report stems from the years 1995–2002. These data were collected, together with relevant supporting information on the data quality, and with an evaluation of whether these data were representative of the country of origin, and thus suitable to estimate national dietary intakes. In 2004, an EFSA panel commented on the data collected in the SCOOP study as part of a review of the evidence that resulted in a report assessing the health risks to consumers associated with exposure to organotins in foodstuffs (EFSA, 2004). Although the SCOOP study represented the largest and most comprehensive overview of organotin compound contamination of food, the panel responsible (EFSA, 2004) still commented that there were large differences in the size of the data sets submitted by the participating countries, and that the data have been obtained in monitoring plans and activities that were not originally designed to evaluate human intake. The data that were collected and evaluated related to the butyltin (MBT, DBT, TBT) and phenyltin (MPhT, DPhT, TPhT) compounds, although occasionally other OTC data were included in the original reports. The national dietary exposures to the six selected OTCs were calculated by the participating countries. Different methods of estimation were used, however. Two of the studies even differentiated intakes of specific population segments – children and high consumers living in coastal municipalities.

Analysis of raw occurrence data in the EU SCOOP report
According to EFSA (2004), during the SCOOP study, a valuable database has been compiled that contains a large and unique set of OTC concentrations (raw data) in many fresh, semi-preserved and fully preserved fish, and fishery products. The latter will hereafter be referred to as 'seafood other than fish' and include molluscs, crustaceans, cephalopods and echinoderms. The accumulated and statistically evaluated data of TBT, DBT, MBT, TPhT, DPhT and MPhT occurrence in fish and fishery products have been published (EFSA, 2004). In the EFSA report (EFSA, 2004) only data officially identified as representative of food contamination have been used.

Despite its uniqueness and enormous versatility, the data set compiled in the SCOOP study has some methodological and statistical shortcomings, as outlined in (EFSA, 2004). There are large differences in the amount, detail and quality of the data from the participating countries. Germany provided by far the largest group of data (N > 5000 or approximately 86% of the entire data set). Furthermore, analytical results appear to have been obtained largely without adequate harmonization of analytical procedures and/or intercalibration processes of the different laboratories, which may limit overall comparability of the data.

Although appropriate statistical tools (nonparametric statistical analysis) were applied for the evaluation of the data set, a number of further observations can still be made: while the OTC data reported spreads over several orders of magnitude, OTC concentration distribution patterns are also severely skewed toward high values. This is readily explained by the large variety of organisms taken into account, including farmed as well as wild fish, fish coming from relatively pristine as well as from more heavily polluted areas, fish belonging to different levels of the trophic web, molluscs, crustaceans, cephalopods and echinoderms. In several cases, arithmetic means are substantially greater than the corresponding medians: these differences are largely associated with the aforesaid distribution high value tails.

The SCOOP report ranks food groups related to their contribution to total intake of OTCs from country to country. These do of course take into account the different food consumption habits in the different countries, but also the large variations in concentrations of OTCs in fish and fishery products. Note that the OTC occurrence levels in seafood other than fish are in general higher than those detected in fish. For instance, calculated median and mean concentration values for TBT in seafood other than fish are 14 and 60 µg/kg fresh weight, respectively, whereas in fish the corresponding estimates are 5 and 17 µg/kg fresh weight. DBT and TPhT concentrations are in general lower than those observed for TBT.

Based on fully aggregated data for fish and fishery products, the estimated concentration medians for TBT, DBT and TPhT are 7.0, 2.5 and 4.0 µg/kg of fresh weight, respectively, and the corresponding mean values being about 28, 17 and 17 µg/kg fresh weight (i.e. 4–7-fold higher). Hardly any data are contained in the EU SCOOP report on DOcT, which was mostly found below the limit of determination. The EFSA report also notes that investigations carried out in the frame of the European Commission Research Project 'OT-SAFE' (QLK1-2001-01437), on sources, consumer exposure and risks of organotin contamination in seafood, have indicated that there is no significant breakdown of organotin compounds during cooking (Willemsen et al., 2004).

Assessment of the European dietary exposures
Average European dietary exposure
According to EFSA (2004), average consumption of fish and fishery products in European countries spans a range between 10 (The Netherlands) and 80 (Norway) g/day/person for adults. The Norwegian people can thus be taken as a conservative paradigm of the exposed European population, as they show a high consumption pattern. With reference to Norway, examples of mean consumption rates of seafood other than fish and fish are 10 and 70 g/day/person, respectively (EFSA, 2004). It is this variability in the composition of food classes considered by the Member-countries participating in the EU SCOOP Task 3.2.13, as well as the species-specific consumption rates, that make a comparative estimation of country-specific intakes based on the average international occurrence data and consumption patterns difficult.

Table 18.6 presents calculations based on the above Norwegian consumption values and the median international concentrations found in fish and fishery products which provide average dietary exposures for three organotin compounds, i.e. TBT, DBT and TPhT: the medians of their individual contributions are within a factor of 3 from each other (that is, between 0.0033 and 0.0093 µg/kg bw/day). The resulting cumulative intake of TBT + DBT + TPhT is 0.018 µg/kg bw/day. When the same calculation is performed with the Norwegian consumption pattern, however, with the mean (instead of the median) international concentrations, then average dietary exposures to TBT, DBT and TPhT of 0.038, 0.022, and 0.023 µg/kg bw/day, respectively, will result. This will result in a cumulative intake for TBT + DBT + TPhT of 0.083 µg/kg bw/day.

Table 18.6 Intakes ($\times 10^{-3}$ µg/kg bw/day) of TBT, DBT and TPhT on the basis of consumption of fish and fish products and median/mean concentrations. Values rounded to three significant digits

Exposure level	Median based				Mean based			
	TBT	DBT	TPhT	Total	TBT	DBT	TPhT	Total
Population (normal consumers)	9.3	3.3	5.3	18.0	37.9	22.4	22.7	83.0
High (95th percentile) consumers	19.2	6.9	11.0	37.1	78.0	46.2	47.0	171.2

456 Persistent organic pollutants and toxic metals in foods

Assessment of the high European dietary exposure
In order to have an estimate for the dietary exposure of high risk groups, the same calculation was repeated for the group of consumers belonging to the 95th percentile figures of the Norwegian consumption distribution. Depending on whether the median or the mean concentrations in fish and seafood other than fish for the individual organotin compounds have been used for the calculations, the cumulative TBT + DBT + TPhT intake is 0.037 μg/kg bw/day in the former case, and 0.17 μg/kg bw/day in the latter case.

The above presented calculations do not include the intake for high consumers eating fish and fishery products with high OTC concentrations.

18.5.3 Further studies of human exposure to organotin compounds from foods

A number of further investigations of organotin compounds in food have been reported since the EFSA study. A recent study (Feinberg *et al.*, 2006) analyzed the contamination by OTC of fish on the French market[2]. It consisted of sampling the fish and seafood mainly consumed by the population studied, taking into account the form of purchase (fresh, frozen, canned, etc.) and provisioning (bought or self-procured). The list of sampled food was based on an analysis of the individual dietary consumptions of the respondents. The final list included 138 products, of which 95 were fish and 43 were molluscs and crustaceans. The sum of DBT, TBT, DOcT and TPhT was about 0.004 μg/g fresh weight for fish and sea food. If a daily intake of 100 g fish were assumed, the total dietary exposure from fish would be 0.4 μg OTC.

In another study, Rantakokko *et al.* (2006) determined the average intake of organotin compounds from foodstuffs in a Finnish market basket. For this purpose, 13 market baskets were collected from various supermarkets and market places, representing altogether 115 different food items. In each basket, foodstuffs were mixed in proportion to their consumption and analyzed for seven organic tin compounds (MBT, DBT, TBT, MPhT, DPhT, TPhT and TOcT). Organotin compounds were detected in only four baskets, mainly in fish. OTCs were not detected in cereals, peas and nuts, milk and milk products, fats and oils, sugar, juices or soft drinks. In contrast, OTCs were detected and quantified in fish and sea foods, potatoes, vegetables and fruits. While in potatoes, vegetables and fruits, the predominant compound was MBT, in fish and sea foods, the predominant compounds were TBT, MBT, TPhT, DBT and DPhT measured at levels up to respectively 2.53, 1.52, 1.11, 0.25 and 0.14 ng/g fresh weight. The overall results of this study indicate that dietary exposure corresponds to a limited fraction of the TDI established by EFSA.

---

[2] Section 18.5.3 is based on SCHER (2006).

It should be mentioned, however, that the methodology used in the Finnish study is relatively optimistic, in that it combines average contamination levels with average food consumption. For consumers eating fish from contaminated areas, the intake may be much higher. The same is true also for high consumers of fish.

In another monitoring project (Sternbeck *et al.*, 2006) Swedish fish samples were analyzed and higher concentrations found than those reported from Finland. The values for fish were nd–2.8 (DBT), nd–7.8 (TBT) and 2.2–12 (TPhT) ng/g fresh weight for background areas, with reporting limits of 0.3–0.4 ng/g ww for food and biota. Fish from Stockholm exceeded even these concentrations, with values up to 14 (DBT), 71 (TBT) and 171 (TPhT) ng/g fresh weight.

18.6 Human exposure to organotin compounds from food packaging material

Human exposure to organotin compounds from food packaging material is discussed in the following subsections.

18.6.1 Introduction and study of migration behavior

In the EU, 27 different organotin compounds or mixtures are authorized for use in plastic-only FCMs (Table 18.7). Specific migration limits vary strongly because they are based on tin content (by weight), not actual molecular weight. In effect, this means that permitted levels in food for most organotins are in the 10–100 ppb range. One exception is the oligomeric organotin additive dibutylthiostannic acid polymer (CAS# 26427-07-6) used in plastics and coatings with no specific migration limit.

For some of the organotin additives the fat consumption reduction factor (FRF) applies. The FRF is a special measure of EU FCM regulation to take into account that products high in fat content will be consumed in lower quantity. For such foods use of the FRF allows higher specific migration. Thus, for highly fatty foods the FRF mirrors the USA's approach to FCM regulation based on consumer exposure (Grob *et al.*, 2007).

In the USA the FDA has authorized the use of 15 different organotins as indirect food additives for various applications. Many of the organotins permitted in the EU are also legally used in the USA, but there are a few compounds only authorized in either economic region (Table 18.7).
Early attempts to study the migration of organotin compounds from PVC packaging materials lacked standardization, and thus affecting comparability with regard to extraction time, solvent and temperature (Crompton, 2007). In order to test the migration behavior of organotin compounds from packaging materials into food, conditions have thus been defined, for example, in FDA 21 CFR 178.2650, to simulate the leaching behavior (Table 18.8).

Table 18.7 Organotin compounds authorized for FCM in the EU and US (Muncke, 2011)

Grouping	CAS number	Chemical name	EU migration limit	FDA ruling
Methyltins	26636–01–1	Dimethyltin bis(isooctyl mercaptoacetate)	$SML(T)^a$ = 0.18 mg/kg (expressed as tin)	21CFR178.2010: antioxidants and/or stabilizers for polymers
	54849–38–6 57583–35–4 57583–34–3 68442–12–6	Monomethyltin tris(isooctyl mercaptoacetate) Dimethyltin bis(ethylhexyl mercaptoacetate) Monomethyltin tris(ethylhexyl mercaptoacetate) Reaction products of oleic acid, 2-mercaptoethyl ester, with dichlorodimethyltin, sodium sulphide and trichloromethyltin	$SML(T)^a$ = 0.18 mg/kg (expressed as tin) FRF applies	21CFR175.300: resinous and polymeric coatings 21CFR177.2420: polyester resins, cross-linked
Butyltins	23850–94–4	MBT tri(2-ethylhexoate)	Not authorized	21CFR177.2420: polyester resins, cross-linked
	818–08–6	Dibutyltin oxide	Not authorized	Not authorized
	22373–43–0 26427–07–6	Hydroxybutyltin oxide Dibutylthiostannoic acid polymer [=Thiobis-(butyltin sulphide), polymer]	Not authorized No SML available	
Monooctyltins	26401–86–5	Mono-n-octyltin tris(isooctyl mercaptoacetate)	$SML(T)^b$ = 1.2 mg/kg (expressed as tin)	21CFR178.2650[d]: organotin stabilizers in vinyl chloride plastics
	27107–89–7 n.a.	Mono-n-octyltin tris(2-ethylhexyl mercaptoacetate) Mono-n-octyltin tris(alkyl(C10–C16) mercaptoacetate		Not authorized
Dioctyltins	10039–33–5	Di-n-octyltin bis-(2-ethylhexyl maleate)	$SML(T)^c$ = 0.006 mg/kg (expressed as tin)	21CFR178.2650[d]: organotin stabilizers in vinyl chloride plastics
	26401–97–8 15571–58–1 33568–99–9 3648–18–8	Di-n-octyltin bis-(iso-octyl mercaptoacetate) Di-n-octyltin bis-(2-ethylhexyl mercaptoacetate) Di-n-octyltin bis-(iso-octyl maleate) Di-n-octyltin dilaurate		Not authorized

	CAS No.	Substance	Restrictions	Remarks
	15571–60–5	Di-n-octyltin dimaleate	Not authorized	21CFR178.2650[d]: organotin stabilizers in vinyl chloride plastics
	69226–44–4	Di-n-octyltin ethyleneglycol bis(mercaptoacetate)		
	15535–79–2	Di-n-octyltin mercaptoacetate		
	n.a.	Di-n-octyltin thiobenzoate 2-ethylhexyl mercaptoacetate		
	n.a.	Di-n-octyltin dimaleate, polymers (N = 2–4)		
	n.a.	Di-n-octyltin dimaleate, esterified		
	n.a.	Di-n-octyltin bis-(ethyl maleate)		
	n.a.	Di-n-octyltin bis-(n-alkyl(C10–C16) mercaptoacetate)		
	n.a.	Di-n-octyltin 1,4-butanediol bis-(mercaptoacetate)		
	n.a.	Dioctyltin maleate		
Dodecyltins	67649–65–4	Mono-n-dodecyltin tris(iso-octyl mercaptoacetate)	0.05 mg/kg food (expressed as sum of mono- and didodecyltin chloride)	21CFR178.2650: organotin stabilizers in vinyl chloride plastics
	84030–61–5	Di-n-dodecyltin bis-(iso-octyl mercaptoacetate)	0.05 mg/kg food (expressed as sum of mono- and didodecyltin chloride) FRF applies	
Further organotins	63397–60–4	Bis(2-carbobutoxyethyl)tin-bis-(iso-octyl mercaptoacetate)	SML = 18 mg/kg FRF applies	Not authorized
	63438–80–2	(2-Carbobutoxyethyl)tin-tris(iso-octyl mercaptoacetate)	SML = 30 mg/kg FRF applies	Not authorized
	83447–69–2	C10–16-alkyl mercaptoacetates reaction products with dichlorodioctylstannane and trichlorooctylstannane	Not authorized	21CFR178.2650[d]: organotin stabilizers in vinyl chloride plastics

n.a., not available. FRF applies: fat (consumption) reduction factor in the EU; an estimate that allows assesment of estimated consumer exposure rather than migration levels in fatty foods.

[a] SML: specific migration level SML(T) in this specific case means that the restriction shall not be exceeded by the sum of the migration levels of the substances mentioned as methyltins.
[b] SML(T) in this specific case means that the restriction shall not be exceeded by the sum of the migration levels of the substances mentioned as monooctyltins.
[c] SML(T) in this specific case means that the restriction shall not be exceeded by the sum of the migration levels of the substances mentioned as dioctyltins.
[d] Migration limit for all organotins listed under 21CFR178.2650 into food simulants: 0.5 mg/kg food simulant.

Table 18.8 Conditions of time, temperature and solvents for the extraction of finished plastics intended for contact with foods of Types II, V, VI-A (except malt beverages), and VI-C (according to FDA 21 CFR 178.2650)

	Food-simulating solvent	Time (hours)	Temperature (degrees Fahrenheit)
Type II	Acetic acid, 3%	48	135
Type V	Heptane	2	100
Type VI-A	Ethyl alcohol, 8%	24	120
Type VI-C	Ethyl alcohol, 50%	24	120

The types of food mentioned in the table refer to Table 18.1 in 21 CFR Section 176.170(c), which enumerates nine Food Types.[1]

21 C.F.R.E 175.300, which lists the components of coatings for metal substrates that FDA has cleared, is another regulation that contains a list of Food Types. The Food Types in Section 175.300 are almost identical to those in Section 176.170, but there are some differences. Since FDA invariably uses the Food Types in Section 176.170 for general reference rather than the Food Types listed in Section 175.300, we have, likewise, listed here the Food Types in Section 176.170.

[1] Food types are as follows:
  I. Nonacid, aqueous products; may contain salt or sugar or both (pH above 5.0).
  II. Acid, aqueous products; may contain salt or sugar or both, and including oil-in-water emulsions of low- or high-fat content.
  III. Aqueous, acid or nonacid products containing free oil or fat; may contain salt, and including water-in-oil emulsions of low- or high-fat content.
  IV. Dairy products and modifications:
    A. Water-in-oil emulsions, high- or low-fat.
    B. Oil-in-water emulsions, high- or low-fat.
  V. Low-moisture fats and oils.
  VI. Beverages:
    A. Containing up to 8% alcohol.
    B. Nonalcoholic.
    C. Containing more than 8% alcohol.
  VII. Bakery products other than those included under Types VIII or IX of this table:
    A. Moist bakery products with surface containing free fat or oil.
    B. Moist bakery products with surface containing no free fat or oil.
  VIII. Dry solids with the surface containing no free fat or oil (no end test required).
  IX. Dry solids with the surface containing free fat or oil.

18.6.2 Consumer exposure assessment

The intake of organotin via PVC food packaging is one topic addressed in SCHER (2006). According to this publication, intake of organotin via PVC food packaging was estimated in a report by Risk & Policy Analysts Ltd. (Norfolk, UK) (RPA, 2007) based on various sources. These included a study by Piringer et al. (2005); however, SCHER (2006) does not consider the values reported in the latter study to be particularly reliable. According to SCHER

(2006), RPA (RPA, 2007) did use values for fatty food which seem to be right (104% and 417% of the TDI for adults and children, respectively) though. RPA (2007) also uses a report from CSL (2005). Since this is also based on the results of the Piringer (2005) study, the values reported there must also be considered with care. The RPA report (RPA, 2007) claims that the level of OTCs in indoor air is below the detection limit (which unfortunately is not specified). Air levels are therefore roughly estimated, assuming that only 1% is going to air. In that case, the investigated material would give an adult exposure from PVC food packaging of about 0.1 µg/kg bw/day of each of DBT and TBT.

According to SCHER (2006), interestingly it is not the diffusion of the organotin compounds within the PVC matrix that determines the emission rate, but rather the removal from the surface. This is well demonstrated by the relative concentrations of MBT, DBT, TBT and TeBT in the octane, water and air extracts from the same PVC matrix. A recent paper (Xu and Little, 2006) also shows that the emission of semi-volatile compounds from polymeric materials is essentially controlled by other factors, such as the partitioning into the gas phase and the convective mass transfer coefficient. Thus the calculations of loss factors based on diffusion rates can be questioned.

18.7 Health risks and toxicity of organotin compounds

This section discusses the health risk and the toxicity of organotin compounds based on the known exposure from food and other sources.

18.7.1 Exposure of adults and children via food

Given the fact that several food items, particularly fish and shellfish, contain organotin compounds, EFSA (2004) recently assessed the possible risks connected to consumption of food containing these substances[3]. The EFSA Panel on Contaminants in the Food Chain focused on TBT, DBT and TPT, primarily found in fish and fishery products, based on data from the SCOOP study (SCOOP, 2003). The results (discussed in Section 18.4) may be summarized so that, particularly in case of high fish and shellfish consumption, TDI levels may be reached for organotin compounds. In comparison to adults, children are generally, because of their lower body weight, considered as a group at risk from food chemicals when the dietary exposure is compared with a TDI expressed in µg/kg bw. For this reason, a factor 3 is commonly used to estimate child dietary exposure when only data on adult consumption are available, i.e. the daily exposure for adults is multiplied by a factor 3. This may be partially offset by the children's lower consumption of certain food categories, such as fish, than for adults. In a recent report, based on

---

[3] Sections 18.7.1 and 18.7.2 are based on SCHER (2006).

a study performed in France by Verger (2006) who assessed the frequency and the quantity of fish consumed by adults and by children, the Scientific Committee on Health and Environmental Risks (SCHER) assumed that the consumption of fish by children corresponds to at least 75% of the consumption of adults (SCHER, 2006). In other EU Member States this percentage could be even higher. Considering both the difference in body weight and food consumption, a factor of 3 can be used to estimate the exposure of children from that of adults.

18.7.2 Effect assessment

In agreement with the CSTEE (Scientific Committee on Toxicity, Ecotoxicity and the Environment) recommendation (EC, 2003a), the RPA (2007) suggests using a group TDI for the four OTCs TBT, DBT, DOT and TPhT based on the immunotoxicity and assuming that the effect of the compounds are additive. A similar approach has been taken by the EFSA study on organotin compounds in foodstuff (EFSA, 2004). The group TDI, corresponding to 0.1 μg Sn/kg bw/day, assumes the same molecular mechanism for the four compounds and the same potency (per μg Sn), although the mechanism of these compounds has not yet been investigated systematically. There are no epidemiological studies on chronic low level exposure to OTCs available for human risk assessment, following either oral, dermal or inhalation exposure. Several experimental studies *in vitro* and *in vivo* have demonstrated various effects after multiple exposures at low dose levels, but most of the positive studies are based upon oral administration.

Inhalation studies
OTCs can be inhaled by humans either directly via the gas phase, or adsorbed on dust particles. Death has been reported in some workers accidentally exposed to OTCs at high levels. There are no quantitative studies regarding absorption of OTCs following inhalation exposure, but several case studies report adverse health effects following exposure to e.g. paint containing TBTO and carpet sprays. However, the level of exposure was not estimated. No studies regarding immunological effects in humans and animals after inhalation exposure to OTCs have been reported.

Persistent neurological changes have been observed in humans following accidental exposure for methyltin compounds, but no neurological effects have been observed in animals. No histopathological changes have been observed in the brains of mice following 6 days of exposure to TBT (ATSDR, 2005).

Dermal exposure studies
No studies regarding absorption in humans or animals after dermal exposure have been reported, but the OECD *in vitro* model (OECD 2004) has been used to compare the effect of different OTCs and species differences in uptake. The uptake of OTCs was significantly lower in human skin than in rat skin, e.g. DOcTC 0.01% and 1.5% respectively over a 24 hrs period (dose 1000 μg/cm^2).

The uptake did depend on the ligand, e.g., DBT dichloride was more easily taken up than dibutyltin-ethylhexylmercaptoacetate (DBT-EHMA), 6.58% and 0.004%, respectively. The relevance of some of these results can, however, be questioned as the dose used in the tests is orders of magnitude higher (in the mg/cm^2 range), while the worst exposure scenarios known to the RPA (RPA, 2007) are in the ng/cm^2 range. The OECD guideline specifically states that the tested dose shall span the realistic range of potential human exposures. Trialkyltin compounds were well absorbed in contact with the rat skin, whereas triphenyltin acetate did not penetrate. Using thus the total amount of OTCs and a fixed uptake rate may give erroneously high estimates of exposure. In turn, there is information from industry that the EHMA ligands are effectively substituted to chlorides after emission, which suggests that the values for the latter have to be used.

Except for direct dermal effects, no adverse health effect was observed in either humans or experimental animals following dermal exposure to OTCs. These compounds are skin irritants in humans, e.g., TPhT produced irritant contact folliculitis and in workers using paint containing TBTO, which is also a severe irritant to the skin in rabbits, whereas TBT and TPhT only produced minimal skin irritation. TBTO induced contact sensitization in mice exposed for the test material for 3 days.

Oral exposure studies
While limited information on OTC toxicity following inhalation and dermal exposure is available, more data exist for oral exposure. OTCs are readily hydrolyzed in human gastric juice and may be converted to the corresponding chlorides, which will more easily be absorbed.

No quantitative data on the absorption of OTCs in humans have been published. In experimental animals, absorption from the intestinal tract of tin compounds with short alkyl chains depends on the chemical compound and shows considerable compound differences, with a higher absorption and with a higher degree of alkyl substitution, e.g. TBT>DBT>MBT. In contrast, TBTO was absorbed incompletely and slowly. Toxicity related to oral exposure has been presented in an EFSA opinion (EFSA, 2004).

The no observed adverse effect level (NOAEL) for OTCs is based upon the immunotoxicity of OTCs. For TBTO, a NOAEL of 0.025 mg/kg bw/day (0.01 mg Sn/kg bw/day) was established based upon resistance to *T. spiralis* and a similar NOAEL was established for TBT chloride (0.00869 mg Sn/kg bw/day). A NOAEL for TPhT was estimated to be equivalent to 0.75 mg/kg bw (0.22 mg Sn/kg bw/day), while for DOcT no NOAEL could be established, but the LOAEL was equivalent to 2.5 mg /kg bw/day (0.68 mg Sn/kg bw/day).

18.7.3 Toxicity and adverse effects of organotin compounds
The most comprehensive summary of organotin compounds' toxicity and biological effects to date is still that presented in the report of the EFSA

Scientific Panel on Contaminants in the Food Chain produced on request of the European Commission (2004). The following effects were discussed for the four organotin compounds TBT, DBT, TPhT, octyltin compounds: (i) genotoxicity, (ii) reproductive and developmental toxicity, and (iii) neurotoxicity.

The report also summarizes and discusses further mechanistic aspects of the immunotoxicity of OTCs (cytotoxicity, cytoskeleton and calcium homeostasis, effects on signaling pathways and proliferation of immunocompetent cells, mitochondrial effects, apoptosis and thymus atrophy). Furthermore, observations in humans are reported. However, these refer to exposure by inhalation, dermal exposure and oral exposure, which all are either occupational or accidental, and do not relate to intake by food or consumer products.

Other reviews that discuss different aspects of toxicity are available for TBT compounds (US EPA, 1997; WHO-IPCS, 1990, 1999b), and TPhT compounds (JMPR, 1999; US EPA, 1999; WHO-IPCS, 1999a).

Mechanism of immunotoxicity
Organotin compounds such as TPhT, TBT, DBT and DOcT have been shown to produce thymus atrophy, which involves a decrease in the number of cortical thymocytes. When exposed for a longer time, T-cell mediated immune responses are suppressed due to both suppression of proliferation of immature thymocytes and apoptosis of mature thymocytes.

Genotoxicity and reprotoxicity
No carcinogenic effect was observed following dermal exposure, but hyperplastic skin changes were observed in mouse skin following 6 weeks of exposure to TBT. More recent studies show that peripubertal exposure to TPhT influences female sexual development in rats (Grote *et al.*, 2006). This endocrine-disrupting effect is mediated indirectly by influencing the expression and activities of the enzyme systems 17-β-HSD1 (17-β-hydroxysteroid dehydrogenase type 1) and 11-β-HSD2 (11-β-hydroxysteroid dehydrogenase type 2), thus influencing the glucocorticoid concentration.

A completely new type of toxic effect was described in 2006 by Grün *et al.* as low doses of OTCs were shown to induce adipogenesis in mice. Based on evaluating a comprehensive set of receptor binding studies, transfection assays, cell culture studies and *in vivo* experiments, this report showed that TBT activates genes promoting adipogenesis, induces differentiation of adipocytes and increases the mass of adipose tissue *in vivo*. The adipogenic activity was observed in offspring of mice exposed to TBT during gestation at maternal dose of 0.5 mg/kg bw/day (and to a lesser extent at 0.05 mg/kg bw/day). It seems that this effect is associated with high affinity and specific binding of TBT to the human retinoid X receptor α (RXRα) and to the human peroxisome proliferator activated receptor γ (PPARγ) at nanomolar concentrations. From comparative studies it seemed that tri-substituted compounds (TBT and TPhT) are most potent among the butyltin and phenyltin compounds, TeBT and DBT moderately potent, whereas MBT lacks the activity.

These findings are important for the risk assessment of OTCs. First, they confirm the validity of the currently used TDI values, which were observed at dose levels similar or even slightly lower than those causing immunotoxicity after chronic administration, representing the most sensitive set of endpoints currently used for organotin risk assessment. There is, therefore, no need to adjust the current TDI value. Second, this paper indicates that there is at least another type of hitherto uncharacterized toxic effect after exposure to low doses of organotin compounds in addition to immunotoxicity. Third, this altogether provides scientific evidence for the use of endpoint specific relative potency factors for different organotin compounds (SCHER, 2006).

Risk characterization

The findings reported above have implications on the risk characterization. As most of the comments from literature indicate an underestimate of the exposure to OTCs, the conclusion must be that the risk estimates may not represent realistic worst case scenarios.

For TBT oxide, a NOAEL for immunotoxicity of 0.025 mg/kg bw/day was identified from chronic feeding studies. Since TBT, DBT, TPhT and DOcT exhibit a similar mode of action and potency in their immunotoxic effects, the Scientific Panel on Contaminants in the Food Chain considered it reasonable to establish a group TDI for these OTC. In the absence of specific studies on combined effects it seemed justified to consider the immunotoxic effects of these compounds as additive. By applying a safety factor of 100, a group TDI of 0.25 µg/kg bw for TBT, DBT, TPhT and DOcT compounds was established (based on TBT oxide molecular mass, this group TDI is 0.1 µg/kg bw when expressed as Sn content or 0.27 µg/kg bw when expressed as TBT chloride (EFSA, 2004).

From the above discussed TDI levels it is possible to determine the maximum residue levels that should be allowed for OTCs in fish and fish products, considering the geographically differentiated average consumption. Based on the TDI of 0.25 µg/kg bw/day for TBT, the maximum TDI for TBT is 15 µg/day for a person with a body weight of 60 kg (Kannan and Falandysz, 1997). The tolerable average residue level (TARL) is defined as the level in seafood that is tolerable for the average consumer with a weight of 60 kg. This indicator value can be calculated according to

$$TARL = \frac{TDI \times 60\,kg\,bw}{average\,daily\,seafood\,consumption}$$

Belfroid *et al.* (2000) have on this basis calculated TARL of TBT in seafood products (Table 18.9). The data for fish consumption in different countries are taken from FAO's food balance sheets for 1996 (FAO, 2009).

Table 18.9 Average seafood consumption per country and calculated TARL of TBT in seafood products (from Belfroid et al., 2000)

Countries	Per capita supply		TARL/day in ng/g seafood product for an average person of 60 kg
	in kg/year	in gram/day	Countries
Australia	19.2	52.6	285
Bangladesh	9.4	25.8	582
Canada	22.7	62.2	241
France	27.9	76.4	196
Germany	15.6	42.7	351
Hong Kong	59.6	163.3	92
India	3.8	10.4	1441
Indonesia	15.2	41.6	360
Italy	23.1	63.3	237
Japan	71	194.5	77
Korea Republic	50.3	137.8	109
Malaysia	53.5	146.6	102
Netherlands	14.6	40	375
Papua New Guinea	26.2	71.8	209
Poland	16.5	45.2	332
Portugal	58.7	160.8	93
Singapore[a]	53.5	146.6	102
Solomon Islands	20	54.8	274
Sweden	30.8	84.4	178
Taiwan[a]	59.6	163.3	92
Thailand	25.9	71	211
UK	20.1	55.1	272
USA	21.6	59.2	253
Vietnam	12.6	34.5	435

[a] Data were unavailable for Singapore and Taiwan; therefore, data for Malaysia and Hong Kong, respectively, were used, which resemble these countries in terms of culture and proximity to the sea.

The most interesting conclusion from the data presented in Table 18.9 is that the TARL may, assuming the same TDI values, vary by a factor of as much as 20 between populations with the highest (Japan) and the lowest (India) consumption of fish and fishery products assessed here.

18.8 Conclusions and future trends

It was remarked, in a recent report on the concentrations of organotin compounds in Portuguese mussels caught along the coast of Portugal after the EU-wide ban of organotin antifouling paints in 2003 (Sousa et al., 2009) that OTC concentration in mussel tissue is strongly influenced by several factors, such as sampling location and distance from hotspots, water temperature, salinity, oxygen content and biological activity. Comparisons with other studies are thus often difficult, and should be critically considered. It is also important to consider the effect of the introduction of the EU Regulation (EC, 2004), which led to significant differences in TBT levels in surveys performed before and after 2003. Typically, a marked decrease in TBT levels is observed. This observation is in line with other studies conducted at the Atlantic coast (Díez and Bayona, 2009), or the Adriatic Sea, which also reported decreasing levels of TBT in mussels collected between 2000 and 2006 (Nemanič et al., 2009). However, despite the decreasing tendency, fresh inputs were still evident (Nemanič et al., 2009). This observation is also in line with similar results from other studies from different locations around the world that also described the continuing TBT inputs into the environment even after the use of this compound had been restricted or banned (Murai et al., 2005; Inoue et al., 2006; Ščančar et al., 2007; Sheikh et al., 2007; Smith et al., 2006).

The levels of phenyltins in mussels collected around Europe are typically below the detection limits of the applied analytical procedures and/or at much lower concentrations than butyltins (Bortoli et al., 2003; Díez et al., 2005; Nemanič et al., 2002, 2009; Ščančar et al., 2007; Zanon et al., 2009). In Europe, the percentage of TPhT used in antifouling paints formulations is relatively low when compared to TBT, which leads to lower environmental levels. It is estimated that contribution of PhTs to total organotins is in the range of 1%. Data on octyltin levels in molluscs are scarce, and the available literature mostly reports concentrations below the detection limit (Sousa et al., 2009). This leads us to assume that the OTC compound profiles are representative, and will continue to be so over the next years.

Related to the toxicological assessment of OTCs, the conclusions and recommendations from the Scientific Panel on Contaminants in the Food Chain (EFSA, 2004) can be summarized. Based on a group TDI of 0.25 µg/kg for TBT, DBT, TPhT and DOcT, the panel noted that intake for the general population is below the TDI. However, in the few cases where seafood contamination with OTC is high, the TDI might be exceeded, for example in the case of consumption of contaminated fish, mussels or other marine animals from the vicinity of harbors and heavily used shipping routes.

In addition to the OTC dietary exposure from fishery products, consumers are also exposed to OTCs from other sources, e.g. pesticides, additives used in plastics, other FCMs and consumer products. Therefore, an integrated risk assessment of OTCs is needed, taking into account human exposure from all possible sources.

Epidemiological studies incorporating biomarkers of exposure and health effects are recommended, particularly in highly exposed populations.

18.9 References

ABALOS M, BAYONA J-M, COMPAÑÓ R, GRANADOS M, LEAL C and PRAT M-D (1997), Analytical procedures for the determination of organotin compounds in sediment and biota: A critical review. *J. Chromatogr. A.* **788**, 1–49.

AGUERRE S, LESPES G, DESAUZIERS V and POTIN-GAUTIER M (2001), Speciation of organotins in environmental samples by SPME-GC: Comparison of four specific detectors: FPD, PFPD, MIP-AES and ICP-MS. *J. Anal. At. Spectrom.* **16**, 263–269.

ALZIEU C, SANJUAN J, DELTREIL JP and BOREL M (1986), Tin contamination in Arcachon Bay: Effects on oyster shell anomalies. *Mar. Pollut. Bull.* **17**, 494–498.

ALZIEU C (2000), Impact of tributyltin on marine invertebrates. *Ecotoxicology.* **9**, 71–76.

ATSDR, Agency for Toxic Substances and Disease Registry. *Toxicological profile for tin and tin compounds.* US Department of Health and Human Services (2005), Available online: http://www.atsdr.cdc.gov/toxprofiles/tp55.pd (Accessed: July 2012).

BACALOGLU R, FISCH MH, KAUFHOLD J and SANDER HJ (2001), *PVC Stabilizers* In ZWEIFEL H (ed.), *Plastics additives handbook*, 5th ed., Hanser, Munich pp. 407–421.

BALTUSSEN E, SANDRA P, DAVID F and CRAMERS C (1999), Stir bar sorptive extraction (SBSE), a novel extraction technique for aqueous samples: Theory and principles. *J. Microcol. Sep.* **11**, 737–747.

BATT JM (2001), The world of organotin chemicals: Applications, substitutes and the environment. Organotin Environmental Program Association (ORTEPA). Available at: http://ortepa.org/WorldofOrganotinChemicals.pdf (Accessed: July 2012).

BELFROID AC, PURPERHART M and ARIESE F (2000), Organotin levels in seafood, *Marine Poll. Bull.* **40**, 226–232.

BENNETT RF (1996), *Industrial Manufacture and Applications of Trubutyltin Compounds*. In DE MORA SJ (ed.), *Tributyltin: A case study of an environmental contaminant*, Cambridge Environmental Chemistry Series, Cambridge University Press, Cambridge, pp. 21–61.

BORTOLI A, TRONCON A, DARIOL S, PELLIZZATO F and PAVONI B (2003), Butyltins and phenyltins in biota and sediments from the Lagoon of Venice. *Oceanologia.* **45**, 7–23.

BRAY S (2006), Tributyltin pollution on a global scale. An overview of relevant and recent research: impacts and issues. Report to Dr. Simon Walmsley WWF UK. Godalming, Surrey. Contract No: FND053998. Available online: http://assets.wwf.no/downloads/tbt_global_review_wwf_uk_oct_2006.pdf (Accessed: July 2012).

BRYAN GW, GIBBS PE, HUMMERSTONE LG and BURT GR (1986), The decline of the gastropod *Nucella Lapillus* around south-west England: Evidence for the effect of tributyltin from antifouling paints. *J Mar Biol Ass UK.* **66**, 611–640.

CARDWELL RD, KEITHLY JC and SIMMONDS J (1999), Tributyltin in US market-bought seafood and assessment of human health risks. *Human Ecol. Risk Assessm.* **5**, 317–335.

CEULEMANS M, WITTE C, ŁOBIŃSKI R and ADAMS FC (1994), Simplified sample preparation for GC speciation analysis of organotin in marine biomaterials, *Appl. Organomet. Chem.* **8**, 451–461.

CHAMP MA (2000), A review of organotin regulatory strategies, pending actions, related costs and benefits. *Sci. Total Environ.* **258**, 21–71.

CHAMP MA and SELIGMAN PF (1996), *An Introduction to Organotin Compounds and Their Use in Antifouling Coatings*, in CHAMP MA and SELIGMAN PF (eds.), *Organotin: Environmental fate and effects*, Chapman & Hall, London, pp. 2–26.

CHAU YK, YANG F and BROWN M (1997), Evaluation of derivatization techniques for the analysis of organotin compounds in biological tissue. *Anal. Chim. Acta.* **338**, 51–55.

CROMPTON TR (2007), *Additive Migration from Plastics Into Foods: A Guide for Analytical Chemists.* iSmithers Rapra Publishing, Shawbury, UK, p. 302.
CRNOJA M, HABERHAUER-TROYER C, ROSENBERG E and GRASSERBAUER M (2001), Determination of Sn- and Pb-organic compounds by solid-phase microextraction-gas chromatography-atomic emission detection (SPME-GC-AED) after in situ propylation with sodium tetrapropylborate. *J. Anal. At. Spectrom.* **16**, 1160–1166.
CSL (2005), *Tin Stabilised Rigid PVC Films Exposure Assessment.* UK Central Science Laboratory, Norwich, UK.
DAVIES AG (2004), *Organotin Chemistry*, 2nd ed., Wiley-VCH, Weinheim.
DÍEZ S, LACORTE S, VIANA P, BARCELÓ D and BAYONA JM (2005), Survey of organotin compounds in rivers and coastal environments in Portugal 1999–2000. *Environ. Poll.* **136**, 525–536.
DÍEZ S and BAYONA JM (2009), Butyltin occurrence and risk assessment in the sediments of the Iberian Peninsula. *J. Environ. Managem.* **90** (1), S25-S30.
DONARD OFX, RAPSOMANIKIS S and WEBER JH (1986), Speciation of inorganic tin and alkyltin compounds by atomic absorption spectrometry using electrothermal quartz furnace after hydride generation. *Anal. Chem.* **58**, 772–777.
DONARD OFX and PINEL R (1989), in HARRISON RM and RAPSOMANIKIS S (eds.), *Tin and Germanium* in *Environmental Analysis Using Chromatography Interfaced with Atomic Spectrometry*, Harrison RM and Rapsomanikis S (eds.), Ellis Horwood, Chichester, pp. 189–257.
EC (1976), Council Directive of 27 July 1976 on the approximation of the laws, regulations and administrative provisions of the Member States relating to restrictions on the marketing and use of certain dangerous substances and preparations (76/769/EEC) *Official J. European Union* **L 262**, of 27.09.1976, pp. 201–203.
EC (EUROPEAN COMMISSION) (2003a), Scientific committee on toxicity, ecotoxicity and the environment (CSTEE). Opinion on assessment of the risks to health and the environment posed by the use of organostannic compounds (excluding use as a biocide in antifouling paints) and a description of the economic profile of the industry. Adopted at 38th Plenary meeting of 12 June 2003.
EC (EUROPEAN COMMISSION) (2003b), *Report on Tasks for Scientific Cooperation (SCOOP), task 3.2.13. Assessment of the dietary exposure to organotin compounds of the population of the EU member states.* European Commission, Directorate-General Health and Consumer Protection, Reports on tasks for scientific co-operation, October 2003.
EC (2004), Regulation (EC) No 782/2003 of the European Parliament and of the Council of 14 April 2003 on the prohibition of organotin compounds on ships. *Official J. European Union* **L 115/11** of 09.05.2003.
EC (2009), Commission Decision of 28 May 2009, amending Council Directive 76/769/EEC as regards restrictions on the marketing and use of organostannic compounds for the purpose of adapting its Annex I to technical progress. *Official J. European Union* **L 138/11** of 04.06.2009.
EFSA (2004), European Food Safety Authority, Opinion of the scientific panel on contaminants in the food chain on a request from the Commission to assess the health risks to consumers associated with exposure to organotins in foodstuffs (Question N° EFSA-Q-2003–110) Adopted on 22 September 2004. *The EFSA Journal.* **102**, 1–119.
EVANS SM, EVANS PM and LEKSONO T (1996), Widespread recovery of dogwhelks, *Nucella lapillus* (L.), from tributyltin contamination in the North Sea and Clyde Sea *Mar. Poll. Bull.* **32**, 263–369.
FAO (2009), FAO (Food and Agriculture Organization of the United Nations) Yearbook 2009. Fishery and Aquaculture Statistics. Food Balance Sheets. Available online: ftp://ftp.fao.org/FI/CDrom/CD_yearbook_2009/navigation/index_content_food_balance_e.htm (Accessed: July 2012).

FEINBERG M, BERTAIL P, TRESSOU J and VERGER P (2006), *Analyse Des Risques Alimentaires*. Sciences et techniques alimentaires; Collection Tec & Doc Lavoisier, Cachan (France).

FØLSVIK N, BREVIK EM and BERGE JA (2002), Organotin compounds in a Norwegian fjord. A comparison of concentration levels in semipermeable membrane devices (SPMDs), blue mussels (*Mytilus edulis*) and water samples, *J. Environ. Monit.* **4**, 280–283.

FRANKLAND E (1854), On a new series of organic bodies containing metals, *Quart. J. Chem. Soc.* London, **6**, 57–71.

FRISTACHI A, XU Y, RICE G, IMPELLITTERI CA, CARLSON-LYNCH H and LITTLE JC (2009), Using probabilistic modeling to evaluate human exposure to organotin in drinking water transported by polyvinyl chloride pipe. *Risk Anal.* **29**, 1615–1628.

GONZÁLEZ-TOLEDO E, COMPANÓ R, PRAT MD and GRANADOS M (2002), Determination of triorganotin species in water samples by liquid chromatography–electrospray-mass spectrometry. *J. Chromatogr. A.* **946**, 1–8.

GONZÁLEZ-TOLEDO E, COMPANÓ R, GRANADOS M and PRAT MD (2003), Detection techniques in speciation analysis of organotin compounds by liquid chromatography. *Trends Anal. Chem.* **22**, 26–33.

GROB K, PFENNINGER S, POHL W, LASO M, IMHOF D and RIEGER K (2007), European legal limits for migration from food packaging materials: 1. Food should prevail over simulants; 2. More realistic conversion from concentrations to limits per surface area. PVC cling films in contact with cheese as an example. *Food Control.* **18**, 201–210.

GROTE K, ANDRADE AJM, GRANDE SW, KURIYAMA SN, TALSNESS CE, APPEL K and CHAHOUD I. (2006), Effects of peripubertal exposure to triphenyltin on female sexual development of the rat. *Toxicology* **222**, 17–24.

GRÜN F, WATANABE H, ZAMANIAN Z, MAEDA L, ARIMA K, CUBACHA R, GARDINER DM, KANNO J, IGUCHI T and BLUMBERG B (2006), Endocrine disrupting organotin compounds are potent inducers of adipogenesis in vertebrates. *Mol. Endocrinol.* **20**, 2141–2155.

HOCH M (2001), Organotin compounds in the environment – an overview. *Appl. Geochem.* **16**, 719–743.

IMO (2008), International Maritime Organization Briefing 40: Harmful ships' paint systems outlawed as international convention enters into force, 16 September 2008. Available online: http://www.imo.org/blast/mainframe.asp?topic_id=1709&doc_id=10131 (Accessed: July 2012).

INOUE H, ABE SI, OSHIMA Y, KAI N and HONJO T (2006), Tributyltin contamination of bivalves in coastal areas around Northern Kyushu, Japan. *Environ. Toxicol.* **21**, 244–249.

JENNINGS TC and STARNES WH JR (1996), *PVC Stabilizers and Lubricants* in SUMMERS JE, and DANIELS CA (eds.), WILKES CE, *PVC Handbook*, Hanser, Munich, pp. 108–116.

JMPR (2009), Priority list of chemical scheduled for evaluation and re-evaluation by JMPR (FAO/WHO Joint Meeting on Pesticides Residues). Available online: http://www.who.int/foodsafety/chem/jmpr/jmpr_schedule_2012to2016.pdf (Accessed: July 2012).

JONES-LEPP TL, VARNER KE and HILTON BA (2001), Speciation and detection of organotins from PVC pipe by micro-liquid chromatography-electrospray-ion trap mass spectrometry. *Appl. Organomet. Chem.* **15**, 933–938.

JUNK GA and RICHARD JJ (1987), Solid phase extraction, GC separation and EC detection of tributyltin chloride, *Chemosphere.* **16**, 61–68.

KANNAN K, CORSOLINI S, FOCARDI S, TANABE S and TATSUKAWA R (1996), Accumulation pattern of butyltin compounds in dolphin, tuna, and shark collected from Italian coastal waters. *Arch. Environ. Contam. Toxicol.* **31**, 19–23.

KANNAN K and FALANDYSZ J (1997), Butyltin residues in sediment, fish, fish-eating birds, harbour porpoise and human tissues from the Polish coast of the Baltic Sea. *Marine Poll. Bull.* **34**, 203–207.

KEITHLY JC, CARDWELL RD and HENDERSON DG (1999), Tributyltin in seafood from Asia, Australia, Europe, North America: assessment of human health risks. *Human Ecol. Risk Assessm.* **5**, 337–354.

LEE M-R and CHEN C-Y (2011), *Organotin compound Analysis*. In NOLLET L (ed.), *Analysis of Endocrine Disrupting Compounds in Food*, (Blackwell, London), pp. 269–288.

LOBINSKI R and SZPUNAR J (eds.) (2003), *Hyphenated Techniques in Speciation Analysis*. Royal Society of Chemistry, Cambridge.

LÖWIG C (1852), *Liebigs Ann. Chem.*, **84**, 308.

MATTHIESSEN P and GIBBS PE (1998), Critical appraisal of the evidence for tributyltin-mediated endocrine disruption in molluscs. *Environ. Toxicol. Chem.* **17**, 37–43.

MIDWEST RESEARCH INSTITUTE (1976), *Manufacture and Use of Selected Alkyltin Compounds: Task II*, Final report. Prepared for: US Environment Protection Agency, Office of Toxic Substances, Washington, DC.

MORABITO R, MASSANISSO P and QUEVAUVILLER P (2000), Derivatization methods for the determination of organotin compounds in environmental samples. *Trends Anal. Chem.* **19**, 113–119.

MUNCKE J (2011), Endocrine disrupting chemicals and other substances of concern in food contact materials: An updated review of exposure, effect and risk assessment. *J. Steroid Biochem. Mol. Biol.* **127**, 118–127.

MURAI R, TAKAHASHI S, TANABE S and TAKEUCHI I (2005), Status of butyltin pollution along the coasts of western Japan in 2001, 11 years after partial restrictions on the usage of tributyltin. *Marine Poll. Bull.* **51**, 940–949.

NAVSEA (US Naval Sea Systems Command) (1984), *Environmental assessment of fleetwide use of organotin antifouling paint*. NAVSEA, Washington, DC, pp. 128.

NAVSEA (US Naval Sea Systems Command) (1986), *Organotin antifouling paint: US Navy's needs, benefit and ecological research. A report to Congress*. NAVSEA, Washington, DC, 38 pp. + appendix.

NEMANIČ TM, LESKOVŠEK H, HORVAT M, VRIŠER B and BOLJE A (2002), Organotin compounds in the marine environment of the Bay of Piran, Northern Adriatic Sea. *J. Environ. Monitor.* **4**, 426–430.

NEMANIČ MT, MILAČIČ R and ŠČANČAR J (2009), A survey of organotin compounds in the Northern Adriatic Sea. *Water Air Soil Poll.* **196**, 211–224.

NÓBREGA JA, SANTOS MC, DE SOUSA RA, CADORE S, BARNES RM and TATRO M (2006), Sample preparation in alkaline media, *Spectrochim. Acta. B.* **61**, 465–495.

OSPAR (2009), *Nucella Lapillus* (Dogwhelk). OSPAR Background Document – First Version. OSPAR Commission, London. Available online: http://www.ospar.org/html_documents/ospar/html/p00408_post%20bdc_nl%20dog%20whelk.pdf (Accessed: July 2012).

PEREIRO IR, SCHMITT VO, SZPUNAR J, DONARD OFX and ŁOBIŃSKI R (1996), Speciation analysis for organotin compounds in biomaterials after integrated dissolution, extraction, and derivatization in a focused microwave field, *Anal. Chem.* **68**, 4135–4140.

PIRINGER O and STEINER K (2005), *Migration Study of Organotin Compounds from Octyltin Stabilized PVC*. FABES Forschungs-GMBH fur Analytik und Bewertung von Stoffubergängen, Munich.

RANTAKOKKO P, KUNINGAS T, SAASTAMOINEN K and VARTIAINEN T (2006), Dietary intake of organotin compounds in Finland: A market-basket study. *Food Add. Contam.* **23**, 749–756.

RAPSOMANIKIS S (1994), Derivatization by ethylation with sodium tetraethylborate for the speciation of metals and organometallics in environmental samples. A review. *Analyst.* **119**, 1429–1439.

RICHARDSON BA (1993), *Wood Preservation*, 2nd ed. E & FN Spon, London, pp. 137–139.

ROSENBERG E (2005), Chapter 2.20 Speciation of tin, in CORNELIS R, CARUSO JA, CREWS H and HEUMANN KG (eds.), *Handbook of Elemental Speciation, II: Species in the Environment, Food, Medicine and Occupational Health*, Wiley & Sons, Chichester, pp. 422–463.

ROSENBERG E, KMETOV V and GRASSERBAUER M (2000), Investigating the potential of high-performance liquid chromatography with atmospheric pressure chemical ionization-mass spectrometry as an alternative method for the speciation analysis of organotin compounds. *Fresenius J. Anal. Chem.* **366**, 400–407.

RPA (2005), *Risk Assessment Studies on Targeted Consumer Applications of Certain Organotin Compounds, Final Report prepared for DG Enterprise and Industry by Risk & Policy Analysts Ltd.*, Norfolk, UK (RPA), 16 September 2005.

RPA (2007), *Impact Assessment of Potential Restrictions on the Marketing and Use of Certain Organotin Compounds. Final Report, prepared for European Commission, Directorate-General Enterprise and Industry*. Risk & Policy Analysts Ltd., Norfolk, UK.

ROWLAND GP JR and REID RJ (1949), *Stabilization of dichlorobutadiene resins*. US. Patent 2,445,739; Application date: Aug 16, 1947; after: *Chem. Abstr.* **43**, 440.

RUIZ JM, BACHELET G, CAUMETTE P and DONARD OF (1996), Three decades of tributyltin in the coastal environment with emphasis on Arcachon Bay, France. *Environ Pollut.* **93**, 195–203.

SADIKI A-I, WILLIAMS DT, CARRIER R and THOMAS B (1996), Pilot study on the contamination of drinking water by organotin compounds from PVC materials. *Chemosphere.* **32**, 2389–2398.

SADIKI A-I and WILLIAMS DT (1999), A study on organotin levels in Canadian drinking water distributed through PVC pipes. *Chemosphere.* **38**, 1541–1548.

ŠČANČAR J, ZULIANI T, TURK T and MILAČIČ R (2007), Organotin compounds and selected metals in the marine environment of Northern Adriatic Sea. *Environ. Monit. Assess.* **127**, 271–282.

SCHER (2006), Scientific Committee on Health and Environmental Risks, *Revised assessment of the risks to health and the environment associated with the use of the four organotin compounds TBT, DBT, DOT and TPT*. Opinion adopted by the SCHER during the 14th plenary of 30 November 2006.

SGS (2009), Safeguards SGS Consumer Testing Services. Hardlines, Softlines, Electrical & Electronic No. 113/09 June 2009, 2 pp. Available online: http://newsletter.sgs.com/eNewsletterPro/uploadedimages/000006/SGS-Safeguards-11309-EU-ban-on-Organotin-Compounds-EN-09.pdf [Accessed: July 2012].

SHEIKH MA, TSUHA K, WANG X, SAWANO K, IMO ST and OOMORI T (2007), Spatial and seasonal behaviour of organotin compounds in protected subtropical estuarine ecosystems in Okinawa, Japan. *Intern. J. Environ. Anal. Chem.* **87**, 847–861.

SIU KWM, GARDNER GJ and BERMAN SS (1988), Atmospheric pressure chemical ionization and ion-spray mass spectrometry of some organotin species, *Rapid Comm. Mass Spectrom.* **2**, 201–204.

SMITH AJ, THAIN JE and BARRY J (2006), Exploring the use of caged *Nucella lapillus* to monitor changes to TBT hotspot areas: a trial in the River Tyne estuary (UK). *Marine Environ. Res.* **62**, 149–163.

SOUSA A, IKEMOTO T, TAKAHASHI S, BARROSO C and TANABE S (2009), Distribution of synthetic organotins and total tin levels in *Mytilus galloprovincialis* along the Portuguese coast. *Mar. Poll. Bull.* **58**, 1130–1136.

STÄB JA, COFINO WP, VAN HATTUM B and BRINKMAN UATH (1994), Assessment of transport routes of triphenyltin used in potato culture in the Netherlands. *Anal. Chim. Acta.* **286**, 335–341.

STERNBECK J, FÄLDT J and ÖSTERÅS AH (2006), Screening of organotin compounds in the Swedish environment. Available online: http://www3.ivl.se/miljo/projekt/dvss/pdf/WSP_TBT.pdf [Accessed: July 2012].
SZPUNAR-ŁOBIŃSKA J, CEULEMANS M, DIRKX W, WITTE C, ŁOBIŃSKI R and ADAMS FC (1994), Interferences in ultratrace speciation of organolead and organotin by gas chromatography with atomic spectrometric detection. *Mikrochim. Acta.* **113**, 287–298.
SZPUNAR J, SCHMITT VO, ŁOBIŃSKI R and MONOD J-L (1996), Rapid speciation of butyltin compounds in sediments and biomaterials by capillary gas chromatography-microwave-induced plasma atomic emission spectrometry after microwave-assisted leaching/digestion. *J. Anal. At. Spectrom.* **11**, 193–199.
TEN HALLERS-TJABBES CC, KEMP JF and BOON JP (1994), Imposex in Whelks (*Buccinum undatum*) from the open North Sea: Relation to shipping traffic intensities. *Marine Poll. Bull.* **28**, 311–313.
TEN HALLERS-TJABBES CC, WEGENER JW, VAN HATTUM BA, KEMP JF, TEN HALLERS E, REITSEMAE TJ and BOON JP (2003), Imposex and organotin concentrations in *Buccinum undatum* and *Neptunea antiqua* from the North Sea: relationship to shipping density and hydrographical conditions. *Mar Environ Res.* **55**, 203–33.
THAIN JE and WALDOCK MJ (1986), The impact of tributyltin (TBT) antifouling paints on molluscan fisheries. *Wat. Sci. Tech.* **18**, 193–202.
THOMAIDIS NS (2007) *Dietary exposure to organotin compounds in Greece from seafood consumption.* Presentation at the AOAC Europe Workshop 2006, 6–7 November 2006, Larnaca.
US EPA (United States Environmental Protection Agency) (1997), *Toxicological Review: Tributyltin Oxide.* Integrated Risk Information System (IRIS). U.S. Environmental Protection Agency, Washington DC.
US EPA (United States Environmental Protection Agency) (1999), *Triphenyltin Hydroxide.* Report EPA 738-R-99-010. U.S. Environmental Protection Agency, Washington DC.
VERCAUTEREN J, PÉRÈS C, DEVOS C, SANDRA P, VANHAECKE and MOENS L (2001), Stir bar sorptive extraction for the determination of ppq-level traces of organotin compounds in environmental samples with thermal desorption-capillary gas chromatography–ICP mass spectrometry. *Anal. Chem.* **73**, 1509–1514.
VERGER P (2006), Consumer Risk Advisory Inquiry Survey. Personal communication to SCHER.
WALDOCK MJ and THAIN JE (1983), Shell thickening in *Crassostrea gigas*: Organotin antifouling or sediment induced? *Marine Poll. Bull.* **14**, 411–415.
WALDOCK MJ, THAIN JE and WAITE ME (1987), The distribution and potential toxic effects of TBT in U.K. estuaries during 1986. *Appl. Organomet. Chem.* **1**, 287–301.
WEINBERG EL and JOHNSON EW (1954), *Tin resin stabilizers.* US. Patent 2,648,650; Application date: Jun 21, 1951; after: *Chem. Abstracts.* **48**, 10056.
WHITE S, CATTERICK T, FAIRMAN B and WEBB K (1998), Speciation of organotin compounds using liquid chromatography–atmospheric pressure ionisation mass spectrometry and liquid chromatography–inductively coupled plasma mass spectrometry as complementary techniques. *J. Chromatogr. A.* **794**, 211–218.
WHO-IPCS (1990), *Enviromental Health Criteria 116: Tributyltin compounds.* World Health Organization, Geneva.
WHO-IPCS (1999a), *Concise International Chemical Assessment 13: Triphenyltin compounds.* World Health Organization, Geneva.
WHO-IPCS (1999b), *Concise International Chemical Assessment 14: Tributyltin oxide.* World Health Organization, Geneva.
WILKES CE, SUMMERS JE and DANIELS CA (eds.) (1996), *PVC Handbook*, Hanser, Munich, pp. 108–116.

WILLEMSEN F, WILLEMSEN, WEGENER J-W, MORABITO R and PANNIER F (2004), *Sources, consumer exposure and risks of organotin contamination in seafood Final report of the European Commission Research Project "OT-SAFE" (QLK1–2001–01437).* Institute for Environmental Studies (IVM), Vrije Universiteit Amsterdam The Netherlands, pp. 143.

XU Y and LITTLE JC (2006), Predicting emissions of SVOCs from polymeric materials and their interaction with airborne particles. *Environ. Sci. Technol.* **40**, 456–61.

YNGVE V (1940), *Stabilized vinyl resins.* US. patent 2,219,463; Application date: Dec 31 1936; quoted after *Chem. Abstracts.* **36**, 1145 (1941).

ZANON F, RADO N, CENTANNI E, ZHAROVA N and PAVONI B (2009), Time trend of butyl and phenyl-tin contamination in organisms of the Lagoon of Venice (1999–2003). *Environ. Monit. Assess.* **152**, 35–45.

18.10 Appendix: abbreviations

AED	atomic emission detector
APCI	atmospheric pressure chemical ionization
bw	body weight
CSL	UK Central Science Laboratory (Norwich, UK)
CSTEE	Comité scientifique sur la toxicité, l'écotoxicité et l'environnement (Scientific Committee on Toxicity, Ecotoxicity and the Environment)
DBT	dibutyltin
DOcT	dioctyltin
DPhT	diphenyltin
EHMA	ethylhexylmercaptoacetate
EFSA	European Food Safety Authority
ESI	electrospray ionization
FAP	free association paints
FCM	food contact material
FDA	(United States) Food and Drug Administration
FPD	flame photometric detector
FRF	fat reduction factor
GC	gas chromatography
HPLC	high-performance liquid chromatography
ICP	inductively coupled plasma
IDL	instrumental detection limit
IMO	International Maritime Organization
JMPR	Joint Meeting on Pesticide Residues
LOA	length overall
LOAEL	lowest observable adverse effect level
MAF	marine antifouling (paints)
MBT	monobutyltin
MOcT	monooctyltin
MPhT	monophenyltin

MS	mass spectrometer, mass spectrometry
NOAEL	no observable adverse effect level
OTC	organotin compound
P-FPD	pulsed flame photometric detector
SBSE	stir-bar sorptive extraction
SCHER	Scientific Committee on Health and Environmental Risks
SIM	selected ion monitoring mode
SML	specific migration limit
SPC	self-polishing copolymer paints
SPME	solid-phase microextraction
TARL	tolerable average residue level
TBT	tributyltin
TBTC	tributyltin chloride
TBTN	tributyltin naphthenate
TBTO	tributyltin oxide
TMAH	tetramethyl ammonium hydroxide
TPhT	triphenyltin
TPhTOH	triphenyltin hydroxide (Fentin hydroxide)
TCyTOH	tricyclohexyltin hydroxide (Cyhexatin)
TCyTT	tricyclohexyltin triazole (Azocyclotin)
TNTO	trineophenyltin oxide (Fenbutatin oxide)
TPhTAc	triphenyltin acetate (Fentin acetate)
WHO	World Health Organization
ww	wet weight

Index

acute reference doses (ARfD), 177, 182
adipose tissue
 polychlorinated napthalenes (PCNs), 385–6
Administration of Quality Supervision, Inspection and Quarantine (AQSIQ), 28
alumina column chromatography, 267
aluminium, 175
analytical methodologies, 373–4
Antifouling System Convention, 437
Aroclor, 217
arsenic
 determination methods in foods, 418–21
  certified reference materials for measuring total arsenic and arsenic species, 421
  determining arsenic species, 419–21
  determining total arsenic content, 418–19
  food analysis certified for arsenic content, 423
 foods and current issues related to analysis, toxicity and metabolism, 414–26
  implications for food industry and policy makers, 425–6
 sources and occurrence in food, 415–17
  origin and distribution of inorganic and organic arsenic in environment, 415–16
 toxicity, 421–5
  dependence on species, 421, 423–4
  toxic action mode for major arsenic species and major toxicity concerns, 424–5
Arsenic Mercury Intake Biometric Study (AMIDS), 396

arsenobetaine, 416–17
arsenolipids, 417
arsenosugars, 417
aryl hydrocarbon hydroxylase (AHH), 371–2
 assay, 53
aryl hydrocarbon receptor (AhR) *see* dioxin responsive elements (DRE)
as low as reasonably achievable (ALARA), 31, 37, 183, 324
Assessing the Toxicology and Hazard of Non-dioxin-like PCBs Present in Food (ATHON), 229–40
 assays in the *in vitro* screening programme, 233
 development and reproduction endpoints, 233–4
 in vivo experiments and studies identified in open literature, 230–2
 neurotoxicity, 234–40
atmospheric pressure chemical ionisation (APCI), 317, 449
atmospheric pressure laser ionisation (APLI), 318
atmospheric pressure photo ionisation (APPI), 317–18
atomic absorption spectrometry (AAS), 96–8, 418
atomic emission detector (AED), 447
atomic fluorescence spectrometry, 98

BDE-99, 8, 16
benchmark dose (BMD), 180–1
benchmark dose level (BMDL), 7–8
benchmark response (BMR), 180–1
Best Management Practices (BMP), 354
bioaccumulation, 291
bioanalytical detection methods (BDM), 53

bioassay, 92–5
  amperometric recording for an acetylthiocholine iodide, 94
  cadmium immunochromatography system, 95
Biobeads S-X3, 351
bioconcentration factors (BCF), 151
biomagnification, 291
biomonitoring, 297
biotransfer factors (BTF), 151–2, 167–9
  schematic diagram, 168
brainstem auditory evoked potentials (BAEP), 239
breastmilk, 384–5
British Retail Consortium (BRC), 115
brominated flame retardants
  food, 261–71
    future trends, 270–1
    implications for prevention and control of contamination, 270
    major incidences of food contamination, 269
    methods of analysis and monitoring, 267–8
    toxicity, 268–9
  sources, occurrence and human exposure, 262–7
  hexabromocyclododecanes (HBCDs) in food, 265–6
  other BFRs in food, 267
  polybrominated diphenyl esters (PBDEs) in food, 262–5

cadmium, 29–30, 31
caffeine complexation, 314
capillary gas chromatography, 314
carcinogen, group 1, 203
carry-over rates (COR), 151–2
CE N333/2007, 83
cell-based assays, 53–4
cell-to-cell communication, 245–6
Center for Food Safety and Applied Nutrition (CFSAN), 28
certified reference materials (CRM), 72
chemical-activated fluorescent protein expression (CAFLUX), 54
chemical-activated luciferase gene expression (CALUX), 14, 53–4, 67
chemical contaminants and residues
  food, 173–85
  risk assessment, 176–82
    exposure assessment, 181–2
    hazard characterisation, 177–81
    hazard identification, 176–7
  paradigm and definitions, 174–5
  risk analysis paradigm, 175
  risk assessment role in risk management, 184–5
  risk characterisation, 182–4
    combined exposure, 184
    health-based guidance values, 182
    margin of exposure (MOE), 182–3
    threshold of toxicological concern (TTC), 183–4
chemiluminescence, 90
Clophen, 217
Codex Alimentarius Code, 27, 314
Codex Committee on Contaminants in Food (CCCF), 32
collision induced dissociation (CID), 56
column chromatography, 314
Commission Decision 2002/657/EC, 74
Commission Directive 2006/13/EC, 50
Commission Directive 2000/60/CE, 352
Commission Directive 2002/70/EC, 51
Commission Directive 2007/19/EC, 352
Commission Directive 2008/105/EC, 352
Commission Recommendation 2011/516/EU, 50
Commission Regulation 10/2011, 352
Commission Regulation 835/2011, 325
Commission Regulation (EC) No. 2065/2003, 325
Commission Regulation (EC) No. 333/ 2007, 39
Commission Regulation (EC) No 152/2009, 48, 51, 64, 67
Commission Regulation (EC) No 178/2002, 27
Commission Regulation (EC) No 252/2012, 12
Commission Regulation (EC) No 315/93, 12
Commission Regulation (EC) No 565/2008, 13
Commission Regulation (EC) No 853/2004, 22
Commission Regulation (EC) No 882/2004, 13
Commission Regulation (EC) No 1259/2011, 9
Commission Regulation (EC) No 1881/2006, 9, 22, 31, 33, 48, 50
Commission Regulation (EC) No 1883/2006, 51, 64, 67
Commission Regulation (EC) No 178/ 2002, 21–2
Commission Regulation (EC) No 199/ 2006, 50

478 Index

Committee on Toxicity of Chemicals in Food Consumer Products and Environment, 353
composite foods, 293
consumer products
   regulation of exposure, 442–4
     summary of new threshold limits for organotin compounds, 443
CONTAM Panel, 7, 28–9, 31–2, 325
Council Directive No 76/769/EEC, 442
Council Directive No 86/278/EEC, 130
Council Regulation 2001/102/EC, 50
Council Regulation (EC) No 2375/ 2001, 50
Crassostrea gigas, 440
crops
   organic pollutants and potentially toxic elements (PTE) uptake, 129–39
     in situ monitoring, 138–9
     plant uptake, 131–5, 135–8
cut-off values, 65–6
   level of confidence calculations, 66
cyhexatin (TCyTOH), 442
cytochrome p-450 system, 244–5
cytotoxicity tests, 424

Danish Environmental Protection Agency, 353
DBT dilaureate, 433
decabromodiphenyl ethane, 267
deconvoluted ion currents (DIC), 60
deodorisation, 314
dietary exposure estimates, 293–5
dietary intake of NDL-PCBs, 220–1
differential pulse anodic stripping voltammetry (DPASV), 85
diode array detector (DAD), 317
dioxin-like polychlorinated biphenyls (DL-PCB), 146–7, 157–69
   BTF, 167–9
   data comparisons, 165–6
     liver and kidney, 165–6
     meat, 165
   highland sheep, 162–5
     residue data, 166
   lowland sheep, 157–62
     residue data, 159
     TEQ values, 160
dioxin responsive elements (DRE), 53, 196–7
dioxins
   biological and physico-chemical tools analytical requirements, 63–6
     Performance-Based Measurement System (PBMS), 64–5
     screening cut-off values, 65–6

biological *vs.* physico-chemical screening, 51–63
dioxin-like PCB, PCDD/F and ICES6, 157–69
experimental rearing, sampling and analysis, 152–7
   daily vegetation consumption estimates, 155
   extraction, purification and analysis, 156
   intake determination, 154–6
   materials, 156
   quality control (QC), 157
   sampling procedures, 153–4
   TEQ, 157
food, 191–213
food contamination, 110–28
   food recall and withdrawal, 117–20
   food traceability, 115–17
   risk analysis, 112–15
   risk communication strategies, 120–5
overview, 47–51
   efficient monitoring, 49
   EU strategy, 49–51
pathways and sources, 147–51
PCDD/F properties and occurrence, 192–5
PCDD/F toxic effect in humans and experimental animals, 198–205
PCDD/F transfer into animal tissues, 151–2
polychlorinated dibenzo-*p*-dioxins and dibenzofurans (PCDD/F), 67–75
QA/QC validation, 67
quantitative *vs.* semi-quantitative approach, 66–7
screening in food, 47–76
structures and labelling of dioxins and PCB, 146
transfer and uptake in sheep, 145–69
Directive 2006/66/EC, 31
Directorate General for Health and Consumers (DG SANCO), 22
donor-acceptor complex chromatography (DACC), 317
dopant-assissted APPI (DA-APPI), 318
duplicate diet method, 293–4

Ecoat, 435
eels, 263
electro-spray ionisation (ESI), 282
electrochemical method, 84–90
electron capture detection, 372
electron impact (EI), 219
electrospray ionisation (ESI), 317
embryotoxicity test, 234
endocrine system, 240–5

Index

cytochrome p-450 system, 244–5
  Venn diagram of CYP-induction pattern of PCBs, 246
  retinoid system, 242, 244
    tissue samples from 28 days and perinatal studies with PCB-180 and 52, 245
  steroid hormones, 242
    human tissue concentrations of PCB 180 and effects seen in serum T3 levels, 243
  thyroid hormones, 240–2
Environmental Health Criteria, 322
environmental pollution, 310
enzyme inhibition biosensor, 92
enzyme-linked immunosorbent assay (ELISA), 95
ethoxyresorufin-O-deethylase (EROD), 371–2
  assay, 53
European Commission Decision 2009/425/EC, 442
European Commission's Scientific Committee for Food, 353
European Council Regulation (EEC) No 315/93, 22
European dietary exposure, 455–6
  intakes of TBT, DBT and TPhT on consumption of fish and fish products, 455
European Economic Commission (EEC), 129–30
European Food Safety Authority (EFSA), 7, 21, 27–9, 31–3, 39, 228–9, 269, 309, 353, 416
European Union Reference Laboratory (EURL), 12
European Union System for the Evaluation of Substances (EUSES), 343
exposure
  NDL-PCB in food and health hazards, 215–49
    ATHON R&D project dedicated to generating NDL-PCB toxicity data, 229–40
    cell regulation and metabolism, 240–6
    future trends, 247–9
    human exposure and tissue levels, 220–3
    NDL-PCB regulatory status, 228–9
    PCB congeners classification, 226–8
    sources, occurrence in food, limit values and monitoring methods, 217–19
    toxicokinetics and metabolism, 223–6
exposure assessment, 338–9

fat consumption reduction factor (FRF), 457
fentin acetate (TPhTAc), 441
fish, 284–5
  mercury concentrations, 396–401
  selected fish and shellfish based on consumption frequency, 402
  top 30 ranked seafood species from US FDA, 397
flame photometric detector (FPD), 447
Florisil, 315
fluorescence spectroscopy (FL), 315
food business operator (FBO), 41
food contamination, 310, 451–2
  dioxins, 110–28
    food recall and withdrawal, 117–20
  food traceability, 115–17
    pathways, 290–3
    risk analysis, 112–15
    risk communication strategies, 120–5
food crises, 48
food drying, 313
food frequency questionnaire (FFQ), 319
food packaging, 341
food packaging material
  human exposure to organotin compounds, 457–61
    consumer exposure assessment, 460–1
    overview and study of migration behaviour, 457–60
food processing, 311, 341
food recall, 117–20
  management, 119
  rapid detection methods, 119–20
  system creation and maintenance, 118–19
  systems, 117–18
food roasting, 311–12
food smoking, 312–13
food traceability, 115–17
  system creation and maintenance, 116–17
food withdrawal, 117–20
  management, 119
  rapid detection methods, 119–20
  system creation and maintenance, 118–19
  systems, 117–18
forced expiratory volume in one second (FEV1), 348
forced vital capacity (FVC), 348
Free Association Paint (FAP), 436

gas chromatography-electron impact ionisation and ion mobility mass spectrometry (GC-EI-Q-IMS-TOF), 76
gas chromatography (GC), 282, 445

480 Index

gas chromatography with electron capture detection (GC-ECD), 351
gas chromatography with electron capture negative ion mass spectrometry (GC-ECNIMS), 267
gas chromatography with flame ionisation (GC-FID), 351
gas chromatography with mass spectrometry (GC-MS), 51–2, 219, 351
GC-HRMS, 51–2, 53, 55
 state-of-the art, 67–72
GC-LRMS, 52
GC-MS TEQ, 54, 55
GC-QISTMS/MS, 56
 parameters optimisation, 57–8
GC-QQQMS/MS, 59, 73–4
 parent and product ions in MRM mode for TCDD, 59
 product ions calculations of TCDD in MRM mode, 59
GC-Triplequad MS *see* GC-QQQMS/MS
GCxGC-TOFMS, 60–2
 chromatographic and mass spectrometric parameters, 61–2
gel permeation chromatography (GPC), 267, 351, 450
genotoxicity
 organotin compounds, 464–5
Good Practice in the Use of Veterinary Drugs (GPVD), 181
Grignard reagents, 447
grilled products, 313
Guidelines for Drinking-water Quality (GDWQ), 32

hanging mercury drop electrode (HDME), 85–6
hazard characterisation, 177–81
hazard identification, 176–7
health-based guidance values, 178, 182
health hazards
 NDL-PCB in food exposure, 215–49
  ATHON R&D project dedicated to generating NDL-PCB toxicity data, 229–40
  cell regulation and metabolism, 240–6
  human exposure and tissue levels, 220–3
  NDL-PCB congeners classification, 247
  NDL-PCB regulatory status, 228–9
  PCB congeners classification, 226–8
  sources, occurrence in food, limit values and monitoring methods, 217–19
  toxicokinetics and metabolism, 223–6

heavy metals
 analytical method suitability, 40–1
 confirmatory methods for detection in food, 95–9
 foodstuff monitoring, 34–7
 industry and enforcement legislation, 37–9
 quality assurance and method validation, 99–102
 regulatory control and monitoring in food, 20–43
 risk assessment and policy making, 21–34
  legislation, 21–7
  risk consideration, 29–30
  risk management, 30–4
  setting maximum limits, 27–9
 screening methods for detection in food, 84–95
hexabromocyclododecanes (HBCDD), 3, 16, 263, 265–6, 268
high-performance liquid chromatography (HPLC), 283, 315, 317, 448–9
high performance liquid chromatography/inductively coupled plasma mass spectrometry (HPLC/ICPMS), 419–21
 anion exchange HPLC separation of arsenic species, 422
high performance liquid chromatography with tandem quadruple mass spectrometry (HPLC-MS/MS), 351
high performance liquid chromatography with UV diode array detection (HPLC-DAD), 351
high resolution gas chromatography coupled with high resolution mass spectrometry (HRGC-HRMS), 374
high resolution gas chromatography (HRGC), 69
high resolution mass spectrometry (HR-MS), 283
human dietary exposure
 organotin compounds, 451–7
 per and polyfluoroalkyl substances (PFASs), 279–99
  analytical methods in food, 281–4
  estimated exposure from food and other exposure media, 293–8
  food contamination pathways, 290–3
  levels in food, 284–90
 polycyclic aromatic hydrocarbons (PAHs), 319–22
human tissue levels, 221–3

ICES6, 157–69
 BTF, 167–9
 data comparisons, 165–6

liver and kidney, 165–6
  meat, 165
  highland sheep, 162–5
  lowland sheep, 157–62
immunoassay, 94
immunotoxicity
  organotin compounds, 464
inductively coupled plasma mass spectroscopy (ICP-MS), 98–9, 100
inductively coupled plasma optical emission spectrometry (ICP-OES), 98–9
ingestion, 336
inhalation, 336–8
  exposure, 310
instrumental detection limits (IDL), 448
internal quality control (IQC), 72
International Agency for Research on Cancer (IARC), 322
International Council for the Exploration of the Sea (ICES), 310
International Food Safety Authorities Network (INFOSAN), 121
International Maritime Organisation (IMO), 436
International Programme for Chemical Safety (IPCS), 322
ion exchange chromatography (IEC), 100
ion-pair extraction (IPE), 281–2
ionisation energies (IE), 318
ISO 17025, 64
ISO/CEN 17025, 72

Joint Expert Committee on Food Additives (JECFA), 27–9
Joint FAO/WHO Committee on Food Additives (JECFA), 309
Joint Meeting on Pesticides Residues (JMPR), 442
Joint Research Centre Institute of Reference Materials and Measurement (JRC-IRMM), 350

Kanechlor, 217
kinase activation, 245–6
Korean National Human Exposure and Biomonitoring Examination, 396

limits of detection (LOD), 49, 82, 97–8
limits of quantification (LOQ), 49, 101
lipophilic organic chemicals, 131
liquid chromatography (LC), 282, 445
liquid chromatography-mass spectroscopy (LC-MC), 317
liquid-liquid microextraction, 91
liquid-liquid partition (LLP), 314, 351

liquid phase separation, 448–9
Lisbon Treaty *see* Treaty on the Functioning of the European Union (TFEU)
liver, 385–6
lowest observed effect level, 228

Margin of Exposure (MoE), 7, 181, 182–3
Marine Antifoulant Paints, 436–41
  regulations controlling the use of TBT paints worldwide up to 1992, 438–9
marine fish, 264
market basket study *see* total diet studies (TDS)
mass spectrometry (MS), 282, 314, 372, 448
matrix-assisted laser desorption ionisation (MALDI), 318
maximum residue level (MRL), 181
MeHg, 393–4, 397, 401, 403
mercaptides, 433
mercury
  concentrations, 396–405
    fish and shellfish, 396–401
    MeHg in rice, 401
  exposures and risks from major food categories, 403, 406
  in foods, 392–406
    analytical techniques, 396
    health effects of mercury exposure, 393–4
    major food-related poisonings, 394–5
    safety thresholds, 395–6
metabolic syndrome, 348
metabolites, 339
micro electron capture detection (micro-ECD), 63
microwave-assisted extraction (MAE), 315
migration behaviour, 457–60
  conditions of time, temperature and solvents for extraction of finished plastics, 460
  organotin compounds authorised for FCM in EU and US, 458–9
migration tests, 292
Minamata, Japan, 394
Ministry of Agriculture (MoA), 28
multiple reaction monitoring (MRM), 56

National Diet and Nutrition Surveys (NDNS), 5
negative chemical ionisation (NCI), 219
neurotoxicity, 234–40
  catalepsy neurotoxicity of PCB 180, 238
  in vivo findings after developmental and subacute NDL-PCB exposures, 235–7

482 Index

no-observed-adverse-effects level (NOAEL), 8, 200, 395, 463
non-cell-based bioassays, 54
non-dietary exposure pathways, 295–6
non-dioxin-like polychlorinated biphenyls (NDL-PCB), 146
   congeners classification, 247
      PCA score plot of the relationship between all tri- to hepta-congeners, 248
   food exposure and health hazards, 215–49
      ATHON R&D project dedicated to generating NDL-PCB toxicity data, 229–40
      cell regulation and metabolism, 240–6
      chemical structure, 216, 219
      human exposure and tissue levels, 220–3
      NDL-PCB regulatory status, 228–9
      PCB congeners classification, 226–8
      sources, occurrence in food, limit values and monitoring methods, 217–19
      total NDL-PCB burden in food samples, 218
      toxicokinetics and metabolism, 223–6
      half-life times of some NDL-PCB congeners plus DL-PCBs 105–118, 224
Nucella lapillus, 440

optical method, 90–2
   absorption profile of thiol-modified gold nanoparticles, 93
   liquid–liquid microextraction preconcentration, 90–2
   thiol-modified gold nanoparticles, 93
organic arsenic, 415–16
organic pollutants
   in situ monitoring
      crop contamination prevention, 139
      metals and metalloids separation, 130–1
      plant uptake, 131–5
      BCF variation, 134
      modelling, 134–5
      particulate deposition, 133
      variation in TSCF with chemical octanol–water partition coefficient, 132
   uptake by crops, 129–39
Organotin Antifoulant Paint Control Act, 436
organotin compounds
   analysis in foods, 445, 447–51
   foods, 430–67
      human exposure from food packaging material, 457–61
   major applications of organotin compounds and quantities used in EU, 431
   health risks and toxicity, 461–6
      exposure of adults and children via food, 461–2
      toxicity and adverse effects, 463–6
   human dietary exposure from foods, 451–7
      database of EU SCOOP Task 3.2.13 report, 453–6
      initial studies of OTC uptake from food, 452–3
      sources, 452
   physical and chemical properties, 444–5
      half-life under different conditions, 444
      physicochemical properties of di- and tri-substituted organotins, 446
   technical, agricultural and industrial usages, 432–44
      biocides in marine antifoulant paints and pesticides for crop protection, 436–42
      catalyst in polymer production, 435
      preservatives to protect wood from fungal attack, 436
      regulation of exposure from consumer products, 442–4
      stabiliser for polyvinyl chloride, 432–5
OT-SAFE, 455

packaging contamination, 265
Panel on Contaminants in the Food Chain, 416, 461
paper chromatography, 315
PCB congeners
   classification, 226–8
      enzyme inducing properties of individual PCB congeners, 225
      summary of WHO TEF values, 227
peak expiratory flow (PEF), 348
per and polyfluoroalkyl substances (PFASs)
   estimated exposure from food and other exposure media, 293–8
      dietary exposure estimates, 293–5
      relative importance of non-dietary exposure pathways, 295–6
      toxicokinetic modelling and exposure reconstruction, 296–7
   human dietary exposure, 279–99
      analytical methods in food, 281–4
      food contamination pathways, 290–3
      levels in food, 284–90
perfluoroalkane sulfonic acid (PFSA), 281
perfluoroalkyl carboxylic acid (PFCA), 281
perfluorooctane sulphonate (PFOS), 3

perfluorooctanoic acid (PFOA), 7, 280, 292
perfluorooctanoic sulfonic acid (PFOS), 280, 284–6, 290
Performance-Based Measurement System (PBMS), 64–5
persistent organic pollutants (POP)
  analytical methods and policy, 14–15
  dietary exposure and total diet studies (TDS), 4–6
  enforcement and implications for food business, 12–14
  food, 3–17
  future trends, 15–17
  risk assessment, policy making and regulations, 6–12
    European review for dioxins, furans and PCB, 12–13
    health risk, 6–8
    limit review, 9, 12
    risk control, 8–9
pesticides
  crop protection, 441–2
  Fenbutatin oxide chemical structure, 441
Phenoclor, 217
phthalates
  carcinogenic effects, 347
  exposure effects in women, 346–7
  food industry and policy making
    implications for prevention and control of contamination, 352–3
    TDI levels established by WHO/CICAD and by EFSA, 353
  foods, 334–54
    future trends, 353–5
    methods analysis and monitoring, 349–52
    sources and occurrence, 339–44
    studies of effects on humans, 344–9
  human exposure, 335–9
    exposure assessment, 338–9
  other effects, 349
  relation to respiratory function, asthma and allergies, 348–9
poisonings
  food-related, mercury, 394–5
polarography, 85
polybrominated biphenyl (PBB), 15–16, 267
polybrominated diphenyl ethers (PBDE), 3, 15–16, 262–5
polychlorinated biphenyls (PCB), 3
  biological and physico-chemical tools
    analytical requirements, 63–6
    Performance-Based Measurement System (PBMS), 64–5

screening cut-off values, 65–6
biological vs. physico-chemical screening, 51–63
  biological screening tools, 53–5
  physico-chemical screening tools, 55–63
dioxin-like PCB, PCDD/F and ICES6, 157–69
experimental rearing, sampling and analysis, 152–7
  daily vegetation consumption estimates, 155
  extraction, purification and analysis, 156
  quality control (QC), 157
  TEQ, 157
food, 191–213
overview, 47–51
  efficient monitoring, 49
  EU strategy, 49–51
  screening strategy, 48–9
pathways and sources, 147–51
PCDD/F properties and occurrence, 192–5
PCDD/F toxic effect in humans and experimental animals, 198–205
PCDD/F toxicity, 195–8
PCDD/F transfer into animal tissues, 151–2
polychlorinated dibenzo-p-dioxins and dibenzofurans (PCDD/F), 67–75
properties and occurrence, 205–9
  chemical structure, 206
  food and human exposure, 207–9
  mean average level in food groups in Europe, 208
  sources, 207
QA/QC validation, 67
quantitative vs. semi-quantitative approach, 66–7
screening and confirmatory methods in food, 47–76
structures and labelling of dioxins and PCB, 146
toxicity, 209–13
transfer and uptake in sheep, 145–69
polychlorinated dibenzo-p-dioxins (PCDD), 3, 50–1, 55–6, 64–5, 146–7, 147–51, 157–69
  BTF, 167–9
  confirmatory methods in food and feed, 67–75
  data comparisons, 165–6
    liver and kidney, 165–6
    meat, 165
  highland sheep, 162–5
  lowland sheep, 157–62
  properties and occurrence, 192–5

484 Index

polychlorinated dibenzo-*p*-dioxins (PCDD) (*cont.*)
  chemical structure, 193
  food and human exposure, 193–5
  mean average level in various food groups in Europe, 194
  toxic effect in humans and experimental animals, 198–205
    carcinogenicity and genotoxicity, 202–4
    case studies, 204
    human toxicity, 200–1
    immunotoxicity, 201–2
    laboratory animals toxicity, 199–200
    maximum regulatory values in food in EU, 205
    regulations, 204
  toxicity, 195–8
    arylhydrocarbon receptor activation pathways, 196
    mode of action, 196–7
    TEF concept, 197–8
    TEF values, 198
    toxicokinetics, 195–6
  transfer into animal tissues, 151–2
polychlorinated dibenzofurans (PCDF), 3, 50–1, 55–6, 64–5, 146–7, 147–51, 157–69
  BTF, 167–9
  confirmatory methods in food and feed, 67–75
  data comparisons, 165–6
    liver and kidney, 165–6
    meat, 165
  highland sheep, 162–5
  lowland sheep, 157–62
  properties and occurrence, 192–5
    chemical structure, 193
    food and human exposure, 193–5
    mean average level in various food groups in Europe, 194
  toxic effect in humans and experimental animals, 198–205
    carcinogenicity and genotoxicity, 202–4
    case studies, 204
    human toxicity, 200–1
    immunotoxicity, 201–2
    laboratory animals toxicity, 199–200
    maximum regulatory values in food in EU, 205
    regulations, 204
  toxicity, 195–8
    arylhydrocarbon receptor activation pathways, 196
    mode of action, 196–7
    TEF concept, 197–8
    TEF values, 198
    toxicokinetics, 195–6
  transfer into animal tissues, 151–2
polychlorinated napthalenes (PCNs)
  food sources, analytical methodology, occurrence and human exposure, 367–87
  methods of analysis, 372–4
  occurrence in food, 375–83
    marine and aquatic biota, 375–7
  occurrence in humans, 383–6
  sources, 368–70
  toxicology, 370–2
    Ah receptor-mediated toxicity, 371–2
    metabolism and half-lives, 370–1
polycyclic aromatic hydrocarbons (PAH), 4, 175
  analysis methods, 314–19
    extraction techniques example, 316
  EU legislation, 324–6
    maximum levels of benzo[*a*]pyrene and PAH4 sum, 326
  food, 308–26
    food scandals, 324
    risk assessment, 322–4
  human dietary exposure, 319–22
    European consumer dietary exposure to benzo[*a*]pyrene, 320
    European consumer dietary exposure to PAH4, 321
  sources and information in food, 309–14
    molecular formula and weight and included in US-EPA16 and EU 15+1, 311
    structure of four PAH included in PAH4, 310
polytetrafluoroethylene (PTFE), 83, 292
polyvinyl chloride
  OTC as stabiliser, 432–5
  stabiliser
    acute toxicity of some organotin chlorides, 434
    scheme of the most common organotin stabilisers for PVC, 434
positive chemical ionisation (PCI), 351
potentially toxic elements (PTE)
  in situ monitoring
    crop contamination prevention, 139
  plant uptake, 135–8
    accumulation in shoot, 136–7
    K_d and BCFs range, 137
    modelling, 137–8
    nutrients dependent on soil pH, 136

particulate deposition, 137
    soil-root transport, 135–6
    uptake by crops, 129–39
potentiometric stripping analysis, 84–5
pressurised liquid extraction (PLE), 315
processed foods, 340
programmed temperature vaporisation (PTV), 317
provisional tolerable monthly intake (PTMI), 7
provisional tolerable weekly intake (PTWI), 28
pulsed-flame photometric detector (P-FPD), 447

quadruple tandem mass spectrometry, 283
quality assurance (QA), 67, 72–3, 99–102
quality control (QC), 67, 72–3, 99, 157
    quality control chart of PCDD/Fs in milk powder, 73
quality manual (QM), 100
quantitative approach, 66–7
quick, easy, cheap, effective, rugged and safe (QuEChERS) method, 315

Rapid Alert System for Food and Feed (RASFF), 35, 48, 121–2
Regulation (EC) No. 782/2003, 437
Regulation (EC) No 178/2002, 112, 114
Regulation (EC) No 1102/2008, 86
relative potency (REP) values, 373
relative response reactor (RRF), 69
reprotoxicity, 464–5
retinoid system, 242, 244
risk characterisation, 465–6
    average seafood consumption and calculated TARL of TBT in seafood products, 466
risk managers
    international risk communication, 120–3
    national risk communication, 123–4
    root concentration factor (RCF), 131

safety thresholds
    mercury exposure, 395–6
saponification, 314
Scientific Co-operation (SCOOP), 453–6
    analysis of raw occurrence data in EU SCOOP report, 454–5
    European dietary exposure assessment, 455–6
Scientific Committee on Food (SCF), 309
Scientific Committee on Health and Environmental Risks (SCHER), 462

Scientific Committee on Toxicity, Ecotoxicity and the Environment (CSTEE), 462
    dermal exposure studies, 462–3
    inhalation studies, 462
    oral exposure studies, 463
screen-printed electrodes (SPE), 88
seafood, 266, 376
selected ion monitoring (SIM), 56, 315
semen, 346
semi-quantitative approach, 66–7
Sephadex, 315
sheep
    dioxins and PCB transfer and uptake, 145–69
        experimental rearing, sampling and analysis, 152–7
        pathways and sources, 147–51
        PCDD/F and ICES6, 157–69
        PCDD/F transfer into animal tissues, 151–2
        structures and labelling of dioxins and PCB, 146
        WHO TEF, 148
shellfish, 284–5
    mercury concentrations, 396–401
    top 30 ranked seafood species from US FDA, 397
silica gel column chromatography, 267
skin absorption, 338
soil-root transport, 131–2
solid-liquid extraction (SLE), 281–2
solid-phase extraction (SPE), 314, 450
solid-phase microextraction (SPME), 450–1
solvent-partitioning method, 419
Soxhlet, 267, 282
specific migration limits (SML), 352
square wave anodic stripping voltammetry (SWASV), 85
squid, 263
State Food and Drug Administration (SFDA), 28
steroid hormones, 242
stir-bar sorptive extraction (SBSE), 450–1
sulphuric acid treatment, 267
supercritical fluid extraction (SFE), 315
sweet preference test, 238

testicular dysgenesis syndrome (TDS), 345–6
tetrabromobisphenol A (TBBP-A), 16, 268–9
2,3,7,8-tetrachlorodibenzo-*para*-dioxin (TCDD), 193, 199–200, 201–4
thin-layer chromatography, 314

Third National Health and Nutrition Examination Survey, 348
threshold of toxicological concern (TTC), 183–4
thyroid function, 347–8
thyroid hormones, 240–2
time-of-flight mass spectrometry (TOFMS), 60, 267, 317
tolerable acceptable residue level (TARL), 465–6
tolerable daily intake (TDI), 7, 218, 353, 376
tolerable weekly intake (TWI), 7
total diet studies (TDS), 4–6, 20–1, 35–7, 43
  food groups used in United Kingdom, 6
  food selection, 36–7
  limitations, 36
toxic equivalence factor (TEF), 50–1, 53, 54, 147, 194, 197–8, 226, 227, 322, 373, 375
toxic equivalents (TEQ), 147, 194–5, 197–8
toxicity
  Ah receptor-mediated, 371–2
  arsenic, 421–5
  brominated flame retardants, 268–9
  organotin compounds, 461–6
  phthalates, 343–4
toxicokinetic modelling, 296–7
trace elements
  analytical method suitability, 40–3
  foodstuff monitoring, 34–7
  industry and enforcement legislation, 37–9
  regulatory control and monitoring in food, 20–43

risk assessment and policy making, 21–34
  legislation, 21–7
  risk consideration, 29–30
  risk management, 30–4
  setting maximum limits, 27–9
transpiration stream concentration factor (TSCF), 132
Treaty on the Functioning of the European Union (TFEU), 27
1,2-bis(2,4,6-tribromophenoxy)ethane, 267
trifluoroacetic acid, 420
two-dimensional gas chromatography, 267

ultraviolet (UV), 315
uncertainty factors, 179–80
  subdivision, 180
United States Environmental Protection Agency (US-EPA), 309
US Food and Drug Administration (US FDA), 396
US National Heath and Nutrition Examination Survey (NHANES), 395
US National Research Council (NRC), 344

vacuum ultraviolet (VUV), 318
vapour generation AAS, 418
Venn diagram, 245

World Health Organization (WHO), 227, 395

xenobiotic response elements (XRE)
  see dioxin responsive elements (DRE)